C0-AUY-836

Proceedings in Life Sciences

{Tübingen. Universität}

Animal Migration, Navigation, and Homing

Symposium Held at the University of Tübingen, August 17–20, 1977

Edited by
K. Schmidt-Koenig and W. T. Keeton

With 237 Figures

Springer-Verlag
Berlin Heidelberg New York 1978

Prof. Dr. Klaus Schmidt-Koenig
Abt. für Verhaltensphysiologie der Universität
Beim Kupferhammer 8, 7400 Tübingen/FRG
and
Dept. of Zoology, Duke University
Durham, N.C. 27706/USA

Prof. Dr. William T. Keeton
Sect. of Neurobiology and Behavior
Langmuir Lab., Cornell University
Ithaca, N.Y. 14853/USA

The Symposium was sponsored by the University of Tübingen, celebrating its 500th anniversary, and held as Internationale Wissenschaftliche Fachkonferenz der Deutschen Forschungsgemeinschaft.

The discussion of this symposium is available directly from the first editor as a mimeographed volume: Animal Migration, Navigation, and Homing: Discussion Volume.
K. Schmidt-Koenig and W. T. Keeton (eds.).

ISBN 3-540-08777-X Springer-Verlag Berlin Heidelberg New York
ISBN 0-387-08777-X Springer-Verlag New York Heidelberg Berlin

Library of Congress Cataloging in Publication Data. Main entry under title: Animal migration, navigation, and homing. (Proceedings in life sciences) Bibliography: p. Includes index. 1. Animal orientation-Congresses. 2. Animal navigation-Congresses. 3. Animal migration-Congresses. I. Schmidt-Koenig, Klaus, 1930-. II. Keeton, William T., 1933-. Tübingen. Universität. IV. Deutsche Forschungsgemeinschaft (Founded 1949). QL 782. 5.A54. 591.1'852.78-8656.

Offsetprinting: Beltz, Offsetdruck, Hemsbach/Bergstr.
Bookbinding: Brühlsche Universitätsdruckerei, Lahn-Gießen.
2131/3130-543210

Preface

Seven years have elapsed between this and the preceding major symposium on animal orientation and navigation, held on Wallops Island, Va., USA in 1970. Never before between two symposia in this field has there been such an enormous increase - more truly an explosion - of new data, new evidence, new ideas, and methods. Moreover, environmental awareness has also increased tremendously. The essential role of animals as bio-indicators, the economic importance of dwindling stocks of fish, whales, and other animals demand increasing efforts to investigate their life cycles and whereabouts, and that means for many of them, their travels between breeding, feeding, or wintering ranges. Although many aspects of research in the various fields covered by this symposium qualify as basic research, environmental and economic considerations cannot be dismissed. Finally, there is a chance to eventually find biologic systems of migration and orientation that may be at least partially useful for human navigation or other undertakings.

The diversity of species involved in animal migration and the methodologies employed in unraveling the modes of animal navigation provide the reader of this volume with a survey of the front line of research in this field, with emphasis on current ideas and with a foretaste of the problems ahead. The reports offer insight into the experimental difficulties, sometimes formidable, that the researcher encounters when dealing with free-ranging animals.

The 500th anniversary of the University of Tübingen was a more than welcome platform and frame for a symposium on animal migration, navigation, and homing. The 500th anniversary fund and the Deutsche Forschungsgemeinschaft contributed to make this symposium possible.

July, 1978 K. SCHMIDT-KOENIG and W.T. KEETON

Contents

Session III
Chairman: A.D. HASLER

Session IV
Chairman: W.T. KEETON

Session V
Chairman: F. PAPI

Session VI
Chairman: J.T. ENRIGHT

Session VII
Chairman: A. CARR

Session I

Chairman:

Franz P. Möhres, Universität Tübingen, Institut für Biologie III
7400 Tübingen, FRG

Second-order Statistical Analysis of Directions

EDUARD BATSCHELET, Universität Zürich, Mathematisches Institut, 8032 Zürich, Switzerland

Abstract

If circular samples of equal size are reduced to their respective mean vectors, a sample of mean vectors is formed. This sample is referred to as a second-order sample. Kolmogorov's test of goodness of fit may be used as a test for concentration. Since second-order samples are essentially bivariate, Hotelling's T^2 test (in case of normality) or a nonparametric competitor may be applied. Here a linear-circular correlation technique is recommended. Also studied is the comparison of two independent second-order samples.

A. Introduction

The analysis of directions is a branch of mathematical statistics which does not belong to the mainstream of statistical research. Rather, it is an area where only few statisticians are active. Mardia (1972) wrote an outstanding book that comprises the theory and some applications of directional analysis. Among the many books on biostatistics there is only one, written by Zar (1974), with a chapter on circular distributions. It is the purpose of this paper to report on newer achievements and applications of circular statistics to animal behavior. Each procedure will be illustrated by a numerical example.

Circular statistics deals with points on the unit circle or, in other words, with vectors of unit length. In this paper we will concentrate on those occasions, where the *vectors to be analyzed are of different lengths*. This calls for suitable tools borrowed from bivariate statistical analysis.

B. First-order Analysis

Assume that an animal is observed either while it is in a cage and tries to escape, or while it moves in its natural surroundings. Let N be the number of angles ϕ_i observed with respect to a zero direction. The angles ϕ_i are not necessarily independent of each other. Hence, the set $\phi_1, \phi_2, \ldots, \phi_N$ need not be a random sample in the strict probabilistic sense. We exclude the case when consecutive angles exhibit a clear trend, e.g., when $\phi_1 < \phi_2 < \ldots < \phi_N$. Under this restriction it is reasonable to reduce the data to the well-known *mean vector* with rectangular components.

$$\bar{x} = (\Sigma \cos \phi_i)/N, \quad \bar{y} = (\Sigma \sin \phi_i)/N. \tag{1}$$

Alternatively, we may calculate the polar coordinates of the mean vector, that is,

$$r = (\bar{x}^2 + \bar{y}^2)^{1/2} \tag{2}$$

and

$$\theta = \begin{cases} \arctan(\bar{y}/\bar{x}) & \text{if } \bar{x} > 0, \\ 180^\circ + \arctan(\bar{y}/\bar{x}) & \text{if } \bar{x} < 0. \end{cases} \tag{3}$$

The length of the mean vector, r, is restricted to the interval from 0 to 1, whereas θ can take on any value in a range of 360° (Fig. 1).

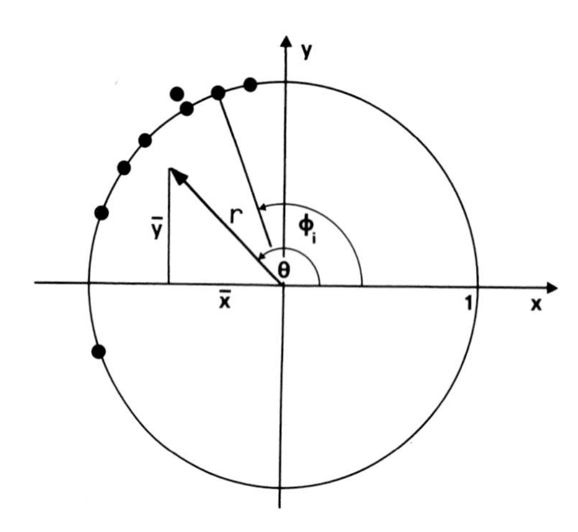

Fig. 1. First-order analysis of a sample of independent or dependent directional observations. The mean vector of length r and polar angle θ points to the "center of mass" with rectangular coordinates $\bar{x}$ and $\bar{y}$

If we knew that the angles ϕ_i were independent, we could test the null hypothesis of a uniform circular distribution, e.g., by application of the widely used Rayleigh test (see, e.g., Zar, 1974). In the opposite case, however, when the angles ϕ_i are dependent on each other, such a test would be of little value.

The calculation of the pair $(\bar{x}, \bar{y})$ or, alternatively, of (r, θ) from a sample of observed angles will be called a *first-order* analysis.

The *sample size*, N, is of some importance. The higher N is, the more precise is the information contained in the pair (r, θ). This is clearly seen if we consider the special case of a *uniform distribution* and independent angular observations. Here the cumulative distribution of r is well-known (see Greenwood and Durand, 1955). Figure 2 depicts the frequency distributions of r for N = 10 and N = 20. Although the theoretical mean vector of a uniform circular distribution is a zero vector, the sample mean vector can range in length from zero to one. For N = 10, the length r takes on values from 0 to practically 0.8, but most frequently it takes on values around the mode at 0.23. For N = 20, the length r ranges from 0 to practically 0.6 with mode at 0.17. In our special case of a uniform circular distribution an increase of sample size results in a decrease of r on the average.

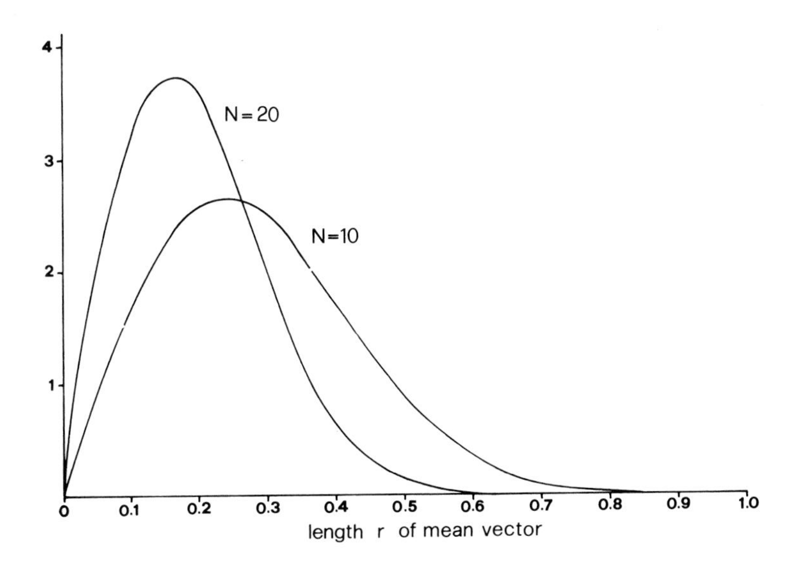

Fig. 2. Distributions of the length r of the mean vector for uniform circular samples of sizes N = 10 and N = 20

C. The Problem

To obtain information on directional behavior, the experimenter rarely works with a single animal. Instead, he tests as many animals of the same kind as he can afford. For each individual animal he collects an equal number N of observations. These observations can be reduced to a pair (r, θ). Let n be the number of animals, then data reduction yields n pairs

$$(r_k, \theta_k), \quad k = 1, 2, \ldots, n. \tag{4}$$

This set may be considered as a new sample, so to say, a "sample of mean vectors." We call the set (4) a *second-order sample*. Since the n animals are expected to be tested independently of each other, the second-order sample is a *random sample* (see Fig. 4).

Second-order samples are shown in various biologic papers (e.g., Ercolini and Scapini, 1976, Fiore and Ioalé, 1973, Herrnkind and McLean, 1971). The term "second-order mean vector" was apparently coined by Wallraff (1970, pp. 331 and 332).

Various questions arise:

1. Do the animals have a preferred direction?

2. If yes, is the preferred direction common to all animals?

3. Is there a significant difference between the directional behavior of "controls" and "experimentals"?

In the following sections we will explain how to solve such questions. We will speak of a *second-order analysis*.

The restriction that N should be equal for all samples of a first-order analysis is essential. The biologist should design his experiments correspondingly. If, however, by some failure N cannot be kept constant, a slight variation of N would hardly damage the statistical analysis.

D. Univariate Test for Concentration

Given is a second-order sample (r_k, θ_k), $k = 1, 2, \ldots, n$, as defined above. To analyze the angles θ_k for themselves, disregarding the lengths r_k, we may apply ordinary tools of circular statistics. They are explained to biologists in Batschelet (1965, 1972) and Zar (1974).

Conversely, we may also consider the values of r_k and disregard the mean angles θ_k. For instance, if a researcher wants to know whether individual animals have a tendency to concentrate on a direction, he should test the univariate second-order sample $r_1, r_2, \ldots, r_n$. For practical reasons, it is simpler to work with the quantity R, the length of the resultant vector calculated for each animal. In our notations

$$R = Nr \tag{5}$$

where N means the number of observations per animal. Thus we consider the second-order sample

$$R_1, R_2, \ldots, R_n.$$

We will also assume that the Rs are already written in ascending order, that is,

$$R_1 \leq R_2 \leq \ldots . \leq R_n. \tag{6}$$

Now we have to confront the sample with the theoretical distribution of R, given N, under the *null hypothesis of a uniform circular distribution*. For this purpose we use a table of the cumulative distribution of R published by Greenwood and Durand (1955). The reader will find it reproduced in Table 1.

To test the null hypothesis, *Kolmogorov's one-sample test* is most suitable. We determine the largest difference in the vertical direction between the cumulative frequency distribution of the sample and the theoretical frequency destribution (cf. Fig. 3). Then we compare this difference with its critical value, given a certain significance level α.

Example 1. To study the possibility of a magnetic orientation in the sandhopper, *Talitrus saltator*, Arendse tested animals individually in the laboratory under isotropic light conditions. For the experimental set-up see Arendse and Vrins (1975). Each animal was observed $N = 12$ times when it jumped from the center of a vessel toward the side wall. In one of these experiments $n = 42$ sandhoppers were tested and a first-order analysis performed. We disregard the mean angles θ_k and consider only the length of all 42 mean vectors r_k or, for the special purpose of the statistical test, the values $R_k = 12r_k$ according to Eq. (5). From the data (pers. comm.) we get the following sample values R_k

Table 1. Cumulative distribution function of R = Nr for uniform circular distributions (from Greenwood and Durand, 1955)

R	N = 6	7	8	9	10	11	12	13	14	15	16	17	18	19	20	21	22	23	24
0.5	.0388	.0323	.0291	.0259	.0235	.0215	.0198	.0183	.0171	.0160	.0150	.0142	.0134	.0127	.0121	.0116	.0110	.0106	.0102
1.0	.1429	.1250	.1111	.1000	.0909	.0833	.0769	.0714	.0667	.0625	.0588	.0556	.0526	.0500	.0476	.0454	.0435	.0417	.0400
1.5	.2932	.2614	.2334	.2118	.1935	.1782	.1651	.1538	.1440	.1353	.1276	.1208	.1146	.1091	.1040	.0994	.0952	.0914	.0878
2.0	.4678	.4178	.3791	.3462	.3188	.2954	.2751	.2575	.2419	.2282	.2159	.2048	.1949	.1858	.1776	.1700	.1631	.1567	.1508
2.5	.6316	.5746	.5280	.4875	.4529	.4227	.3963	.3729	.3521	.3334	.3167	.3015	.2877	.2751	.2635	.2529	.2431	.2340	.2256
3.0	.7672	.7140	.6642	.6213	.5828	.5487	.5181	.4906	.4658	.4434	.4229	.4042	.3871	.3713	.3568	.3433	.3308	.3192	.3083
3.5	.8702	.8228	.7782	.7368	.6987	.6638	.6318	.6025	.5756	.5508	.5279	.5068	.4872	.4691	.4522	.4364	.4217	.4079	.3950
4.0	.9376	.9004	.8646	.8287	.7945	.7619	.7312	.7023	.6752	.6499	.6262	.6039	.5831	.5635	.5452	.5279	.5116	.4962	.4818
4.5	.9732	.9507	.9240	.8964	.8680	.8401	.8128	.7864	.7610	.7368	.7136	.6916	.6707	.6509	.6320	.6141	.5971	.5809	.5654
5.0	.9912	.9786	.9615	.9420	.9207	.8985	.8759	.8533	.8310	.8092	.7880	.7674	.7476	.7284	.7100	.6923	.6753	.6590	.6434
5.5	.9986	.9920	.9828	.9702	.9556	.9393	.9220	.9039	.8855	.8669	.8485	.8302	.8122	.7947	.7775	.7608	.7446	.7289	.7136
6.0	1.0000	.9979	.9933	.9863	.9770	.9660	.9536	.9400	.9257	.9108	.8956	.8801	.8646	.8492	.8339	.8188	.8039	.7893	.7751
6.5		.9998	.9978	.9944	.9892	.9823	.9740	.9645	.9540	.9426	.9307	.9182	.9055	.8925	.8794	.8663	.8532	.8401	.8272
7.0		1.0000	.9995	.9980	.9954	.9915	.9864	.9801	.9728	.9647	.9558	.9462	.9362	.9257	.9149	.9040	.8928	.8816	.8703
7.5			1.0000	.9994	.9983	.9963	.9934	.9895	.9848	.9792	.9729	.9659	.9583	.9502	.9417	.9329	.9238	.9144	.9049
8.0				.9999	.9995	.9986	.9970	.9948	.9920	.9884	.9841	.9792	.9738	.9678	.9613	.9544	.9472	.9396	.9318
8.5				1.0000	.9999	.9995	.9988	.9977	.9960	.9938	.9911	.9879	.9841	.9798	.9751	.9699	.9644	.9585	.9524
9.0					1.0000	.9999	.9996	.9990	.9982	.9969	.9953	.9932	.9907	.9878	.9845	.9808	.9767	.9722	.9675
9.5						1.0000	.9999	.9996	.9992	.9986	.9976	.9964	.9948	.9929	.9906	.9881	.9852	.9819	.9784
10.0							1.0000	.9999	.9997	.9994	.9989	.9982	.9972	.9960	.9946	.9928	.9908	.9886	.9860
10.5								1.0000	.9999	.9997	.9995	.9991	.9986	.9979	.9970	.9958	.9945	.9930	.9912
11.0									1.0000	.9999	.9998	.9996	.9993	.9989	.9984	.9977	.9968	.9958	.9946
11.5										1.0000	.9999	.9998	.9997	.9995	.9992	.9988	.9982	.9976	.9968
12.0											1.0000	.9999	.9999	.9998	.9996	.9994	.9990	.9986	.9981
13.0												1.0000	1.0000	.9999	.9999	.9998	.9997	.9996	.9994
14.0														1.0000	1.0000	1.0000	.9999	.9999	.9998
15.0																	1.0000	1.0000	1.0000

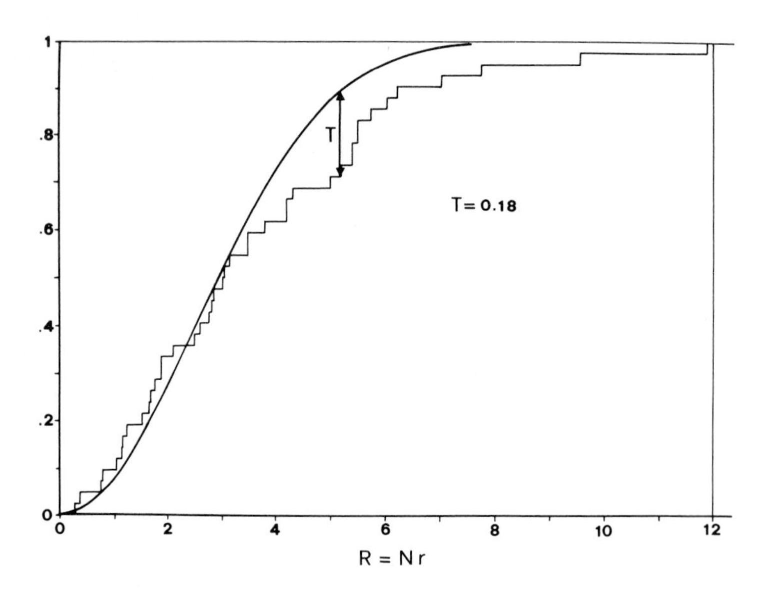

Fig. 3. A second-order analysis of r or R = Nr. Kolmogorov's one-sample test is applied to discover whether the r values are on the average higher than expected from the null hypothesis of uniform circular distributions

0.28	1.18	1.90	2.77	3.43	5.17	6.07
0.38	1.25	1.90	2.81	3.78	5.33	6.20
0.74	1.52	2.12	3.16	4.16	5.36	7.04
0.79	1.63	2.50	3.20	4.16	5.42	7.70
1.04	1.69	2.53	3.28	4.21	5.47	9.56
1.16	1.75	2.74	3.42	4.92	5.76	11.83

The cumulative frequency distribution (or sample distribution function) is plotted in Figure 3.

The null hypothesis states that the animals were not oriented at all, that is, that the values R_k correspond to sample values based on the uniform circular distribution. Under the null hypothesis we can plot a graph of the theoretical distribution function using Table 1 for N = 12. The largest difference, denoted by T, turns out to be T = 0.18. At a significance level of α = 5% and n = 42, the critical value is $T_{0.05}$ = 0.205 (see, e.g., Conover, 1971). Since $T < T_{0.05}$, the null hypothesis cannot be rejected. There is no statistical evidence that the animals were oriented.

E. Hotelling's One-sample Test

Suppose we are given a second-order sample

$$(r_k, \theta_k), \; k = 1, 2, \ldots, n$$

as described previously. A graphic presentation consists of n vectors of variable length r_k and of polar angles θ_k (cf. Fig. 4). We may also use rectangular coordinates

$$(x_k, y_k), \; k = 1, 2, \ldots, n$$

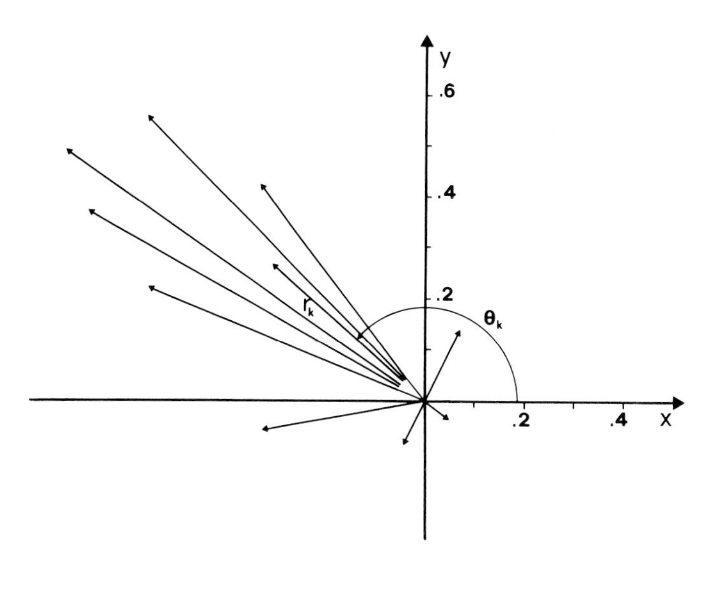

Fig. 4. A second-order sample of directions. Each vector is given by a pair (r_k, θ_k)

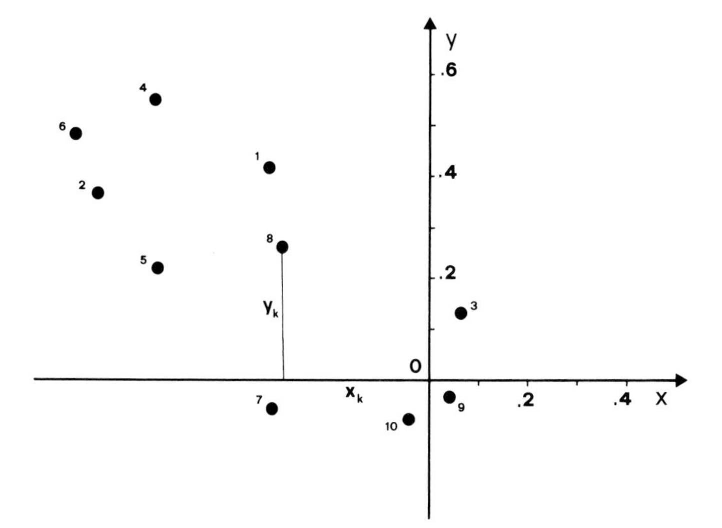

Fig. 5. A scatter diagram equivalent to the second-order sample of directions depicted in Figure 4. Each sample point is given by a pair (x_k, y_k)

and plot only dots with these coordinates (cf. Fig. 5). Remember that x_k and y_k are arithmetic means calculated in a first-order analysis, but we will no longer write $\bar{x}_k$ and $\bar{y}_k$.

The n dots form a scatter diagram with center

$$\bar{x} = (\Sigma\ x_k)/n, \quad \bar{y} = (\Sigma\ y_k)/n. \tag{7}$$

We may ask whether this center deviates significantly from the origin O of the coordinate system. If yes, this would indicate that the n vectors in Figure 4 are more or less concentrated around a certain direction. Therefore, the question whether a group of animals shows an orientation toward a common direction reduces to the other question-whether the center of a scatter diagram differs from a reference point, such as O.

Hotelling solved this problem by extending the well-known one-sample t test. For this purpose we calculate

$$SS_x = \Sigma(x_k - \bar{x})^2, \quad SS_y = \Sigma(y_k - \bar{y})^2 \tag{8}$$

and

$$s_1 = [SS_x/(n-1)]^{1/2}, \quad s_2 = [SS_y/(n-1)]^{1/2} \tag{9}$$

as well as

$$S_{xy} = \Sigma(x_k - \bar{x})(y_k - \bar{y}), \quad r = S_{xy}/(n-1)s_1 s_2 \quad . \tag{10}$$

Here, s_1 and s_2 are the standard deviations of x_k and y_k, respectively, and r is the correlation coefficient of the pairs (x_k, y_k). The test statistics of the univariate t test are

$$t_1 = \frac{\bar{x} - \mu_1}{s_1} n^{1/2}, \quad t_2 = \frac{\bar{y} - \mu_2}{s_2} n^{1/2} \tag{11}$$

where μ_1 und μ_2 are population mean values of x_k and y_k, respectively.

Instead of using confidence intervals for μ_1 und μ_2, Hotelling (1931) introduced a bivariate confidence region for the population center (μ_1, μ_2). The borderline is given by the equation

$$t_1^2 - 2rt_1t_2 + t_2^2 = (1 - r^2)T^2 \tag{12}$$

where Hotelling's T^2 is given by

$$T^2 = 2\,\frac{n-1}{n-2}\,F_{2,n-2} \tag{13}$$

with the critical F value taken for 2 and n-2 degrees of freedom. $F_{2,n-2}$ depends of course on the chosen significance level α. The theory is based on the *assumption that the sample is taken from a bivariate normal distribution.*

The confidence region defined by Eq. (12) turns out to be an ellipse with center in $(\bar{x}, \bar{y})$. With probability $1-\alpha$ this ellipse covers the unknown population center (μ_1, μ_2). Now we have to distinguish between two cases:

1. *If the confidence ellipse covers the origin O, the population center* (μ_1, μ_2) *could be identical with O.* Therefore, there is no significant difference between $(\bar{x}, \bar{y})$ and O. Consequently, *a concentration of the vectors* (r_k, θ_k) *around a certain direction cannot be proven statistically.*

2. *If the confidence allipse does not cover the origin O, the population center* (μ_1, μ_2) *is different from O at the chosen significance level.* Therefore, $(\bar{x}, \bar{y})$ differs significantly from O. Consequently, the vectors (r_k, θ_k) are somewhat concentrated around a mean direction.

Hotelling's test may be performed without the use of a confidence ellipse. For this purpose we combine Eqs. (12) and (13) and calculate the quantity

$$F_{2,n-2} = \frac{n-2}{2(n-1)}\,\frac{1}{1-r^2}\,(t_1^2 - 2rt_1t_2 + t_2^2) \tag{14}$$

with $\mu_1 = 0$ and $\mu_2 = 0$ in Eqs. (11). If $F_{2,n-2}$ exceeds the critical value for a given level of significance, we conclude that the center $(\bar{x}, \bar{y})$ is significantly different from the origin of the coordinate system.

Hotelling's test is widely used in chronobiology. The method was introduced by F. Halberg and co-workers (see, e.g., Halberg et al., 1965).

The drawing of the confidence ellipse requires special attention. Without any further explanation we indicate all consecutive steps:

Using $\bar{x}$, $\bar{y}$, s_1, s_2, r as calculated from Eqs. (7) through (10), determine

$$\left.\begin{aligned} &A = s_1^{-2}, \quad B = -r/s_1 s_2, \quad C = s_2^{-2}, \\ &D = \frac{1-r^2}{n} \cdot \frac{2(n-1)}{n-2} F_{2,n-2} \end{aligned}\right\} \qquad (15)$$

where $F_{2,n-2}$ is the critical F value with 2 and n-2 degrees of freedom.

Case 1: $B \neq 0$. Calculate

$$\left.\begin{aligned} R &= [(A-C)^2 + 4B^2]^{1/2}, \\ \lambda_1 &= (A+C-R)/2D, \quad \lambda_2 = (A+C+R)/2D, \\ a &= \lambda_1^{-1/2}, \quad b = \lambda_2^{-1/2}, \\ \theta_0 &= \arctan - \frac{A-C+R}{2B}. \end{aligned}\right\} \qquad (16)$$

Case 2: $B = 0$.

$$\left.\begin{aligned} &\text{Subcase 2.1: } A < C \\ &\quad a = (D/A)^{1/2}, \quad b = (D/C)^{1/2}, \quad \theta_0 = 0^0 \\ &\text{Subcase 2.2: } A > C \\ &\quad a = (D/C)^{1/2}, \quad b = (D/A)^{1/2}, \quad \theta_0 = 90^0 \\ &\text{Subcase 2.3: } A = C \\ &\quad a = b = (D/A)^{1/2}, \quad \theta_0 = 0^0. \end{aligned}\right\} \qquad (17)$$

The confidence ellipse has center $(\bar{x}, \bar{y})$, semi-axes a and b $(a \geq b)$, and angle θ_0 between major axis and the positive x axis (cf. Fig. 6).

If computer and plotter are available, we define a sequence of angles

$$\phi_j = j \cdot \delta, \quad j = 1, 2, \ldots, 360^0/\delta$$

where the increment δ is suitably chosen, e.g., $\delta = 4^0$. Then we calculate and plot the points (x_j, y_j) with coordinates

$$\begin{aligned} x_j &= \bar{x} + a \cos\theta_0 \cos\phi_j - b \sin\theta_0 \sin\phi_j, \\ y_j &= \bar{y} + a \sin\theta_0 \cos\phi_j + b \cos\theta_0 \sin\phi_j. \end{aligned} \qquad (18)$$

These points form the confidence ellipse.

For applications of the confidence ellipse see the paper by Wallraff in this volume.

Example 2. A first-order analysis resulted in ten mean vectors as shown in Figure 4. We want to test whether there exists a preferred direction or, at least, whether there is some concentration around a mean direction. The diagram may be interpreted as a scatter diagram of ten dots, numbered 1, ..., 10 in Figure 5. The coordinates are

k	x_k	y_k
1	-0.36	+0.45
2	-0.73	+0.40
3	+0.08	+0.16
4	-0.60	+0.58
5	-0.60	+0.22
6	-0.80	+0.52
7	-0.35	-0.06
8	-0.34	+0.29
9	+0.04	-0.06
10	-0.04	-0.10
total	-3.70	+2.40

The center of the ten points is located at

$$\bar{x} = -0.37, \quad \bar{y} = +0.24.$$

We obtain further from Eqs. (8) through (10)

$$SS_x = 0.9012, \quad SS_y = 0.5686, \quad S_{xy} = -0.5252,$$

$$s_1 = 0.3164, \quad s_2 = 0.2514, \quad r = -0.7336.$$

Formula (15) yields

$$A = 9.990, \quad B = 9.223, \quad C = 15.83, \quad D = (0.1039)F_{2,8}.$$

With $\alpha = 0.05$ we read the critical F value from a table:

$$F_{2,8} = 4.46.$$

Hence,

$$D = 0.4634.$$

Since $B \neq 0$, Case 1 is applicable. From Formulas (16) we get

$$R = 19.34, \quad \lambda_1 = 6.981, \quad \lambda_2 = 48.74$$

$$a = 0.3785, \quad b = 0.1432, \quad \theta_0 = -36.2^0$$

Thus the center of the confidence ellipse is at $\bar{x} = -0.37$, $\bar{y} = +0.24$. The major axis is inclined by $\theta_0 = -36.2^0$ versus the positive x axis. The semi-axes are $a = 0.3785$, $b = 0.1432$. The confidence ellipse is depicted in Figure 6. Since the origin O does not fall into the confidence ellipse, we conclude that $(\bar{x}, \bar{y})$ is significantly different from O and, therefore, that the vectors of the first-order analysis are somewhat concentrated around a mean direction; there exists a preferred direction.

We could have avoided the confidence ellipse by applying Eqs. (11) with $\mu_1 = 0$, $\mu_2 = 0$, and Eq. (14). We get

$$t_1 = -3.70, \quad t_2 = +3.02, \quad F_{2,8} = 6.17.$$

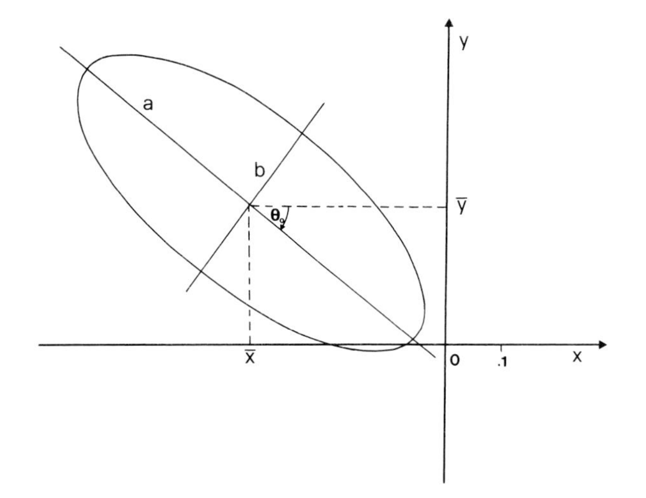

Fig. 6. Confidence ellipse for the bivariate mean (μ_1, μ_2) associated with the second-order samples depicted in Figures 4 and 5. The center is at ($\bar{x}$, $\bar{y}$), the principal axis inclined by an angle θ_o. The half-axes are denoted by a and b ($a \geq b$)

Since the F value from the sample exceeds the critical value 4.46 for $\alpha = 0.05$, significance is proven.

Example 3. Herrnkind and McLean (1971) studied the orientation in the western Atlantic spiny lobster *(Panulirus argus)*. Each animal was tested N = 8 times, and a first-order analysis resulted in a mean vector (r_k, θ_k) for the *k*th animal. From their Table 3 we copy the data for n = 11 unblinded lobsters (with all release angles equivalent) and add the rectangular coordinates of the mean vectors:

k	r	θ	x	y
1	.97	22°	+.899	+.363
2	.74	349°	+.726	-.141
3	.84	118°	-.394	+.742
4	.80	41°	+.604	+.525
5	.98	356°	+.978	-.068
6	.56	357°	+.559	-.029
7	.63	35°	+.516	+.361
8	.90	23°	+.828	+.352
9	.91	59°	+.469	+.780
10	.58	16°	+.558	+.160
11	.64	337°	+.589	-.250
total	-	-	6.332	2.795

The eleven mean vectors are plotted in Figure 7.

For the center of the confidence ellipse we get

$$\bar{x} = 0.576, \qquad \bar{y} = 0.254.$$

It follows from Eqs. (8) through (10) that

$$SS_x = 1.309, \qquad SS_y = 1.224, \qquad S_{xy} = -0.657$$

$$s_1 = 0.362, \qquad s_2 = 0.350, \qquad r = -0.519$$

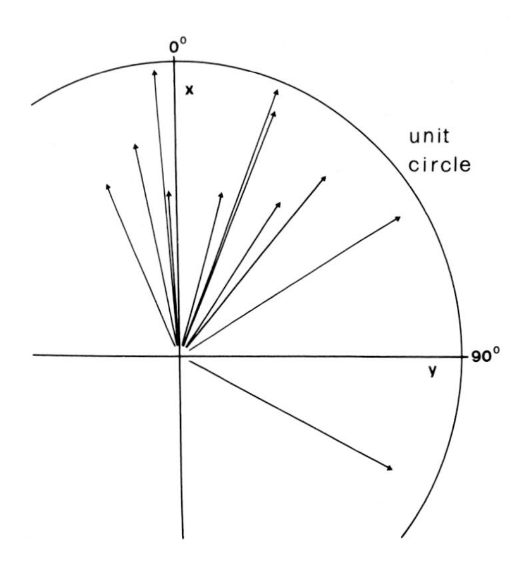

Fig. 7. A second-order sample from Herrnkind and McLean (1971)

Formula (15) leads to

$$A = 7.637, \quad B = 4.102, \quad C = 8.171, \quad D = 0.1476 \cdot F_{2,9}.$$

Given $\alpha = 0.05$, the critical F value is

$$F_{2,9} = 4.26.$$

Hence,

$$D = 0.629.$$

Since $B \neq 0$, we proceed with Case 1. Eqs. (16) yield

$$R = 8.22, \quad \lambda_1 = 6.04, \quad \lambda_2 = 19.11$$

$$a = 0.407, \quad b = 0.229, \quad \theta_o = -43.1^\circ.$$

Thus the semi-axes of the confidence ellipse are $a = 0.407$, $b = 0.229$. The center is at $\bar{x} = 0.576$, $\bar{y} = 0.254$, and the inclination of the major axis versus the positive x axis is given by $\theta_o = -43.1^\circ$. The confidence ellipse is depicted in Figure 8.

Since the confidence ellipse does not cover the origin O, the lobsters were significantly oriented around a common direction.

F. A Nonparametric Competitor of Hotelling's One-sample Test

Hotelling's T^2 test is based on the assumption that the sample is taken from a bivariate normal distribution. This assumption is not always justified. Particularly, in second-order analyses of directions it occurs sometimes that the assumption of normality is seriously violated. Therefore, it is desirable to have a nonparametric alternative.

We will explain the basic idea using Figure 4. Starting from a circular plot we will change to a linear plot, that is, we will plot the angles θ_k

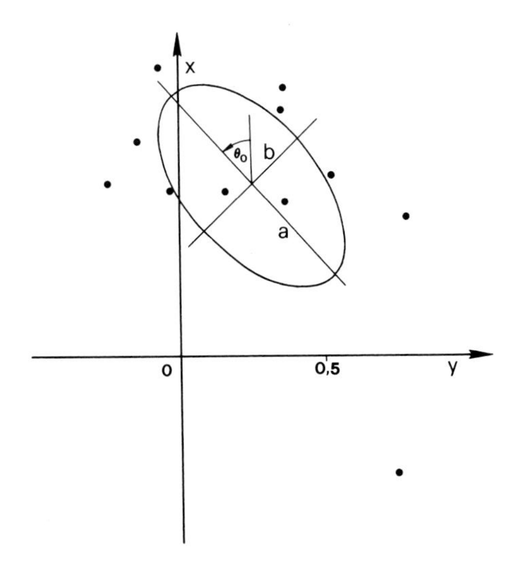

Fig. 8. Scatter diagram and confidence ellipse corresponding to Figure 7

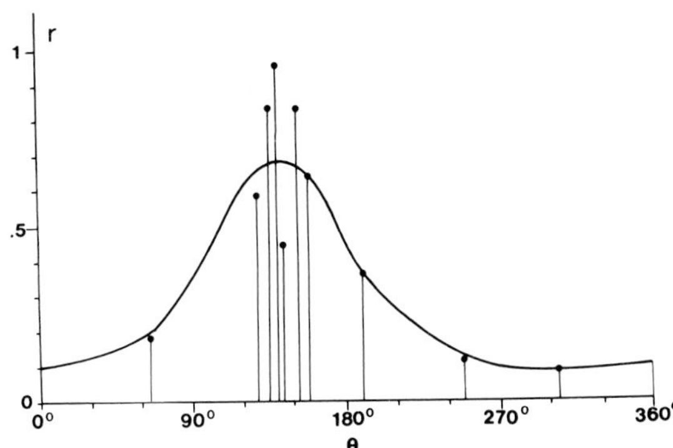

Fig. 9. A linear-circular regression model applied to the second-order sample shown in Figure 4

on a horizontal axis and the vector lengths r_k as ordinates. The result is shown in Figure 9. The figure suggests some trend: an increase in r from $\theta = 0^\circ$ to a maximum at about 150° and a decrease in r when θ approaches 360°. This calls for a correlation analysis. The scale of r is linear (nonperiodic) whereas the scale of θ is circular (periodic). Therefore, we have to consider the case of a *linear-circular correlation*. Mardia (1976) defined a linear-circular rank correlation coefficient D_n, which is useful in this context. If D_n is significantly different from zero, we conclude that the trend is not spurious and, therefore, that the second-order sample has a "peak direction." In other words, there is a significant concentration of directions around a mean direction.

The calculation of Mardia's correlation coefficient proceeds as follows:

Arrange the vector lengths r_k in ascending order and assign ranks $k = 1, 2, 3, \ldots, n$. Denote the corresponding ranks of θ_k by $q_1, q_2, \ldots, q_n$. Calculate the auxiliary angles

$$\psi_k = (q_k/n) \cdot 360^\circ, \quad k = 1, 2, \ldots, n \tag{19}$$

and the quantities

$$T_C = \sum_{k=1}^{n} k \cos \psi_k,$$

$$T_S = \sum_{k=1}^{n} k \sin \psi_k. \tag{20}$$

Table 2. Table of coefficients and of critical values $U_n(\alpha)$ for Mardia's linear-circular correlation coefficient (adapted from Mardia, 1976)

n	a_n	c_n	$\alpha = 0.01$	$\alpha = 0.05$	$\alpha = 0.10$
5	4.033×10^{-2}	3.968	$U_n(\alpha) =$ ——	——	3.968
6	1.923×10^{-2}	4.952	4.952	4.667	4.572
7	1.032×10^{-2}	5.934	5.747	4.896	4.296
8	6.023×10^{-3}	6.918	6.148	5.170	4.488
9	3.750×10^{-3}	7.901	6.639	5.339	4.503
10	2.456×10^{-3}	8.885	6.679	5.480	4.524
11	1.675×10^{-3}	9.870	7.2	5.5	4.55
12	1.181×10^{-3}	10.854	7.5	5.6	4.57
15	4.828×10^{-4}	13.808	7.9	5.7	4.59
20	1.525×10^{-4}	18.733	8.3	5.8	4.60
30	3.009×10^{-5}	28.586	8.7	5.9	4.60
40	9.518×10^{-6}	38.440	8.8	5.9	4.60
50	3.898×10^{-6}	48.295	8.9	6.0	4.61
100	2.435×10^{-7}	97.570	9.1	6.0	4.61

Take a_n from Table 2. Then the linear-circular correlation coefficient is

$$D_n = a_n(T_C^2 + T_S^2), \quad 0 \leqq D_n \leqq 1. \tag{21}$$

As a test statistic use

$$U_n = c_n D_n \tag{22}$$

where c_n is given in Table 2. Choose a significance level α. Determine the critical value $U_n(\alpha)$ from Table 2. Significance is established if U_n exceeds $U_n(\alpha)$.

Example 4. We use the same data as in Example 2. We only calculate the polar coordinates (r_k, θ_k) from (x_k, y_k). At the same time we rearrange the vectors in ascending order of r_k. We also determine the ranks of θ_k and the auxiliary angles ψ_k:

k	r_k	θ_k	ranks	ψ_k
1	0.072	304°	q_1 = 10	360°
2	0.108	248°	q_2 = 9	324°
3	0.179	63°	q_3 = 1	36°
4	0.355	190°	q_4 = 8	288°
5	0.447	140°	q_5 = 4	144°
6	0.576	129°	q_6 = 2	72°
7	0.639	160°	q_7 = 7	252°
8	0.832	151°	q_8 = 6	216°
9	0.835	136°	q_9 = 3	108°
10	0.954	147°	q_{10} = 5	180°

From Eq. (20) we get

$$T_C = -17.33, \quad T_S = +2.629.$$

With $a_n = a_{10} = 0.002456$ from Table 2, Eq. (21) yields

$$D_n = D_{10} = 0.754.$$

To test whether D_n differs significantly from zero we calculate the test statistic U_n from Eq. (22) with $c_n = c_{10} = 8.885$ from Table 2:

$$U_n = U_{10} = 6.70.$$

With $\alpha = 0.05$ we find the critical value

$$U_{10}(0.05) = 5.48.$$

Since U_{10} as calculated from our sample exceeds the critical value, significance is established. We conclude that in Figure 4 there is some concentration of the vectors around a mean direction.

Example 5. We return to the data of Example 3, which were taken from Herrnkind and McLean (1971). Is there a nonparametric test for orientation? At a glance we can see that the test proposed in this section would fail in our example. There are two reasons for the failure. First, the mean vector lengths fall within a relatively small range. Second, a large sector of the unit circle does not contain any mean vector.

However, we may apply the bivariate sign test by Hodges (see Batschelet 1972, p. 66). All mean vectors are on the same side of a suitably chosen straight line through the origin. Hence, the test statistic is $K = 0$. With $n = 11$ we obtain significance at a level of $p = 0.011$. We conclude that all eleven lobsters are oriented around a common direction.

G. Hotelling's Two-sample Test

An investigation may result in two independent second-order samples, and the question then arises: Can the two samples be considered as being taken from the same population (null hypothesis), or is there a significant difference between the two samples?

In Figure 10 two samples of vectors are depicted, a first sample consisting of seven vectors (triangles at tip) and a second sample consisting of nine vectors (circles at tip). If we only look at the tips, we are faced with scatter diagrams of two bivariate samples. Also shown in Figure 10 is the common center of mass, S.

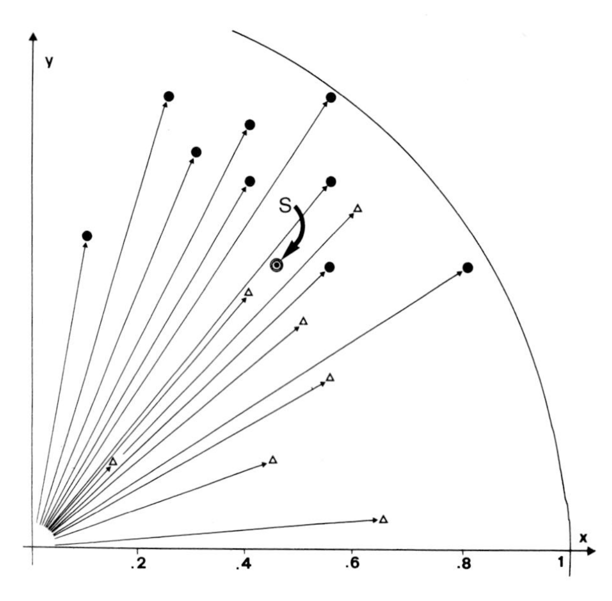

Fig. 10. Two independent second-order samples. The common "center of mass" is denoted by S

For a statistical test of the two samples it does not matter which point we choose as reference point. It is convenient to shift the x,y coordinate system so that S becomes the new origin. This is done in Figure 11. If x, y are the old and x', y' the new coordinates, then

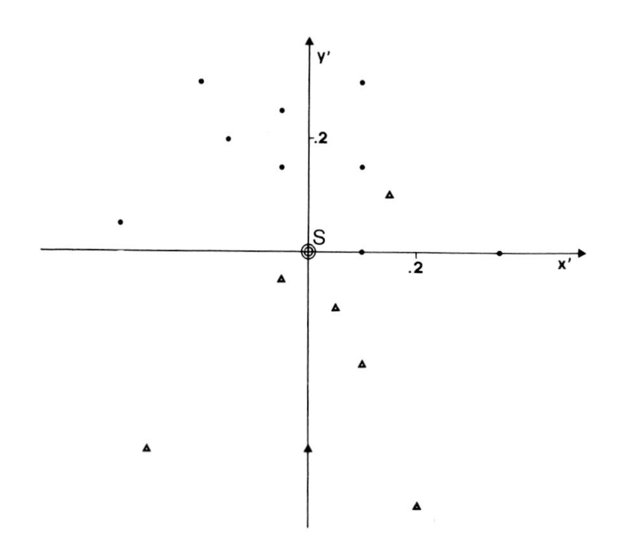

Fig. 11. A scatter *diagram* of the two second-order samples shown in Figure 10. The new origin is S

$$x' = x - \bar{x}_S, \quad y' = y - \bar{y}_S \tag{23}$$

where $\bar{x}_S$ and $\bar{y}_S$ are the original coordinates of S.

Hotelling considered the case where both samples can be assumed as being taken *from bivariate normal distrimutions with the same covariance matrix*. Under these conditions the null hypothesis states that the two parent populations, have the same location parameters. If the null hypothesis can be rejected, it simply means that the parent populations differ only by a translation. Figure 10 suggests a displacement of about 0.4 in the y direction.

Hotelling's test resembles somewhat the familiar two-sample t test. We denote the abscissas and ordinates of the first sample of size n by x_{i1} and y_{i1}, respectively. The corresponding coordinates of the second sample are x_{i2} and y_{i2} $(i = 1, 2, \ldots, m)$.

Now we calculate

$$\left.\begin{aligned} \bar{x}_1 &= \frac{1}{n}\sum_{i=1}^{n} x_{i1}, \quad & \bar{y}_1 &= \frac{1}{n}\sum_{i=1}^{n} y_{i1}, \\ \bar{x}_2 &= \frac{1}{m}\sum_{i=1}^{m} x_{i2}, \quad & \bar{y}_2 &= \frac{1}{m}\sum_{i=1}^{m} y_{i2}, \end{aligned}\right\} \tag{24}$$

$$\left.\begin{aligned} SS_{x1} &= \Sigma(x_{i1} - \bar{x}_1)^2, \quad SS_{y1} = \Sigma(y_{i1} - \bar{y}_1)^2, \\ S_{xy1} &= \Sigma(x_{i1} - \bar{x}_1)(y_{i1} - \bar{y}_1), \\ SS_{x2} &= \Sigma(x_{i2} - \bar{x}_2)^2, \quad SS_{y2} = \Sigma(y_{i2} - \bar{y}_2)^2, \\ S_{xy2} &= \Sigma(x_{i2} - \bar{x}_2)(y_{i2} - \bar{y}_2), \end{aligned}\right\} \tag{25}$$

$$\left.\begin{aligned} s_x^2 &= (SS_{x1} + SS_{x2})/(n + m - 2), \\ s_y^2 &= (SS_{y1} + SS_{y2})/(n + m - 2), \\ s_{xy} &= (S_{xy1} + S_{xy2})/(n + m - 2), \\ r &= s_{xy}/s_x s_y \end{aligned}\right\} \tag{26}$$

$$t_x = \frac{\bar{x}_1 - \bar{x}_2}{s_x\left(\frac{1}{n}+\frac{1}{m}\right)^{1/2}}, \quad t_y = \frac{\bar{y}_1 - \bar{y}_2}{s_y\left(\frac{1}{n}+\frac{1}{m}\right)^{1/2}}. \tag{27}$$

Notice that Eq. (27) defines the t values for two-sample t tests performed on the x and y values separately (cf. Hald, 1952, p. 618).

Hotelling's T^2 is defined by

$$T^2 = \frac{1}{1-r^2}\left(t_x^2 - 2rt_xt_y + t_y^2\right). \tag{28}$$

Under the null hypothesis of no shift of the parent populations the quantity

$$\frac{n+m-3}{2(n+m-2)}\, T^2 = F_{2,n+m-3} \tag{29}$$

has an F distribution with 2 and n+m-3 degrees of freedom. Therefore, Hotelling's bivariate two-sample test is reduced to an ordinary F test.

Example 6. We use the data of Figure 10 or, having performed the translation indicated by Eqs. (23), the data of Figure 11. Written without primes, these data are:

Sample 1 (n = 7)		Sample 2 (m = 9)	
x_{i1}	y_{i1}	x_{i2}	y_{i2}
-0.30	-0.35	+0.35	0
0	-0.35	+0.10	0
+0.20	-0.45	-0.35	+0.05
+0.10	-0.20	+0.10	+0.15
+0.05	-0.10	-0.05	+0.15
-0.05	-0.05	-0.15	+0.20
+0.15	+0.10	-0.20	+0.30
		-0.05	+0.25
		+0.10	+0.30

Application of Eqs. (24) through (29) yields

$\bar{x}_1 = 0.0214$, $\bar{y}_1 = -0.2000$, $\bar{x}_2 = -0.0167$, $\bar{y}_2 = 0.1656$

$SS_{x1} = 0.1643$, $SS_{y1} = 0.2300$, $SS_{x2} = 0.3400$, $SS_{y2} = 0.1122$

$S_{xy1} = 0.0375$, $S_{xy2} = -0.0592$,

$s_x^2 = 0.0360$, $s_y^2 = 0.0244$,

$s_{xy} = -0.00155$, $r = -0.0522$,

$t_x = 0.398$, $t_y = -4.51$,

$T^2 = 20.4$, $F_{2,13} = 9.47$.

If we choose $\alpha = 0.01$, the critical F value is

$$F_{2,13}(0.01) = 6.70.$$

$F_{2,13}$ is larger than the critical value. Hence, the two samples deviate significantly from each other.

H. Nonparametric Competitors to Hotelling's Two-sample Test

Hotelling's two-sample test is based on the assumption that the two samples are taken from parent populations which are both *normal and coincide in the covariance matrix*. This assumption is a heavy burden and hardly ever satisfied. To be on the safe side it is wise, therefore, to apply nonparametric tests.

We will list a few such alternatives, which will all be based on the same idea. As in Section G, we shift the coordinate system in such a way that the new origin falls into the common center of gravity, S. The result of such a shift is illustrated in Figure 11. As a rule, one sample lies on one side of S, the other sample on the opposite side. Some overlapping is possible. Now we consider the directions from S to the sample points and disregard the distances. The directions to the points of one sample are somewhat different from the directions to the points of the other samples. To test significance we may apply a circular two-sample test. Figure 12 depicts two circular samples that originate from Figure 11 if we only consider the directions.

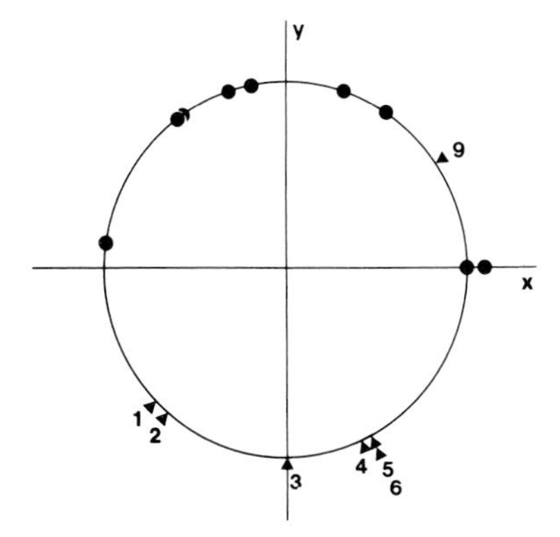

Fig. 12. Two independent circular samples derived from Figure 11 by disregarding the distances. By this method testing of bivariate samples is reduced to a problem of circular statistics according to an idea by Mardia

The idea to reduce the problem of testing bivariate samples to circular samples was conceived by Mardia (1967).

As soon as this reduction is performed, any exact circular two-sample test can be applied, e.g., the now famous Mardia-Watson-Wheeler test, the runs test, or a modification of the Wilcoxon test (cf. Batschelet 1965, 1972, Zar 1974).

Example 7. We refer to Figures 11 and 12. The two samples are represented by the following angles:

Sample 1 (n = 7)	Sample 2 (m = 9)
229°	0°
270°	0°
294°	172°
297°	56°
297°	108°
225°	127°
34°	124°
	101°
	72°

When applying the Mardia-Watson-Wheeler test we have to rank all 16 directions, whereby we may start at an arbitrary location (Batschelet, 1972). Starting with the angle 225° of Sample 1 and turing counterclockwise, Sample 1 obtains the following ranks

$$q_i = 1, 2, 3, 4, 5, 6, 9.$$

The corresponding auxiliary angles are

$$\beta_i = [360°/(n+m)](q_i) = 22.5°, 45°, 67.5°, 90°, 112.5°, 135°, 202.5°.$$

It follows

$$V = \sum_{i=1}^{n} \cos \beta_i = 0, \quad W = \sum_{i=1}^{n} \sin \beta_i = 4.262,$$

$$B = R^2 = V^2 + W^2 = 18.2.$$

At a level of $\alpha = 0.05$ the critical value is $B(0.05) = 12.66$. Our B exceeds the $B(0.05)$. Hence, significance is established. We would also obtain significance at the level $\alpha = 0.01$.

We conclude with listing the results of two other circular two-sample tests applied to the same data (cf. Batschelet, 1965):

Test	Test statistic	Significance at level
Runs	$h = 4$	0.035
Modified Wilcoxon	$S = 30$, $U = 2$	0.01

Example 8. Heiligenberg (1974, Fig. 2) compared two samples of vectors in a plane. The vector length represents the gain in electromotor response of fish and the polar angle the phase lag. In Figure 13 only the tips of the vectors are shown. Do the samples differ significantly from each other?

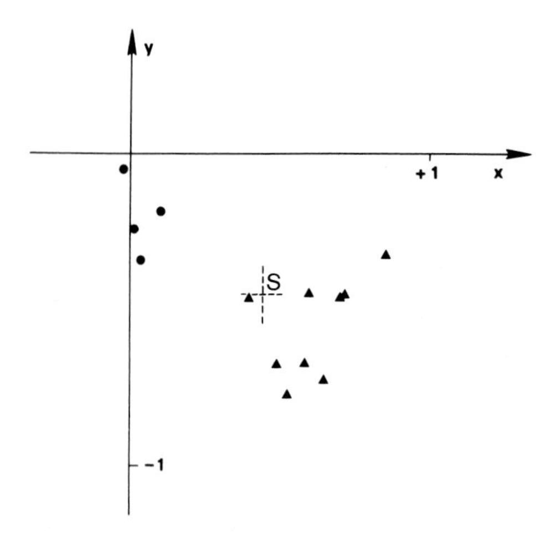

Fig. 13. A scatter *diagram* of two independent bivariate samples studied by Heiligenberg (1974)

Hotelling's parametric two-sample test is hardly applicable since the covariance matrices of the two populations are likely to differ considerably from each other. Therefore, we apply a bivariate nonparametric test. We shift the origin to the common center, S, of the two samples and disregard the distances from S. Finally, we proceed with a circular two-sample test.

The rectangular coordinates of the sample points are

Sample 1 (dots)		Sample 2 (triangles)	
x	y	x	y
-0.03	-0.05	+0.87	-0.32
+0.11	-0.19	+0.41	-0.46
+0.02	-0.24	+0.60	-0.44
+0.03	-0.34	+0.70	-0.46
(n = 4)		+0.72	-0.45
		+0.50	-0.67
		+0.59	-0.67
		+0.65	-0.72
		+0.54	-0.78
		(m = 9)	

The center S, common to both samples, has coordinates

$$\bar{x} = 0.439, \qquad \bar{y} = -0.445.$$

Choosing S as the new origin and disregarding the distances from S we obtain two circular samples as shown in Figure 14. The two samples are entirely separated. Applying some circular two-sample tests we get the following results:

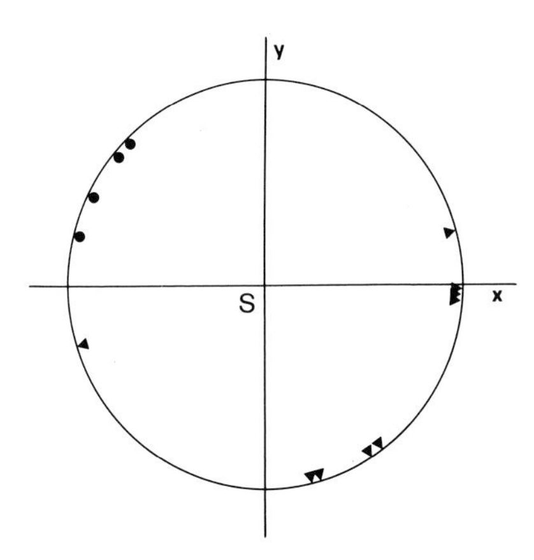

Fig. 14. A circular test applied to the data of Figure 13

Test	Test statistic	Significance at level
Mardia-Watson-Wheeler	B = 11.8	< 0.05
Runs	h = 2	0.018
Modified Wilcoxon	U = 0	0.018

All three tests result in significance at a level below 5%.

References

Arendse, M.C., Vrins, J.C.M.: Magnetic orientation and its relation to photic orientation in *Tenebrio Molitor* L. Netherl. J. Zool. 25, 407-437 (1975)

Batschelet, E.: Statistical Methods for the Analysis of Problems in Animal Orientation and Certain Biological Rhythms. Washington, D.C.: Am. Inst. Biol. Sci., 1965

Batschelet, E.: Recent statistical methods for orientation data. In: Animal Orientation and Navigation. Galler, S.R. et al. (eds.). NASA-Symposium. Washington, D.C.: U.S. Government Printing Office, 1972

Conover, W.J.: Practical Nonparametric Statistics. New York and London: Wiley, 1971

Ercolini, A., Scapini, F.: Fototassia negativa e orientamento astronomico solare in due specie di stafilinidi ripari. Redia 59, 135-153 (1976)

Fiore, L., Ioalè, P.: A method for studying orientation in pigeons by operant conditioning. Rev. Comp. Animal 7, 275-278 (1973)

Greenwood, J.A., Durand, D.: The distribution of length and components of the sum of n random unit vectors. Ann. Math. Stat. 26, 233-246 (1955)

Halberg, F., Tong, Y.L., Johnson, E.A.: Circadian system phase - An aspect of temporal morphology. In: Cellular Aspects of Biorhythms. Symposium on Rhythmic Research in Wiesbaden. Berlin - Heidelberg - New York: Springer, 1965

Hald, A.: Statistical Theory with Engineering Applications. New York: Wiley, 1952

Heiligenberg, W.: Electrolocation and jamming avoidance in a *Hypopygus* (Rhamphichthyidae, Gymnotoidei), an electric fish with pulse-type discharges. J. Comp. Physiol. 91, 223-240 (1974)

Herrnkind, W.F., McLean, R.: Field studies of homing, mass emigration, and orientation in the spiny lobster, *Panulirus Argus*. Ann. New York Acad. Sci. 188, 359-377 (1971)

Hotelling, H.: The generalization of Student's ratio. Ann. Math. Stat. 2, 360-378 (1931)

Mardia, K.V.: A nonparametric test for the bivariate two-sample location problem. J. R. Stat. Soc., Ser. B. 29, 320-342 (1967)

Mardia, K.V.: Statistics of Directional Data. London - New York: Academic Press, 1972

Mardia, K.V.: Linear-circular correlation coefficients and rhythmometry. Biometrika 63, 403-405 (1976)

Wallraff, H.G.: Über die Flugrichtungen verfrachteter Brieftauben in Abhängigkeit vom Heimatort und vom Ort der Freilassung. Z. Tierpsychol. 27, 303-351 (1970)

Zar, J.H.: Biostatistical Analysis. Englewood Cliffs, N.J.: Prentice-Hall, 1974

Sensory Mechanisms for Animal Orientation - Can Any New Ones be Discovered?

MELVIN L. KREITHEN, Division of Biological Sciences, Cornell University, Ithaca, NY 14853, USA

Abstract

Several new sensory mechanisms usable for animal orientation have recently been discovered. Homing pigeons can detect polarized light, ultraviolet light, and atmospheric infrasounds (0.1 → 10 Hz). Each of these sensory mechanisms had been tested previously without success or had been rejected as impossible. This paper presents an analysis of *why* earlier tests were unsuccessful and looks at why more recent tests succeeded.

A. Introduction

The past few years have produced an unprecedented level of active research on animal migration, navigation, orientation, and homing. Many new sensory capabilities have been discovered. Homing pigeons can detect polarized light, ultraviolet (UV) light, barometric pressure changes, and atmospheric infrasounds (Kreithen and Keeton, 1974a, b; Delius et al., 1976; Kreithen and Eisner, 1978; Delius, 1978; Yodlowski et al., 1977; Kreithen and Quine, manuscript in preparation).

Other vertebrates have been tested and found to be sensitive to cues such as polarized light, ultraviolet light, and electric and magnetic fields (Adler and Taylor, 1973; Groot, 1965; Dietz, 1972; Huth and Burkhardt, 1972; Kimeldorf and Fontanini, 1974; Moehn, 1974; Kalmijn, 1971, 1978).

Many of these sensory channels had been tested and rejected earlier. Out of failures and false interpretations grew generalizations about what animals could and could not do. As a new biologist, I was trained on such doctrines as: Birds cannot hear below 200 Hz; vertebrates cannot see UV light. The effect of these generalizations was to inhibit further studies.

Yet many of these barriers have now been scaled, and on the other side is the rich world of animal sensory capabilities and a revival of the old tradition that the sense organs of animals can be as sensitive as our best physical instruments.

I would like to offer three of my own experiments as case studies of how barriers can be crossed and to offer some tentative explanations of why earlier experiments failed to show what is now easy to accomplish.

Case Study 1: Polarized Light. Polarized light has been an accepted cue for animal orientation ever since the pioneering work of von Frisch (1949, 1967). Although nearly every invertebrate is sensitive to polarized light, most studies with vertebrates were negative or inconclusive. Pigeons had been tested, without success, by Montgomery and Heinemann in 1952. Yet it was an important enough environmental cue that I thought it was worth trying again.

My own tests showed that pigeons could be trained to respond to polarized light (Fig. 1). Successful polarized light tests were also performed by Delius et al., (1976). Why were these tests successful when

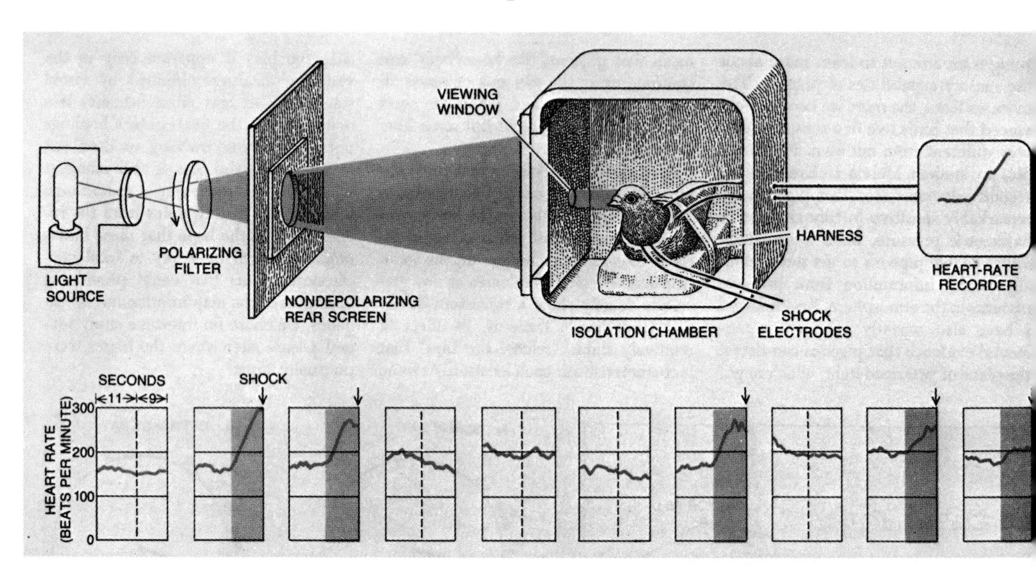

Fig. 1. Polarized light testing apparatus and cardiac responses. The distance between the pigeon's eye and the screen is 1.9 m, which allows the pigeon to use its lateral visual field (see text). EKG and shock electrodes allow simple cardiac conditioning procedures to be used. Each frame in the lower half of the figure is the heart rate obtained in one trial. The *shaded areas* are times when the polarization axis is turning and signals the approach of a mild electric shock *(arrows)*. The heart rate increases dramatically in anticipation of the shock, but does not increase during the corresponding time in other trials when the polarization axis remains stationary (*not shaded*, no shock follows). (Copyright Sci. Am. Dec. 1974 All rights reserved)

others had failed? The differences in results can be accounted for by examining some fundamental differences between the pigeon's visual system and our own visual system. In the middle of building the apparatus for these experiments, I read, quite by accident, an important series of papers by Blough, Catania, and others (Blough, 1971; Catania, 1964; Chard, 1939). These studies revealed that the pigeon's visual system is quite different from ours. The retina has specialized regions each with its own peculiar neural organization, colored oil droplet distribution, and spectral and spatial sensitivity (Fig. 2). Humans tend to center and focus all images on the fovea regardless of distance to the object. Pigeons, on the other hand, use one portion of the retina for close objects (peripheral retina, red oil droplets), and a different part of the retina for distant targets (central and lower retina, yellow oil droplets). Binocular vision is only used for nearby objects; for distant targets, pigeons use each eye separately with independent, monocular fields of view. The polarization of the sky is both distant and overhead, and falls on the region of the yellow retinal field. Montgomery and Heinemann (1952) presented their

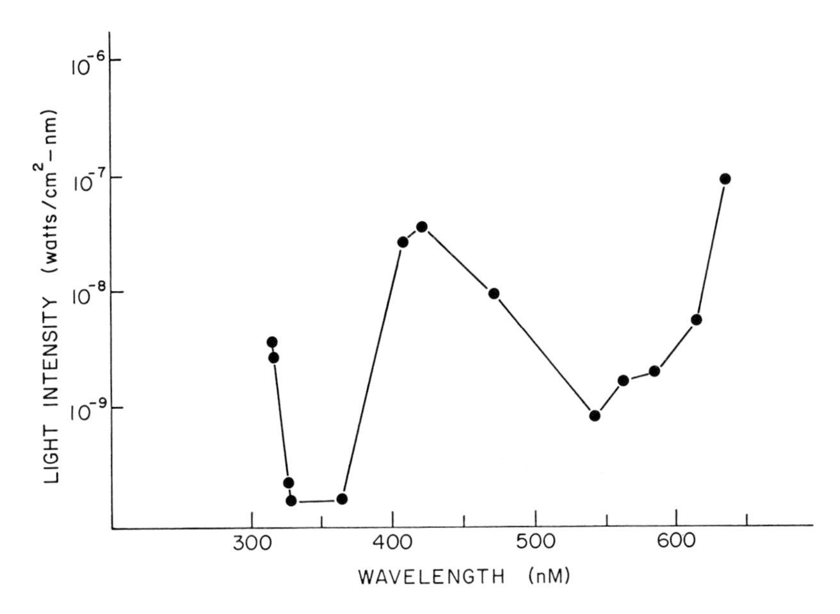

Fig. 2. Ultraviolet and visible light sensitivity of pigeon #3075. Each *dot* is a behavioral threshold (50%) obtained with cardiac conditioning methods. (From Kreithen and Eisner, 1978)

light source close up in the binocular red field of the retina and did not get positive results.

After reading about the pigeons' unique visual system, I rebuilt my experimental apparatus to present the polarized light source at a distance of 1.9 m from the bird, and the birds did respond to the polarizers! Delius et al., (1976) presented his polarized targets overhead and also got behavioral responses. As further confirmation, Delius has repeated Montgomery and Heinemann's tests with near-field illumination and found that the birds still do not respond to polarized targets in the binocular field.

The message seems clear; pigeons can detect polarized light only if it falls on the part of the retina where the sky is normally projected. The problem in this case consisted of thinking of the pigeon's eye in terms of our own eye, which has only a single viewing axis and is always binocularly organized. Distance and position make little difference to our visual system, but it is a completely different matter to a pigeon's eye. We must constantly be on our guard for such differences.

Polarized light vision is an important navigation cue. It has not been conclusively shown that pigeons actually use the sky polarization in the field. More work still needs to be done to determine exactly how pigeons and other birds actually use polarized light.

Case Study 2: Ultraviolet Light. Ultraviolet sunlight wavelengths between 310 - 400 nm provide a well-known source of illumination for insects and other arthropods. Insects routinely use UV to (1) locate the sun behind thin clouds, (2) orient to patterns of polarized UV light in the sky, (3) detect UV flower patterns invisible to us, and (4) locate UV color patches on other insects (von Frisch, 1967; Ghiradella et al., 1972).

Since humans cannot see these UV wavelengths, it is often assumed that most vertebrates are also UV blind. However, in my lab, pigeons have shown a remarkable ability to detect UV light. They have a maximum sensitivity to UV wavelengths between 325 - 360 nm (Kreithen and Eisner, 1978) (Fig. 2.) UV sensitivity has recently been shown in hummingbirds (Huth and Burkhardt, 1972), toads (Dietz, 1972), newts (Kimeldorf and Fontanini, 1974), and lizards (Moehn, 1974), It appears that UV vision may be more widespread among the vertebrate groups than was previously assumed.

Why did it take so long to discover that the pigeon could see UV light? People have been studying the visual system of the pigeon for years; the color vision of the pigeon is the most well-known system of any vertebrate animal besides man. Why was it so difficult to test the UV wavelengths along with the other colors? I believe there were two reasons for the delay; one mental and one physical reason. The mental barrier was that for most of the spectra studied the pigeon's visual sensitivity is much like our own, with a maximum spectral sensitivity between 500 and 600 nm. We have UV-absorbing pigments in our lens that exclude UV wavelengths from the retina. Since light of less than 400 nm wavelength does not reach our retina and the spectral sensitivity is diminishing rapidly also, there appeared to be little reason to test at short wavelengths. The second barrier was the physical reason that glass optics are of little use for UV wavelengths. The performance of ordinary glass lenses and windows is not satisfactory for 300 - 400 nm light. Quartz lenses and optics are required, but the price of these lenses is often exorbitant. The extra cost of UV testing apparatus was a sufficient deterent to further examination of color vision given that the spectral sensitivity was failing rapidly anyway.

As it turns out, there is a less expensive solution to the optical problem. For wavelengths longer than 300 nm, pyrex is a relatively inexpensive glass that transmits UV wavelengths and is about one-tenth the price of quartz optics. When I fitted the testing apparatus with pyrex optics, it was then a simple matter to demonstrate the UV spectral responses of the pigeons (Fig. 2). Notice that the birds' UV sensitivity cannot be predicted from an examination of the responses to visible light, since in this region, the pigeons' vision is much like our own. A second sensitivity maximum is not expected from the behavior at light wavelengths greater than 400 nm. This is an important lesson; sensitivity curves are not always smooth with only one maximum; multiple peaks can occur. Care must be exercised in interpreting data, lest we assume too much.

The lenses and ocular media of the pigeon are clear in the UV rather than being yellow like our own. UV light can reach the retina of many species of birds and possibly many other vertebrate species as well (Govardovski and Zueva, 1977; Dietz, 1972).

We do not know in what ways pigeons use UV light cues. Nor do we know the exact mechanisms of UV transduction. There are many unanswered questions. Can pigeons form sharply focused images at UV wavelengths? Can they see and use the polarized UV light in the sky the way ants and bees can? There is still much left to discover.

Case Study 3: Atmospheric Infrasounds. There is a loud and complex world of low frequency sounds (infrasounds) that we cannot hear. Infrasounds have the capability of traveling thousands of kilometers with little attenuation. With suitable transducers, these sounds can provide navigational and meteorologic information.

Fig. 3. Acoustic tracking of a severe thunderstorm from the four infrasonic monitoring stations operated by the U.S. National Weather Service. (Modified from Bowman and Bedard, 1971)

Geophysicists and meteorologists, using special microphones, have identified numerous infrasound sources. Thunderstorms, magnetic storms, earthquakes, ocean waves, jetstreams, and mountain ranges all produce coherent acoustic signals that can be identified thousands of kilometers from their sources. Figures 3, 4, and 5 show some infrasound sources and their localization by acoustic tracking stations. It is tempting to think that infrasounds might be used for navigation and orientation since the signals arrive from all parts of the world and can be identified and localized. To test if birds could detect infrasounds, I began a long series of laboratory experiments to see if pigeons could respond to artificially generated infrasounds. Much to my surprise, the homing pigeons were very sensitive to low frequency sounds. The birds could respond to the lowest frequency sounds that I could produce - 0.06 Hz! I have no idea what the actual lower limit is - this is all I can produce at the present time. Figure 6 shows the results of behavioral testing of the low frequency sensitivity of two Cornell homing pigeons, #3075 and #7876. Also included in the Figure are an estimate of some natural infrasound intensities and a low frequency audiogram for humans (Yeowart and Evans, 1974; Whittle et al., 1972). Unlike the pigeons, our own auditory thresholds are above the natural infrasound levels and thus we cannot hear these sounds.

Sonic booms from the transatlantic flights of the Concorde produce infrasound levels at Cornell that are within the hearing capabilities of our pigeons (Balachandran et al., 1977). Perhaps our pigeons could be trained to inform us if the SST flights are behind schedule. The data shown on Figure 6 extend the known auditory range 11 octaves lower than any previously published work (Heise, 1953; Harrison and Furomoto, 1971).

Fig. 4. Areas of continuous infrasound production by the interaction of winds and mountains in N.W. America. Another mountain-associated source is located in Argentina. (Modified from Larson et al., 1971)

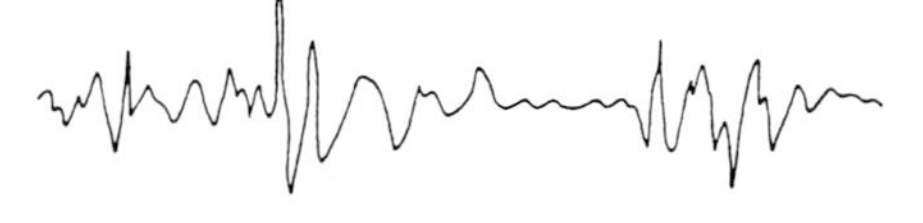

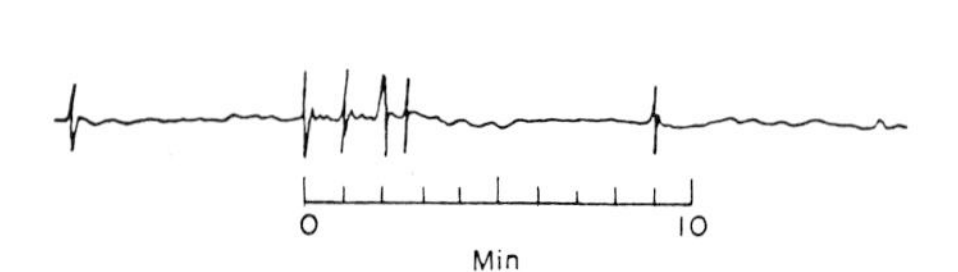

Fig. 5. Infrasound pulses in the 1-16 Hz range, *upper trace*, produced during visible aurora borealis displays. Infrasound associated with geomagnetic activity can also be detected at lower frequencies, *bottom trace*. Data from Procunier, 1971. Peak amplitudes are approximately 1-10 dynes/cm^2

We do not know where the infrasound signals are converted to nerve impulses, but the following two experiments suggest that the mechanism is probably located in the inner ear organ system. (1) Birds with surgically removed cochleas and lagenas do not respond at all to infrasounds (Yodlowski et al., 1977). (2) Other birds, with their collumellas removed, responded to infrasounds but with a 50-dB reduction in sensitivity (Kreithen and Quine, MS). The collumella in birds is a single bone, homologous to our three middle ear bones, that links the tympanic membrane with the inner ear. These findings are only the results of laboratory studies. We do not know if pigeons can use outdoor infrasounds for navigation and orientation. Two major obstacles that the birds must overcome are (1) confusion with pseudosound signals, and (2) the difficulty of localizing long wavelength sounds.

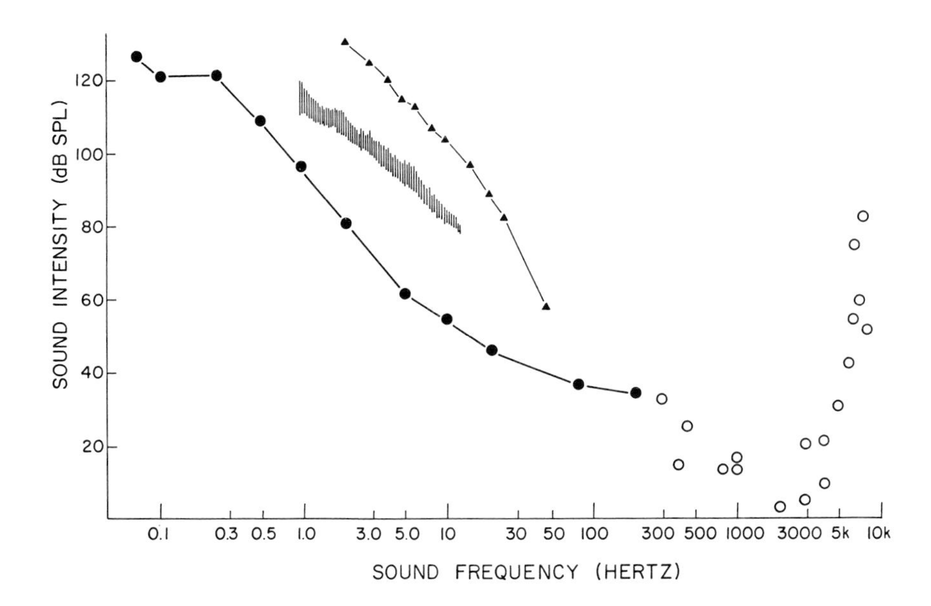

Fig. 6. Infrasound sensitivity of the homing pigeon. Each *filled circle* is the behaviorally determined threshold (50%) at that frequency. *Open circles* are from other laboratory studies. *Triangles* are human thresholds; *vertical bars* indicate natural infrasound levels

Pseudosounds are nonpropagating pressure variations due to local winds and turbulence. On windy days, pseudosounds can mask true infrasounds unless special wind-filtering devices are used. These wind screens are very large - sometimes as much as 3 km in length. The birds need a more compact wind filter, but we do not know if they have one. Perhaps if the birds fly through the air, they can use the spatial patterns of pressure to sort out true infrasounds from local turbulence. It is a serious problem and the bird must have a solution if infrasounds are to be detected on windy days.

Localization of the direction of the sound source is essential for navigational purposes. The bird would need to know where it is on the "acoustic map." Normally, sounds are localized by using binaural differences in signal strength, phase, or time of arrival. None of these parameters is usable at the wavelengths of infrasounds. For example, a sound of 0.1 Hz has a wavelength of 3.4 km in air. There are virtually no differences in infrasound signal quality between the two ears of the pigeon. Microphone arrays used for infrasound research space multiple microphones several kilometers apart. They can then use the time of arrival differences to locate the direction of the sound source. Bird cannot use this method - they are too small and their ears are too close together.

Quine and I are investigating a new method of sound localization based on the doppler shift principle. Regardless of sound frequency, a sound will appear higher in pitch if a bird flies toward the source, and it will appear lower in pitch if the bird flies away from the sound. Pigeons fly at 20 m/s and therefore should encounter a 14% shift in all sound frequencies as they fly toward or away from the source. In a preliminary study by Quine in my laboratory, pigeons have been trained to discriminate shifts in sound frequencies of less than 5%

at 5 Hz. This means that doppler shifts are a possible mechanism for sound localization, although we are still a long way from a definite answer.

The major lesson I would like to offer from this case study is the fact that previous studies of bird hearing failed to look at frequencies below 200 Hz because the thresholds were becoming increasingly less sensitive (see Fig. 6, open circles). What was overlooked was the simple fact that the world is becoming increasingly louder at the lower frequencies, and therefore less sensitive thresholds are completely adequate to hear the low frequency sounds. As I pointed out in the case study of UV light detection, it is dangerous to assume that a sensitivity function is smooth with a single maximum, or that a curve that is increasing will continue to increase smoothly. We must force ourselves to ask empirical questions where there are no data, and not yield to the simpler temptation of extending the results of other studies. In the case of infrasound detection, it is the comparison of the spectral sensitivity *with the ambient sounds* that is important, not just the absolute thresholds of sound pressure levels.

The major task ahead is to see if pigeons can filter and sort out some relevant information from the mass of infrasound signals that surround them. In my most optimistic moments I envision a kind of acoustic topographic map that gives a wealth of navigation cues. In my more pessimistic hours, I think of all of the unsolved problems and obstacles. Regardless of my moods, however, the fact remains that pigeons can detect very low frequency sounds, and it is our job to find out how they use this capability.

B. Concluding Remarks

The title of this paper asks a question: Can any new sensory mechanisms be discovered? The three case studies given here showed how three new sensory capabilities were discovered and briefly indicated how existing barriers, mostly mental, had to be crossed. The question is still unanswered - will any *new* sensory capabilities be discovered? I can answer that question with a statement of logical necessity: If we are going to understand animal navigation, we *must* discover a new sensory channel. The ones we now have are not sufficient to explain the animals' behavior. Since we must discover a new cue, we can proceed to search with the optimism that it exists and that it is our job to find it. We must beware of our own limitations and try to overcome mental barriers about what can and what cannot be accomplished.

One final *caveat* that I am compelled to mention: Although we can proceed with optimism, excitement, and the faith that animals can be as good as any of our instruments, we must nevertheless maintain the highest standards of rigor in our experimental procedures. When we announce a new discovery, it must really be the case, supported with proper controls and checks against false interpretations. I hope I will be present at the next symposium on animal orientation, when important new discoveries will surely be revealed.

References

Adler, K., Taylor, D.H.: Extraocular perception of polarized light by orienting salamanders. J. Comp. Physiol. 87, 203-212 (1973)

Balachandran, N.K., Dunn, W.L., Rind, D.H.: Concorde sonic booms as an atmospheric probe. Science 197, 47-49 (1977)

Blough, P.M.: The visual acuity of the pigeon for distant targets. J. Exp. Anal. Behav. 15, 57-67 (1971)

Bowman, H.S., Bedard, A.J.: Observations of infrasound and subsonic disturbances related to severe weather. Geophys. J.R. Astr. Soc. 26, 215-242 (1971)

Catania, A.C.: On the visual acuity of the pigeon. J. Exp. Anal. Behav. 7, 361-366 (1964)

Chard, R.D.: Visual acuity in the pigeon. J. Exp. Psychol. 24, 588-608 (1939)

Delius, J., Emmerton, J.: Visual Performance of Pigeons. Accepted for publication in: Brain, Behavior, Evolution (1978)

Delius, J., Perchard, R., Emmerton, J.: Polarized light discrimination by pigeons and an electroretinographic correlate. J. Comp. Physiol. Psychol. 90(6), 560-571 (1976)

Dietz, M.: Erdkröten können UV-Licht sehen. Naturwissenschaften 59(7), 316 (1972)

Frisch, K. von: Die Polarisation des Himmelslichtes als orientierender Faktor bei den Tänzen der Bienen. Experientia 5, 142-148 (1949)

Frisch, K. von: The dance language and orientation of bees. Cambridge: Belknap Press, 1967, 566pp.

Georges, T.M., Beasley, W.H.: Refraction of infrasound by upper atmospheric winds. J. Acoust, Soc. Am. 61, 28-34 (1977)

Ghiradella, H., Aneshansley, D., Eisner, T., Silberglied, R., Hinton, H.: Ultraviolet reflection of a male butterfly: Interference color caused by thin-layer elaboration of wing scales. Science 178, 1214-1217 (1972)

Govardovski, V.I., Zueva, L.V.: Visual pigments of chicken and pigeon. Vision Res. 17, 537-543 (1977)

Groot, C.: On the orientation of young Sockeye Salmon *(Oncorhynchus nerka)* during their seaward migration out of lakes. Behavior Suppl. 14, 198pp. (1965)

Harrison, J.B., Furomoto, L.: Pigeon audiograms: comparison of evoked potential and behavioral thresholds in individual birds. J. Auditory Research 11, 33-42 (1971)

Heise, G.A.: Auditory thresholds in the pigeon. Am. J. Psychol. 66, 1-19 (1953)

Huth, H., Burkhardt, D.: Der spektrale Sehbereich eines Violettohr-Kolibris. Naturwissenschaften 59, 650 (1972)

Kalmijn, A.J.: The electric sense of sharks and rays. J. Exp. Biol 35, 371-383 (1971)

Kimeldorf, D.J., Fontanini, D.F.: Avoidance of near ultraviolet radiation exposures by an amphibious vertebrate. Environ. Physiol. Biochem 4, 40-44 (1974)

Kreithen. M.L., Eisner, T.: Detection of ultraviolet light by the homing pigeon. Nature (London) 272(5651), 347-348 (1978)

Kreithen, M.L., Keeton, W.T.: Detection of polarized light by the homing pigeon, *Columba livia*. J. Comp. Physiol 89, 83-92 (1974a)

Kreithen, M.L., Keeton, W.T.: Detection of changes in atmospheric pressure by the homing pigeon, *Columba livia*. J. Comp. Physiol 89, 73-82 (1974b)

Kreithen, M., Quine, D. (MS): Infrasound detection by the homing pigeon: A behavioral audiogram. Submitted to J. Comp. Physiol.

Larson, R.J., Craine, L.B., Thomas, J.E., Wilson, C.R.: Correlation of winds with production of certain infrasonic signals in the atmosphere. Geophys. J. Roy. Astr. Soc. 26, 201-214 (1971)

Moehn, L.: The effects of quality of light on agonistic behavior of iguanid and agamid lizards. J. Herpetology 8, 175-183 (1974)

Montgomery, K.C., Heinemann, E.G.: Concerning the ability of homing pigeons to discriminate patterns of polarized light. Science 116, 454-456 (1952)

Procunier, R.W.: Observations of acoustic aurora in the 1-16 Hz range. Geophys. J. Roy. Astr. Soc. 26, 183-189 (1971)

Whittle, L.S., Collings, S.J., Robinson, D.W.: Audibility of low frequency sounds. J. Sound. Vibr. 21, 431-448 (1972)
Yeowart, N.S., Evans, M.J.: Thresholds of audibility for very low frequency pure tones. J. Acoust. Soc. 55, 814-818 (1974)
Yodlowski, M.L., Kreithen, M.L., Keeton, W.T.: Detection of atmospheric infrasound by homing pigeons. Nature (London) 265(5596), 725-726 (1977)

Sensory Mechanisms Related to Homing in Pigeons

JUAN D. DELIUS and JACKY EMMERTON, Ruhr-Universität Bochum, Psychologisches Institut, 4630 Bochum-Querenberg, FRG

Abstract

Besides having a well-developed vestibular system, pigeons possess a gravity and inertia sensitive system mediated by displacement of the viscera, presumed to stimulate mesenteric mechanoreceptors. Classical conditioning experiments confirm a sensitivity to atmospheric pressure changes. Failure to demonstrate a magnetic field orientation sensitivity in an instrumental conditioning situation is reported. Behavioral and physiological experiments establishing the pigeon's ability to detect and orient itself with respect to an overhead polarized light are referred to. Failure to demonstrate discrimination of such a plane orientation on other experiments suggests that mechanisms connected with the upper field of vision are specialized for polarization plane detection. Evidence that pigeons can perceive and discriminate ultraviolet light is mentioned. Finally reference is made to the widespread forebrain projections of olfactory input.

Accurate information about position or position changes with reference to gravity and inertia must be an important prerequisite for navigation. The vestibular sense organs of pigeons are accordingly well developed. The cerebellum, the main processing projection of this sensory complex in vertebrates, is comparatively large in birds. The vestibular projections to the thalamus of pigeons also seem to be relatively more extensive than those found in mammals (Vollrath and Delius, 1976). Moreover some of the units seem to be engaged in performing a compensating function that must be postulated if vestibular information is to be used for spatial orientation of the body in an animal with a neck as mobile as that of birds. Neck mechanoreceptor afferences and labyrinthine input are integrated by these units in a way that is qualitatively compatible with a compensation of head movements relative to the body (Mittelstaedt, 1964). Birds do not seem to rely on this mechanism alone. It has been known for some time that wing, tail and leg righting-reflexes persist in pigeons whose spinal cord has been transected at thoracic levels (Biederman-Thorson and Thorson, 1973). Recording from the sacral dorsal root fibers of such pigeons we found that a proportion of the units responded to pitch, roll, and yaw even after we had transected the somatic branches of the dorsal roots. Mechanical movement of the viscera with an implanted balloon or lever suggests that this activity was due to the stimulation of mesenteric mechanoreceptors whose axons reach the dorsal roots through the rami communicantes. The viscera thus function in a manner analogous to otoliths (Delius and Vollrath, 1973).

Even though pigeons seem unusually well oriented with respect to gravity, they still have some striking difficulties in certain situations in which we might expect these orientation abilities to be used. We have found that they have considerable difficulty in learning to discriminate an oblique cross (x) from a vertical/horizontal one (+)

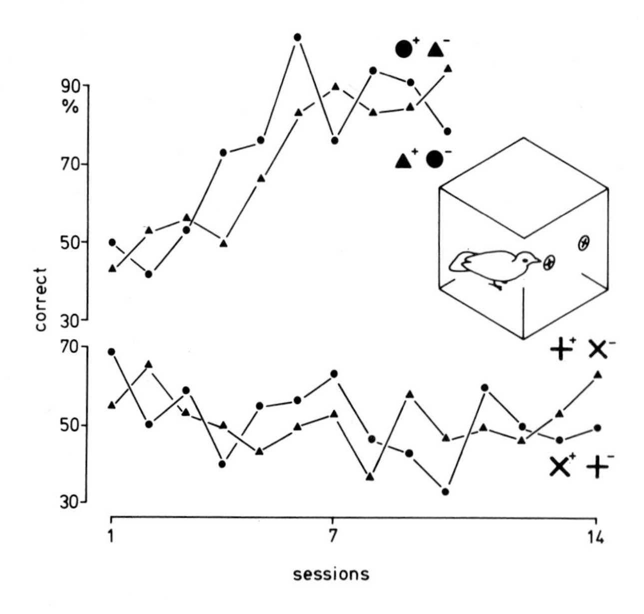

Fig. 1. Learning curves of the same two pigeons discriminating an *upright* and an *oblique cross* and a *triangle* and *circle* in separate experiments. Each session consisted of 30 trials

projected on the pecking keys of a Skinner box (Fig. 1). Reports suggesting that pigeons have some difficulty in recognizing varying orientations of other visual patterns displayed on Skinner box keys can also be found in the literature (Corballis and Beale, 1976). The error seems to arise from the fact that when observing the figures the pigeon seems to align its head along the vertical or oblique axis of the figures and obviously does not compensate this alignment with vestibularly available information about the true vertical. It remains to be established whether this difficulty is restricted to the pigeon's binocular frontal field of vision, served by the red field of the retina, that appears somewhat specialized for food gathering (Galifret, 1968, Catania, 1964, Romeskie and Yager, 1976 but see Friedman, 1975). An inability to learn to discriminate grains by their orientation seems affordable, perhaps even desirable, for a pigeon, if it can spontaneously align its head to enable a good bill grip on them.

We have established that pigeons know up from down. It would be useful for them to have information about how far up or down they were. This can be provided by barometric sensitivity, i.e., by a sense organ responsive to atmospheric pressure levels (see however Bilge, 1973). Such a sensitivity was demonstrated by Kreithen and Keeton (1974a). We just wish briefly to report experiments that were begun independently (Watson, 1974) and that fully support their findings. They were initiated by an incidental observation that pigeons performing steadily in certain behavioral experiments were disturbed by the opening and closing of remote laboratory doors even when it was highly unlikely that they could have heard anything. We used an aversive classical heart rate conditioning technique similar to that used by Delius and Tarpy (1974). Pigeons equipped with chronically implanted electrocardiogram and shock electrodes were placed in a through-ventilated pressure chamber and, at random intervals, experienced 10-s pressure increases (or decreases) preceding the administration of a 50-Hz 0.8 mA 0.1-s shock. Heart beats were counted during pairs of 10-s periods that were quasirandomly alternated and separated by an approximately 2-min variable interval. During half of these pairs (control), chamber pressure remained equal to atmospheric pressure. In the remaining pairs (experimental), atmospheric pressure was again used in the first 10-s period but during the second period a pressure change, followed by shock, was instituted.

From these 25 control and 25 experimental pairs of periods we derived separate signed mean changes in heart beats per 10-s where the control yielded an estimate of the baseline heart-rate trend over the session (usually a slight downward trend) and the experimental pairs gave an estimate of the conditioned heart-rate increases superimposed on this trend. The signed difference between the two means gave the net conditioned heart-rate increases occurring in the presence of the pressure wave.

For two pigeons the conditioned stimulus was a pressure increase, for one it was a pressure decrease. During the initial sessions with each animal the 10-s pressure pulses had an amplitude of 10 cm H_2O. Then each animal had two sessions with 2 cm H_2O pulses and finally they had three control sessions incorporating shocks but without pressure changes.

Since the results for the subjects receiving pressure increases and decreases did not differ, they were lumped together for Figure 2, which presents the mean net conditioned responses for all sessions in the form of histograms. It is clear that pigeons can detect pressure changes and that their threshold lies below 2 cm H_2O. This means that they should be able to detect altitude changes of at least 20 m. Yodlowski et al. (1977) showed that the pressure sensitivity can be considered as an extension of the hearing capabilities into the very low frequency range and speculated on whether they might be able to extract navigational cues from naturally occurring ultra-low frequency sound phenomena. It is, however, unclear whether pigeons can detect steady pressure levels.

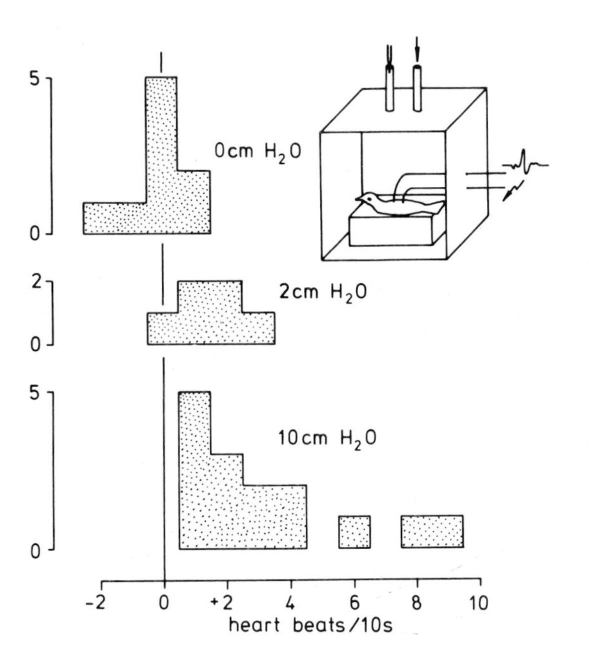

Fig. 2. Distribution of mean conditioned heart-rate changes for 3 subjects tested with pressure changes of 10 cm H_2O (5 sessions), 2 cm H_2O (2 sessions) and in control sessions (0 cm H_2O, 3 sessions)

Our present effort is directed at identifying the sensory organ responsible for this pressure sensitivity, a likely candidate being Vitali's organ, a closed, hair-cell lined vesicle located just below the middle ear lining (Benjamins, 1926), but so far we have not been able to produce definitive evidence because surgical access to the vesicle is difficult.

Needless to say, like many others, we have also attempted to condition pigeons to magnetic field changes. While there can be no doubt that pigeons, among other birds, are sensitive to magnetic forces (e.g., Larkin and Keeton, 1976; Walcott and Green, 1974; Wiltschko and Wiltschko, 1972), it has proved difficult to demonstrate this in the laboratory (Kreithen and Keeton, 1974c; Beaugrand, 1976, 1977; but see Bookman, 1977), a prerequisite for attempting to identify the sensory process by which they are able to do so. We can only report a further failure (Freeman, 1974). We surrounded an octagonal Skinner box bearing a pecking key on every alternate wall with three pairs of orthogonally collocated Helmholtz coils. Automatic programming controlled the sequence of events in each session and recorded the subjects' responses. In particular it activated the Helmholtz coil pairs in a manner adequate to produce the desired net magnetic fields, cancelling where necessary natural earth magnetic field components.

Two animals were presented with alternative 0.20 Oe magnetic fields devoid of a vertical component and at right angles to each other. One subject was rewarded for choosing either of the pair of keys lined up with the relevant magnetic field. The other subject had to choose the keys at right angles to the momentary field orientation. A third pigeon had the same task as the first one but the fields had a downward vertical component of 0.40 Oe (corresponding to the normal local vertical component). None of the animals showed any sign of discriminating between the alternative fields during 17 sessions. In a similar situation described below, but involving polarized light, discrimination was observed after one to seven sessions. This also applied to two of the above pigeons that were subsequently used in that experiment. Eight to 11 further sessions, in which the horizontal component intensities were doubled (0.40 Oe), with only two pecking keys available, and in which a response was demanded to the key coinciding with the magnetic north (first and third subjects) or west (second subject), did not yield discrimination either (Fig. 3). Five to eight additional sessions, with only two keys, in which the horizontal fields were twice the strength as before (0.40 Oe), all without a vertical component and with reversed polarity at a rate of 0.5 Hz (first two subjects) or 10 Hz (last subject), were similarly ineffectual.

The fact that pigeons are capable of orienting by the sun compass method, much as bees do, suggested that they might use the sun-coupled polarization pattern of skylight to supplement this capability. This presupposes that they can detect the polarization plane of light. Kreithen and Keeton (1974b) showed that pigeons could differentiate between a light source with a stationary and rotating plane of polarization. Independently we found (Delius et al., 1976) that pigeons placed in a Skinner box similar to that described earlier for the magnetism experiment, could learn to choose keys aligned in specified ways with the randomly rotated polarization plane of an overhead light source. In additional experiments we found that the time course of the bioelectric response of the pigeon's eye to a light flash varied with the orientation of its polarization plane. The effect was maximal when the light fell on the lower retina. This would agree with the results of two other behavioral experiments that yielded weak or no evidence of discrimination of polarization plane orientation in pigeons. Montgomery and Heinemann (1952) could not get pigeons to perform such dis-

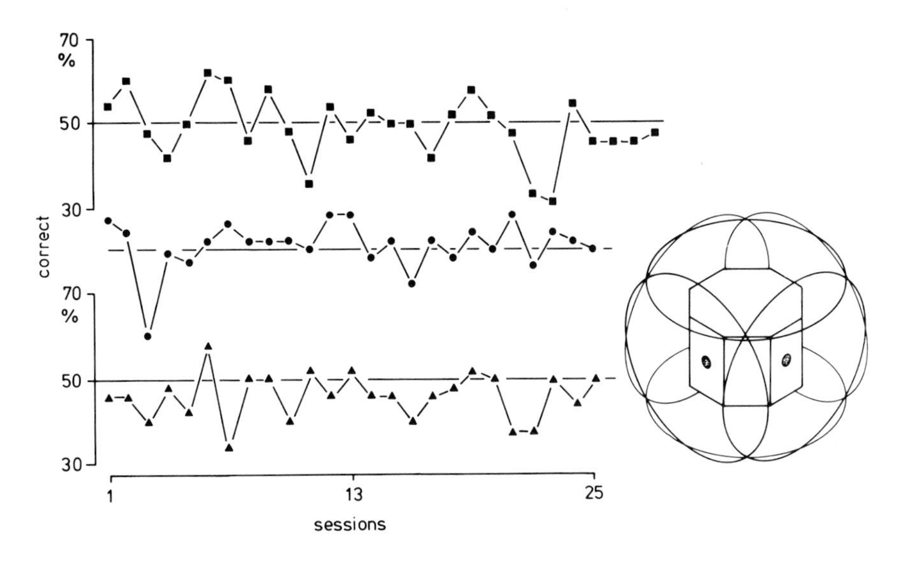

Fig. 3. Discrimination performance of three birds tested under various static magnetic field intensity conditions (see text). Four response keys were used during the first 17 sessions; only 2 keys were available during subsequent sessions

criminations when the stimuli were shown on pecking keys. In another of our experiments (Delius and Emmerton, in press), using an automated Y maze with stimuli displayed on the end walls, we found at best weak evidence of discrimination even after using various procedures designed to aid the animals in this discrimination task (Fig. 4). In both cases

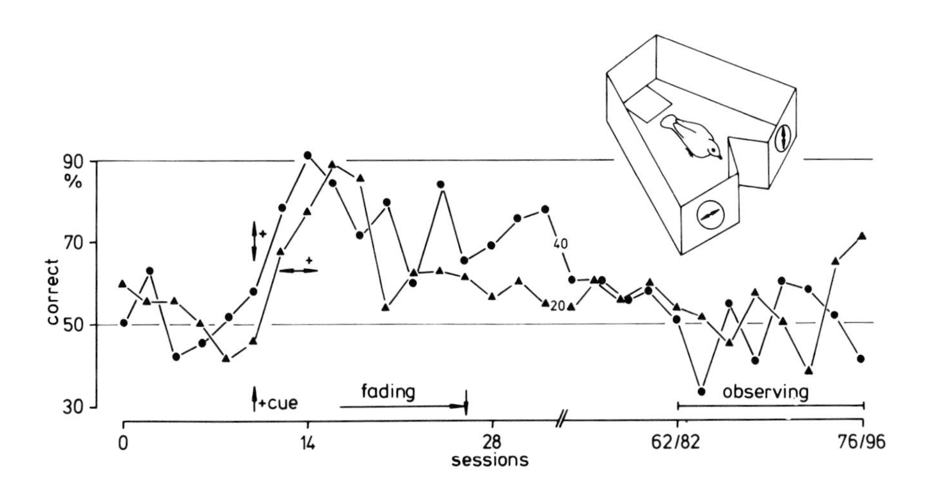

Fig. 4. Learning curves of two pigeons discriminating the polarization plane orientation of a pair of stimuli in a Y maze. In an effort to aid the discrimination task a correction procedure was used throughout; a temporary light cue, subsequently faded off, marked the correct stimulus; later, subjects had to fulfil a 5-s stimulus observing period prior to making a choice. Means of pairs of sessions are plotted. For simplification 20 (40) sessions have been omitted

the pigeons can be expected not to have viewed stimuli with their upper field of vision. We have referred before to possible functional differences between the various parts of the pigeon's visual space that are reflected, to some extent at least, in differentiation of various areas of the retina (Galifret, 1968). The results of another experiment using the same tall octagonal Skinner box, but a bright light bar as an orientational cue instead of the polarization plane, suggest that the upper field of vision might not be as adept at the utilisation of brightness patterns for orientation (Delius and Emmerton, in press).

Although we know that pigeons, unlike bees, can detect the polarization plane orientation of nonultraviolet light, the fact that the polarization of skylight is maximal in the ultraviolet range raises the question of whether pigeons can see this short-wavelength light. Wright (1972) produced good evidence that they can by showing that nonexclusion of ultraviolet cues can contaminate the discrimination of longer wavelength light. One of us has shown electroretinographically that the eye's sensitivity extends at least to 360 nm, where sensitivity is, however, still increasing (Emmerton, 1975, Delius and Emmerton, in press; see also Kreithen, this volume). In a behavioral experiment it was also shown that pigeons can discriminate well between ultraviolet stimuli of different wavelengths. This experiment also suggested that their color vision within the human visible range is more complex than ours. In view of the pigeon's wavelength detection and discrimination abilities, we are at present examining the wavelength-dependent responses to polarized light, being particularly interested to see if polarization sensitivity is enhanced when ultraviolet stimuli are used.

Olfaction is said to provide cues for homing (Papi et al., 1972; Benvenuti et al., 1973), which would seem to be contrary to the general belief that birds have a poorly developed sense of smell. This generalization has been found to need correction in several instances (Wenzel, 1973). As far as pigeons are concerned, we can report from some current electrophysiological work (Delius and Middler, in prep.; see also Wenzel and Rieke, 1975) that the plurality and complexity of olfactory projections found in the pigeon's forebrain is surprising, suggesting that this bird cannot be an olfactory slouch!

Acknowledgements. The work reported here has been supported by the Science Research Council and the Deutsche Forschungsgemeinschaft. It was partially carried out while the authors were at the Department of Psychology, University of Durham, England. We are grateful to the technical staff and many students in Durham and Bochum for their efficient assistance and collaboration.

References

Beaugrand, J.P.: An attempt to confirm magnetic sensitivity in the pigeon, *Columba livia*. J. Comp. Physiol. 110, 343-355 (1976)

Beaugrand, J.P.: Test of magnetic sensitivity in seven species of European birds using a cardiac nociceptive conditioning procedure. Behav. Proc. 2, 113-127 (1977)

Benjamins, C.E.: Y a-t-il une relation entre l'organe paratympanique de Vitali et le vol des oiseaux? Arch. Physiol. 11, 216-222 (1926)

Benvenuti, S., Fiaschi, L., Fiore, L., Papi, F.: Homing performances of inexperienced and directionally trained pigeons subjected to olfactory nerve section. J. Comp. Physiol. 83, 81-92 (1973)

Biederman-Thorson, M., Thorson, J.: Rotation compensating reflexes independent of the labyrinth and the eye: neuromuscular correlates in the pigeon. J. Comp. Physiol. 83, 103-122 (1973)

Bilge, M.: A visual altimeter system with four windows in the optic tectum of the pigeon. IRCS (73-9) 16-2-28 (1973)

Bookman, M.A.: Sensitivity of the homing pigeon to an earth strength magnetic field. Nature (London) 267, 340-341 (1977)

Catania, A.C.: On the visual acuity of the pigeon. J. Exp. Analysis Behav. 7, 361-366 (1964)

Corballis, M.C., Beale, I.L.: The Psychology of Left and Right. Hillsdale, New Jersey: Lawrence Erlbaum Associates, 1976

Delius, J.D., Emmerton, J.A.: Visual performance of pigeons. In: Neural Mechanisms of Behavior in the Pigeon. Granda, A.M., Maxwell, J.H. (eds.) Plenum Press (in press)

Delius, J.D., Middler, L.: Evozierte olfaktorische Potentiale im Hirn der Taube (in prep.)

Delius, J.D., Perchard, R.J., Emmerton, J.A.: Polarized light discrimination by pigeons and an electroretinographic correlate. J. Comp. Physiol. Psychol. 90, 560-571 (1976)

Delius, J.D., Tarpy, R.: Stimulus control of heart rate by auditory frequency and auditory pattern in pigeons. J. Exp. Analysis Behav. 21, 297-306 (1974)

Delius, J.D., Vollrath, F.W.: Rotation compensating reflexes independent of the labyrinth: neurosensory correlates in pigeons. J. Comp. Physiol. 83, 123-134 (1973)

Emmerton, J.A.: The colour vision of the pigeon. Ph. D. Thesis, Durham Univ., U. K. (1975)

Freeman, R.J.: The perception of magnetic fields by pigeons. Unpublished Honours dissertation, Durham Univ., U. K. (1974)

Friedman, M.B.: How birds use their eyes. In: Neural and Endocrine Aspects of Behaviour in Birds. Wright, P., Caryl, P.G., Vowles, D.M. (eds.) Amsterdam: Elsevier, 1975, pp. 181-204

Galifret, Y.: Les diverses aires fonctionelles de la rêtine du pigeon. Z. Zellforsch. Mikrosk. Anat. 86, 535-545 (1968)

Kreithen, M.L., Keeton, W.T.: Detection of changes in atmospheric pressure by the homing pigeon, *Columba livia*. J. Comp. Physiol. 89, 73-82 (1974a)

Kreithen, M.L., Keeton, W.T.: Detection of polarized light by the homing pigeon, *Columba livia*. J. Comp. Physiol. 89, 83-92 (1974b)

Kreithen, M.L., Keeton, W.T.: Attempts to condition homing pigeons to magnetic stimuli. J. Comp. Physiol. 91, 355-362 (1974c)

Larkin, T.S., Keeton, W.T.: Bar magnets mask the effect of normal magnetic disturbances on pigeon orientation. J. Comp. Physiol. 110, 227-231 (1976)

Mittelstaedt, H.: Basic control patterns of orientational homeostasis. Symp. Soc. Exp. Biol. 18, 365-385 (1964)

Montgomery, K.C., Heinemann, E.G.: Concerning the ability of homing pigeons to discriminate patterns of polarized light. Science 116, 454-456 (1952)

Papi, F., Fiore, L., Fiaschi, V., Benvenuti, S.: Olfaction and homing in pigeons. Monit. Zool. Ital. (N.S.) 6, 85-95 (1972)

Romeski, M., Yager, D.: Psychophysical studies of pigeon color vision. I Photopic spectral sensitivity. Vision Res. 16, 501-505 (1976)

Vollrath, F.W., Delius, J.D.: Vestibular projections to the thalamus of the pigeon. Brain Behav. Evol. 13, 58-68 (1976)

Walcott, C., Green, R.: Orientation of homing pigeons altered by a change in direction of an applied magnetic field. Science 184, 180-182 (1974)

Watson, H.: Pressure sensitivity in the pigeon. Unpublished Honours dissertation, Durham Univ. U. K. (1974)

Wenzel, B.M.: Chemoreception. In: Avian Biology. Farner, D.S., King, J.R. (eds.) New York: Academic Press, 1973, vol. III, pp. 389-415

Wenzel, B.M., Rieke, G.K.: The ipsilateral olfactory projection in the pigeon. In: Olfaction and Taste. Denton, D.A., Coghlan, J.P. (eds.) New York: Academic Press, 1975, vol. V, pp. 361-368

Wiltschko, W., Wiltschko, R.: Magnetic compass of European robins. Science 176, 62-64 (1972)

Wright, A.A.: The influence of ultraviolet radiation on the pigeon's colour discrimination. J. Exp. Analysis Behav. 17, 325-337 (1972)

Yodlowski, M.L., Kreithen, M.L., Keeton, W.T.: Detection of atmospheric infrasound by homing pigeons. Nature (London) 265, 725-726 (1977)

A Role for the Avian Pecten Oculi in Orientation to the Sun?

JOHN D. PETTIGREW, 216-76 Beckman Laboratories, California Institute of Technology, Pasadena, CA 91125, USA

Abstract

In addition to its well-accepted nutritive role, I propose a function for the pecten that is a synthesis of the views of Griffin (that it might play a role in migration), Menner (that it casts a shadow), and Barlow (that it blocks intraocular light rays scattered from the sun's image).

New measurements, supported by observations on a scale model, show that the pecten is precisely placed to produce a sharply defined shadow on the field of light scattered within the eye when the sun is imaged on one edge of the retina. The precision of the shadow-casting function becomes evident when one considers the sun's *image* as the source of light, rather than direct rays from the sun itself, as Menner had done. The sharp definition of the shadow, its extended linear contour, its proximity to the image of the horizon, and its position within the area of highest retinal resolution would all facilitate precise observation of the sun.

Key words: Avian pecten - celestial navigation - eye - bird migration.

A. Introduction

The elegant structure of the avian pecten oculi has intrigued scientists since it was first described by Borrichius in 1674 (see quotation in Wingstrand and Munk, 1965). A large number of possible functions has been proposed for the pecten but none can be said to be generally accepted, with the exception of the nutritive function proposed by Wingstrand and Munk (1965), who also provide a comprehensive review of functions proposed in the past. However, the nutritive function does not account for a number of the extraordinary features of the pecten such as its heavy pigmentation, its extraordinary shapes in different species (Wood, 1917), and its precise orientation within the eye.

I propose an additional role for the pecten that takes into account some of these extraordinary features. I was stimulated to think about this problem after finding a remarkable degree of both precision and reproducibility in the visual coordinates of the pectens of a number of owls (Pettigrew and Konishi, 1976a, b). This finding led me to test the theory of Barlow and Ostwald (1972) that the pecten is positioned to intercept light rays that might be scattered inside the eye from the sun's image. Scattered light would act to reduce the sensitivity of retinal regions receiving dimmer parts of the image, and one imagines that it would be a particular problem for the owl, with its highly reflectile choroidal coat, on those not infrequent occasions when it is forced to venture out in sunlight. As Barlow and Ostwald (1972) had done for the pigeon, I was able to verify in the owl that the pecten does intercept light scattered from a bright image, such as the sun's. Indeed the heavily pigmented pleats seem ideally suited for this. On the other hand the orientation of the pecten did not appear very appropriate for this function, only a small fraction of the retina being shielded from the positions of the sun's image in the inferior retina, which one would expect to occur in the natural situation. The temporal fovea appeared to have the least shielding of all, since the pecten would present the smallest of its possible profiles to rays scattered in the fovea's direction. This last observation led me to wonder whether the bird might actually be able to utilize the visual information contained in the pattern of scattered light and shade.

I find that the pecten casts a well-defined shadow that should be readily visible to the bird. This is reminiscent of Menner's (1938) suggestion, but it is important to stress the difference in our formulations. Menner proposed that the pecten cast a shadow by directly intercepting the sun's rays entering the eye. Such a shadow is confined to a small area at the base, because the pecten is oriented along the optical axis of the eye and in any case would be largely invisible to the bird because of its position overlying the "blind spot" of the optic nerve head. My proposal hinges on Barlow's insight (Barlow and Ostwald, 1972) that a significant source of light is the *image* of the sun within the eye. This study concerns the nature of the shadow produced when the pecten intercepts rays scattered within the eye from the sun's image.

B. Materials and Methods

Observations were made on the eyes of a variety of bird species. Nine *Tyto alba*, two *Falco sparverius*, two *Speotyto cunicularis*, two *Bubo virginianus*, one *Columba livia*, and one *Larus argentatus* were studied relatively completely in the laboratory situation. Additional ophthalmoscopic observations in the field and later laboratory work on fixed eyes were available from four *Creagrus furcatus* and two *Steatornis caripensis*.

The owls were anesthetized with Ketamine as part of another study on the physiology of the visual pathway (Pettigrew and Konishi, 1976a, b). The other bird species were anesthetized with Ketamine (3 - 6 mg/kg), sometimes with additional thiopentone (2 - 5 mg/kg), and paralyzed with d-tubocurarine (0.25 - 1 mg). Respiration was maintained in paralyzed birds with a continuous flow of warm, moist air (5 - 20 cm H_2O pressure) into the posterior air sac on one side. Corneas were protected with plastic contact lenses of zero power whose radium of curvature (5.0 mm - 12 mm) was chosen to bring the temporal retina into focus on the tangent screen at 57 cm. Visible landmarks within each eye, such as the margins of the base of the pecten and foveal pits, were projected to the tangent screen in front of the bird with a reversible ophthalmoscope. The limits of the visual field for each eye and the boundaries of the region of binocular overlap were also determined ophthalmoscopically, and the latter boundaries were also plotted onto the tangent screen. Photographs of the retinal landmarks were taken with a fundus camera (Kowa II).

At the termination birds were given an overdose of thiopentone and perfused with fixative (Karnovsky's of buffered formalin) and the eyes dissected out. The anterior half of the eyeball was carefully removed so as to include the vitreous, and photographs were taken of the posterior globe with the pecten and retina in situ. The full extent of the pecten was best seen if photography was carried out with the posterior globe under water to avoid the collapse that tended to occur because of surface tension and partial drying.

The retina was then dissected free and flat mounted on a freshly gelatinized slide with the ganglion cell layer uppermost. When dry, the flat mount was stained with cresyl violet to reveal ganglion cells (Hughes, 1975). The area centralis could thereby be defined in those species like *Tyto* and *Steatornis* where there is no visible fovea. The relationship between the pecten and the area centralis on the retinal whole mount could then be used to infer the projection of the area centralis to the screen from the previously plotted pecten's projection. This will be dealt with in full in a subsequent paper (Pettigrew and Wathey, in preparation).

In some cases the eye was rapidly excised, carefully cleaned of muscle and fat, fitted with a contact lens and sealed into a light-tight tube so that only the corneal surface projected and the posterior globe could be seen by looking down the tube. The shape of the eye was maintained with 20 - 40 cm H_2O pressure from a cannula containing Ringer's solution and attached via a 26 gauge needle inserted into the posterior chamber near the equator of the eye. The whole arrangement could then be taken outside and used to observe the retinal image of the sun through the sclera if care was taken to exclude all light except that entering through the corneal surface. When the preparation was favorable it was possible to see the shadow cast by the pecten in the light scattered from the sun's image, but deterioration of the optics was too fast in bright sunlight for much accurate observation or photography to be carried out. Henceforward, when I speak of a "shadow" in this paper I shall be referring to this "secondary shadow," so-called by Barlow and Ostwald (1972) to draw attention to its relation with the light rays coming from the sun's *image* and to contrast it with the "primary shadow" associated with direct rays entering the eye from the sun itself. As already pointed out by others (e.g., Wingstrand and Munk, 1965, pp. 12 and 15; Barlow and Ostwald, 1972), the "primary shadow" is a somewhat theoretical concept, being more or less coextensive with the base of the pecten and lacking sharply defined boundaries. I had the greatest difficulty in visualizing it. For these reasons the rest of the paper deals only with the "secondary shadow" to which I shall refer simply as "shadow" to avoid repetition.

Because of the difficulties of working with a real eyeball in sunlight, I used the information gained about relative dimensions from the ophthalmoscopy and anatomy to make two scale models of the eye and pecten. In the first I used a Ping-pong ball to form the back of the eye and a 20-mm focal length lens for the optics in front. The pecten was made of folded black paper and was carefully modeled after the pecten of *Tyto*. The total length of this model pecten for *Tyto* subtended 27° (see Fig. 1) at its base (4.8 mm) and the height of the highest point was 80% (3.8 mm) of the length of the base. These relative proportions, in addition to the shape, were similar in both the other species of owl studied. The angular subtense and position of the base of the pecten within the eye of each species can be derived from the data of Figures 1 - 5. The shape of the pectens of the owls, all of which are closely similar in profile, can be derived from Figure 6 and also from Figure 9b (p. 53) of Wood (1917). The model was sealed, with optics facing out, into a light-tight tube mounted on a camera so that the "scleral" surface could be photographed from behind while the sun was imaged upon it via the optics in front.

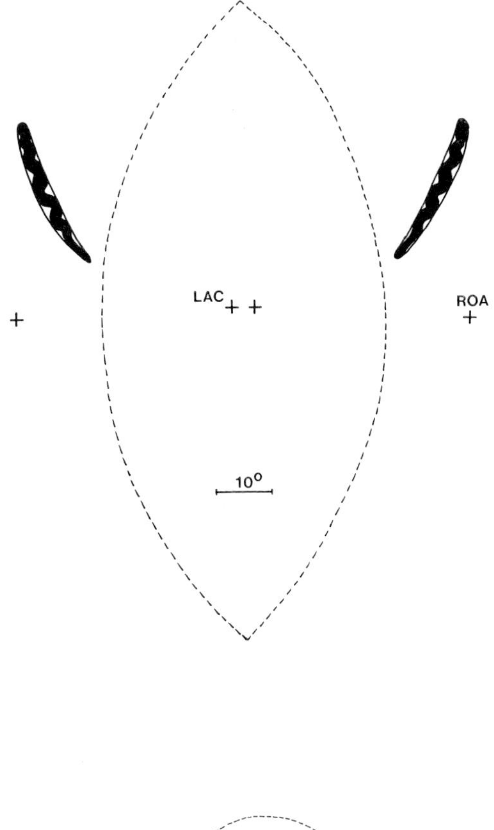

Fig. 1. Binocular visual field *(dotted lines)* and retinal landmarks of *Tyto alba* projected to a tangent screen 57 cm in front of the animal. *LAC*, left area centralis (as determined from retinal whole mounts stained for ganglion cells). *ROA*, right optical axis (determined by centering a light reflex in the pupil). *Crescent shapes* are projections of the bases of the pectens. The position was remarkably constant and exactly as shown in nine different *Tyto alba*. Note in this figure and also in Figures 2 - 5, that the *long axis* of the base of the pecten indicates the direction of the temporal area of retinal specialization. In *Tyto* this area is vertically elongated with a plateau of high ganglion cell density extending into the retina superior to the point of peak density (marked by *Ac*). In *Speotyto* (Fig. 2), *Bubo* (Fig. 3), and *Falco* (Fig. 5) this specialization reaches an extreme form with the development of a fovea and in each case the pecten bases "point" fairly accurately toward the temporal fovea. The significance of this is that the pecten's shadow is narrowest and has a pointed termination when it is directed toward the temporal fovea (see text)

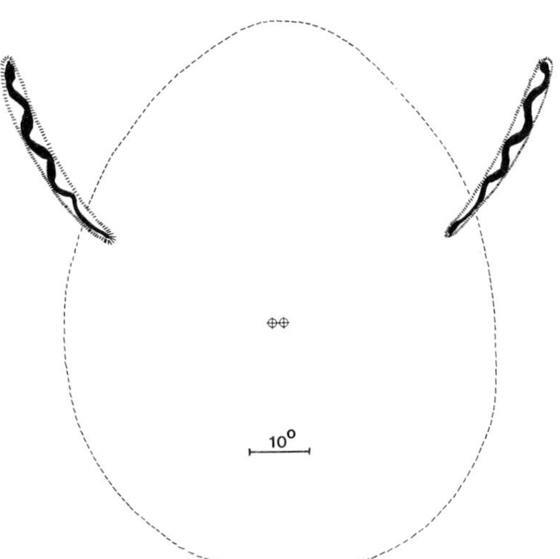

Fig. 2. Binocular visual field and retinal landmarks of the burrowing owl *(Speotyto cunicularia)*. Conventions as for Figure 1. Projection of right and left foveas (visible ophthalmoscopically as dark pits with a surrounding light reflex) are shown as *crossed circles*

The second model was a plastic, translucent hemisphere, 60 cm in circomference. In this case, instead of using optics to image the real sun upon the hemisphere, I used a 1-mm incandescent bulb moved around on the inside of the hemisphere to mimic the sun's image. This scaled-up version facilitated an examination of the effects of changes in the

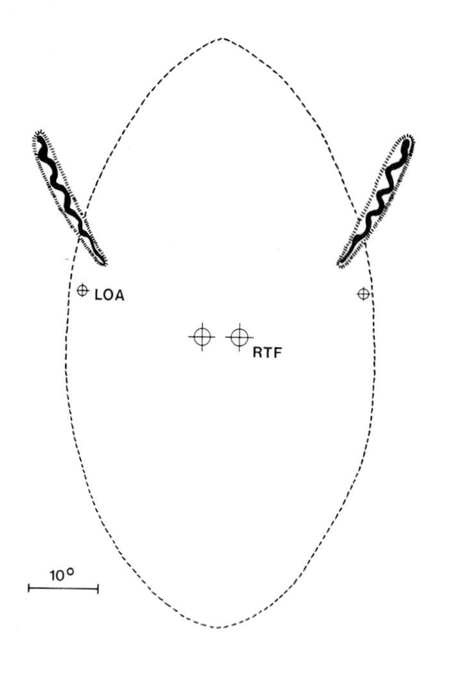

Fig. 3. Binocular visual field and retinal landmarks of a Great Horned Owl *(Bubo virginianus)*. *RTF*, *right* temporal fovea; *LOA*, *left* optical axis

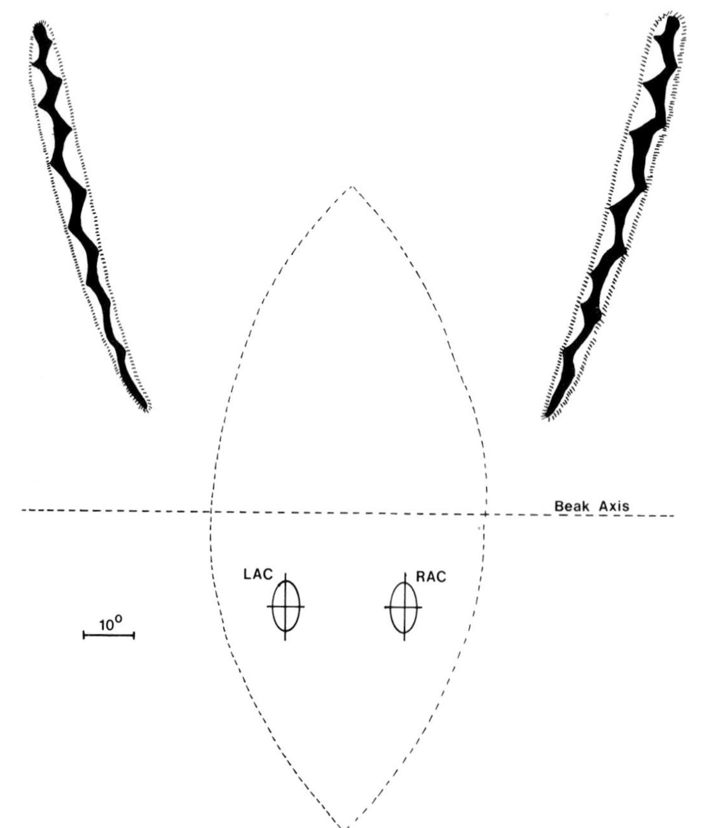

Fig. 4. Binocular visual field and retinal landmarks of the nocturnal oilbird, *Steatornis caripensis*. Note the extensive pecten, which is like that of a diurnal bird and quite uncharacteristic of other Caprimulgiformes, all of which have small pectens like that of *Bubo* in Figure 3. This observation suggests that *Steatornis* may spend more time in daylight than commonly supposed. The positions of the areae centrales *(AC)* are shown in paralysis. In the natural situation they would be overlapped to a greater extent and the binocular field thereby increased

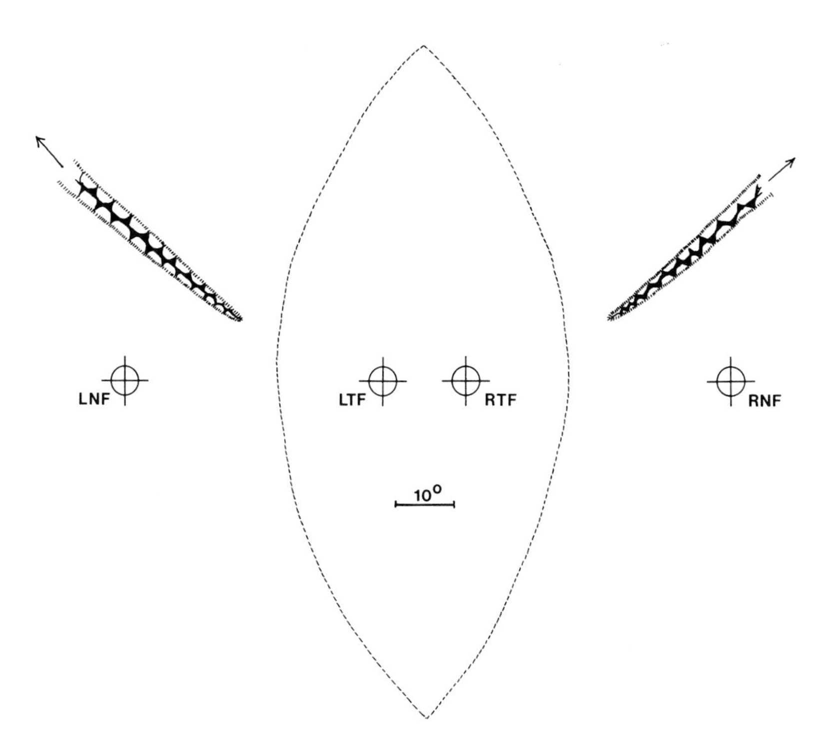

Fig. 5. Binocular visual field and retinal landmarks of the kestrel *Falco sparverius* projected to the tangent screen at 57 cm. *RNF* and *LNF*, right and left nasal foveae; *RTL* and *LTF*, right and left temporal foveae. All four foveae were visible ophthalmoscopically as dark pits surrounded by light reflexes in the shape of blue concentric rings

shape and orientation of pecten upon the properties of its shadow. Measurements could also be carried out more easily upon the geometric transformations brought about between the path taken by the sun's image and the concurrent changes in the path of the shadow. These spherical models only approximate the posterior globes of birds, many of which are to some degree aspherical. The drawings taken from these models (Figs. 7 and 8) could therefore be in error by as much as 30% with respect to the position of the shadows shown.

C. Results

I. Position of Pecten Base on Retina

In all species studied the elongated base of the pecten pointed in the direction of the temporal foveal region (see Figs. 1 - 5). In those species where a temporal fovea is not well developed, such as *Tyto*, *Steatornis*, and *Creagrus*, a temporal area centralis of increased ganglion cell density could be defined on retinal whole mounts. The base of the pecten was directed toward the temporal area centralis in each of these cases also.

This finding seems likely to be a general one since it also applies to most of the birds with temporal foveae studied ophthalmoscopically

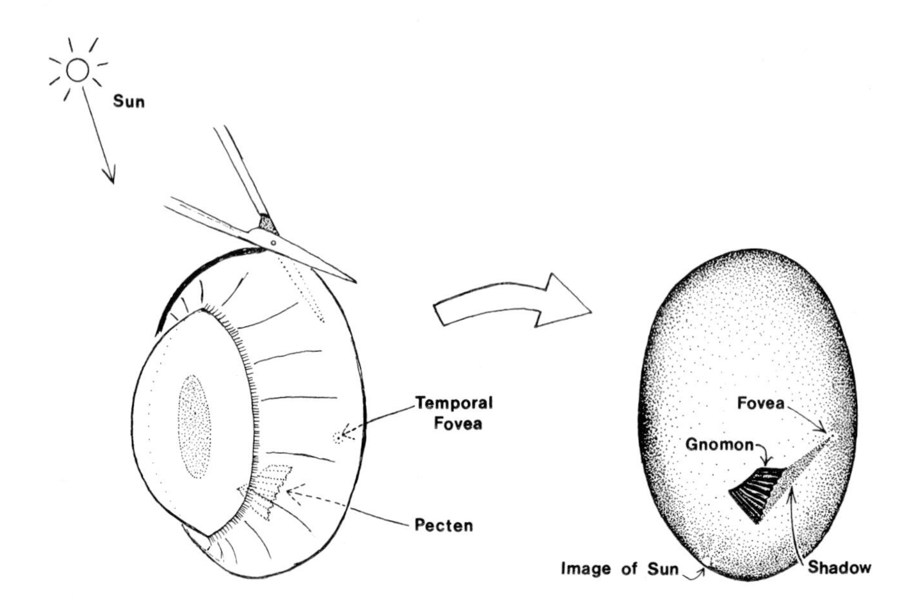

Fig. 6. Diagram to illustrate formation of a shadow by the pecten in the field of light scattered from the sun's image. The *right-hand view* is a representation of what might be seen in the posterior globe if it were possible to view the retinal image produced in the situation shown at *left*. If the sun is imaged on the edge of the inferior retina, light will be scattered to other parts of this spherical surface. The retina on the opposite side of the pecten is shielded from these rays scattered from the sun's image and a shadow is formed there. The shape of the pecten ensures that this shadow has sharp boundaries. There is a point on the edge of the retina where the sun's image produces a shadow with a sharp termination at the fovea. This special case is shown in the illustration at right

by Wood (1917), e.g., penguin, night parrot, swift, bittern, and kingfisher in addition to some other species of owls not studied by me. The California shrike and mockingbird appear to be exceptions to this rule.

II. Projection of the Pecten in Space

In most birds, the presence of eye movements makes it difficult to infer the projections of ocular landmarks in life from ophthalmoscopic observations made when the bird is held. However, in owls such inferences can readily and accurately be made from a knowledge of head position because eye position is fixed relative to the head with an accuracy of from one to three degrees, depending on the species [~1°, *Bubo virginianus*, Steinbach and Money (1962); 2 - 3°, *Tyto alba*, Pettigrew and Konishi (1976a)].

In all the species of owls which I studied, the positions of the pectens in space were remarkably constant. For example in nine *Tyto alba* whose heads had been held in a standard position with a beakholder in a stereotaxic frame, the projection of both pecten tips lay in the same horizontal plane 56° apart. The variance in their positions was 0.8°, which is close to the limits of accuracy of the reversible ophthalmoscope method used. In both great horned owls studied the pecten tips were found in the same position, 37° apart on the same horizontal plane. The corresponding values for 2 *Speotyto* were 54° and 56°.

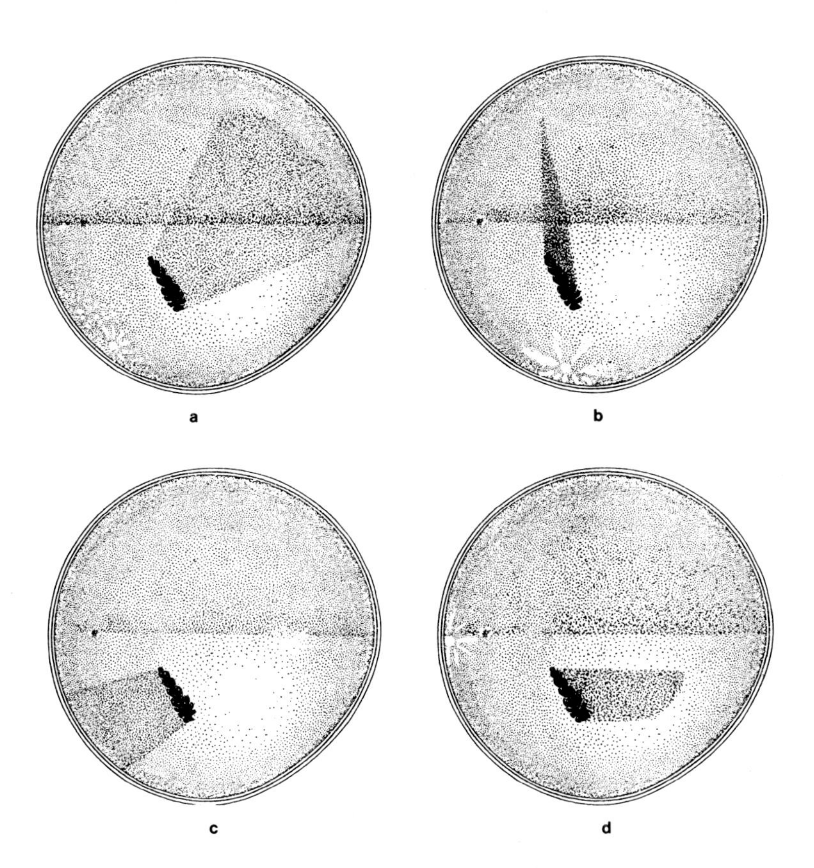

Fig. 7a - d. Different shapes and orientations of the pecten's shadow seen with different positions of the sun's image. Drawn from the model eye described in text.

For each case the eye is fixated upon the horizon, whose image is shown passing through the temporal fovea (the *point* shown on the equator of the eyeball near its left-hand margin).

The various shapes and orientations could be produced by different sets of conditions. The simplest case to consider is that where the bird remains at one location and continues to look in the same azimuthal direction. In this case the changes seen would be simply related to the movement of the sun across the sky with time; (c) representing sunrise, (d) sunset, (b) noon, and (a) early afternoon.

There are alternative conditions that produce the same shadow patterns. For example (b) can also be obtained earlier (or later) than noon at a different latitude (closer to the equator) if the bird changes the azimuth of its head or eye appropriately

III. The Shadow

While it did prove possible to observe the pecten's shadow directly through the sclera of excised eyeballs as described in Methods, the optics deteriorated rapidly in the bright sunlight necessary for this experiment. Consequently most of what follows is derived from observations of scale models based on the eyes of *Tyto*, *Bubo*, *Speotyto*, and *Columba*. It is possible to say on the basis of direct observations on excised eyeballs, that the shadow is sharply defined and that, when the sun is imaged on the far periphery of the retina, the shadow has a pointed termination.

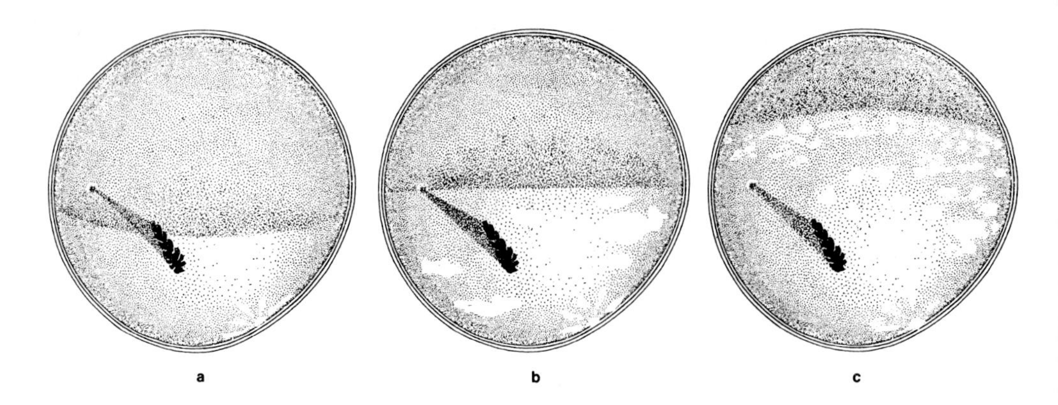

Fig. 8a - c. Illustration of a possible strategy by which a bird might estimate the elevation of the sun by fixating the tip of the shadow and observing the position of the horizon. Such a determination could be used either as an additional measure of time (if the bird stayed in one place and applied a gradually changing, seasonal correction to its estimate of the horizon's position), or as a measure of latitude (if the estimate could be extrapolated to, or obtained exactly at, noon).

For this strategy the bird orients its head/eye so that the sun's image falls on the point on a line between gnomon and fovea. In other words, the bird fixates upon its fovea the shadow's tip.

In the example shown an owl has adopted this strategy at noon, at the time of the equinox at three different latitudes (a) 50^0, (b) 30^0, and (c) 10^0

The scale models allowed more detailed observations of the shadow, which can be seen in Figures 6 - 8. Remarkably, in all the species examined, the shadow is thinnest and its tip most obvious at precisely that point where the sun's image is positioned so that the shadow falls upon the temporal fovea or area centralis. This may apply to the majority of birds, since as already pointed out, there is a general trend for the temporal fovea to lie in the direction of the long axis of the base of the pecten.

The most important shadow-casting edge of the pecten appears to be the sharp superior limb that slopes down to meet the retina (see Fig. 6). This is because the shadow of this edge is straight. I found this to be true for all of the different pecten shapes (eight species) that I tested in models, in addition to the direct observations which I made upon excised eyes of *Columba*, *Tyto*, and *Bubo*. This observation is in apparent conflict with Barlow and Ostwald (1972) who found a slight S-shaped curvature which they thought might be attributable either to the shape of the pecten or to the projection system they employed. The latter explanation seems the more likely in view of the straightness of this edge of the shadow which I observed in all cases. The shadow of the superior limb also radiates from the point where the superior limb of the pecten meets the retina. It would therefore be readily visible, whatever the shadow's orientation, in contrast to the shadow of the inferior edge, which would often be obscured by the base of the pecten itself. I shall therefore refer to the superior edge of the pecten as the *gnomon*, the important shadow-casting edge of a sundial (Dolan, 1975). In addition to its visibility and straight edge, the shadow of the pecten's *gnomon* appears to have another important feature: Its orientation upon the retina will tend to change in a manner which is relatively linear with time. This is because the gnomon is oriented roughly parallel to the axis of rotation of the sun's image within the eye (that is, if we imagine that the eye is fixed and the temporal fovea is looking at the horizon). This relationship cannot be exact, of course, since it depends upon the season and upon the latitude. However, it is intriguing to observe that the angle that the gnomon makes with the retina varies somewhat from one species to another and may possibly be related to latitude just as sundials of different latitudes have to have different gnomon angles (Dolan, 1975). For example, the snow goose, from circumpolar latitudes, has a very small gnomon angle [Fig. 93, p. 52, Wood (1917)] whereas *Tyto*, from temperate latitudes, has a large one (Fig. 6). (Note that latitude compensation of the gnomon's angle in the eye is complementary to that operating for sundials because the sun's image is *below* the gnomon in the former case. A sundial operating near the pole has to have a very large gnomon angle.)

Another remarkable feature of all species studied was that the gnomon was situated half way on the straight line between the temporal fovea and the edge of the retina. This means that if the sun were imaged on the edge of the retina such that the tip of the shadow lay on the fovea, then movement of the sun across the sky would be faithfully and accurately reflected in movement of the tip of the shadow, without magnification or reduction. This may be an important point if the pecten is being used for celestial observation, since the source of light is not at infinity as it is for a sundial or a sextant. It is not inconceivable that a bird could track the sun-produced movement of the shadow across the retina with reference to some nearby stationary contour such as the horizon. [The ON-type of direction selective ganglion cells of the rabbit retina (Oyster, 1968) respond to such low velocities that they could easily manage this task; thus it is not empty speculation to propose that birds could do as well.] If such tracking is carried out, one might expect the bird to use its highest resolution apparatus in the fovea for the task. It is therefore of some interest to find that the shadow formed on the temporal fovea is exactly yoked to the sun without magnification.

The larger model permitted some observations to be made on the geometric transformations brought about between changes in the position of the sun's image and changes in the shadow. Some of these observations are shown in Figures 7 and 8, where it can be seen, for example, that when the sun's image has an elevation close to that of the fovea (as for example when the horizon is being fixated around the time of sunrise or sunset) the orientation of the shadow is most sensitive to changes in the sun's elevation and rather insensitive to changes in its azimuth (which produce only changes in the shadow's length). Conversely, at moderate to high elevations of the sun, the shadow's orientation more faithfully reflects changes in the sun's azimuth and is insensitive to elevation changes (which now produce changes in the shadow's length). The elevation of the sun at which the shadow's orientation switches from being more elevation-sensitive to more azimuth-sensitive depends upon the orientation of the whole pecten with respect to the horizon. This varies from species to species according to the "infulapapillary angle" of Wood (1917, p. 66), which is the angle of intersection of the long axis of the base of the pecten with the horizon. Perhaps the marked variation in this angle [see Figs. 1 - 5, and also Figs. 110 - 142 of Wood (1917)] can be related to the kinds of solar observations being made by different species of birds and the times of day when each species tends to carry them out.

D. Discussion

The observations presented here provide preliminary evidence to support the view that the shape and precise position of the pecten within the eye would provide a bird with the means for indirect observations upon the sun. A sharply defined shadow is formed by the pecten in the field of stray light scattered within the eye from the sun's image and this shadow could readily be detected by the orientation-sensitive mechanisms described in the forebrain visual Wulst of owls and hawks (Pettigrew and Konishi, 1976a; Pettigrew, 1977). The visual Wulst is analogous to the striate cortex of mammals where high resolution visual tasks like vernier and stereoscopic acuity (accurate to 2 s arc in man) are though to be carried out. The visual Wulst is also primarily concerned with visual processing in temporal retina (Pettigrew, 1977), where the shadow is both well defined and exactly yoked to the sun's movement. Taken together, these considerations suggest that the pecten might enable the literally astronomical accuracy of visual resolution to be brought to bear upon the sun.

The first question which arises about this possible role for the pecten is the following: Why have such an elaborate mechanism for carrying out indirect observations upon the sun when the same information might be obtained from direct visualization of the sun's image itself? There are a number of possible answers to this question.

For example, photic damage which would follow any prolonged direct observation of the sun can be avoided by observing the yoked shadow. Even high resolution apparatus like the fovea can then be safely brought to the task. It is even possible that there exists a specialized area in the posteror globe (e.g., at the ora serrata or upon the ciliary body) where the albedo and resistance to photic damage are higher than usual and would thereby facilitate prolonged fixation of an even more distinct shadow.

In addition, the shadow is an elongated edge whose orientation might be more accurately estimated in relation to a reference mark than the position of the sun, which very often has no nearby stationary contour. A particularly important contour in this respect is the horizon. Like the sextant, the pecten arrangement effectively brings the sun (in the form of the shadow yoked to its image) in close relation to the horizon. Both the position of the horizon and of the sun could thereby be simultaneously viewed by the same retinal region. In fact, in a given bird's eye, there is a particular altitude of the sun when both the image of the horizon and the tip of pecten's shadow can be placed precisely on the center of the temporal fovea. As pointed out in Results, the shadow in this position will be exactly yoked to that of the sun, without magnification or distortion and would therefore provide an ideal opportunity for estimating the sun's trajectory if the bird were using the strategy for celestial navigation proposed by Matthews (1955). While there appears to be little experimental evidence at the moment to support Matthews' proposal (Emlen, 1975), at least it cannot be ruled out on the ground that birds are incapable of performing the necessary measurements. The measurement in this case will be like the vernier task of judging the separation of two nearby contours, whose accuracy approaches 2 s arc in man and is not likely to be much less in birds.

A fourth possible advantage in viewing a shadow, rather than the sun itself, may be associated with the geometric transformations involved in the production of the shadow. For example, as described in Results, for moderate elevations of the sun the angle made by the shadow faithfully reflects changes in the sun's azimuth but is relatively insensitive to elevation changes, which produce changes in the shadow's length but not its angle on the retina. The pecten may thereby be viewed as simplifying the task of observation by reducing the effects of unwanted variables to enable a more direct reading of the desired one (e.g., azimuth).

E. Sun Compass, Sundial, or Sextant?

If it is accepted for the moment that the pecten is used as an aid in solar orientation, we can then ask which of the three possible kinds of information from the sun it is designed to gather, viz., north-south direction, time of day, or latitude. In other words, is the pecten designed as a sun-compass, sundial, or sextant? All three possibilities exist, since there are different strategies that could be used by the bird to achieve all three goals. Just which strategy is chosen might depend upon what kinds of other information were available and may possibly vary with the species.

The sun-compass function follows most directly since the orientation of the pecten's shadow gives the sun's azimuth, which then requires only some sort of time compensation to yield compass direction. Since the gnomon is roughly parallel to the axis of rotation of the sun's image on the retina, and may even be corrected for latitude in some cases (see Results), this will ensure a fairly linear relation between the orientation of the shadow upon the retina and time. The pecten's design plus an internal clock of the kind known already to be found in birds are all that are needed to account for the sun-compass ability that has been shown in pigeons and starlings (e.g., Hoffman, 1960).

If the bird were already aware of north-south direction, for example, because it was at a familiar place, then the azimuth given by the pecten's shadow could be converted to time if the bird kept looking in a particular direction. Local time could also be obtained from the pecten without knowledge of north-south direction by a more elaborate strategy involving visual fixation and estimation of the horizon with seasonal compensation. This strategy is illustrated in Figure 6. The same strategy can be used to estimate latitude if the bird estimates the trajectory of the shadow tip as well as the position of the horizon.

This last strategy of fixating the tip of the shadow is more elaborate than that required for the use of the pecten as a sun-compass. It is therefore less likely that it is actually used by birds for navigation. Moreover, such a strategy may not be equally feasible in the eyes of all birds. For it to be used successfully the sun must be imaged on a point in the eye which is on the line joining the temporal fovea and that part of the pecten furthest from the retina. Such a point may not exist for some species where the pecten extends well forward and may even touch the posterior surface of the lens [see p. 44, Wood (1917)]. Direct examination of the eyes in question would be necessary to settle this point.

Acknowledgments. This work was supported by the Spencer Foundation and U.S.P.H.S. grants MH 25852 and EY 01909-01. Drs. Horace Barlow, Donald Griffin, and Andrew Ingersoll provided helpful discussion of some of the ideas presented. The data on *Steatornis* were gathered with the help of a grant from the National Geographic Society to M. Konishi. The study on *Creagrus* was carried out with the assistance of the Charles Darwin Research Station, Isla Santa Cruz.

References

Barlow, H.B., Ostwald, T.J.: Pecten of the pigeon's eye as an intra-ocular eye shade. Nature (London) 236, 88-90 (1972)

Dolan, W.W.: A Choice of Sundials. Portland, Ore: Stephen Greene Press, 1975

Emlen, S.T.: Migration, navigation and orientation. In: Avian Biology. Vol. V. London - New York: Academic Press, 1975

Hoffman, K.: Experimental manipulations of the orientational clock in birds. In: Cold Spring Harbor Symp. Quant. Biol. Vol. 25, 379-388 (1960)

Hughes, A.: The distribution of ganglion cells in the cat retina. J. Comp. Neurol. 163, 197 (1975)

Matthews, G.V.T.: Bird Navigation. Cambridge: Univ. Press, 1955

Menner, E.: Die Bedeutung des Pecten im Auge der Vögel für die Wahrnehmung von Bewegungen. Zool. Jb., Abt. 3 (Allgem. Zool. v. Physiol.) 58, 481-538 (1938)

Oyster, C.W.: The analysis of image motion by the rabbit retina. J. Physiol. 199, 613-635 (1968)

Pettigrew, J.D.: A comparison of the visual field representations in the visual Wulst of *Falco* and of *Tyto*, with a note on the evolution of frontal vision. In: Frontiers in Visual Science Cool, S., Smith, E. (eds.), Berlin - Heidelberg - New York, Springer, 1978

Pettigrew, J.D., Konishi, M.: Neurons selective for orientation and binocular disparity in the visual Wulst of the barn owl *(Tyto alba)*. Science 193, 675-678 (1976a)

Pettigrew, J.D., Konishi, M.: Effect of monocular deprivation on binocular neurones in the owl's visual Wulst. Nature (London) 264, 753-754 (1976b)

Steinbach, M.J., Money, K.E.: Eye movements in the great horned owl, *Bubo virginianus*. Vision Res. 13, 889-890 (1972)

Wingstrand, K.G., Munk, O.: The pecten oculi of the pigeon with particular regard to its function. Biol. Skr. Dan. Vid. Selsk. 14, 1-64 (1965)

Wood, C.A.: The fundus oculi of birds, especially as viewed by the ophthalmoscope. Chicago, Illinois: Lakeside Press, 1917

Session II

Chairman:

DONALD R. GRIFFIN, Biol. Laboratories, The Rockefeller University, 1230 York Avenue, New York, NY 10021, USA

Do Pigeons Use Their Eyes for Navigation? A New Technique!

KARL-LUDWIG KÖHLER, Flugshofer Straße 14, 8535 Emskirchen, FRG

Abstract

A simple camera was attached to a pigeon's head. A shutter closed the aperture 60 s after release. Each time the pigeon held its head steady for a few seconds, the sun formed a spot on the horizontal film at the bottom of the camera. With the aid of the sun's position, known from nautical tables, the direction of fixation of the pigeon could be calculated and associated with landmarks at the horizon. Preliminary results show that many sun spots are indeed produced on the film upon release of the bird.

A. Introduction

In investigating the homing mechanism of pigeons the most important question to be asked concerns the sensory modalities involved. Which sensory inputs give navigational information of the displaced position to the animal? Which information is stored in the animal enabling it to correlate release site and home? From preliminary results and the hypothesis I advanced earlier (Köhler, 1975), I conclude that the visual system is very important for navigation.

In contrast to previous suggestions, my hypothesis requires utilization of a vertical rather than a bi-coordinate system. If the vertical from the home position can be established at the place to which the bird has been displaced, it will differ from the local vertical (Fig. 1a). The direction in which the home vertical deviates from the local vertical indicates the home direction. The magnitude of the angular difference between the two verticals corresponds to the distance from home. The same results can be arrived at if the corresponding horizontal planes are used instead of the verticals (Fig. 1b): Both are parallel at home; their differences at displaced positions indicate direction and distance from home.

Distance and direction from home are given with precision at any locality if the vertical, which always points to the center of the globe, is used. However, this is not always the case if the horizontals are used instead of the verticals, and deviations may result, especially at close range. Since pigeons do, in fact, frequently deviate from the home direction upon release, usage of the horizontal seems to have preference over usage of the vertical.

What does the horizontal mean in terms of the hypothesis? It is the plane as determined by the visible horizon and the altitude of the site. This plane may be different from the imaginary, ideal plane based solely on the courvature of the earth at that site. Different visibility may result in differently inclined planes. Experimental

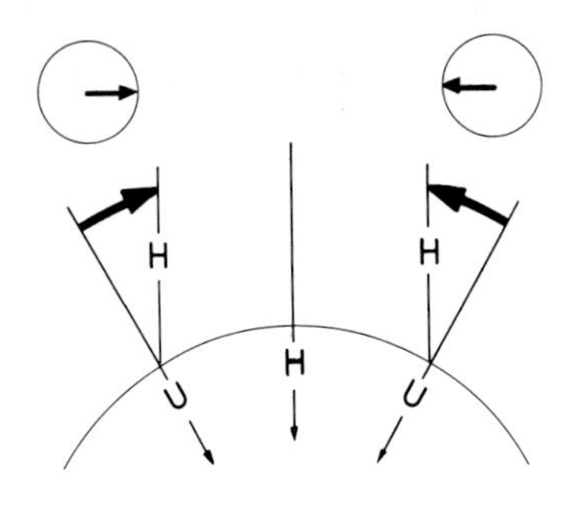

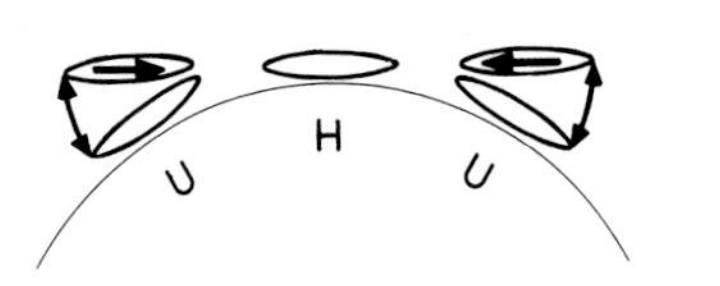

Fig. 1. Diagrammatic presentation of the navigation hypothesis (a) using the verticals and (b) using the horizontals
H indicates the home position, *U* a position unknown to the bird

results show that slight changes in the layer of haze alter the influence of mountain ridges on the direction of the initial orientation of pigeons.

How and when do the pigeons collect information on the plane? It has been shown that already 20 s after take-off pigeons are roughly oriented in the final vanishing direction (Schmidt-Koenig, 1965). It follows that the first minute after take-off is the critical interval in which essential navigational processes are performed. If visual stimuli have an important part in this process, the first seconds after take-off should be the most interesting moments.

B. Methods

To determine where the pigeons are looking immediately upon departure, I designed a camera, weighing 2 g, to be fixed on top of the pigeon's head (Fig. 2). A circular disc of orthocromatic film AGFA HA 68 is horizontally aligned with a notch indicating the head's axis. An aperture of 0.05 mm diameter centrally above the film permits light to enter. A shutter closes the aperture after 60 s. If the camera is held horizontally the sun forms a spot on the film. After rotation of the camera around its vertical axis, a spot is formed on the film at a different position, a faint line connecting both spots.

The perpendicular axis is taken to represent the bird's axis of fixation. Experiments are currently under way to find out whether or not the directions of fixation of both eyes are, indeed, precisely perpendicular to the long axis of the head, as given in simplified fashion in Figure 2, or whether some angular correction is necessary. Also, possible orbital movements of the eyeball are being investigated. We can frequently watch birds holding their heads very still when fixating objects in some direction. When changing the object of fixation they turn their heads. These turns also turn the camera relative to the sun. Since the azimuth and altitude of the sun are known, the "sun spots"

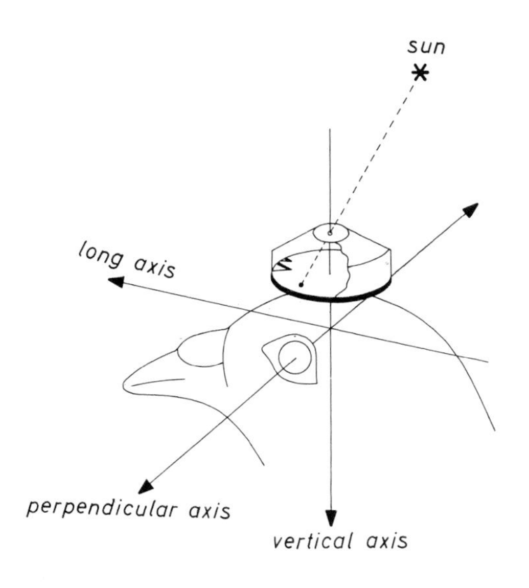

Fig. 2. The camera on the pigeon's head. Note the deep notch on the film pointing toward the beak

on the film permit calculation of the positions of the head. Figure 3 shows the position of the eyes relative to the film.

Figure 4 presents a frontal view of a pigeon and a sample film. The sun spots are circularly arranged if the bird turns its head only around the vertical axis. If objects of different altitudes are fixated at the horizon, a broader circle of sun spots is produced. Improper attachment of the camera (Fig. 5) yields a noncircular arrangement of sun spots.

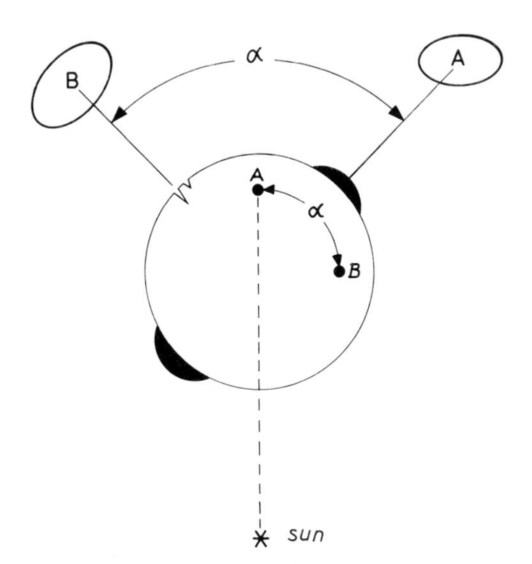

Fig. 3. The angular relation between a sun spot formed on the film and the assumed direction of fixation by the right eye *(black half disc)*. The deeper of the two notches at the periphery of the film is pointing toward the beak and is aligned with the long axis of the head. The right eye is aiming at object *A*; the left eye has no object to fixate. The sun is due south and produces the spot *A* on the film. By turning the head to the left, object *B* can be fixated with the same eye. The sun then produces spot *B* on the film. Angle α subtended between spot *A* and spot *B* corresponds to the angle subtended between object *A* and object *B*. If more objects were fixated, correspondingly more spots and angles would be produced

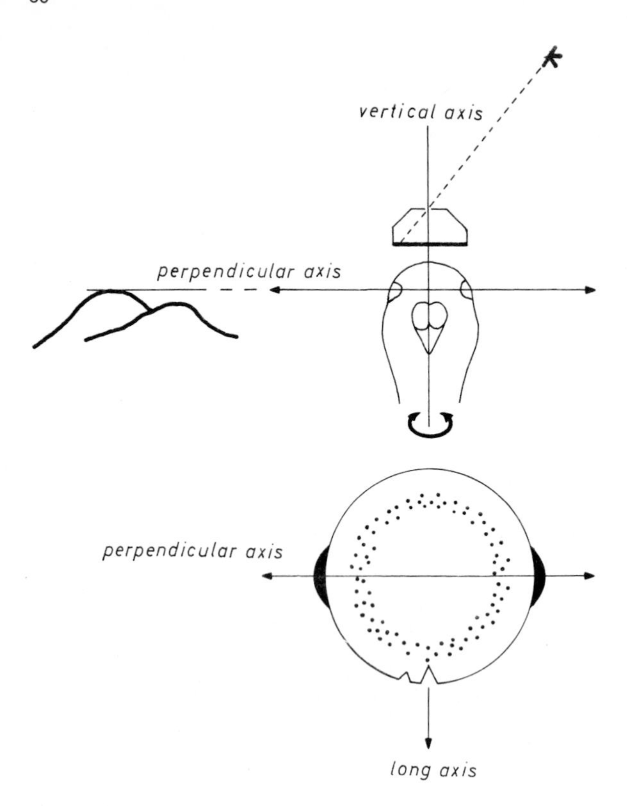

Fig. 4. A circle of sun spots formed on the film when the bird turns its head around the vertical axis

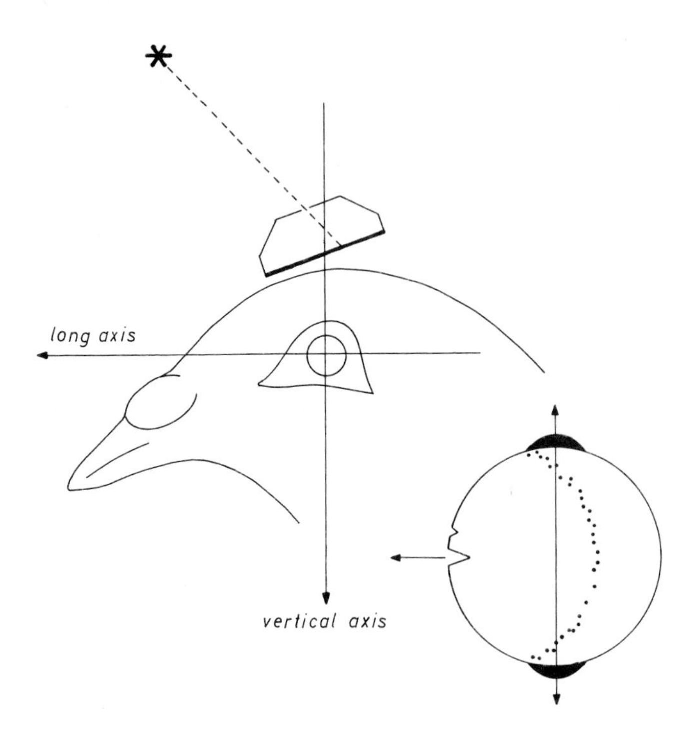

Fig. 5. Improper placement of the camera produces noncircular sun spots on the film

C. Results

I first used this method for recording the direction of fixation in pilot studies in 1976. Pending the solution of partial problems, I can present the following preliminary results:

1. Vision is used during the initial portion of a homing flight.
2. The pigeon does not randomly look around; rather it systematically scans, as can be deduced from the circular arrangement of sun spots.
3. Pigeons fixate objects with their heads kept amazingly stable. Even a slight movement during fixation would yield a blurred spot or a streak.
4. Poor or good atmospheric visibility produces different results. With ca. 12 km visibility, more sun spots are produced (Fig. 6) from the same site than with ca. 30 km visibility (Fig. 7). At this site, the horizon appears without elevated points. This may mean that without distinct visual cues the bird tries harder to find some, whereas

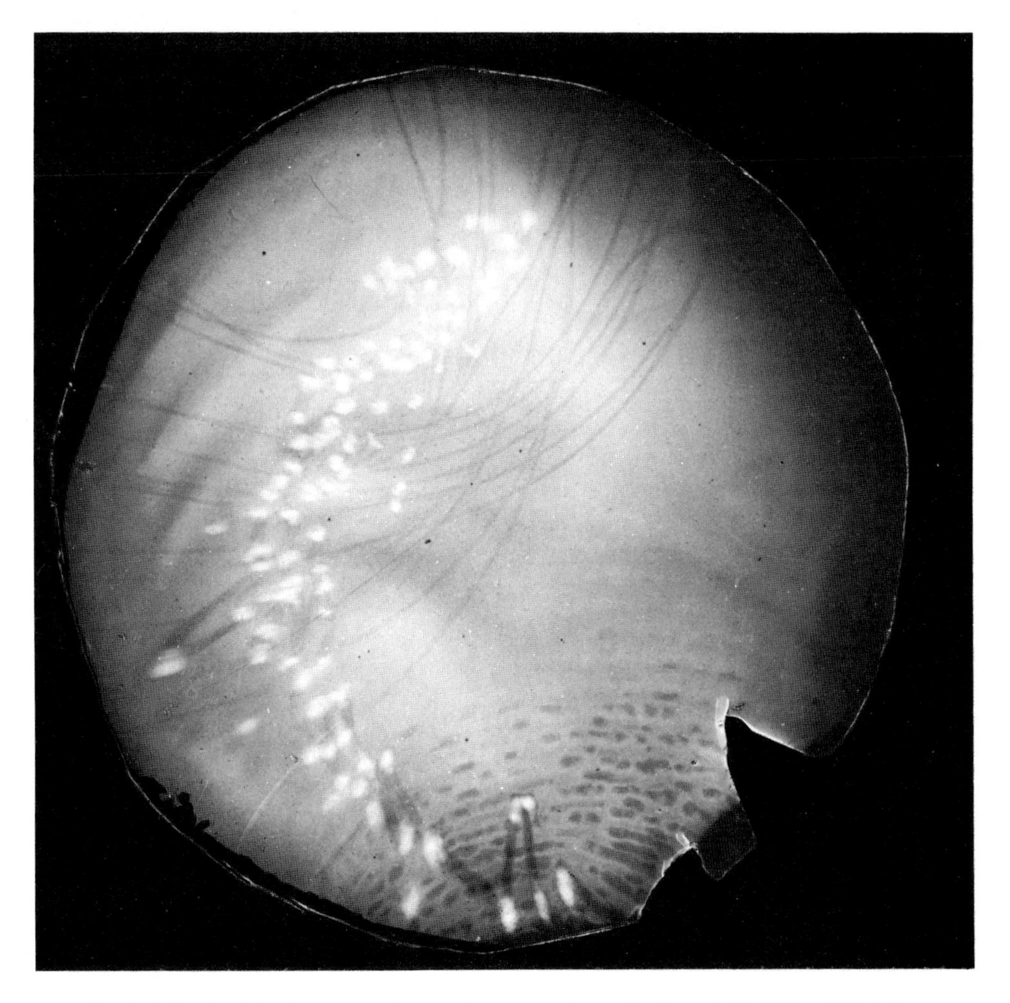

Fig. 6. Original film from a release in poor atmospheric visibility (12 km)

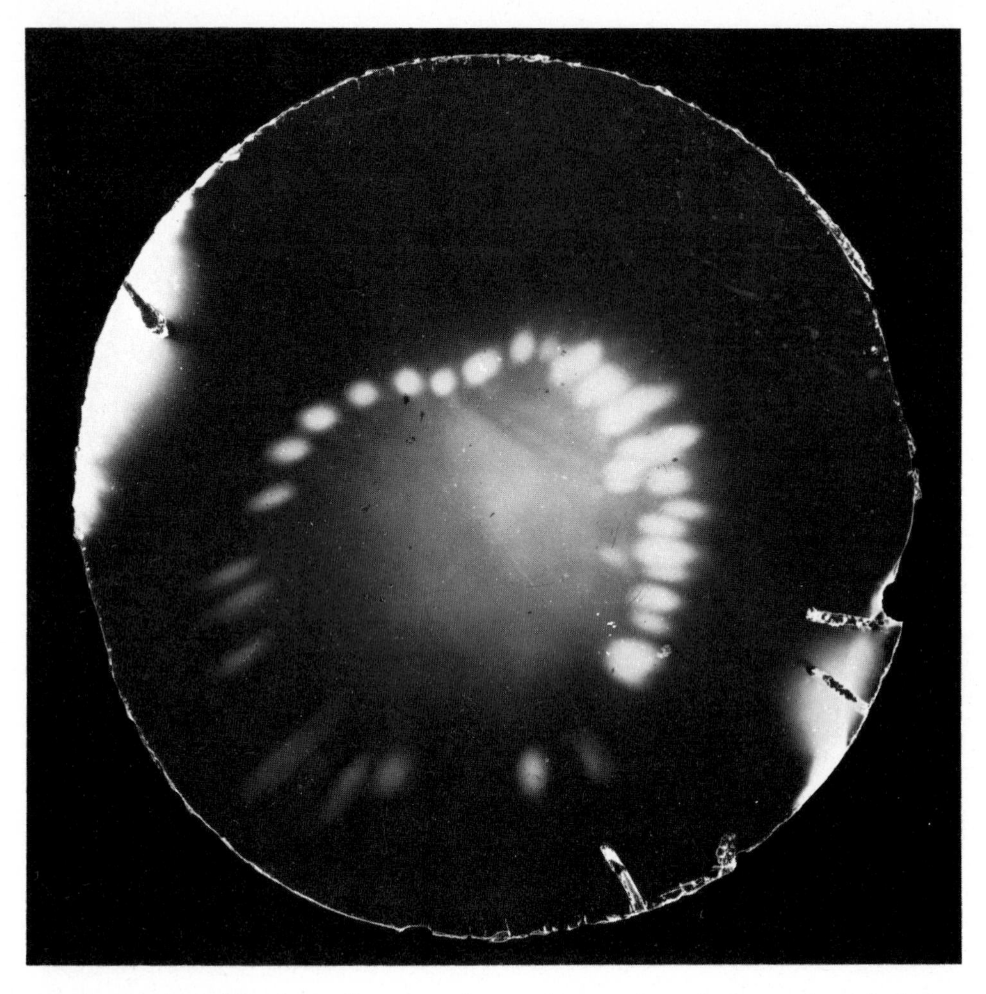

Fig. 7. Original film from the same release site as that of Figure 6, with 30 km visibility

under good atmospheric visibility mountains and ridges can easily be seen and fixated.

5. The birds fixate mainly horizontally. There are no sun spots inside or outside the circle, indicating that objects above or below the horizon are not fixated.
6. Some portions of the horizon are of more interest to the bird than others, resulting in an uneven distribution of spots (Figs. 7 and 8).
7. Some objects may be fixated only once, others repeatedly, as may be concluded from clumped arrangements of sun spots on the film.
8. Figure 8 presents a preliminary analysis relating the sun spots on a film to higher elevations at the horizon (assuming the axis of fixation to be perpendicular to the long axis of the head). The bird was released 10 km west of the loft and disappeared to the northwest. The elevation of the release site was 375 m above sea level. The bird flew an estimated 25 m above the ground, thereby attaining a total altitude of 400 m above sea level. All elevations exceeding 400 m above sea level, visible at the time of take-off, (Fig. 8) relate sun spots to

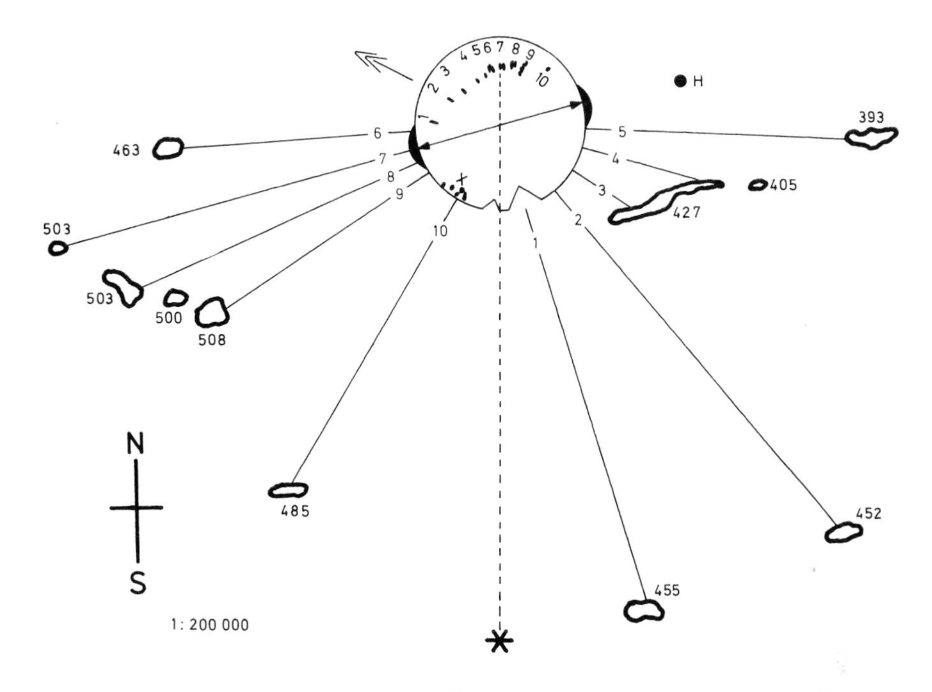

Fig. 8. Relation (designated by *numbers*) between sun spots on the film and elevations (in m above sea level) as calculated from the sun's azimuth position. *H* indicates home; the *double arrow* indicates the flight direction of the bird

the various elevations. For example No. 7 relates fixation of elevation 503 by the right eye to the sun spot No. 7 produced by the sun in the south etc. To obtain all ten sun spots on the film the pigeon, flying in the direction of the double-headed arrow in Figure 8, must have turned its head rather extensively. The spots marked by designate those produced without a turn of the head away from the direction of departure. There is an abvious agreement between sun spots and the highest elevations at the horizon. The highest peaks scored two and four dots.

D. Discussion

If we succeed in decoding the sun spots, a thus far untapped source of information will be available. The pigeon eye, freely moving, within certain limits, seems to rest in a sort of zero position during fixation. Otherwise agreement between sun spots and landmarks could hardly be found. Application of frosted lenses (Schmidt-Koenig and Walcott, 1973) reducing vision considerably would permit checking the present interpretaion. The experimenter does not know which of the bird's eyes is actually fixating, but this ambiguity in the interpretation could be eliminated with additional experiments.

The interpretation of nonsense-orientation may also be facilitated by this approach. I could imagine a final solution that integrates nonsense-orientation into a navigational concept.

There is no evidence for recognition of landmarks. The present results indicate that the pigeon scans the horizon only. The horizon and the horizontal plane derived from it are, however, essential parts of the navigation hypothesis advanced recently (Köhler, 1975).

I do not advocate a purely visual navigation system. More likely, vision is one factor in a highly complex system. Vision should, however, be more emphasized than has recently been the case, especially since it seems to be used in a dimension that emerged as important from several experiments, namely, detection of the horizontal plane. Kramer (1959) suggested a role of the horizontal plane; Wallraff (1966) excluded visual stimuli in his palisade experiments, but suggested a horizontally structured source of information.

This is a short summary of my privately run orientation research - fun to me despite high costs in terms of money and time. I have a large number of experiments pending for the reasons mentioned. I especially plan to further improve and modify the camera for successive picture taking and for vertical arrangement of the film. This would permit answering several questions. Current techniques require the experimenter to watch the pigeon disappear upon releasse; this new technique enables the experimenter to see, in a certain sense, through the eyes of the pigeon.

Acknowledgments. I am grateful to Prof. Dr. G. Rüppell, Prof. Dr. K. Schmidt-Koenig, Dr. D. Haarhaus, and Mr. K. Büchner for their assistance and encouragement.

References

Köhler, K.-L.: Eine neue Navigationshypothese für Nah- und Fernorientierung von Vögeln. J. Ornithol. 116, 357-368 (1975)

Kramer, G.: Über die Heimfindeleistung unter Sichtbegrenzung aufgewachsener Brieftauben. Verh. Deut. Zool. Ges. Frankfurt a.M. 1958, 168-176 (1959)

Schmidt-Koenig, K.: Über den zeitlichen Ablauf der Anfangsorientierung von Brieftauben. Verh. Deut. Zool. Ges., Kiel 1964, 407-411 (1965)

Schmidt-Koenig, K., Walcott, C.: Flugwege und Verbleib von Brieftauben mit getrübten Haftschalen. Naturwissenschaften 60, 108-109 (1973)

Wallraff, H.G.: Über die Heimfindeleistungen von Brieftauben nach Haltung in verschiedenartig abgeschirmten Volieren. Z. Vergl. Physiol. 52, 215-259 (1966)

Pigeon Homing: Cues Detected During the Outward Journey Influence Initial Orientation

F. Papi, P. Ioalé, V. Fiaschi, S. Benvenuti, and N. E. Baldaccini,
Istituto di Biologia Generale, Università di Pisa, Via A. Volta 6, 56100 Pisa, Italy

Abstract

The initial orientation of homing pigeons released from the same site can be significantly different according to the odorous stimuli perceived during the outward journey. In fact, differences in orientation of pigeons transported to the same release site by different routes ("detour effect") resulted only when the birds were allowed to perceive the odors of the areas crossed during the outward journey. Moreover, birds carried in the same vehicle were different in orientation when subjected to a different pattern of olfactory stimulation during the journey.

Birds previously trained from a direction opposite to that of the test releases and transported in an iron container showed a dramatic difference in their initial orientation with respect to the control birds transported in an aluminum container. This suggests that magnetic stimuli perceived during the outward journey may also be involved in orientation.

Introduction

More than a century ago, Darwin (1873) wondered if the navigational capability of certain animals might be based on a kind of dead reckoning system on which they could rely even during an artificial displacement. Since then this idea has often been discussed and has even been developed into a complete theory (Barlow, 1964). For the homing pigeon, one of the most celebrated navigators, the use of information detected during the outward journey for navigational purposes does not seem improbable if one considers the behavior of its ancestor, the rock pigeon. The latter performs daily flights of distances up to tens of kilometers to and from foraging sites (Alleva et al., 1975). During the outward flight, the birds probably memorize landmarks and physical parameters useful to orientation during the return. When comparing the behavior of the two forms, the homeward flights of both can be considered as corresponding, whereas the outward flight of the rock pigeon has been replaced by the forced displacement of the homing pigeon. Therefore, one may wonder if the homing pigeon has preserved the ability to collect navigational cues, as far as they are still detectable, while being transported to the release site in a closed basket.

Until a few years ago, experiments performed with birds failed to produce evidence for the existence of inertial navigation or even of an influence of the outward journey on the homing behavior (see for ref. Keeton, 1974a). However, our group decided to perform new experiments devoted to this problem, also because we realized that the olfactory hypothesis of pigeon homing, which we had recently proposed (Papi et al., 1972), would have been in better agreement with the available evidence if one assumes that pigeons, besides the odors of the release area, also use those perceived during the outward journey. Indeed, as we often emphasized, a mechanism based on an olfactory map can operate only within a certain distance from the loft. Therefore correct initial orientation and successful homing from long distances would require that birds use alternate nonolfactory mechanisms and/or deduce the direction of displacement by smelling the familiar odors perceived during the initial part of the outward journey. If this last hypothesis is true, birds should be wrong in orientation when transported to the release site by a circuitous route. The detour experiments originated from this assumption.

Four years ago we published a short report on a first series of detour experiments (Papi et al., 1973). The positive results encouraged us to perform a further long series of experiments. Most of these were made with birds unfamiliar to the release site and are summarized in the first part of this paper. The second part is devoted to the influence of nonolfactory stimuli perceived en route, probably magnetic in nature, which can be demonstrated by transporting the birds in an iron container.

A. The Detour Effect

Each detour experiment was performed by carrying two groups of pigeons of the same (or balanced) age and experience to the same release site by different routes, with the first leg of the two routes strongly divergent. According to our hypothesis, each group was expected to fly roughly toward a direction opposite to that of the first part of the journey, thus one group deflecting clockwise (C birds), the other group counterclockwise (CC birds) with respect to the straight home course.

I. Methods

The two groups were transported by two vehicles starting simultaneously from the loft; the baskets were provided with a rich ventilation during transit. Birds were released singly, sometimes just after the arrival of both vehicles at the release site, sometimes on the following morning (in one case, some of the birds were released two days after the journey). No differences attributable to these variations in procedure were observed.

Birds were tossed singly by hand, alternating the release of one to three birds from one group with an equal number of birds from the other group. Each bird was observed with 10 × 40 binoculars until it disappeared from sight. From the individual bearings the direction and length of the mean vector were calculated. The bearing distribution was tested for randomness by the Rayleigh test and the difference between the samples by the Watson U^2 test (Batschelet, 1965, 1972).

The experiments with birds unfamiliar with the release site, 27 in all, were performed using pigeons of four lofts located near Pisa (S. Piero and Arnino lofts), Florence (Antella loft), and Siena (Catignano loft). Experiments were numbered according to the decreasing distance of the release site from the loft (max 135.8 km, min 78.4 km, airline). The home directions from the release sites were as follows: 0° - 90°, 14 cases; 91° - 180°, 5 cases; 181° - 270°, 5 cases; 271° - 360°, 3 cases. Twelve different kinds of detours were used. The experiments 1 - 3, 4 - 5, 6 - 7, 10 - 12, 14 - 20, 21 - 22, and 25 - 27 are repetitions of seven kinds of detours performed with different birds. From the 54 groups of birds used in the 27 experiments, 609 bearings were recorded (309 from the C, 300 from the CC birds; min. Sample size 5 bearings, max. 18).

II. Results

The effect of the detour was very variable, probably depending on the release site and on the routes used for transportation. The results of four single experiments are shown in detail to give an idea of the variability in the results obtained (Figs. 1-4).

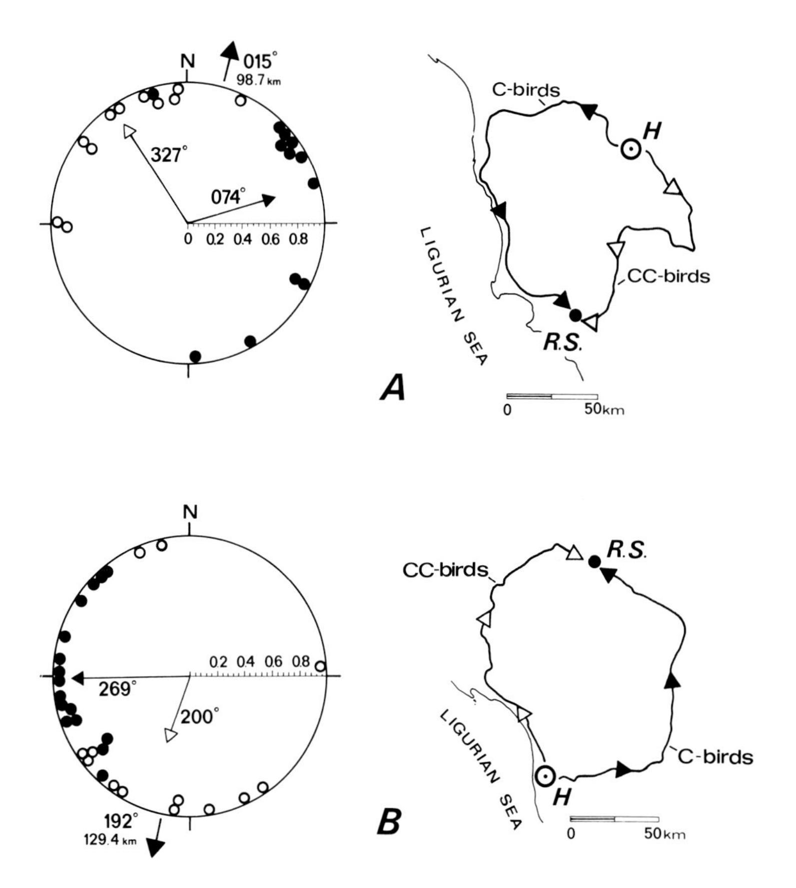

Fig. 1 A and B. Two detour experiments shown in detail. Experiment (A) refers to Florentine birds released from the South, experiment (B) to Pisa birds released from the North. In the maps, the routes from the loft (*H*) to the release site (*R.S.*) are shown. The birds expected to deflect clockwise (*C birds*) were transported on the road marked with *solid arrowheads;* the birds expected to deflect counterclockwise (*CC birds*) on the road marked with *open arrowheads*. In the circular diagrams, showing initial orientation, *solid symbols* are again used for the C birds, and *open symbols* for the CC birds. Each *small circle* is the vanishing bearing of one bird, the *inner arrows* represent the mean vectors of the two sets of bearings. The length of the vectors can be read using the scale. The *outer arrow* indicates home, the direction and distance of which are given.

In (A) each group of birds deflected as predicted with respect to the home direction, whereas in (B) both groups deflected in the same way. However, the geometric relationship of the two mean bearings was as predicted. In both experiments the two sets of bearings were nonrandom and the difference between them was significant ($p < 0.001$)

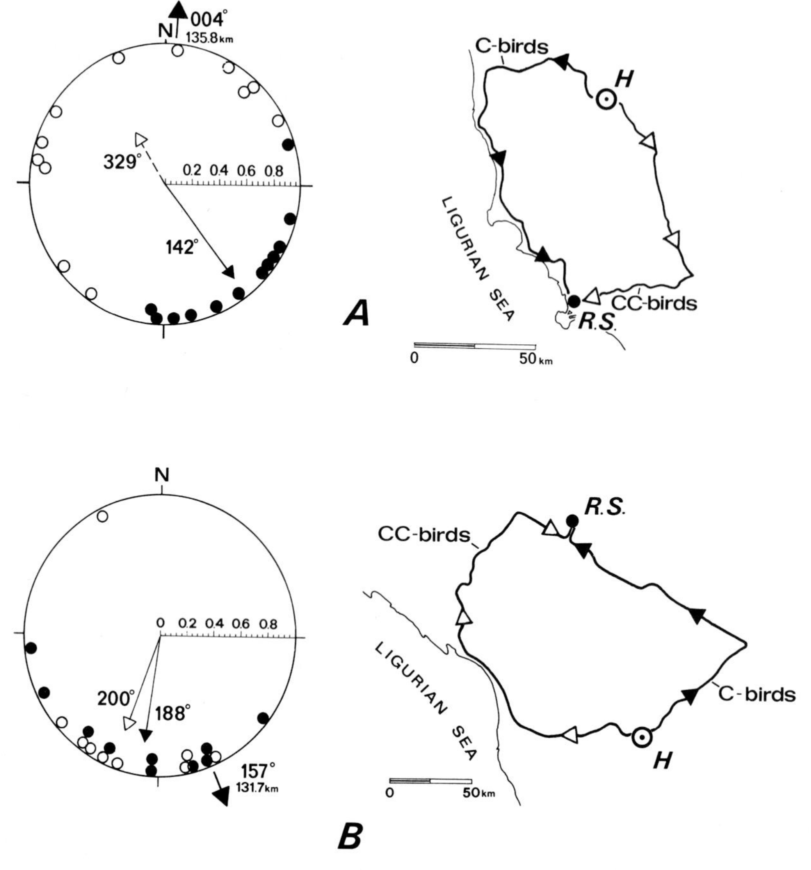

Fig. 2 A and B. Two further detour experiments in detail, both performed with Florentine birds released from the south (A) or from the north (B). In experiment A the bearings of CC birds were random (*dashed arrow*), whereas those of C birds were not and deflected very much in the predicted way. The difference between the two bearing sets was significant ($p < 0.001$). In B both groups were nonrandom but no difference in orientation resulted. The geometric relationship of the two mean bearings was in fact contrary to the prediction. Other explanations as in Figure 1

Considering all the experiments, it was found that one group of birds was oriented randomly in 13 experiments and both groups in one experiment. From the comparison of the bearing sets of each experiment, a significant difference resulted in 17 of the 27 experiments (in 9 cases at the 0.01 level, in the other 8 cases at the 0.05 level). As predicted, most C groups deflected clockwise from the home direction (21 cases) and most CC groups deflected counterclockwise (20 cases, Fig. 3).

Pooling the mean bearings (as in Fig. 3) and considering them as unit vectors, a second order analysis was possible. For the C birds there was a mean deflection from the home direction of 66° clockwise, a vector length of 0.43, and a nonrandom distribution of the bearings ($p < 0.01$). For the CC birds there was a mean deflection from the home direction of 24° counterclockwise, a vector length of 0.78, and a nonrandom distribution of the bearings ($p < 0.001$). The amount of deviation

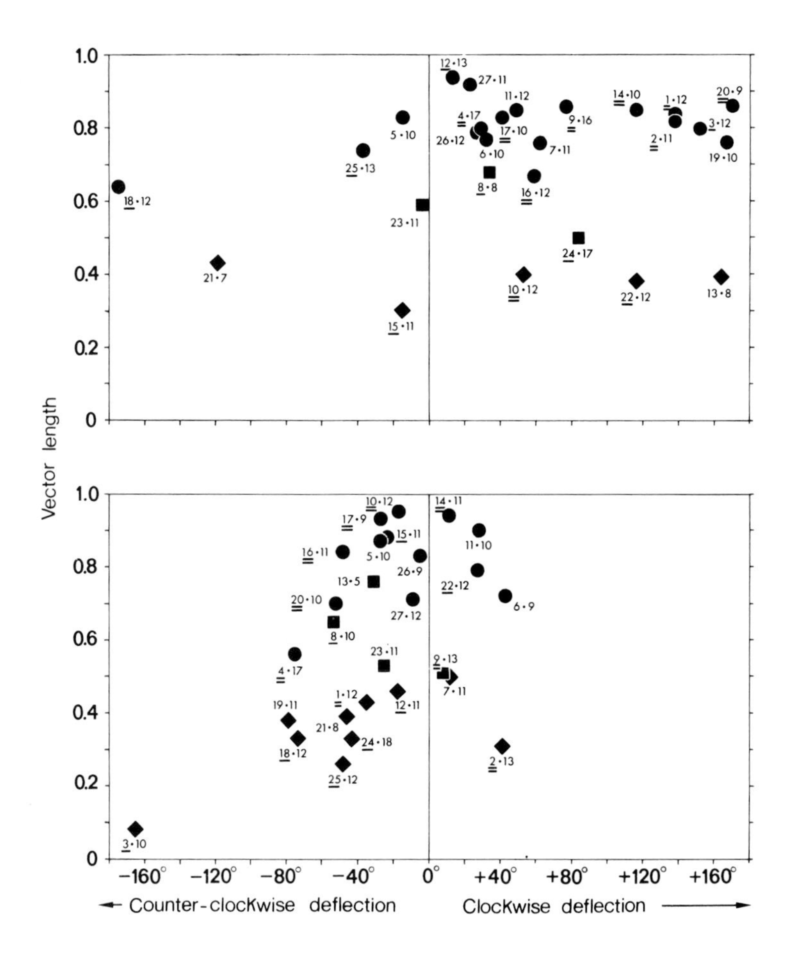

Fig. 3. Detour effect. Mean deflection from the home direction (*abscissa*) and length of the mean vector (*ordinate*) in each experiment. *Top*, C birds; *bottom*, CC birds. Whether the bearings of the set are nonrandom at the 0.01 level, nonrandom at the 0.05 level, or random, *circles*, *squares*, or *diamonds* are used respectively. The number of the experiment and the sample size (*second number*) are indicated near each symbol. The *number* of the experiment is underlined once or twice to denote when the difference between the C birds and the corresponding group of the CC birds is significant at the 0.05 and 0.01 levels, respectively

between the two grand mean bearings amounted to 90°, and the difference between the two samples was significant ($p < 0.001$). Also pooling all the vanishing bearings of the C and CC birds and comparing the two distributions, one obtains similar results (amount of deviation between the two grand mean bearings 83°, difference between the two samples still significant at the 0.001 level).

If one assumes that some detours always or mostly produce a significant difference in the birds' behavior, whereas other detours do not, the value of the results obtained by pooling our data are biased by the fact that some experiments were repeated exactly several times. For example, we performed the same detour experiment seven times (experiments No. 14 - 20) obtaining a significant difference between the two groups six times (among the 20 remaining experiments only 11 produced

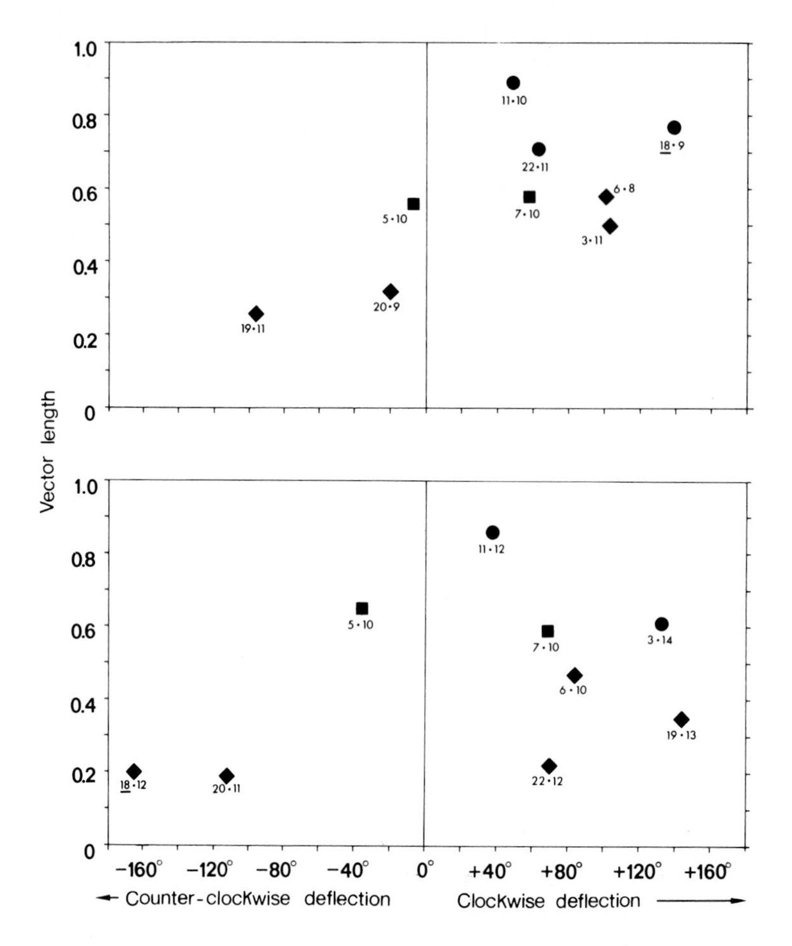

Fig. 4. Detour effect. Birds transported with their nostrils plugged: mean deflection from the home direction (*abscissa*) and length of the mean vector (*ordinate*). *Top*, CO birds; *bottom*, CCO birds. Other explanations as in Figure 3

a significant result). However, it was not our aim to establish the frequency of positive effects in a series of randomly choosen detours, but to test whether the detour effect is so frequent that it cannot be attributed to chance.

Further experiments were made to test if olfactory cues were involved in the detour effect. In 9 of the preceding 27 experiments, two additional groups of birds, transported with their nostrils plugged, were also tested (O pigeons). One group was transported with the C and the other with the CC birds. We called these new groups CO and CCO birds, respectively. They were aged and experienced as the C and CC birds, and housed in the same pen. During the test releases, tosses of the four groups of birds were alternated. To minimize the differences in treatment, the nostrils of the C and CC birds were plugged just after their arrival at the release site and remained plugged for an amount of time equal to that of the other two groups.

From the 18 O groups 193 bearings were recorded (89 from CO, 104 from CCO birds; min. sample size 8, max. 14). One group was oriented at random in three experiments and both groups in three other experiments. From the comparison of the two sets of bearings, a significant difference resulted in one experiment only (No. 18). Comparing each CO and CCO group with the corresponding group of C and CC birds, the vector length of the O birds was smaller in 13 cases and greater in 5. This difference was significant (binomial test, $p < 0.05$) and showed that the scattering of the bearings was greater in the O birds.

Pooling the mean bearings of CO and CCO groups, a tendency of both groups to deflect clockwise from the home direction appeared (Fig. 4). The difference between the two samples of mean bearings was not significant ($p > 0.50$), whereas that between the corresponding samples of C and CC birds was significant ($p < 0.02$). Similar results are obtained by pooling the vanishing bearings (Fig. 5).

In a few further detour experiments, instead of plugging the birds' nostrils, we transported the birds in containers through which bottled air flowed. Again, no detour effect resulted. These experiments will be reported in a subsequent paper.

Non consistent differences in homing performance were found between C and CC birds or between CO and CCO birds. Also a comparison between the homing performances of the O birds and those of their controls gave a negative result.

B. The "Iron Container" Effect

We recently began a series of experiments to test the possible influence of magnetic stimuli perceived during the outwart journey by transporting the birds in either iron or aluminum containers. In the first two tests the olfactory experience of the birds was also manipulated.

I. Methods

The birds of the different treatment groups were transported without detour to the release site in the same vehicle, but were enclosed in different containers. The iron container had the shape of a big cooking pot with lid, 63.5 cm in diameter, 30 cm high, and 1.6 mm thick. Inside the container the strength of the earth's magnetic field was reduced to 1/200 of its natural value. The aluminum containers used in experiments No. 1, 2, 3, and 7 were rectangular, 60 cm × 50 cm × 25 cm high, whereas the aluminum container used in the remaining experiments had the same shape and size of the iron container. All the aluminum containers were 2.0 mm thick. Differences in initial orientation between birds transported in an aluminum or the iron container persisted also after having painted the inner surface of the containers with the same paint in order to avoid differences in odors between the containers. Air sucked in from the outside or bottled air was used to provide a forced ventilation of the containers.

Just after the arrival at the release site, the birds were put into wicker baskets and kept there until the release. The test was usually performed on the same day as the outward journey, but on two occasions (experiments No. 6 and 9) it was made on the following morning.

Birds from the Arnino loft near Pisa (PA and PB birds) and from the Antella loft near Florence (FA and FB birds) were used. The age of the birds when the test releases began was: PA birds 1 - 2 years, PB birds 2 - 4 years, FA birds 1 year, FB birds 2 - 3 years. Before beginning the test releases, Pisa birds had been directionally trained by releasing them 18 - 20 times from SSE at distances of 8 - 93 km. Between

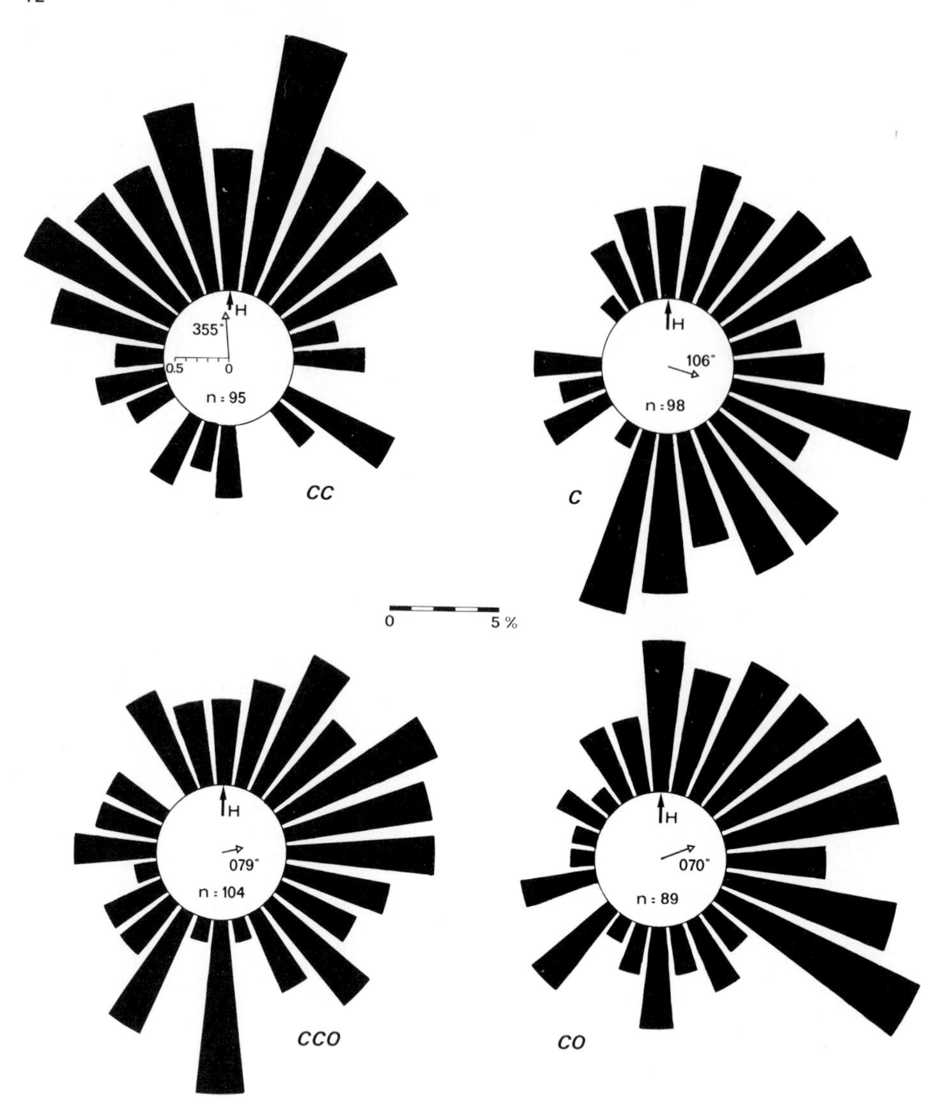

Fig. 5. Detour effect. Initial orientation of the birds transported with their nostrils plugged (*CCO* and *CO*, *bottom*) and of the control birds simultaneously transported with their nostrils free (*CC* and *C*, *top*). The length of the bars is proportional to the number of vanishing bearings in each 15^0-sector. The home direction (*H*) is set to 0^0. The mean vectors are represented inside the circles, and *n* is the number of bearings.

The difference between C and CC birds is significant ($p < 0.005$), whereas that between CO and CCO birds is not

two successive tests they were always released for training 1 - 3 times from 93 km SSE. Florentine birds had been subjected to a training from W, once from 30 km and four times from 73 km. Between each test they were released for training 2 - 4 times from 73 km W. If not otherwise stated, birds were assigned to the treatments by lots.

The V test was used to test if birds were homeward oriented (Batschelet, 1972). For other methods see Section 1.

II. Experiments and Results

The first two experiments were performed with three groups of birds: (1) Birds transported in the iron container and ventilated with bottled air (E1 birds); (2) birds transported in an aluminum container and ventilated with bottled air (E2 birds); and (3) birds transported in a second aluminum container and ventilated with air sucked in from outside during the journey (C birds).

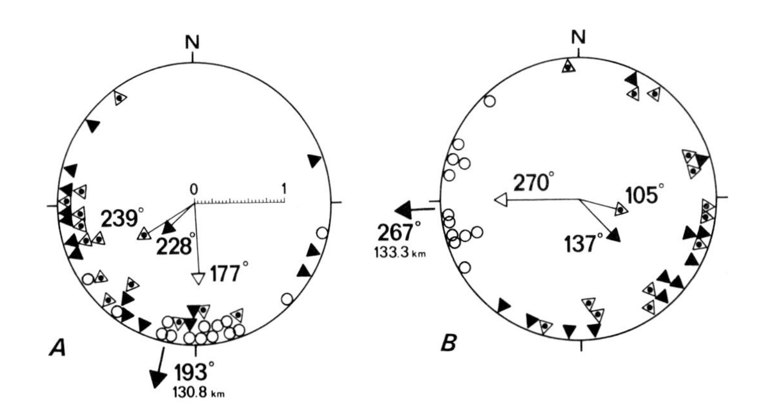

Fig. 6 A and B. Iron container effect. Initial orientation of E1 birds (*solid triangles*, iron container and bottled air), E2 birds (*dotted triangles*, aluminum container and bottled air), and C birds (*open triangles*, aluminum container and air sucked in from outside). Each *triangle* represents the vanishing bearing of one bird, the *inner arrows* represent the mean vectors. The *outer arrow* indicates home, the direction and distance of which are given. (A) experiment No. 1; (B) experiment No. 2

Experiment No. 1, Figure 6A. April 6, 1977. Pisa birds (PA) released from Bagnolo, 130.8 km NNE. All the groups were nonrandom and homeward oriented, but the orientation of C birds was more precise and accurate than that of both other groups. Comparing C with both E1 and E2 birds, a significant difference was evident ($p < 0.01$), whereas E1 were not different in orientation from E2 birds.

Experiment No. 2, Figure 6B. April 13, 1977. Florentine birds (FA) released from Fano, 133 km E. All the groups were nonrandom, but only C birds were homeward oriented. They were different in orientation from both E1 and E2 birds ($p < 0.001$), which were not different each other.

The results of both of these experiments seemed to indicate that only the quality of the air breathed by the birds influenced their orientation, whereas the kind of container did not. This was contradicted by the results of the subsequent experiments. They were all performed with two groups of birds, one transported in the iron container (E birds), and one in the aluminum container (C birds). Both groups were ventilated with air sucked in from outside during the journey.

The following four tests were performed with Pisa birds released from Bagnolo, 130.8 km NNE.

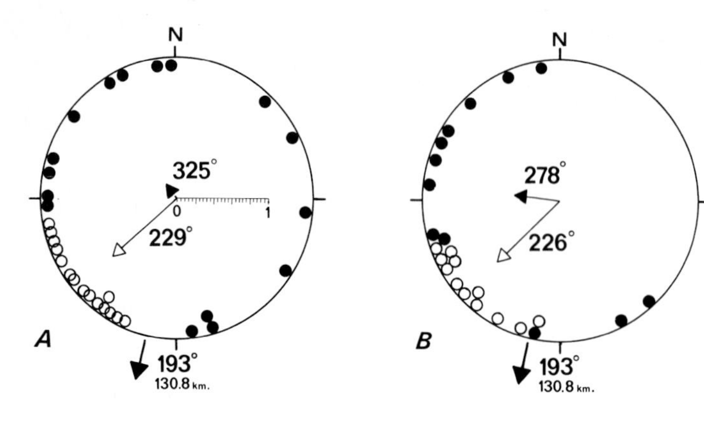

Fig. 7 A - D. Iron container effect. Initial orientation in experiments No. 3 - 6. *Solid symbols* for the E birds, *open symbols* for the C birds. Each *small circle* represents the vanishing bearing of one bird. For other explanations see Figure 6

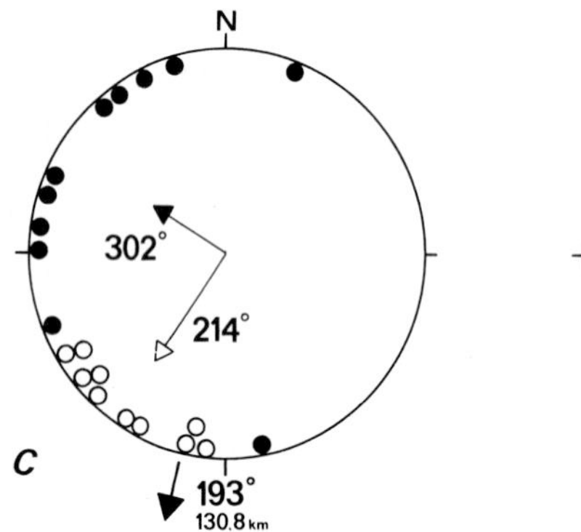

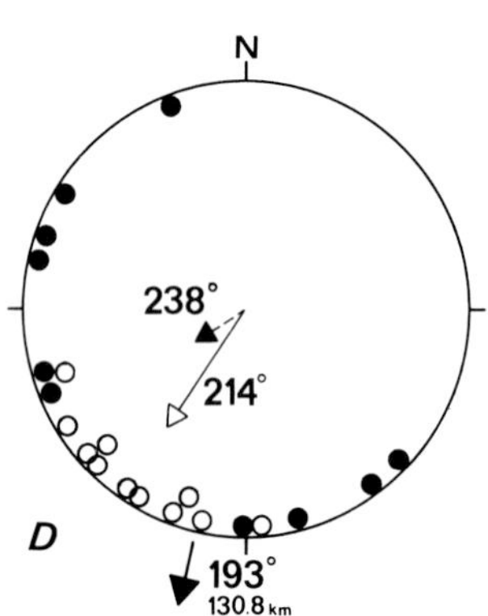

Experiment No. 3, Figure 7A. May 6, 1977. PB birds were used. E birds were random and not homeward oriented; C birds were nonrandom and homeward oriented. The difference between the two groups was significant ($p < 0.001$).

Experiment No. 4, Figure 7B. May 12, 1977. Veterans from experiment No. 1 (PA birds) were used. Both groups were nonrandom, but only C birds were homeward oriented. The difference in orientation was significant ($p < 0.001$).

Experiment No. 5, Figure 7C. May 23, 1977. Half of the PA birds already used in experiments No. 1 and 4 and half of the PB birds already used in experiment No. 3 were released. The result was the same as in the previous experiment.

Experiment No. 6, Figure 7D. May 24, 1977. PA and PB birds already used in experiments No. 1, 3, and 4, but not in experiment No. 5, were released. The result was the same as in experiment No. 3.

The remaining five experiments were performed with Florentine birds. The first three releases were made at Fano, 133.3 km E, and the last two at Bagnolo, 131.7 km NNW from the loft.

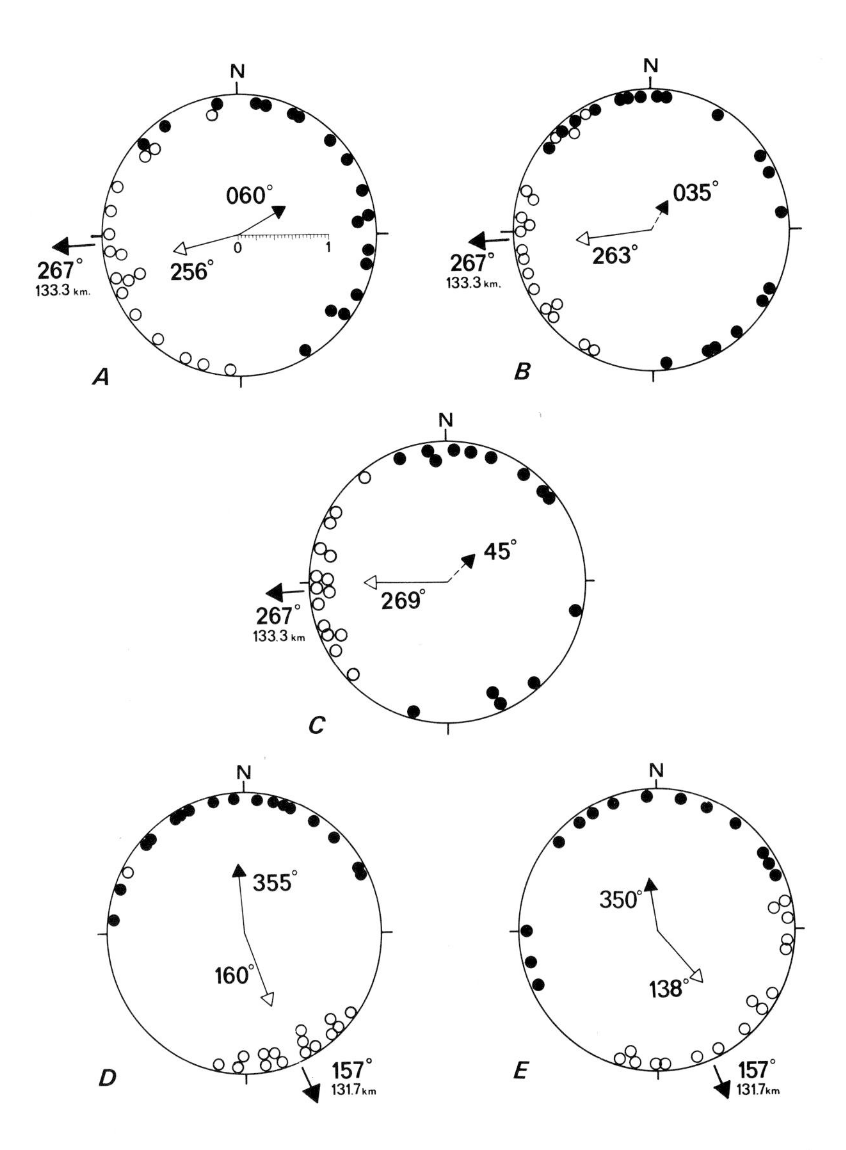

Fig. 8 A - E. Iron container effect. Initial orientation in experiments No. 7 - 11. For explanations see Figures 6 and 7

Experiment No. 7, Figure 8A. April 21, 1977. FB birds were used. Both groups were nonrandom but only C birds were homeward oriented. The difference in orientation was significant ($p < 0.001$).

Experiment No. 8, Figure 8B. May 10, 1977. FA birds veterans from experiment No. 1 were used. E birds were random and not homeward oriented; C birds were nonrandom and homeward oriented. The difference between the two groups was significant ($p < 0.001$).

Experiment No. 9, Figure 8C. June 2, 1977. The same two groups of FA birds of experiment No. 9 were used, but their roles were reversed. The result was as in the previous experiment.

Experiment No. 10, Figure 8D. June 10, 1977. FB birds already used in experiments No. 7 and 9 were released. The result was as in experiment No. 7.

Experiment No. 11, Figure 8E. July 25, 1977. FA birds veterans from experiments No. 2 and 8 were released again. The result was as in experiments No. 7 and 10.

Despite the dramatic differences in initial orientation between the treatments, no significant differences in homing performance resulted from the experiments, taken either singly or as a whole.

C. Discussion

The detour experiments performed with birds transported with their nostrils plugged show that the detour effect is due to the odors sensed during the outward journey. This conclusion is supported also by the results of the first two experiments reported in the second section and by further unpublished experiments in which the birds, carried in the same vehicle but subjected to different olfactory stimuli during the journey, differed in their initial orientation. These last experiments will be described in a subsequent paper, which will include also a full report of all our detour experiments.

In many instances, the initial bearing of detour birds appears to be intermediate between the home direction and the direction opposite to that of the first leg of the journey. This indicates that the olfactory cues detected during the initial part of the journey form only part of the information on which the deduction of the initial bearing is based. The reason why birds expected to deflect clockwise did so more than those expected to deflect counterclockwise, and why both bird groups transported with their nostrils plugged deflected clockwise, can be, at present, only a matter of speculation.

In some of the detour experiments on pigeons transported with their nostrils free, no difference in initial orientation resulted. The reason for this is again unknown, but this fact explains the apparent discrepancy with the first detour experiments performed in the United States on Cornell pigeons (Keeton, 1974b). Very recently, indeed, Keeton and Brown, in a cooperative research program together with two of us (F.P. and S.B.), were able to find a kind of detour which produced on Cornell birds a significant divergence, both in the original experiment and in its repetition with the birds conveyed by the opposite leg of the detour (unpublished data). In Germany, also, the effect of the detour is very variable and Hartwick et al. (this volume) found only one release site out of 12 at which the detour birds persistently revealed the expected deflection. On the other hand, both the detour experiments performed in Switzerland produced positive results (Fiaschi and Wagner, 1976).

The results reported in the second section of this paper suggest that homing pigeons also use magnetic stimuli detected during the outward journey. This conclusion is supported by the results reported by Kiepenheuer and by Wiltschko et al. at this Symposium.

It is too early even to speculate about the role of the magnetic information detected en route on the navigational system of the homing pigeon. The fact that the birds transported in the iron container had the same homing performances as the control birds indicates that the lack of information produced by the iron container can be quickly compensated. The magnetic cues detected during the outward journey cannot be considered as alternative to the olfactory cues, especially to those which the birds detect while in flight. In fact, both earlier as well as more recent experiments (Papi et al., 1971, 1972; Benvenuti et al., 1973; Baldaccini et al., 1975; Snyder and Cheney, 1975; Hartwick et al., 1977) demonstrate that the elimination of olfactory ability, effected in whatever manner, disastrously impairs pigeon homing from unfamiliar localities, thus confirming the unreplaceable role of olfaction (Papi, 1976; Hartwick et al., 1977).

Acknowledgments. This work was supported by the Consiglio Nazionale delle Ricerche. We thank Professor Adriano Gozzini for stimulating discussions and for measuring the strength of the earth's magnetic field inside the iron container. P.J. Roach kindly improved the English text.

References

Alleva, E., Baldaccini, N.E., Foà, A., Visalberghi, E.: Homing behaviour of the rock pigeon. Monitore Zool. Ital. (N.S.) 9, 213-224 (1975)

Baldaccini, N.E., Benvenuti, S., Fiaschi, V., Papi, F.: New data on the influence of olfactory deprivation on the homing behavior of pigeons. In: Olfaction and Taste. Denton, D.A., Coghlan, J.P. (eds.), New York: Academic Press pp. 351-353 (1975)

Barlow, J.S.: Inertial navigation as a basis for animal navigation. J. Theor. Biol. 6, 76-117 (1964)

Batschelet, E.: Statistical methods for the analysis of problems in animal orientation and certain biological rhythms. Washington, D.C.: Am. Inst. Biol. Sci. 57 p. (1965)

Batschelet, E.: Recent statistical methods for orientation data. In: Animal orientation and navigation. Galler S.R. et al. (eds.) NASA SP-262, U.S. Gov. Printing Office, Washington, D.C. pp. 61-91 (1972)

Benvenuti, S., Fiaschi, V., Fiore, L., Papi, F.: Homing performances of inexperienced and directionally trained pigeons subjected to olfactory nerve section. J. Comp. Physiol. 83, 81-92 (1973)

Darwin, C. Origin of certain instincts. Nature (London) 7, 417-418 (1873)

Fiaschi, V., Wagner, G.: Pigeons homing: some experiments for testing the olfactory hypothesis. Experientia 32, 991-993 (1976)

Hartwick, R.F., Foà, A., Papi, F.: The effect of olfactory deprivation by nasal tubes upon homing behavior in pigeons. Behav. Ecol. Sociobiol. 2, 81-89 (1977)

Keeton, W.T.: The orientational and navigational basis of homing in birds. Adv. Study Behav. 5, 47-132 (1974a)

Keeton, W.T.: Pigeon homing: no influence of outward-journey detours on initial orientation. Monitore Zool. Ital. (N.S.) 8, 227-234 (1974b)

Papi, F.: The olfactory navigation system of the homing pigeon. Verh. Dtsch. Zool. Ges. 1976: 184-205. Gustav Fischer Verlag, Stuttgart, 1976

Papi, F., Fiaschi, V., Benvenuti, S., Baldaccini, N.E.: Pigeon homing: outward journey detours influence the initial orientation. Monitore Zool. Ital. (N.S.) 7, 129-133 (1973)

Papi, F., Fiore, L., Fiaschi, V., Benvenuti, S.: The influence of olfactory nerve section on the homing capacity of carrier pigeons. Monitore Zool. Ital. (N.S.) 5, 265-267 (1971)

Papi, F., Fiore, L., Fiaschi, V. Benvenuti, S.: Olfaction and homing in pigeons. Monitore Zool. Ital. (N.S.) 6, 85-95 (1972)

Snyder, R.L., Cheney, C.D.: Homing performance of anosmic pigeons. Bull. Psychonomic Soc. 6, 592-594 (1975)

Investigation of Pigeon Homing by Means of "Deflector Cages"

N. E. Baldaccini, S. Benvenuti, V. Fiaschi, P. Ioalé, and F. Papi,
Istituto di Biologia Generale, Università di Pisa, Via A. Volta 6, 56100 Pisa, Italy

Abstract

Homing pigeons, kept from fledging time in cages supplied with screens to deflect the winds clockwise (CW) or counterclockwise (CCW), show a corresponding deflection in their initial orientation (Baldaccini et al., 1975). New experiments produced the following main results:
1. The acquired deflection in initial orientation was modified by subsequently reversing the arrangement of the deflectors, i.e., by subjecting CW-deflected birds to CCW-deflected wind and vice versa.
2. Birds reared in control cages, which had shown a normal initial orientation in test releases, deflected as expected three months after the deflectors had been applied to their cages. 3. The orientational bias of deflector-cage birds was not eliminated by repeated homing experience.

A. Introduction

According to the olfactory hypothesis of pigeon navigation, homing pigeons acquire, during their first months of life, the "map component" of their navigational mechanism by associating the odors carried by the winds with the direction from which they come (Papi et al., 1972). Therefore, pigeons should acquire a rotated map and be wrong in their orientation, if at their loft they perceive the winds deflected with respect to their actual direction. In fact, if kept from fledging time in large aviaries fitted with screens to deflect the winds clockwise or counterclockwise, pigeons show a corresponding deflection in their initial orientation (Baldaccini et al., 1975). So far, keeping birds in such "deflector cages" is the only treatment that yielded birds with a rotated map.

Partial aim of the experiments reported here was to ascertain whether the acquired knowledge that we call "map" is unmodifiable or whether, once it has been determined, it can be rotated by modifying the wind direction through the inside of the cages. Finally, a series of experiments was made to determine if the orientational bias of the deflector-cage birds could be eliminated by repeated homing experience.

B. Materials and Methods

I. The Cages

All cages were installed on the ground of the Arnino laboratory near Pisa. Besides the three cages - two for experimental and one for control birds used in previous experiments (Baldaccini et al., 1975) -

two new deflector cages were used in the third series of experiments: one cage with clockwise (CW) and one with counterclockwise (CCW) deflecting screens. The new cages, installed close to the old ones (Fig. 1), were similar to these, but larger (2.60 × 2.60 × 2.10 m high). The deflectors were made entirely of glass whereas the old cages had deflectors made of panes of glass in the upper part only.

Fig. 1. View of the deflector cages at Arnino with the new CW cages in the *foreground*. In the *background*, from the *left*, the new CCW cage, the old CCW cage, the control cage, and the old CW cage

II. Release Procedures

The birds were carried to the release sites inside a van. From their baskets the birds could not view the outside but were well ventilated by two broad pipes conveying air to them from the outside. Many experiments were carried out in two days, with a daily release of nearly equal numbers of birds of each group. Because of changing weather one experiment was made on three different days. The releases were made under clear skies or at least with the sun disk visible. Bird were tossed singly and observed with 10 × 40 binoculars until they vanished from sight. Mostly the number of birds released in each test is usually somewhat larger than that of bearings recorded in the corresponding vanishing diagrams because some birds landed in the vicinity, joined other pigeons, or disappeared prematurely behind the vegetation or other objects and had to be omitted.

III. Treatment of the Birds and Test Releases

1. First Experimental Series. The experiments were performed with two groups of birds put into the old experimental cages at the age of 1 or 2 months. They had never been kept in open air before. The emprisonment in the cages begun for most birds the last week of April

1975 but a few birds, eight in all, had been put into the cages at different dates between November, 1974 and March, 1975. Shortly before the beginning of the test releases, the birds were allowed free flight during three consecutive days and then subjected to three training releases from distances of 150, 400, and 700 m. During all these flights the birds wore masks preventing them from breathing through the nostrils (Papi et al., 1973).

Three test releases were performed between July 11 and August 4, 1975. On August 7 and 8 the arrangement of the screens of both cages was reversed so that the CW birds were from then on subjected to CCW-deflected winds and became CCW-deflected birds, and vice versa. Twenty out of 39 birds were lost or died between the first test and the releases to test the effect of deflectors' inversion. Afraid that the losses had been nonrandom we took into account only the birds that survived until the second release series. The data collected on the orientation of all the birds tested before the inversion of the deflectors were reported by Papi (1976, Fig. 7C-D).

After the inversion of the deflectors the birds were tested eight times between November 14, 1975 and January 16, 1976.

2. Second Experimental Series. The birds, about one month old, were divided by lots into three groups (α, β, and γ birds) at the beginning of April 1976 and each group was housed in a similar cage without deflectors. In fact we used the three old cages after having removed the deflectors from the two experimental cages. On May 15 and 20 we allowed the birds to go out for flight exercise. Birds that did not fly out spontaneously were forced to do so. Later, in the same month, the pigeons were released for training from 150, 400, and 700 m distance. Between June 3 and 11 three test releases were made from three different sites. Contrary to the procedure used in the first experimental series we took into account the bearings of all the birds tested before attachment of the deflectors without excluding the birds that died or were lost (only 8 out of 67) before being tested again for deflectors' effect.

On June 16 - 18, the deflectors were attached to the cages of the α and γ birds. The arrangement of the deflectors was such that α birds were subjected to clockwise-deflected winds (thus becoming CW birds) and the γ birds to counterclockwise-deflected winds (CCW birds), whereas β birds were destined to act as control birds. The releases aimed at testing the effect of the deflectors now attached began on September 6, 1976. The birds remained imprisoned in their cages until this date. The first three releases were made at the same sites already used by testing the pigeons before the attachment of the deflectors. Six more releases were performed, four from unfamiliar sites.

3. Third Experimental Series. The birds, all about one month old, were housed in two new deflector cages on April 16, 1976. On July 22 the birds were allowed or forced to fly for exercise. At the beginning of August two training releases from 150 and 300 m were made. The test series, begun on August 12, 1976, included nine releases from the same site and four releases from other sites.

IV. Statistical Methods

The mean vector for each group of bearings was calculated by vector analysis. Bearings were tested for nonrandomness by the Rayleigh test

(Batschelet, 1965). Bearing sets were compared by the U^2 Watson test (Batschelet, 1972), and by the Mardia multisample test (Mardia, 1972). Homing performances were tested by the Mann-Whitney-U test and by the sign test (Siegel, 1956).

C. Results

1. First Experimental Series. In the first two releases before inversion of the deflectors the predicted deflection of the two groups was observed. In the third release, made from over the sea, the CCW birds were random (Fig. 3, left). However, the difference between CW and CCW birds was always significant at the 0.001 level. Pooling all the bearings (Fig. 2 A - B), two nonrandom distributions resulted. CW and CCW birds were deflected in the expected direction 59° and 35°, respectively, from the home direction. The difference between the two groups was significant at $p < 0.005$.

In the eight releases after inversion of the deflectors, two of which were made over the sea, one group was random on three occasions (releases 4, 5, and 9). In four out of the five remaining releases the geometric relationship between the mean vectors was as expected according to the new arrangement of the deflectors (Fig. 3, right). The difference between the two bearing sets was significant only in test releases 9 and 10. Pooling the bearings (Fig. 2 C - D) two nonrandom distributions resulted. With respect to the home direction the mean vector of the CW (formerly CCW) birds and of the CCW (formerly CW) birds deflected 12° and 20° in the expected direction. However, the difference between the two distributions was not significant.

There were no differences in homing performances between CW and CCW birds, either before or after inversion of the deflectors.

2. Second Experimental Series. The three groups of birds (α, β, and γ) oriented nonrandomly and in a similar way in the first three releases made before the deflectors were attached (Fig. 4). Mardia's multisample test did not lead to significance in any of the three tests.

The first three test releases after attachment of the deflectors were performed from the same release sites, but the results obtained were different (Fig. 5). Nonhomogeneity was revealed in all three tests ($p < 0.02$ in the first, $p < 0.001$ in the other two tests). Therefore, we tested the distributions in pairs by the Watson U^2 test. The result of the first test release (Fig. 5 A_{1-3}) was unusual because both experimental groups were randomly oriented whereas the control group was not. The difference between CW and control birds was significant ($p < 0.001$) while the differences both between CCW and CW birds and between CCW and control birds were not.

In the two subsequent releases the three groups were nonrandomly oriented with the mean vectors of the experimental groups deflected as expected with respect to both the home direction and the mean vector of control birds (Fig. 5 B_{1-3}, 5 C_{1-3}). Differences were always significant at least at the 0.005 level.

In all subsequent releases birds were always nonrandomly oriented. Because the first test after attachment of the deflectors had produced an unusual result, we first tested the birds again at the site. The

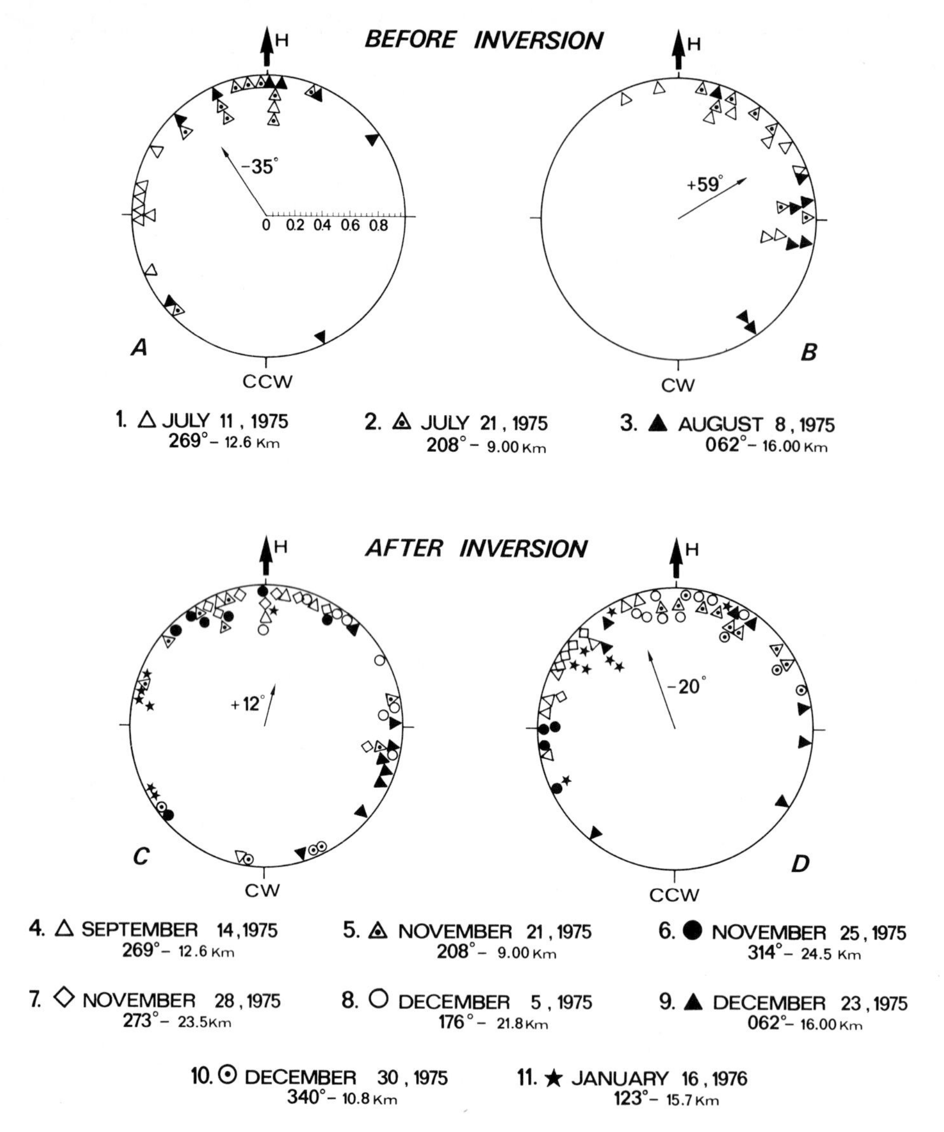

Fig. 2 A - D. First experimental series. Initial orientation of deflector cage birds before (*top*) and after (*bottom*) inversion of the deflectors. Diagrams (A) and (C) refer to the same birds which were expected to deflect counterclockwise (*CCW*) before inversion of the deflectors, and clockwise (*CW*) after inversion; correspondingly, diagrams (B) and (D) refer to a second group of birds which were expected to deflect clockwise (*CW*) before inversion, counterclockwise (*CCW*) after inversion. Home direction (*H*) is set to 0°. The *inner arrow* represents the mean vector, the length of which can be read according to the scale in the first diagram. In the four diagrams each symbol on the periphery of the circle represents the vanishing bearing of one bird. Different symbols refer to different releases as indicated below in each pair of diagrams. For each release home bearing and distance are given below the dates

three treatments produced homogeneous results (Fig. 6 A_{1-3}). We then released the birds from an unfamiliar site 11 km further out on the same line (Fig. 6 B_{1-3}), and subsequently from a familiar site 9 km

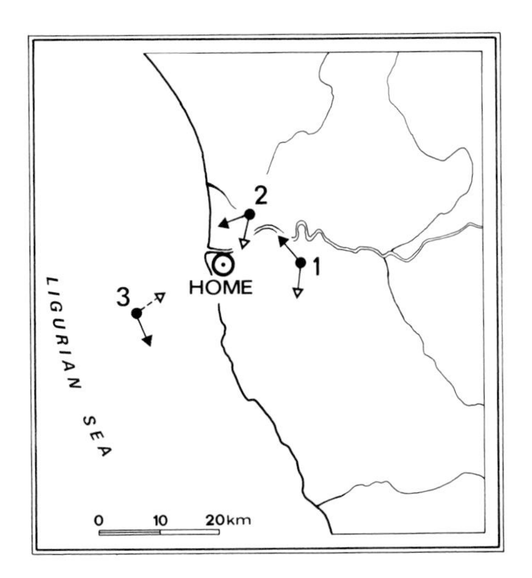

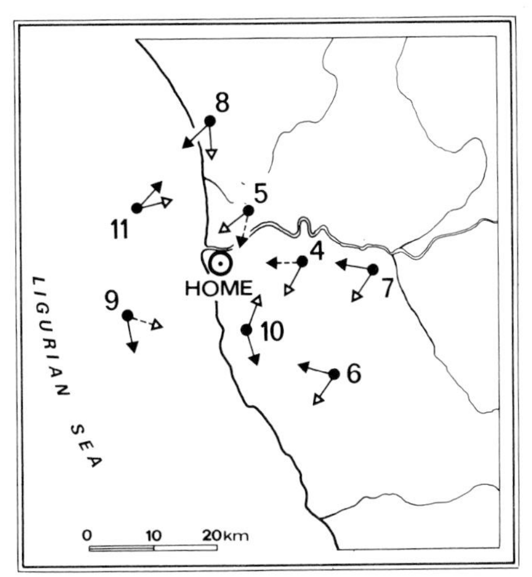

Fig. 3. First experimental series. Maps showing the release sites and the mean directions of the birds expected to deflect clockwise (*solid arrow head*) and counterclockwise (*open arrow head*). Releases before the inversion of the deflectors at the *left*, after inversion at the *right*. The length of arrows is not proportional to the vector length, but *dashed arrows* are used for random distributions (p >0.05). Releases are numbered as in Figure 2. See also Figure 2 and text

NNE (Fig. 6 C_{1-3}) from which the pigeons had already been released both before and after attachment of the deflectors (Fig. 4 B_{1-3}, and 5 B_{1-3}). No differences among the groups resulted.

Looking at the negative results of the last three releases, we were going to conclude that the effect of attaching the deflectors had been transient. However, before giving up, we decided to perform three further releases (24.5 km SE, 21.8 km N, and 94.1 km E). The three sets were never homogeneous and from comparing them in pairs always a significant difference at the 0.001 level resulted (Fig. 7).

No significant differences in homing performances resulted before attachment of the deflectors. On the contrary, control birds performed better than CW birds in the 1st, 2nd, and 9th experiments ($p < 0.05$) after attachment of the deflectors and better than CCW birds in the 7th ($p < 0.05$) and 9th ($p < 0.01$) experiments. Comparing the homing performances of the pairs that could be formed of birds of different groups released one immediately before (or after) the other, we obtained significant differences after attachment of the deflectors. Control birds performed better than both CW birds (91 cases out of 137, $p < 0.01$) and CCW birds (82 cases out of 135, $p < 0.01$). No differences resulted, however, between CW and CCW birds, the former having been faster in 70 cases, the latter in 61 cases (p >0.20).

3. Third Experimental Series. The first three out of the nine test releases performed from the same site yielded unusual results. At the first release CCW birds vanished randomly, CW birds were not homeward oriented (Fig. 8 A), and no significant difference between the two groups resulted. In the second experiment (Fig. 8 B) both groups were nonrandom and deflected counterclockwise from home, the geometric relationship between the two mean bearings being contrary to predic-

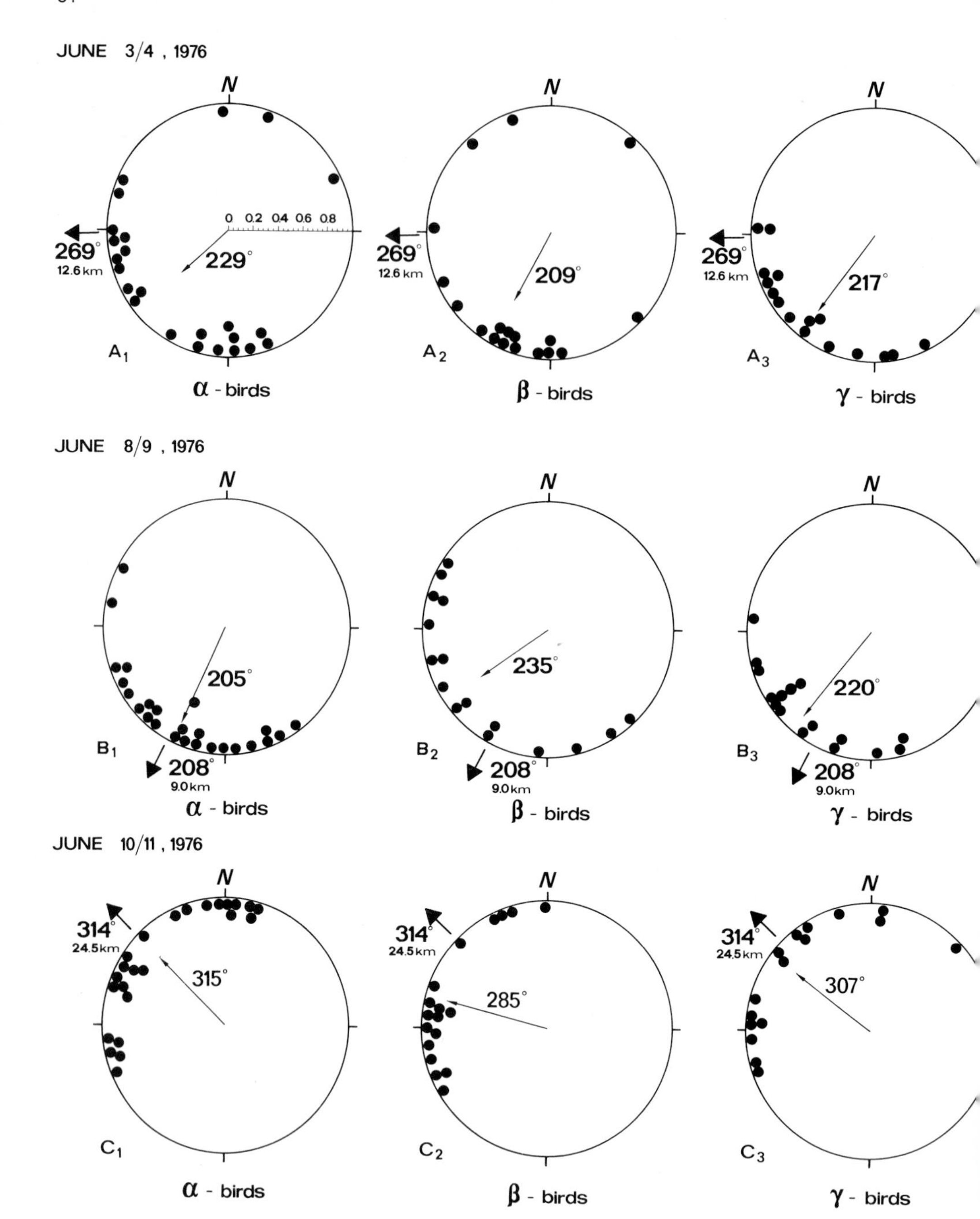

Fig. 4 A - C. Second experimental series. Initial orientation of α, β, and γ birds in the three test releases before attachment of the deflectors. In the diagrams each *dot* represents the vanishing bearing of one bird, the *inner arrow* is the mean vector, the *outer arrow* designates the home direction

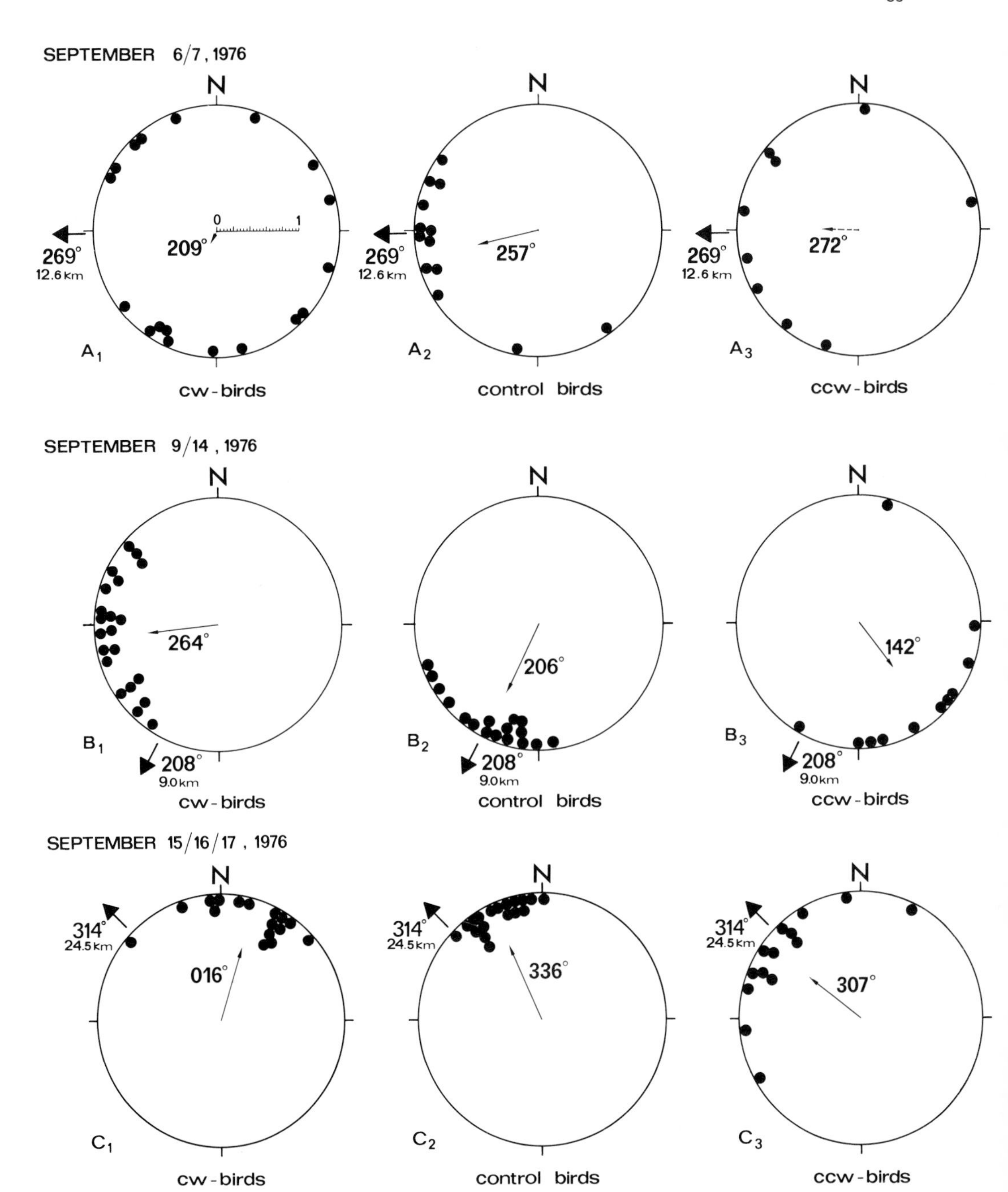

Fig. 5 A - C. Second experimental series. Initial orientation of CW (formerly α), control (formerly β), and CCW (formerly γ) birds in the first three releases after attachment of the deflectors. Other explanations as in Figure 4

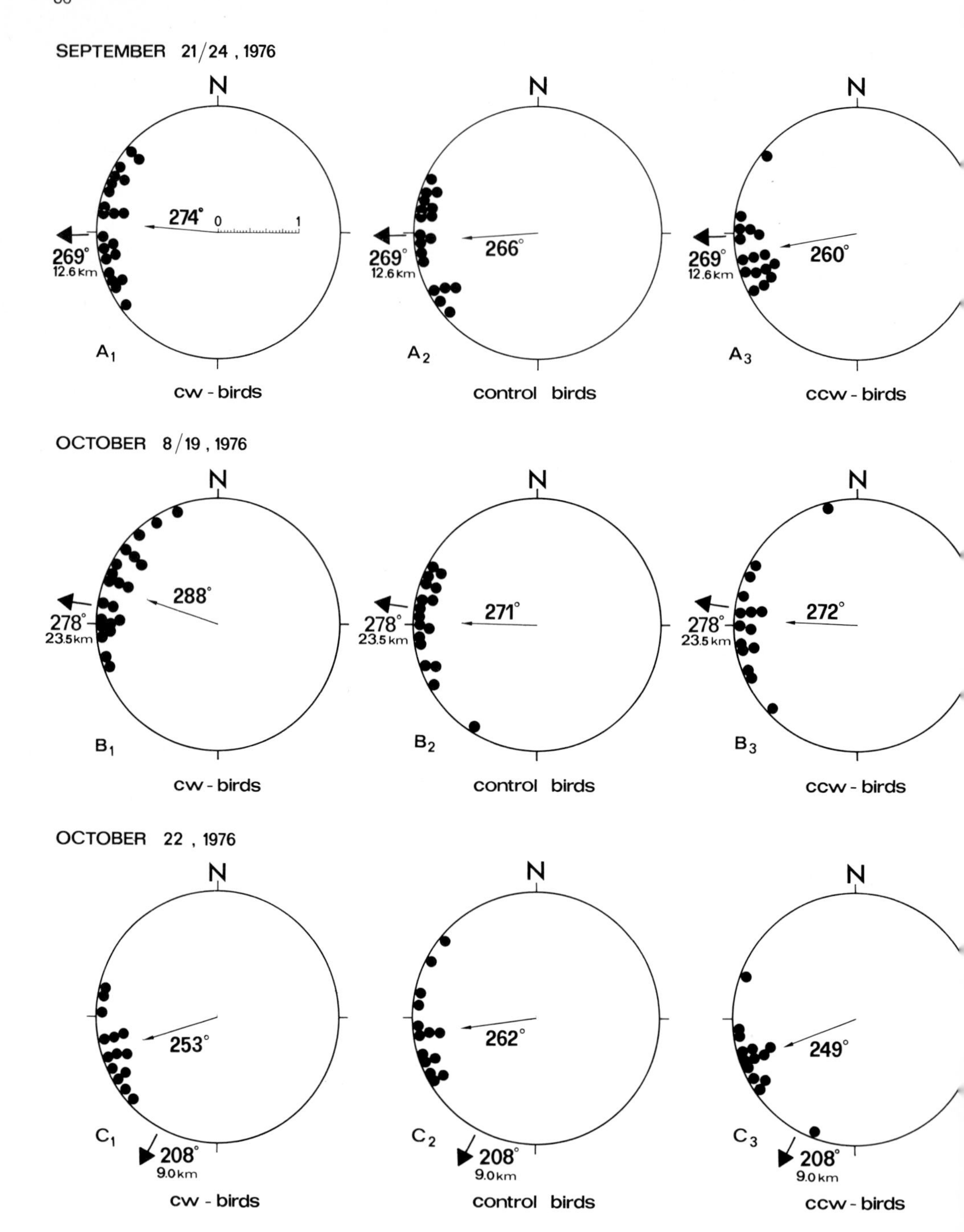

Fig. 6 A – C. Second experimental series. Initial orientation of the birds after attachment of the deflectors in three further releases. Other explanations as in Figures 4 and 5

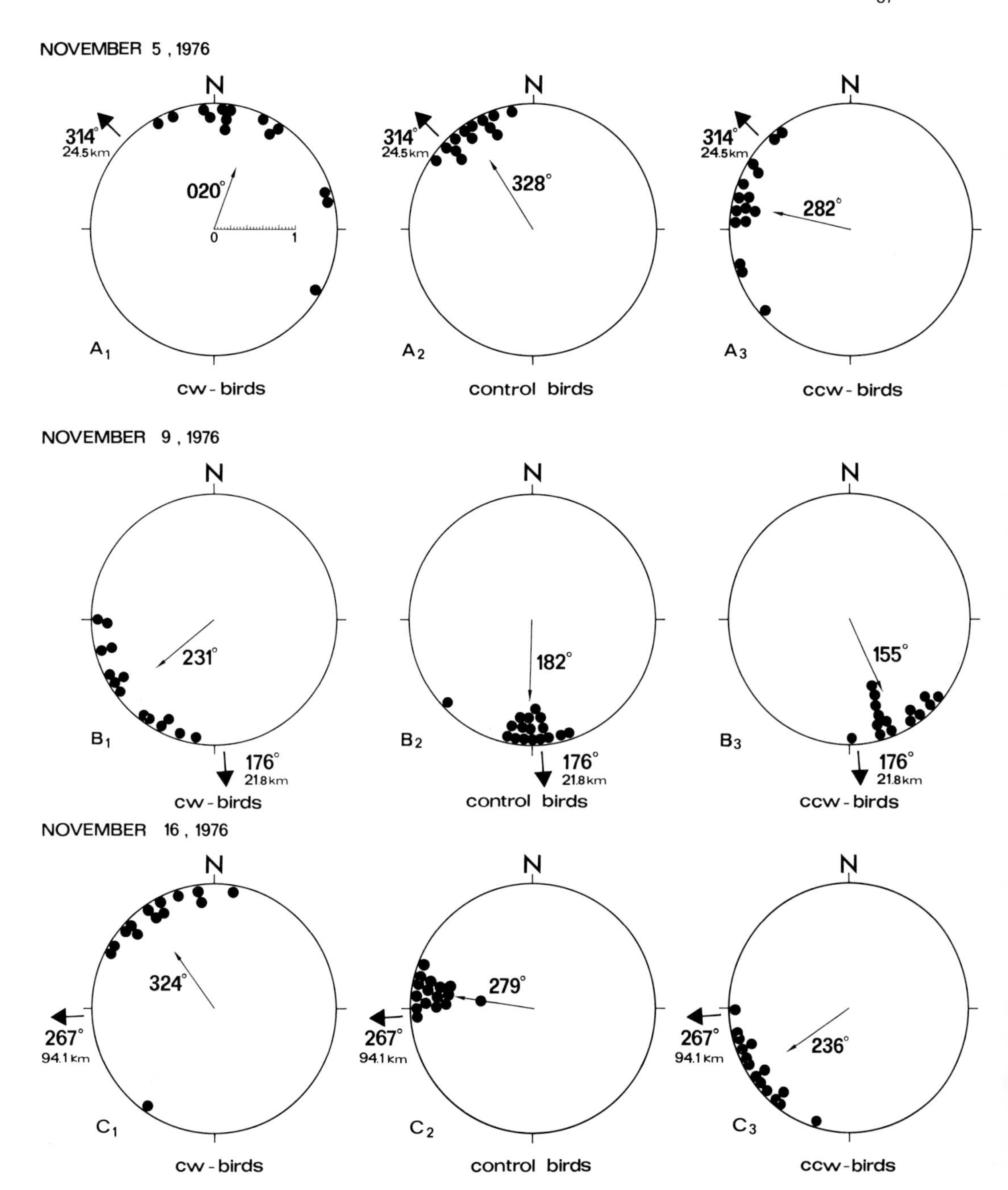

Fig. 7 A - C. Second experimental series. Initial orientation of the birds in the last three releases after attachment of the deflectors. Other explanations as in Figures 4 and 5

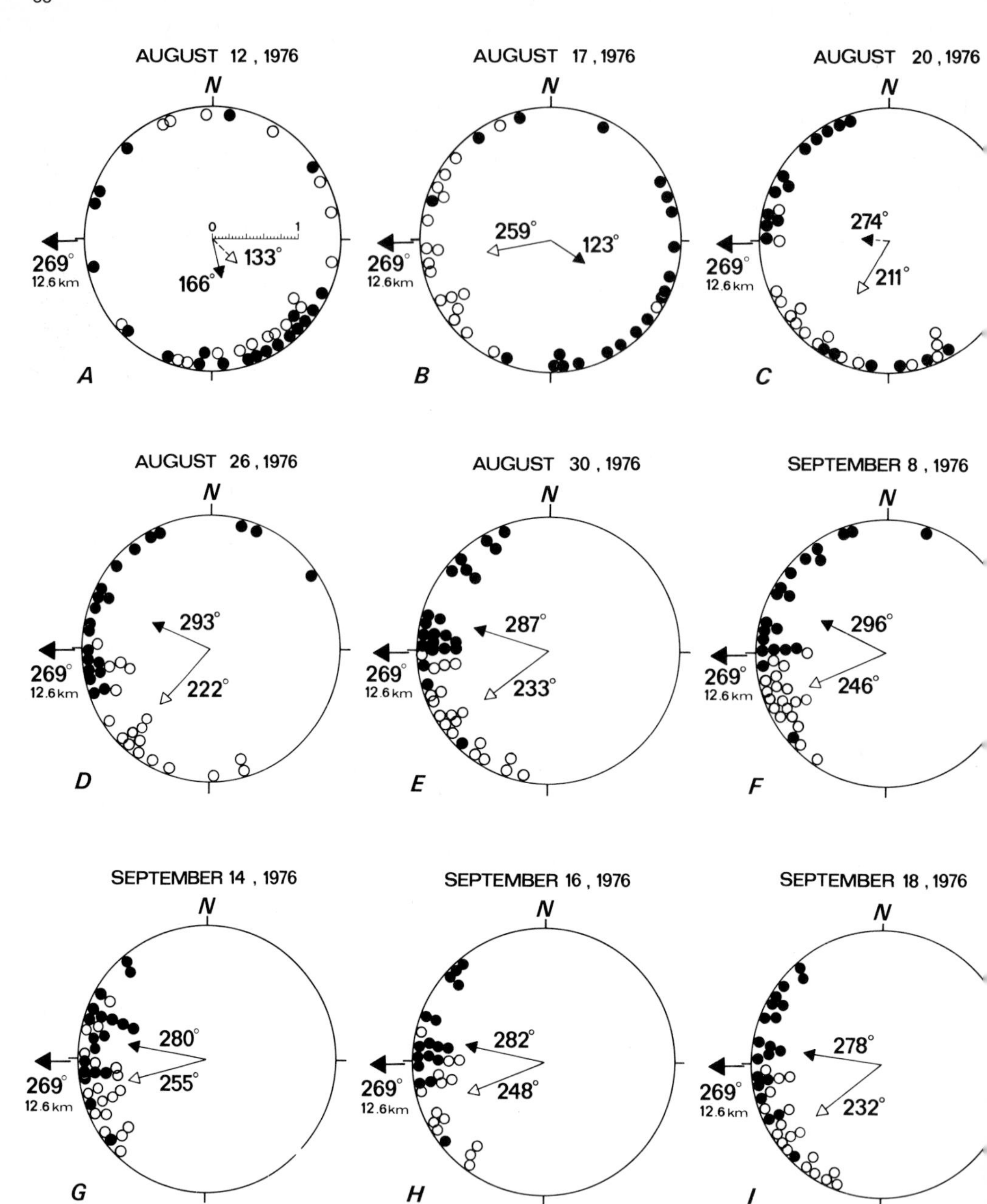

Fig. 8 A - I. Third experimental series. Each diagram refers to one experiment and shows the bearings both of the birds expected to deviate counterclockwise (*open symbols*) and of the birds expected to deviate clockwise (*solid symbols*). This figure includes the first nine releases performed at the same site. Other explanations as in Figure 4

tion. The difference between the two groups was significant ($p < 0.005$). In the third experiment (Fig. 8 C) the CW birds were randomly oriented, while the CCW birds performed nonrandomly and were deflected as expected with respect to the home direction. The difference was again significant at the 0.005 level.

From the fourth experiment on, the results became consistent. All releases produced similar results (Fig. 8 D - I). Both groups were always nonrandomly oriented and deflected as predicted with respect to the home direction. The divergence between the mean vectors, initially decreasing (from 71° in the 4th experiment to 25° in the 7th experiment) increased in the eight and nineth experiments (34° and 46° respectively). The differences between the two sets were always significant ($p < 0.05$ in the 7th experiment, $p < 0.005$ in all the others).

Four more experiments were performed from other sites (Fig. 9). The first was done from a site along the line of the previous releases but at a farther distance. The result was similar (Fig. 9 A). The mean vectors were deflected as expected, the deviation attaining 91° and the difference being significant ($p < 0.005$). The subsequent release was performed from a site NNE but both groups of birds headed westward

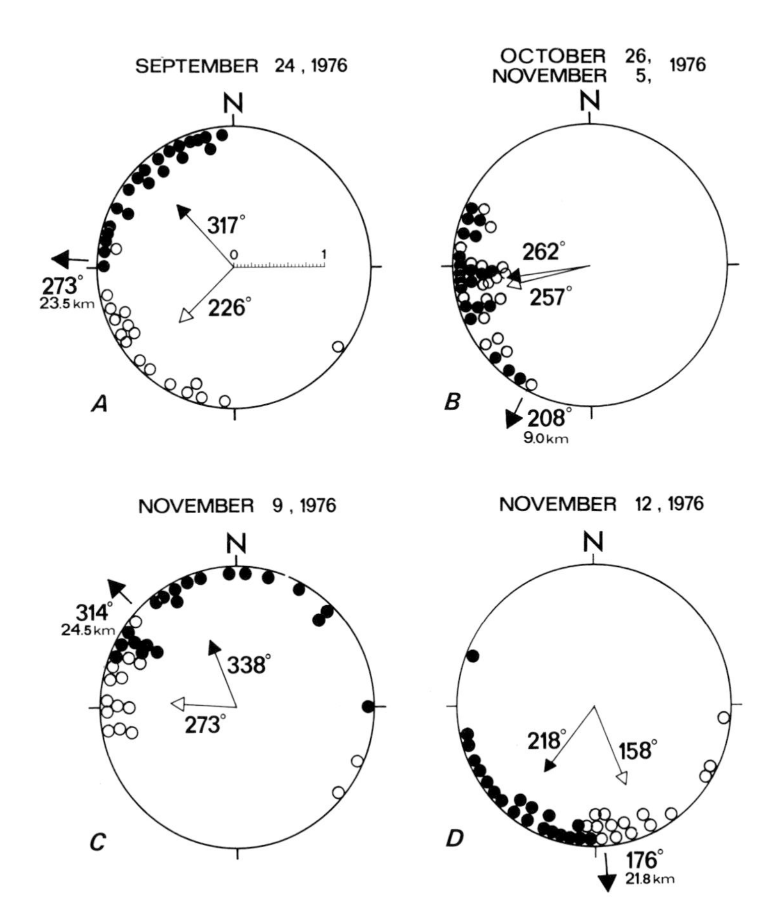

Fig. 9 A - D. Third experimental series. Initial orientation in the last four releases. Other explanations as in Figures 4 and 8

and no significant difference resulted (Fig. 9 B). However, positive and similar results were produced in the last two experiments (Fig. 9 C - D). The mean vectors were deflected as expected and the differences were in both cases significant at the 0.005 level. The value of the deviation of the mean vectors was 65° and 60°, respectively.

In the nine releases from the same site, CCW birds always homed faster than CW birds. The difference was significant in the first ($p < 0.01$), second ($p < 0.01$), fifth ($p < 0.001$) and seventh releases ($p < 0.05$). Comparing the homing performances of each CW bird with that of the CCW bird released immediately before or after we found in 114 out of 168 pairs that the CCW bind had shorter homing times ($p < 0.01$). In the four subsequent releases made from other sites, CCW birds performed better again in the tenth experiment ($p < 0.005$). The CW birds homed faster in the other experiments, but the difference was significant in the 13th experiment only ($p < 0.05$).

D. Discussion

The birds of the first experimental series modified their orientation after the inversion of the deflectors. In fact, in the single releases, they often showed a tendency to invert their bias, and no significant difference between the two groups of pooled bearings was subsequently observed. However, considering this result separately, it would be premature to conclude that the change in orientation was due to the inversion of the deflectors, because homing experience or other factors might have been involved. The result of the second experimental series, however, clearly shows that the map component of the navigational mechanism can be rotated after it has been determined by modifying the wind direction inside the cages. Finally, we concluded from the results of the third experimental series that the orientational bias of the deflector-cage birds cannot be corrected, at least not entirely, by a large amount of homing experience. However, we would like to emphasize that the birds between one release and the next continued to be housed in the deflector cages so that the sensory experience at home may have contributed to retaining the bias and to neutralizing the effect of the homing experience.

Two minor questions arising from our results are presently without a satisfactory explanation. In three releases of the second series after attachment of the deflectors, the deflector effect was lacking or it was not significant. Similarly, no effect resulted in the 11th release of the third series. In this case, both groups were oriented westward instead of southward. A training effect might be involved here because the ten previous releases had been performed from the east, but the lack of deviation between CW and CCW birds remains unexplained.

The second question concerns the puzzling behavior of the third-series birds in the first tests. There is a discrepancy between their behavior and that of the old-deflector-cage birds as reported by Baldaccini et al. (1975) and Papi, (1976). This difference in the results cannot be attributed to the kind of deflector cage (old or new) because other birds reared in the new cages in 1977 did not show the anomalous behavior of their 1976 companions (unpublished data).

A problem of more general interest is how experimental birds find their way home despite their wrong initial orientation. They were often slower than control birds but usually successful in homing. It should be emphasized that in the releases of the third series, performed from the same site, the pigeons of one group were consistently faster in homing. Therefore one should conclude that these birds corrected their course in a different and more effective way than their companions. Presently we have no suggestion as to which mechanism is used to correct initially false headings. A few flocks of experimental birds have been tracked by helicopter but the data are, so far, too scarce for any conclusions. Further work is in progress.

Switching to final conclusions, our present results confirm that homing pigeons establish the map component of their navigational mechanism during the first months of life by associating the odors carried by winds with the direction from which they come. In determining initial orientation upon release, the olfactory information acquired at the loft seems to play a more important role than the experience collected during homing flights. The acquisition of the map is not an imprinting process; pigeons easily modify their map according to new information they receive. The adaptive value of this may be that pigeons must continuously update their map under natural conditions if the odorous cues they use to navigate vary according to the seasons.

Acknowledgements. This work was supported by the Consiglio Nazionale delle Ricerche. We thank the Command of the Brigata Paracadutisti "Folgore", Italian Army, which kindly allowed us to use a power boat for the releases from over the sea. Drs R.F. Hartwick and J. Kiepenheuer helped to carry out several releases reported here.

References

Baldaccini, N.E., Benvenuti, S., Fiaschi, V., Papi, F.: Pigeon navigation: effects of wind deflection at home cage on homing behavior. J. Comp. Physiol. 99, 177-186 (1975)

Batschelet, E.: Statistical methods for the analysis of problems in animal orientation and certain biological rhythms. Washington, D.C. Am. Inst. Biol. Sci. 57 p. (1965)

Batschelet, E.: Recent statistical methods for orientation data. In: Animal Orientation and Navigation. Galler, S.R. et al. (eds.) NASA SP-262, U.S. Gov. Printing Office, Washington D.C., pp. 61-91

Mardia, K.V.: A multi sample uniform scores test on a circle and its parametric competitor. J. R. Stat. Soc. (B) 34, 102-113 (1972)

Papi, F.: The olfactory navigation system of the homing pigeon. Verh. Deut. Zool. Ges. 1976: 184-205. Gustav Fischer Verlag, Stuttgart (1976)

Papi, F., Fiore, L., Fiaschi, V., Benvenuti, S.: Olfaction and homing in pigeons. Monitore Zool. Ital. (N.S.) 6, 85-95 (1972)

Papi, F., Fiore, L., Fiaschi, V., Benvenuti, S.: An experiment for testing the hypothesis of olfactory navigation of homing pigeons. J. Comp. Physiol. 83, 93-102 (1973)

Siegel, S.: Nonparametric statistics for the behavioral sciences. New York: Mc Graw-Hill Book Co. 1956 XVII, 312 pp.

An Apparent Lunar Rhythm in the Day-to-Day Variations in Initial Bearings of Homing Pigeons

TIMOTHY LARKIN and WILLIAM T. KEETON, Section of Neurobiology and Behavior, Cornell University, Ithaca, NY 14853, USA

Abstract

In view of the suggestion from earlier experiments that detection of magnetic and gravitational stimuli may be linked, we investigated the possibility that gravitational changes across the synodic lunar month might influence the initial bearings of homing pigeons under conditions when naturally occurring fluctuations in the geomagnetic field have been shown to influence the birds' orientation behavior. Six separate series of releases were conducted during four different years at three different sites. In all six, a nearly linear correlation between the pigeons' mean vanishing bearings (MVB) and the day of lunar month (DLM) was found. The monthly oscillation of MVB could apparently exist stably in either of two modes, each 180° out of phase with the other; in some series the oscillation was from new moon to new moon and in others from full moon to full moon. Statistical analyses of several types support the notions that the best frequency of the oscillations of MVB is indeed almost exactly that of the synodic lunar month and that the turning points of the oscillations are the syzygies (new moon and full moon). However, attempts to tie the variations in MVB directly to the gravitational changes occurring during the lunar month have so far been unsuccessful; hence it is not yet known what environmental stimuli are acting as the immediate causes of the birds' changing orientation behavior.

As has been reported previously, most release sites are characterized by a site-specific bias in the initial bearings chosen by pigeons released there (e.g., Keeton, 1973). Though these release-site biases are relatively stable, they do show some day-to-day variation (Fig. 1). For several years we have been attempting to discover the causes of this variation, in hopes that in so doing we would learn more about the pigeons' navigational "map" and about how the birds integrate the various components of their navigational system.

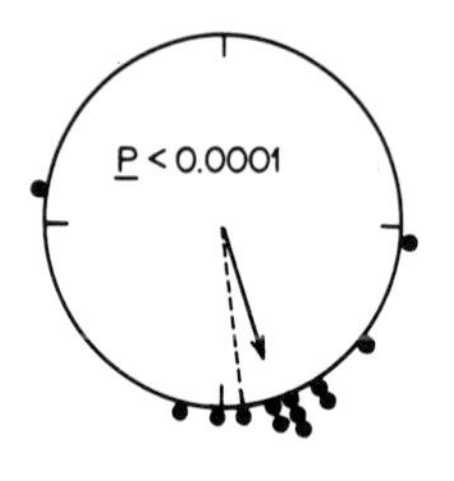

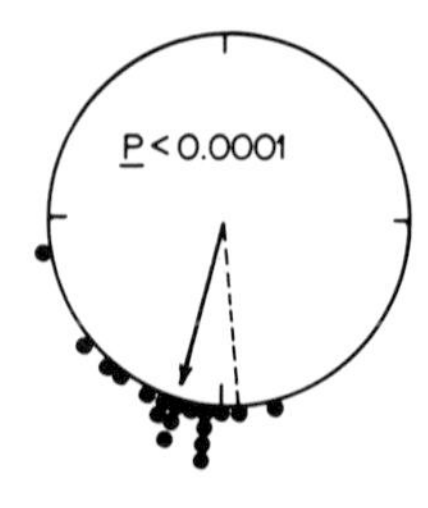

Fig. 1. Bearings of the same group of pigeons on two different days at Weedsport in 1972, illustrating the type of day-to-day variation seen. The mean vanishing bearing (*arrow*) usually deviates clockwise from the home direction (*dashed line*) at this site, so the MVB for 16 June (*left*) is unusual

In earlier publications, we reported evidence that naturally occurring fluctuations of the earth's magnetic field (as measured by the K index) probably constitute one of the environmental factors contributing to the day-to-day variation in pigeons' orientational behavior (Keeton et al., 1974; Larkin and Keeton, 1976). Now, in many other instances where magnetic fields have been found to influence animal orientation, the animals have been simultaneously responding to gravitational cues (see Keeton, 1974 for summary), which raises the possibility that detection of magnetic and gravitational stimuli may be linked. Having found that our pigeons exhibit slight but consistent changes in their vanishing bearings as a function of K-type magnetic fluctuations, we wondered if this response, too, might be in any way related to gravity.

Being unable to alter gravity in a laboratory test of orientation, we chose to determine whether the natural monthly gravitational cycle caused by the changing relative positions of the earth, sun, and moon, might influence the orientation of pigeons in the type of test releases we had used in studying the "K effect." We here report evidence from six separate release series that the mean bearings of a group of pigeons released repeatedly at a single release site do indeed vary in an orderly fashion across the lunar synodic month.

A. Methods

We first reanalyzed the data from our initial three series of releases (1970, 1972, 1973) in the K-effect study (Keeton et al., 1974), to see if the mean vanishing bearings exhibited any relationship to the day of the lunar month. As a follow-up, we performed three additional series of releases, all in 1974: One was from Weedsport, New York (73.5 km N of the loft; home direction 173°), which was the same site used in the series of 1970, 1972, and 1973; another was from Orwell, Pennsylvania (65.5 km S; home direction 348°); and the third was from Campbell, New York (70.2 km W; home direction 72°).

In each of the six series, a single group of pigeons was repeatedly released throughout a period of several months from the designated test site and from no other site. In this way we controlled as well as possible for differences in the birds' ages, experience, or individual idiosyncrasies, and for direction of previous release, special properties of release site, etc. - all of which are variables whose effects on the birds' orientation behavior might otherwise have obscured the relationships we were seeking to examine. All releases were under sunny skies, when winds were weak or moderate.

The mean vanishing bearings (MVB) for the individual releases of each series were graphed as a function of day of lunar month (DLM); since the distributions appeared to approximate linearity, the statistics for linear regression of MVB on DLM were calculated. For our purposes, a lunar "day" was arbitrarily defined as one-fifteenth of the period between successive syzygies (i.e., of the period between new moon and the next full moon, or between full moon and the next new moon). Thus a complete synodic lunar month (which averages 29.53 solar days) was assigned 30 lunar days, although the 15 days of the first half month (new moon to full moon) might not be of exactly the same duration in solar minutes as the 15 days of the second half month (full moon to new moon) because of variations in the syzygetic periods, and although a given lunar day (e.g., day 12) might differ slightly in duration from one month to the next because of synodic variation. (Our method of defining lunar days was a modification of a method used by Lieber and Sherin, 1972.)

The special statistical methods used to try to answer the two questions - (1) whether emphasis on the syzygies as the starting points for regression is justified, and (2) whether the best period of oscillation of MVB is really the same as the period of the synodic lunar month - will be briefly described in the Results section when those questions are addressed.

B. Results

I. Evidence for a Lunar Rhythm in Orientation

Significant positive regression of MVB on DLM was found in each of the three original release series (Fig. 2). When the data were plotted with the day of full moon as the starting point on the X axis, the

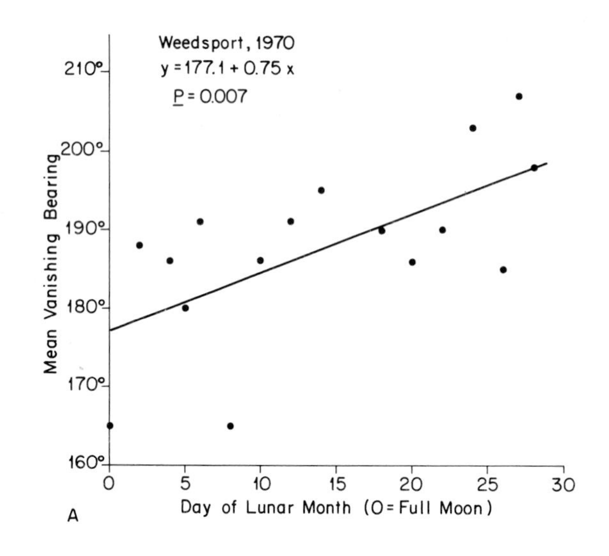

Fig. 2 A - C. Graphs showing MVB as a function of DLM for the original three series of releases at Weedsport

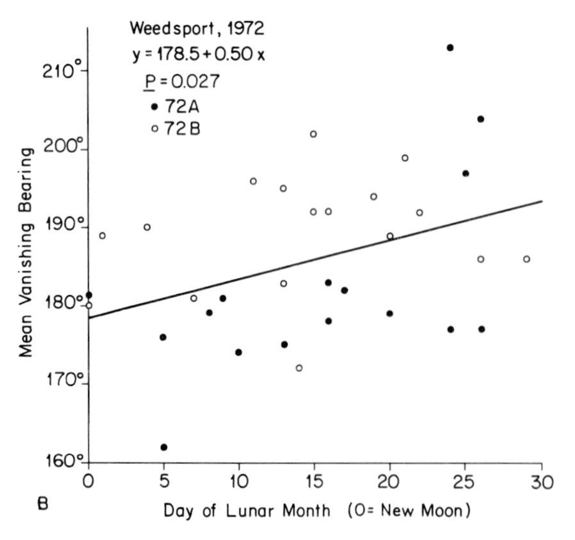

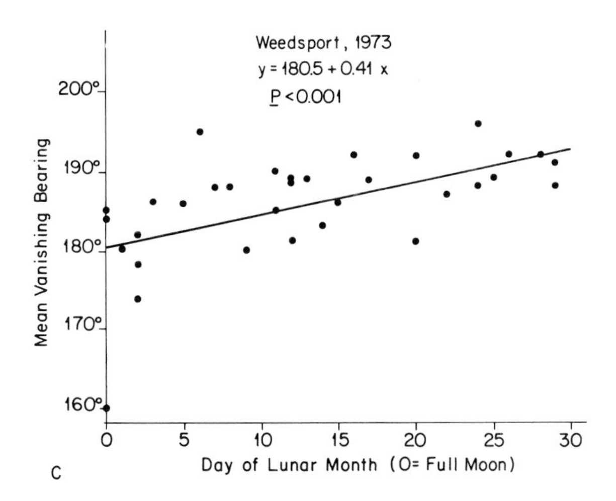

regression coefficient for 1970 was significant at $p = 0.007$ ($n = 16$), and the coefficient for 1973 was significant at $p < 0.001$ ($n = 33$). When the data for 1972 were plotted with the day of new moon as the starting point, the regression coefficient was significant at $p = 0.027$ ($n = 33$). (All p values are for two-tailed tests.) Note that, if MVB really varies in a regular manner across the lunar month, that variation can apparently oscillate in either of two modes 180° out of phase with each other - i.e., the oscillation can be from full moon to full moon (as in 1970 and 1973) or from new moon to new moon (as in 1972).

The graphs of Figure 2 include the pooled data for all test months of each year. However, if MVB is oscillating as a function of day of lunar month, we should be able to see this on a month-to-month basis also. Figure 3 shows the data for the first two months of the 1972 series, as an example. There are relatively few data points in any single month, and consequently the separate regressions are not significant. However, month-by-month analysis of this sort yields a picture consistent with the one given by the yearly pooled data, namely, that MVB tends to "rise" (i.e., move clockwise) throughout the month until just before the end of the month (day of new moon in 1972, or day of full moon in

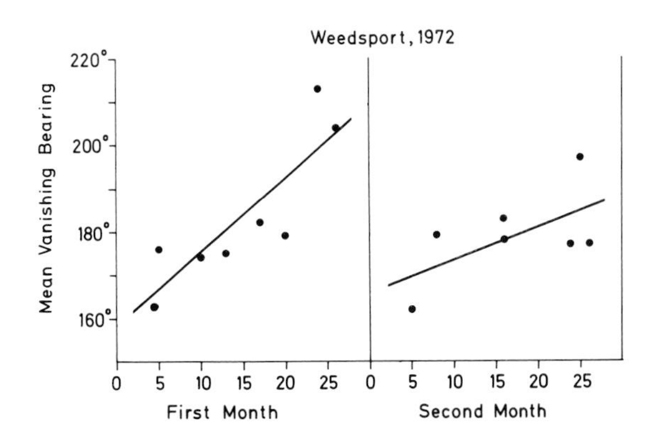

Fig. 3. Graphs of MVB as a function of DLM for the first two months of the 1972 series

1970 and 1973), when it abruptly swings back counterclockwise and the cycle begins again. Thus MVB does not vary as a sinusoidal function, as might be expected, but rather it traces a strongly skewed saw-toothed curve. It is because of this rather peculiar shape of the monthly oscillation that we have been able to apply linear regression techniques to cyclical data.

One may legitimately ask whether we are justified in choosing the syzygies as the beginning and ending points for our analyses. Obviously any true cycle could be taken to "begin" at any point, but what we are here seeking is the beginning point that results in the best approximation of a linear function. One way to determine the best beginning point is to fit 30 different regression curves to the same data, with each curve utilizing a different one of the 30 days of the lunar month as the beginning point, and then to see which curve gives the best linear fit (i.e., which leaves the smallest residual sum of squares). In other words, using the data from our 1970 series as an example, one first calculates the regression with the day of the full moon (day 0) as the starting point; then one moves this day to the end of the month and calculates a new regression with the day after full moon (day 1) as the starting point; next one moves this day to the end of the month and again calculates a new regression, this time with the second day after full moon (day 2) as the starting point; and so on through day 29 (days 0 and 30 are the same). If the t values derived from these regressions are plotted as a function of DLM, curves like those shown in Figure 4 result.[1] Such curves help us gain a better understanding of the way MVB varies across the synodic lunar month.

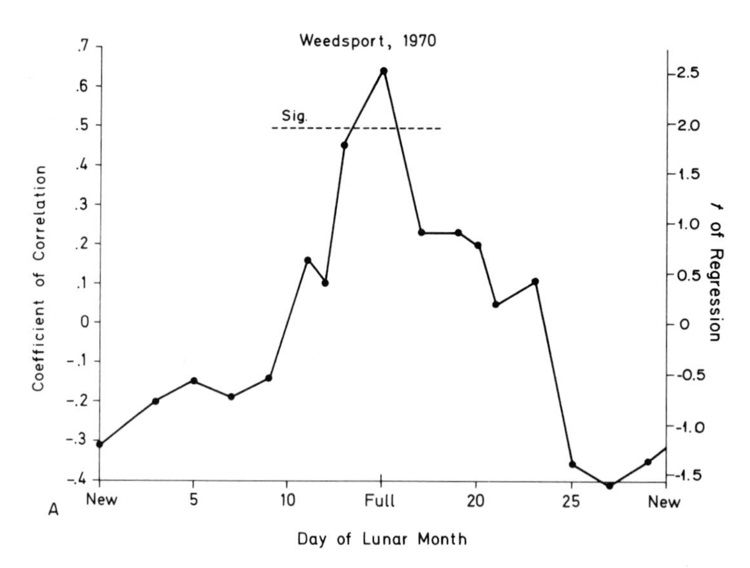

Fig. 4A

[1]Because most people find it easier to think in terms of a coefficient of correlation (r), ranging from -1 to +1, rather than in terms of t, we show values for both t and r on the vertical axis of our graphs. These two ways of treating the data yield identically shaped graphs and the same significance levels, even though purists may object that correlation analysis is not properly applied to cases like this, where one variable (i.e., DLM) can be measured exactly.

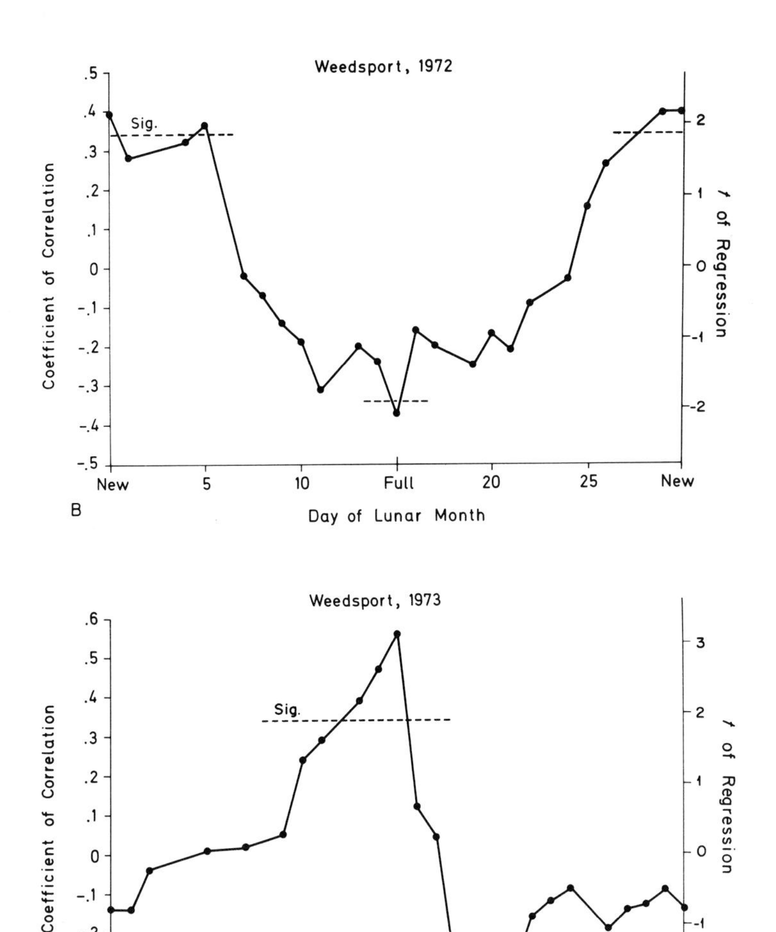

Fig. 4 A - C. Graphs of the coefficient of correlation (and of the t value of linear regression) as a function of day of lunar month for the original three series of releases at Weedsport. Note that the 1970 and 1973 curves peak precisely at full moon (day 15), whereas the 1972 curve has its highest point on the day of new moon and its lowest point at full moon

As Figures 4 A and 4 C clearly show, the day of full moon (day 15) is indeed the best starting point for positive linear regression of the data from 1970 and 1973 - i.e., the curves for these years peak precisely at day 15, in both cases well above the two-tailed significance level. Similarly, Figure 4 B shows that the day of new moon (day 0) is the best starting point for positive linear regression of the data from 1972 (alternatively, we may think of day 15 as the best starting point for a negative regression). It seems, then, that our choice of the syzygies is not an arbitrary one based on our human perception of new and full moons as the important events of the lunar month, but rather these events must truly mark turning points in whatever environmental parameters are influencing our pigeons' orientation behavior.

In our earlier study of the K effect (Keeton et al., 1974), we found that the first and second halves of the 1972 series differed greatly; there was far less consistency in the length of the mean vector during the first half, and there was a significant correlation of MVB with K_{12} only during the second half. In view of these differences, we also examined the two halves of 1972 for possible differences in their relationship to DLM. We found that MVB was well correlated positively with DLM on a new moon-new moon basis during the first half of the series ($y = 169.3 + 0.856\ \chi$; $r = 0.586$; $p = 0.017$) but not during the second half ($p = 0.44$). MVB during the second half was, however, negatively correlated ($r = -0.539$; $p = 0.026$) with DLM on a full moon-full moon basis; indeed, a plot of r for the second half as a function of DLM shows its most extreme negative point precisely on day 15. In short, as in the case of the K effect, the two halves of the 1972 series showed important differences in the relationship between MVB and DLM. We call special attention to the fact that the influence of DLM seemed to be greatest during the half year when the influence of K_{12} was least.

Having found such intriguing results from analysis of the data from the 1970, 1972, and 1973 series, we conducted three additional series of releases, all in 1974, from three sites at comparable distances N, S, and W of the loft. On the basis of the previous tests, we now predicted positive regression and hence used one-tailed tests of significance. The results were generally consistent with those of the original three series, though not so clear-cut. We shall examine each of these three additional series in turn.

About a third of the way through the 1974 Weedsport series, we had a distinct impression that MVB was oscillating in the full moon-full moon mode. However, about mid-June there appeared to be a reversal, and the oscillation became new moon-new moon. Hence, when the r values from the 30 successive regressions were plotted against DLM, a curve with a major peak at new moon and a second smaller peak at full moon were seen. As Figure 5 A shows, division of the field season

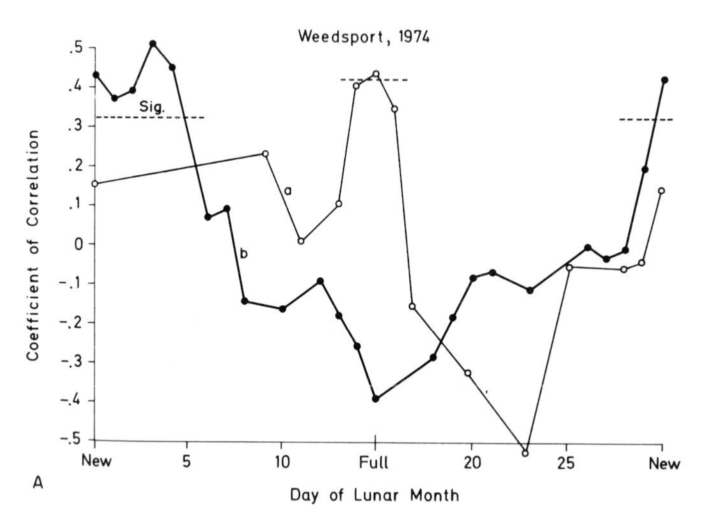

Fig. 5A

into two parts - one before 20 June and one after - yields curves conforming to the two modes, namely, a curve (a) for the early part of the season that peaks at day 15 (the shape of the curve is irregular because there are no data points between days 0 and 9), and a curve (b) for the later part of the season that peaks around new moon and has its minimum at day 15. We believe, therefore, that in this series we witnessed in mid-June a shift from the full moon mode to the new

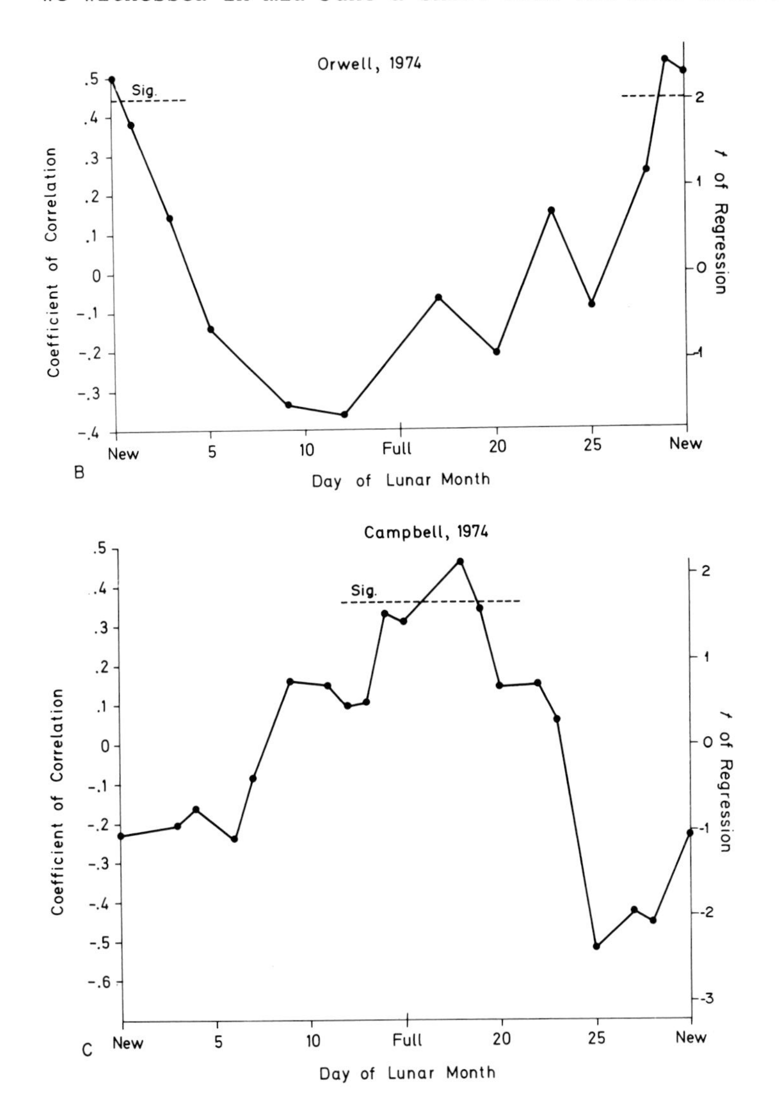

Fig. 5 A - C. Graphs of the coefficient of correlation as a function of day of lunar month for the three series of releases in 1974. (A) Weedsport - curve *a*, which peaks at full moon (day 15), is for the early part of the series, prior to 20 June; curve *b*, which has its lowest point at full moon, is for the latter part of the series (curve *b* is smoother than *a* because it is based on more data points - 27 versus 16). (B) Orwell - note that there is no datum point for full moon, so we cannot tell whether this curve would have agreed with all the other new moon-new moon curves in having its lowest point at day 15. (C) Campbell - this is the only curve of the full moon-full moon mode that does not peak precisely at day 15

moon mode. Particularly striking is the fact that all the birds appeared to undergo this shift in behavior simultaneously.

The 1974 Orwell series, all of which was conducted after 20 June, yielded data that clearly fit the new moon-new moon mode (Fig. 5 B). Unfortunately, no release was performed on day 15, so we cannot say whether the minimum of the curve would have been on day 15, as in Weedsport 1972 and Weedsport 1974b, but the minimum does occur on the last day a release was run prior to day 15.

The 1974 Campbell series yielded data that clearly conform to the full moon-full moon mode, although the peak is not precisely at full moon but rather at the first day a release was run after full moon (Fig. 5 C). Had we had data for days 16 and 17, the peak might well have been shifted closer to day 15, or even to day 15 itself.

II. Periodogram Analysis for Best Period

One can legitimately ask whether the synodic lunar month of 29.53 solar days is really the best period of oscillation of MVB, or whether we merely forced the data into that period because of some preconceived bias. In an attempt to check ourselves, we tried using the periodogram analysis suggested by Enright (1965). This technique involves systematically regrouping the data to form a series of Buys-Ballot tables with successive periodicities, each of which yields a statistic A_p, the root mean square amplitude. Comparisons of A_p for the different possible periodicities allows one to determine which is the best.

Unfortunately, this approach depends on the availability of a large number of data points for averaging. It was therefore essential, if we were to use periodogram analysis, that we pool the data from all our Weedsport test series. But this posed a problem because we believed the data from some series were oscillating 180° out of phase with other series, hence straight pooling would lead to mutual canceling. One solution would have been to subtract 15 days from the new moon-new moon series, so as to adjust them to the full moon-full moon mode. However, the selection of a 15-day adjustment assumes a 30-day periodicity, which is precisely the assumption we were trying to test. We chose, therefore, to make a slight change in the mathematics of the periodogram: We calculated the grand mean of all the Weedsport data, and then replaced each point with the square of its deviation from the grand mean. It was these values which we then subjected to periodogram analysis.

Figure 6 shows the results of analysis for all possible half-day periods between 22.5 and 36.5 days, i.e., half-day periods within the two-week interval centered on the 29.53-day synodic lunar month. The highest peak is at day 30, which is seductively close to the 29.53-day synodic month, but we caution that our sample was so small and so unevenly distributed across days that removal of only one or two data points could appreciably alter our periodogram results. We feel this analysis should be looked at with considerable skepticism and considered as suggestive only. However, we can certainly say that the results of this analysis are not inconsistent with the assumption of a synodic lunar month cycle.

III. Analysis of the Data in Terms of Lunar Hours

Because lunar and solar days are of unequal duration, a given solar hour will correspond to a different lunar hour each day of the synodic

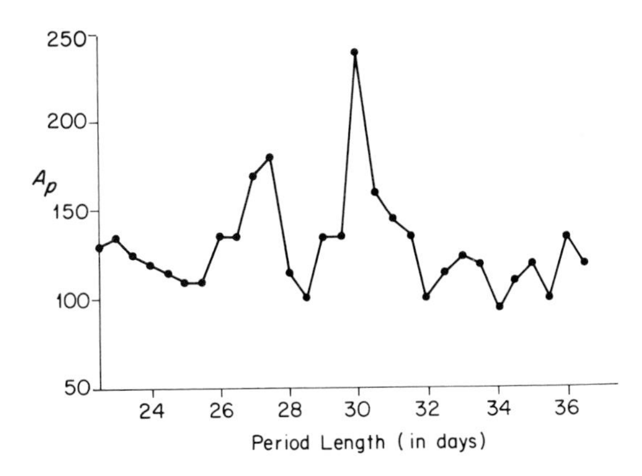

Fig. 6. The results of periodogram analysis by half days, beginning with a period one week shorter than the synodic lunar month and ending with a period one week longer than the synodic month. The highest peak is at 30 days, which is very close to the 29.53 days of the synodic month. We note with interest that a second peak is between 27 and 27.5 days, perhaps suggestive of the 27.32-day sidereal lunar month

month. Since we tended to conduct our releases at more or less the same solar time each day (the mean release time in 1972 was 11:49 EST, with a standard deviation of 2:36), we automatically conducted them at regularly varying lunar times. This means that the periodicity in MVB could be examined as a function of lunar hour, in a manner analogous to our earlier examination using lunar days. In assigning "hours" to the lunar day, we used the moments of nadir and zenith transits as analogs of the syzygies, and divided the lunar day into 30 hours. We then determined the lunar hour in which each vanishing bearing was taken, and calculated the mean for each of the 30 hours. We eliminated all nonsignificant means (i.e., $p > 0.05$ under both the Rayleigh and the V tests) and also all means where $N < 4$.

As expected, this alternative way of grouping and analyzing the data yielded evidence of periodicity in MVB as a function of lunar hour (Fig. 7). The crucial turning point, where the essentially linear variation of MVB "resets" (thereby providing the best coefficient of correlation), occurs one hour before the nadir transit in both 1970 and 1972. However, the curves for the two years are in antiphase with each other, as they were for lunar days.

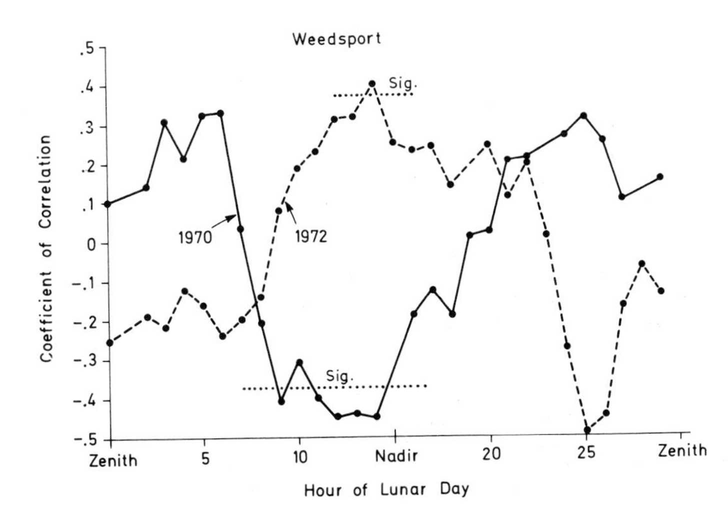

Fig. 7. Graphs of the coefficient of correlation as a function of hour of lunar day for the Weedsport series of 1970 and 1972

Having evidence for interpreting our data in terms of two different cycles, one based on the moon's synodic monthly motion and the other on its diurnal motion, we have no convincing way of deciding which interpretation is the more correct. Because of the relationship between lunar month and lunar day, either of the birds' apparent orientation rhythms may be an artifact of the other. In the absence of better grounds for a choice, we elect to deal with our data in terms of the lunar month because this yields the clearest and most consistent pattern across all six release series.

IV. Attempts to Relate the Lunar Rhythm of Orientation Changes to Gravity

We began this study because we suspected a linkage between magnetic and gravitational detection in animals. Therefore, once we found evidence for a lunar rhythm in our pigeons' choice of initial bearings, we attempted to relate this rhythm to moon and sun-induced tidal acceleration changes. For this study, we modified a computor program written by John Kuo for calculating the vertical components of gravitation, and we wrote a new program for calculating horizontal accelerations. Both programs were based on equations in Longman (1959) and in Schureman (1924), updated with information from The American Ephemeris and Nautical Almanac. With these programs, we could compute gravitational values for any place and any time with considerable precision.

We first attempted to correlate MVB with the actual values of vertical tidal acceleration, which combine solar and lunar effects. To do this, we tried several different ways of defining the time base, inasmuch as a single release always spanned several hours, during which there was substantial change in the tidal acceleration. Amongst other approaches, we tried using the noon tidal value, the average of a series of values spanning the release period, and a value representing the mean rate of change. The results of these correlations were not consistent. For almost all series we found some significant correlations, but the series did not agree with each other. This failure is understandable when we consider the shape of the gravitational changes across the lunar month. First, gravity varies along a modified sinusoidal curve (Fig. 8) whereas MVB does not, and second, gravity cycles twice each month (viewed in terms of a given solar hour, such as noon) whereas MVB cycles only once. Clearly, then, there could not be a point-for-point correspondence between MVB and gravity, even if variation of the former were triggered by variation of the latter.

In view of this difficulty, we attempted to find some derivative parameter of acceleration that varied across the lunar month in a manner more like that of MVB. Our approach was to search for a parameter that correlated linearly with DLM. This search was partially successful, in that we discovered that the noon value of a considerably modified version of the homeward component of horizontal acceleration did indeed correlate significantly with DLM. Moreover, the correlation changed phases between years just as the pigeons' orientation did, so that in 1970 and 1973 the correlation was in the full moon-full moon mode and in 1972 it was in the new moon-new moon mode. Furthermore, the correlation was very high during the first half of 1972 and non-existant during the second half, just as in the case of the correlation of MVB with DLM.

In short, we had found a component of acceleration - and, of particular interest, one related to the home direction - that varied statistically with lunar days in a way remarkably similar to the way our orientation data varied with lunar days. Unfortunately, however, we could not make the transition between the two sets of correlations. MVB did not cor-

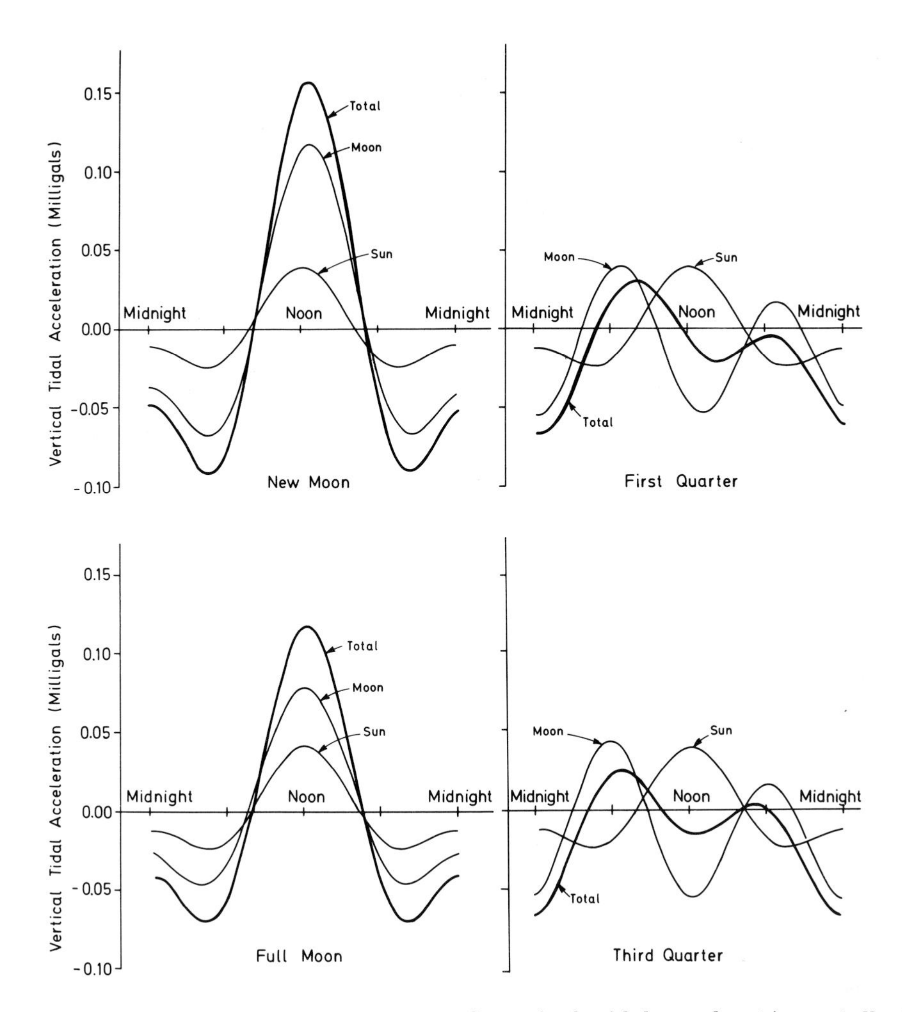

Fig. 8. Graphs showing the pattern of vertical tidal accelerations at Weedsport for four days in June, 1973. In addition to the total values, the contributions of the sun and moon are shown separately. Note that new moon and full moon are very much alike. Similarly, the first and third quarters closely resemble each other

relate significantly with the modified homeward component of horizontal acceleration in 1970 or 1973, and it correlated only weakly in 1972. Hence we came to an impasse that we have never been able to overcome. We cannot demonstrate a direct relationship between MVB and gravitational changes.

C. Discussion

We are well aware of the past history of skepticism concerning lunar rhythms in biologic systems. For example, the numerous reports of such rhythms by Frank Brown and his colleagues (e.g., Brown, 1954, 1965; Bennett, 1954; Bennett et al., 1957; Brown et al., 1960; Brown and Park, 1964, 1965, 1967a, 1967b; Brown and Webb, 1968; Brown and Chow,

1973, 1975) have met with a generally negative reception (e.g., Enright, 1963, 1965; Aschoff, 1965), though some advocates of lunar periodicites (e.g., Hauenschild, 1960) have been viewed somewhat more favorably. We do not wish to engage in polemics about such rhythms, and especially we wish, at this point, to take no position on the question of whether such rhythms do or do not involve an endogenous component. We merely report, on the basis of the general agreement between our six separate series of test releases covering four different years and three different release sites, that we think it reasonable to think a synodic lunar periodicity contributes to the day-to-day variation in the initial bearings chosen by pigeons, and that the syzygies appear to be major inflection points in the observed periodicity.

Although we do not think our difficulty in explaining the lunar periodicity of MVB in terms of gravitational changes definitely rules out gravity as the causal factor, we do acknowledge that the possibility is doubtful. It is entirely possible that the pigeons respond to some other environmental factor that varies across the lunar month, such as atmospheric movements or even nocturnal illumination (even though the birds spent the nights in buildings with limited openings to the outside, some slight difference in light intensity between, say, full moon and new moon would have been apparent). We feel we can rule out direct view of the moon at the release site as the cause, because omission from the analyses of releases when the moon was visible did not fundamentally alter the results.

We are especially interested in the fact that MVB shows a unimonthly period, although many physical events (such as oceanic tides) caused by the moon's synodic movements exhibit bimonthly periods. There are, of course, many aquatic organisms that have also been reported to have unimonthly rhythms (e.g., Lang, 1967; Korringa, 1947; Enright, 1972), and some of these are said to sometimes undergo relatively sudden 180^{0} phase shifts (from new moon-new moon cycles to full moon-full moon cycles, or the reverse), just as our pigeons do. We would very much like to discover what triggers such a shift of the cycle in our birds, because that should help show us what environmental factors underlie the lunar rhythm in the birds' orientation behavior.

We must add that, although the pigeons' orientational rhythm appears to be stable in either of the two modes (which would be consistent with the very similar gravitational characteristics of new and full moons; see Fig. 8), we think it possible the birds' orientation rhythms in the two modes are somewhat different. We are struck by the observation that the form of the curves in Figures 4 B and 5 Ab are quite different from the forms of the curves in Figures 4 A, 4 C, 5 Aa, and 5 C, even if these are inverted. We note, for example, that the new moon-new moon curves reach significance not only at new moon (with a positive value) but also usually at full moon (with a negative value). By contrast the full moon-full moon curves (which reach positive significance at or near full moon) never reach negative significance at or near new moon. Moreover, the full moon curves (Figs. 4 A, 4 C, 5 Aa, 5 C) always reach their greatest negative value sometime during the period between day 18 and day 27, well before the syzygy, whereas the new moon curves do not have a corresponding high point during that period.

Perhaps the most perplexing question we face concerns the importance of this lunar effect (or "lunacy", as we ourselves sometimes call it) in our understanding of pigeon navigation. We admit to having no good answer. We may have been looking at a small perturbation in the birds' behavior that has no great importance. On the other hand, the periodicity we have discovered may be a hint to us that information from geophysical parameters of the environment hitherto largely ignored in studies of avian orientation are contributing to pigeons' remarkable navigational abilities. In this regard, we feel the possible role of gravity in bird orientation, whether as a link to magnetic cues or as a cue system in its own right, is worthy of further investigation.

Finally, we can say that our studies of release-site biases, and of the factors that appear to influence their day-to-day variation, have convinced us that the navigational "map" of pigeons will eventually be found to be no single cue but a complex of cues, probably partially redundant, that can be integrated in a variety of different ways depending on both external conditions and the internal state of the birds (as influenced by age, experience, motivation, etc.). Certainly our studies suggest that the navigational map is neither simple nor obvious. But from what we have learned about the pigeon navigation system during the past ten years, complexity and redundancy should no longer be surprising to us.

Acknowledgements. We especially thank Irene Brown and Marilyn Yodlowski for their help in conducting the releases, and Melvin Kreithen for stimulating discussions. John Kuo of Columbia University wrote the original version of the program for computing tidal accelerations; Clifford Frohlich made the program available to us. This research was supported by National Science Foundation Grants GB-35199X and BMS 75-18905 A02.

References

Aschoff, J.: Diurnal rhythms. Ann. Rev. Physiol. 25, 581-600 (1965)
Bennett, M.F.: The rhythmic activity of the quahog, *Venus mercenaria*, and its modification by light. Biol. Bull. 107, 174-191 (1954)
Bennett, M.F., Shriner, J., Brown, R.A.: Persistent tidal cycles of spontaneous motor activity in the fiddler crab, *Uca pugnax*. Biol Bull. 112, 267-275 (1957)
Brown, F.A.: Persistent activity rhythms in the oyster. Am. J. Physiol. 178, 510-514 (1954)
Brown, F.A.: Propensity for lunar periodicity in hamsters and its significance for biological clock theories. Proc. Soc. Exp. Biol. Med. 120, 792-797 (1965)
Brown, F.A., Chow, C.S.: Lunar-correlated variations in water uptake by bean seeds. Biol. Bull. 145, 265-278 (1973)
Brown, F.A., Chow, C.S.: Differentiation between clockwise and counterclockwise magnetic rotation by the planarian, *Dugesia dorotocephala*. Physiol. Zool. 48, 168-176 (1975)
Brown, F.A., Park, Y.H.: Seasonal variations in sign and strength of gamma-taxis in planarians. Nature (London) 202, 469-471 (1964)
Brown, F.A., Park, Y.H.: Phase-shifting a lunar rhythm in planarians by altering the horizontal magnetic vector. Biol. Bull. 129, 79-86 (1965)
Brown, F.A., Park, Y.H.: Association-formation between photic and subtle geophysical stimulus patterns - a new biological concept. Biol. Bull. 132, 311-319 (1967a)
Brown, F.A., Park, Y.H.: Synodic monthly modulation of the diurnal rhythm of hamsters. Proc. Soc. Exp. Biol. Med. 125, 712-715 (1967b)
Brown, F.A., Webb, H.M.: Some temporal and geographic relations of snail response to very weak gamma radiation. Physiol. Zool. 41, 385-400 (1968)
Brown, F.A., Webb, H.M., Brett, W.J.: Magnetic response of an organism and its lunar relationships. Biol. Bull. 118, 382-392 (1960)
Enright, J.T.: Endogenous tidal and lunar rhythms. Proc. XVI Intern. Cong. Zool. 4, 355-359 (1963)
Enright, J.T.: The search for rhythmicity in biological time-series. J. Theoret. Biol. 8, 426-468 (1965)

Enright, J.T.: A virtuoso isopod: Circa-lunar rhythms and their tidal fine structure. J. Comp. Physiol. 77, 141-162 (1972)
Hauenschild, C.: Lunar periodicity. Cold Spring Harbor Symp. Quant. Biol. 25, 491-497 (1960)
Keeton, W.T.: Release-site bias as a possible guide to the "map" component in pigeon homing. J. Comp. Physiol. 86, 1-16 (1973)
Keeton, W.T.: The orientation and navigational basis of homing in birds. Adv. Study Behav. 5, 47-132 (1974)
Keeton, W.T., Larkin, T.S., Windsor, D.M.: Normal fluctuations in the earth's magnetic field influence pigeon orientation. J. Comp. Physiol. 95, 95-103 (1974)
Korringa, P.: Relations between the moon and periodicity in the breeding of marine animals. Ecol. Monogr. 17, 347-381 (1947)
Lang, H.-J.: Über das Lichtrückenverhalten des Guppy *(Lebistes reticulatus)* in farbigen und farblosen Lichtern. Z. Vergl. Physiol. 56, 296-340 (1967)
Larkin, T.S., Keeton, W.T.: Bar magnets mask the effect of normal magnetic disturbances on pigeon orientation. J. Comp. Physiol. 110, 227-231 (1976)
Lieber, A.L., Sherin, C.R.: Homicides and the lunar cycle: Toward a theory of lunar influence on human emotional disturbance. Am. J. Psychiat. 129, 69-74 (1972)
Longman, I.M.: Formulas for computing the tidal accelerations due to the moon and the sun. J. Geophys. Res. 64, 2351-2355 (1959)
Schureman, P.: A manual of the harmonic analysis and prediction of tides. U.S. Coast and Geodetic Survey, Spec. Publ. 98 (1924)
United States Government: The American ephemeris and nautical almanac. U.S. Nautical Almanac Office, Washington, D. C., 1973

Further Experiments on the Olfactory Hypothesis of Pigeon Homing*

ROBERT HARTWICK[1], JAKOB KIEPENHEUER[2], and KLAUS SCHMIDT-KOENIG[2],
[1] Department of Marine Biology, James Cook University of North Queensland, Townsville, QLD 4811, Australia. [2] Abteilung Verhaltensphysiologie der Universität Tübingen, 7400 Tübingen, FRG

Abstract

Given the inconsistent results of experiments testing the olfactory hypothesis of pigeon homing, we have undertaken a variety of experiments with pigeons from several lofts in Germany. In a series of detour experiments we found a consistent effect at one of 12 release sites. Results at the other sites showed great variability. A more direct test of the role of olfaction in homing involved application of odorous nasal pouches to pigeons' beaks just prior to release. This usually resulted in increased scatter of vanishing bearings, but had only marginal effects on homing velocity. Some of the limitations in these experiments and possible sources of inconsistency are discussed.

A. Introduction

Since the first formulation by Papi and his co-workers (Papi et al., 1972) of the olfactory hypothesis of pigeon homing, a succession of varied experiments, supporting and elaborating on it, has been reported by the Italian researchers and, more recently, by workers in other countries (Snyder and Cheney, 1975; Fiaschi and Wagner, 1976; Kiepenheuer et al., in press). Relatively slow acceptance of the Papi hypothesis as a universal explanation for pigeon homing can be attributed to a variety of factors. These include the startling emphasis on a sensory modality hitherto viewed as being of minor significance to most birds, as well as the persistent uncertainty shrouding the actual olfactory cues employed. Undoubtedly, a third major factor arose from the attempts by Keeton and co-workers at Cornell to repeat certain of Papi's early experiments. This work includes detour experiments to determine whether pigeons gather guidance information during the outward journey to the release site (Keeton, 1974 - compare with Papi et al., 1973 and Papi, 1976), and several experiments eliminating or interfering with the birds' powers of olfaction: (1) severing of the olfactory nerves (Hermeyer and Keeton, manuscript in preparation - cf. Papi et al., 1971, 1972; Benvenuti et al., 1973b; Baldaccini et al., 1975a), (2) insertion of plastic tubes through the nostrils (Keeton et al., 1977 - cf. Hartwick et al., 1977), and (3) application of odorous substances to the nostrils (Keeton and Brown, 1976 - cf. Benvenuti et al., 1973a; Papi, 1976). In most of the Cornell experiments the results either did not show an effect conforming to the olfactory hypothesis or, when an effect was present, were viewed as arising indirectly from an artifact of the technique. Such contradictory results, emerging from research groups in different countries, could be attributed to the use of different strains of pigeons with different genetic backgrounds and experience, although the possible effects of subtle differences in research emphasis and technique need also to be considered, as has been shown in other orientation research

*Supported by the Deutsche Forschungsgemeinschaft.

(Hartwick, 1976). In an effort to clarify some of these uncertainties, we have used pigeons from lofts in Germany in a number of experimental series testing the olfactory hypothesis. We report here on a series of detour experiments and releases of birds with odorous nasal pouches applied to the beak.

B. Materials and Methods

I. Detour Experiments

According to Papi's interpretation of the detour effect (Papi et al., 1973), pigeons on the outward journey to the release site acquire the most familiar and, therefore, most relied-upon olfactory information during the initial kilometers nearest the loft, and, presumably, make less use of information obtained at greater distances, near the release site. In our first experiments we sought to accentuate this distinction and, thereby, the detour effect, by giving pigeons useful olfactory information only during the initial detour leg of the outward journey.

Pigeons were randomly assigned to either of two detour groups. They were placed in open cages fixed to the roof of one of the two vehicles, fully exposed to the surroundings above and to the sides and rear. An opaque windscreen was attached forward. The two groups were driven simultaneously away from the loft along opposite paths directed about 90° to either side of the loft - release site line (Fig. 1a). This initial 'detour leg' comprised an approximate straight-line course for a distance at least equal to the release site range. During this leg 3 stops of 5 min. duration were made at equidistant points to al-

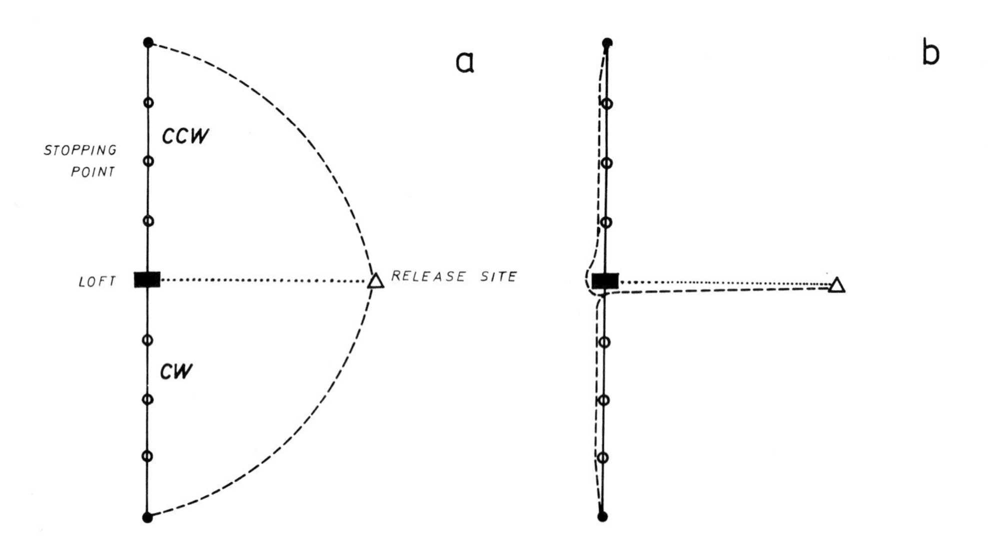

Fig. 1a and b. Schematic representation of the types of detour journeys employed. *Solid lines* indicate the "exposed" detour leg. *Dashed lines* indicate the closed portion of the outward journey. *Dotted lines* represent the loft - release site line in a normal release. (a) Detour plan followed in Series I and the first release in Series II. A similar plan but with open transport throughout, was followed in Series III - V. (b) Detour plan followed in the last five release of Series II. Sealed birds were transported back toward the loft on the same highway used for the detour leg. Further details in text

low the birds to receive and process information in relatively still air. At the terminal points of the detour leg, the birds were removed from the cages and placed in individual cardboard cartons, which were, in turn, sealed in air-tight plastic bags. To prevent an asphyxiating build-up of carbon dioxide in the bags, a small sachet of CO_2 absorbent (Sodasorp Atemkalk, W.R. Grace & Co., New York) was enclosed in each carton. The cartons were placed inside the vehicles and driven without further delay on the long path to the release site. Thus, until the moment it was removed from its carton and released, the pigeon was deprived of any further olfactory (or visual) contact with the outside environment. Birds from the two detour groups were released alternately, followed with binoculars, and the vanishing bearing recorded.

The five experimental series with their variations in technique were as follows:

Series I. Short-range releases of first-flight young birds maintained in a loft in Göttingen (1975). Two releases were performed at a site 15 km north of the loft, and one at a site 12 km west. The birds were all 4 - 5 months old and had flown free only in the immediate vicinity of the loft.

Series II. Medium-range releases of experienced birds from a loft in Frankfurt (1975). Three releases were performed at 2 adjacent sites 70 km east and one release each from sites 70 km to the north, south, and west. It was found after the first release 70 km east that, because of the distribution of highways around Frankfurt, the experiment could be carried out most efficiently by returning the birds toward Frankfurt on the same highway and then transporting them directly to the release site (Fig. 1b). It was assumed that, once the birds were sealed in cartons at the end of the detour leg, subsequent variations in the route would not affect their orientation. This 'T-route' technique was employed in the final 5 releases in the series.

Series III. Long-range releases of experienced Frankfurt birds (1976). A total of 4 releases were made at sites 130 km south, 150 km west, 130 km east, and 160 km north. A group of 40 birds with similar age and experience (average 12 releases each, maximum ranges: 100 km north and 70 km in the other three directions) was randomly divided into 2 detour groups that were maintained intact through the 4 releases - one group was always transported so as to predict a clockwise deflection, the other a counterclockwise deflection. Because of the much longer driving time involved, the technique of sealing birds in plastic bags was abandoned. The birds were confined in open cages and transported inside well-ventilated vehicles over the full detour course, which necessarily had the normal triangular rather than T shape. Outward transport occurred during one day, and the birds were released on the following day.

Series IV. Medium-range releases of experienced birds east of Frankfurt (1976). Because of evidence in Series 2 for a detour effect at the 70 km east site (Glasofen), one further release was made here with Frankfurt birds, as well as releases at 2 other sites (Autobahn and Steinfeld) within a few kilometers of it. These were one-day releases which otherwise followed the technique in Series III.

Series V. Long-range release of experienced Tübingen birds east of Frankfurt (1976). As a further test of detour effect at the Glasofen site, a single release was made with experienced birds from a loft in Tübingen, 152 km south of the release site. The Series III technique was followed.

II. Nasal Pouch Experiments

Nasal pouches were constructed of 2 trapezoidal pieces of cloth adhesive tape pressed together around a small wad of absorbent cotton which protruded slightly from the open base. The pouches were glued with branding cement (Allweather Tag Cement, Lincoln, Nebraska, U.S.A.) down over the pigeon's upper beak such that the exposed border of cotton was parallel with and 2 - 3 mm anterior to the slit opening of the external nares. A one-to-one mixture of the odorous substance with Vaseline was applied to the cotton border with a plastic syringe just prior to release. Control birds were equipped with nasal pouches and pure Vaseline.

Experimentals and controls were released alternately at sites 25 km north, south, east, and west of the loft. They were tracked in the usual manner and the homing time recorded by an observer at the loft. Experienced Frankfurt birds were used, most having had a number of prior releases in each of the 4 compass directions - a few in the vicinity of the release sites and others well beyond them.

Two experimental series were performed as follows:

Series I. Birds were transported to the release sites in open cages. After attachment of the nasal pouch, a mixture of Vaseline and menthol was applied to experimentals. Four releases were performed.

Series II. Birds were placed in individual cardboard cartons, as in the early detour series, and sealed in plastic bags before leaving the loft. Experimental boxes contained a small open vial of the same odorous mixture - Vaseline and α-pinene - which was later applied to their nasal pouches immediately after each was removed from its carton for releasing. It was hoped that this technique would minimize the possibility of the bird obtaining useful olfactory information during the trip to the release site or while awaiting release. Three experiments were performed.

Data analysis followed standard techniques. Vanishing bearings were summed by means of vector analysis, evaluated for nonrandomness by the Rayleigh test, for homeward directedness by the V test, and for homogeneity between treatments (detour paths) by Watson's U^2 test. Vanishing times and homing velocities were compared using the Mann-Whitney U test, birds failing to home being counted as the slowest returners.

C. Results

I. Detour Experiments

The results of the individual detour releases are summarized in Table 1. Detour groups are arranged according to whether their initial departure from the loft would predict a clockwise or counterclockwise deflection of their vanishing bearings from the true homeward direction.

The distribution of vanishing bearings showed a great deal of variability. In 13 of the 17 releases one or both detour groups made directional choices not distinguishable from randomness. Although the mean vectors of the detour groups deviated from the home direction in the predicted semicircle in 20 cases, while in the wrong semicircle in only 14 cases, fully half of the vectors were of nonsignificant length. In view of such variability, it was deemed impractical to analyze in-

Table 1. Results of detour experiments

Experiment number	Date	Loft distance (km)	Loft direction	Clockwise deviating group: No. released and (bearings)	Clockwise: Vanishing bearing mean vector Bearing	Clockwise: Length	Clockwise: p (Rayleigh)	Counterclockwise group: No. released and (bearings)	Counterclockwise: Vanishing bearing mean vector Bearing	Counterclockwise: Length	Counterclockwise: p
Series I. Short-range releases of first-flight birds (Göttingen - 1975)											
1	25. 7.75	15	172	8 (7)	057	0.329	>*.05*	7 (7)	039	0.949	<.01
2	31. 7.75	12	092	14 (14)	029	0.214	>*.05*	14 (13)	346	0.600	<.01
3	13. 8.75	15	172	14 (12)	112	0.435	>*.05*	14 (10)	121	0.660	<.01
Series II. Medium-range releases of experienced birds (Frankfurt - 1975)											
4	2. 9.75	70[a]	297	13 (13)	355	0.604	<.01	13 (11)	105	0.566	<.05
5	17. 9.75	70[a]	297	18 (15)	353	0.729	<.01	18 (17)	170	0.122	>*.05*
6	27. 9.75	70	115	16 (12)	355	0.096	>*.05*	17 (10)	349	0.279	>*.05*
7	29. 9.75	70	018	15 (14)	309	0.188	>*.05*	16 (15)	304	0.535	<.01
8	2.10.75	70	205	15 (12)	191	0.549	<.05	16 (14)	125	0.251	>*.05*
9	10.10.75	70[b]	297	16 (16)	247	0.583	<.01	15 (14)	261	0.682	<.01
Series III. Long-range releases of experienced birds (Frankfurt - 1976)											
10	25. 5.76	130	042	20 (14)	144	0.326	>*.05*	20 (16)	220	0.445	<.05
11	20. 6.76	150	088	20 (18)	149	0.654	<.01	20 (15)	157	0.628	<.01
12	3. 7.76	130	273	17 (12)	236	0.288	>*.05*	17 (16)	178	0.034	>*.05*
13	9. 7.76	160	172	16 (14)	249	0.423	>*.05*	15 (13)	263	0.849	<.01
Series IV. Medium-range releases of experienced birds east of Frankfurt (1976)											
14	12. 7.76	70[a]	297	16 (12)	091	0.267	>*.05*	16 (14)	218	0.747	<.01
15	28. 7.76	70[b]	297	9 (4)	274	0.769	>*.05*	10 (6)	182	0.192	>*.05*
16	6. 8.76	74[c]	288	18 (14)	016	0.620	<.01	18 (12)	357	0.529	<.05
Series V. Long-range release of experienced birds east of Frankfurt (Tübingen - 1976)											
17	20. 8.76	152[a]	197	19 (14)	237	0.334	>*.05*	21 (12)	146	0.090	>*.05*

[a]Marktheidenfeld-Glasofen site. [b]Marktheidenfeld-Autobahn site. [c]Marktheidenfeld-Steinfeld site.
N.B. Release 15 curtailed by rain.

dividual experiments for evidence of a detour effect. Those detour groups expected to deviate clockwise could, however, be pooled by superimposing home bearings, and then compared with the counterclockwise groups, similarly pooled (Fig. 2). The resulting mean vectors were both significantly nonrandom by the Rayleigh test, although considerable scatter remained evident. Both vector bearings were deflected from the home direction in the expected quadrants. The two distributions were significantly different from each other according to Watson's U^2 test ($U^2 = 0.541$, $p < .005$).

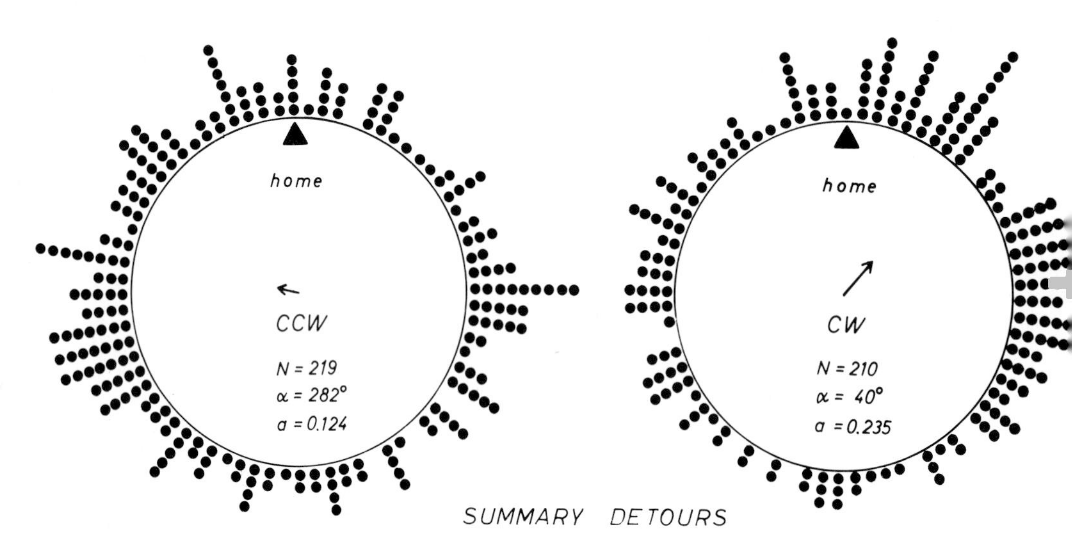

Fig. 2. Vanishing bearings from the 17 detour experiments. Detour groups were pooled according to whether a clockwise (*CW*) or counterclockwise (*CCW*) deviation from the home direction is expected. *Triangles* represent the home direction in each release. *Arrows* in the center are the overall mean vectors. Both are significantly nonrandom (Rayleigh test, $p < .05$)

This apparent detour effect was not uniformly present among all releases, however. Pooled summaries of the 5 experimental series revealed a significant difference between CW and CCW groups only when these included data from one particular release site, Marktheidenfeld-Glasofen, located 70 km east of Frankfurt. A summary of the 4 releases made here with birds from the Frankfurt and Tübingen lofts (Fig. 3) indicates the clear, repeatable deviation between CW and CCW groups ($U^2 = 0.605$, $p < .005$). The mean bearings correspond closely to the paths of the detour legs. Control birds transported directly to the Glasofen site on another occasion were homeward oriented, intermediate to the deflected detour groups (Fig. 3c). Evidence for a detour effect at the other 11 release sites was less convincing. The deviations of the pooled mean vectors (CW = +23°, CCW = -56°) were again in the expected quadrants, but the vector lengths (CW = 0.184, CCW = 0.109) were smaller, and nonrandom only for the clockwise birds. The smaller deviation was less than significant according to Watson's U^2 test ($U^2 = 0.166$, m = 156, n = 165, $.05 < p < .10$).

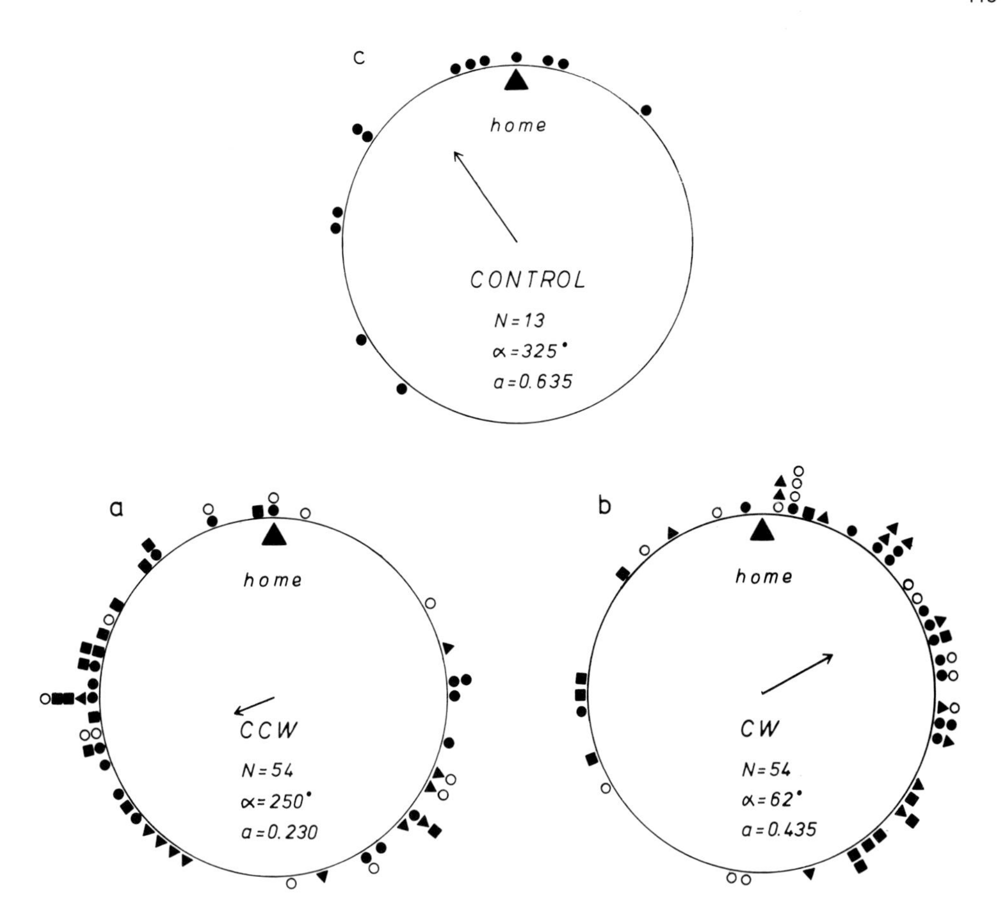

Fig. 3 a - c. Vanishing bearings of pigeons from the Frankfurt and Tübingen lofts used in detour releases Nos. 4, 5, 14, and 17 at the Marktheidenfeld-Glasofen site 70 km east of Frankfurt. Symbols refer to releases as follows: *Closed triangles*, No. 4; *closed circles*, No. 5; *closed squares*, No. 14; *open circles*, No. 17. (a) Orientation of groups expected to deflect counterclockwise (*CCW*), (b) orientation of groups expected to deflect clockwise (*CW*), (c) control orientation of birds transported directly to the release site on 16.7.76. All mean vectors are significantly nonrandom (Rayleigh test, $p < .05$)

II. Nasal Pouch Experiments

The results for the 7 releases are summarized in Table 2. In terms of homing performance, experimental birds did only marginally poorer than controls. In one of the 7 releases the Mann-Whitney U test revealed a significant decrement in experimental homing velocity (Exp. 15: $U_{18,18} = 108$, $p < .05$), while there was no significant difference in the other 6. Although 20 experimentals out of 102 failed to home during the normal observation day (up until 2 h after the last release), while only 10 out of 99 controls were late, the difference is not significant ($\chi^2 = 2.87$, $p > .05$). No significant differences in vanishing times occurred.

The most obvious difference between controls and experimentals appears in the greater scatter among the latter. Experimental mean vectors

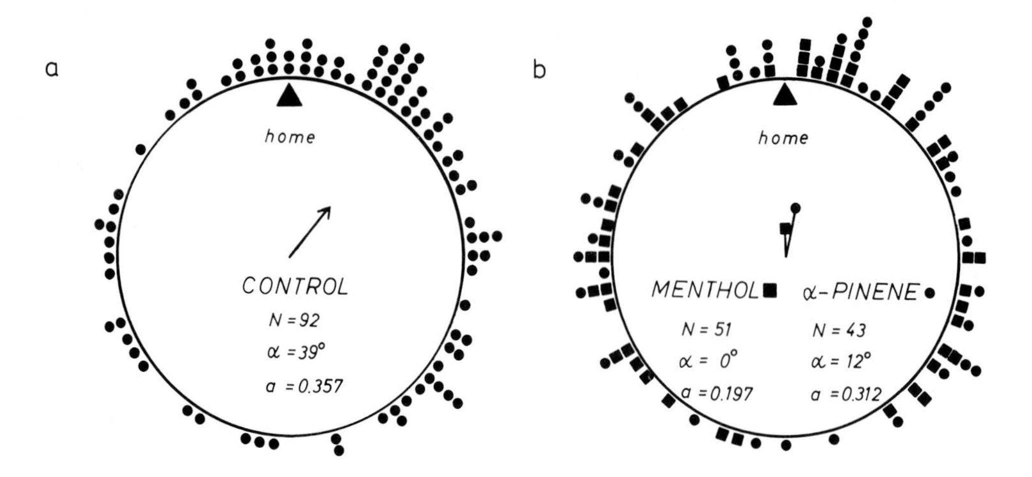

Fig. 4 a and b. Vanishing bearings of birds with odorous nasal pouches attached to the beak (b) as compared with controls carrying nonodorous pouches (a). Control and α-pinene mean vectors are significantly nonrandom, but the menthol birds were not distinguishable from random scatter (Rayleigh test, $p > .05$)

were shorter than controls in 6 of 7 releases, and, in the pooled menthol series, experimental bearings were not distinguishable from randomness. In terms of home-directedness (V test - Table 2), differences were less clear. Experimentals were home oriented in only 2 of 7 releases, but controls were home oriented in only one additional experiment.

D. Discussion

I. Detour Experiments

Although the implications of the detour effect are not central to the olfactory hypothesis, they do provide an important ancillary mechanism for explaining pigeon homing from distances well beyond those regions, adjacent to the loft, with which the birds could reasonably be expected to have an olfactory familiarity. The sharp discrepancy between the findings of Papi et al. (1973) and those of Keeton (1974) is scarcely clarified by the results reported here, in which a consistent effect is evident at only one of 12 release sites, while any possible effect at the other 11 is obscured in random scatter and inexplicable deviations. Similar variability has been reported in a significant proportion of Papi's releases as well (Papi, 1976), and it thus may be worthwhile to consider some of the possible sources of variability in detour experiments.

One immediate limitation arises from the postulate that pigeons can obtain guidance information during two different phases of the release procedure, the outward journey to the release site and following release. If one of these sources is lacking or unequivocably dominated by the other, then consistent orientation should be maintained. If, however, both sources are available in significant measure, then a conflict will arise when they do not closely agree. In detour experiments it may be assumed that they do not agree, and Papi (1976) has speculated that scattered or partially deflected orientation may re-

Table 2. Results of nasal pouch experiments

Experiment	Date	Loft distance (km)	Loft direction	Treatment	No. released and (bearings)	Mean vector vanishing bearings: Bearing	Length	p (Rayl.)	p (V Test)	No. birds homing[a]	Mean homing speed (km/h)	No. birds late	Mean vanishing time (min)
Series I. Exposed transport - menthol experiments (Frankfurt - 1974-1975)													
1	9.7.74	25.4	100	C	19 (17)	168	0.560	<.01	>.05	18	22.4	1	4.1
				E	20 (16)	190	0.487	<.05	>.05	17	22.0	3	4.0
2	28.5.75	29.1	185	C	13 (11)	186	0.694	<.01	<.001	12	33.5	1	5.0
				E	14 (13)	178	0.688	<.01	<.001	14	34.2	0	3.6
3	6.6.75	23.2	296	C	11 (11)	309	0.471	>.05	<.05	11	40.6	0	3.5
				E	11 (8)	321	0.230	>.05	>.05	9	37.6	2	3.4
4	12.6.75	26.9	020	C	15 (11)	315	0.473	>.05	>.05	12	26.5	3	2.6
				E	15 (14)	286	0.543	<.05	>.05	13	24.6	2	3.1
Series II. Sealed transport - α-pinene experiments (Frankfurt - 1975)													
5	4.7.75	26.9	020	C	18 (16)	033	0.737	<.01	<.001	16	26.8	2	2.9
				E	18 (16)	002	0.638	<.01	<.001	12	23.6	6	2.6
6	28.7.75	23.2	296	C	16 (11)	031	0.753	<.01	>.05	16	29.9	0	3.4
				E	19 (13)	345	0.390	>.05	>.05	16	29.1	3	3.3
7	4.8.75	29.1	185	C	15 (15)	316	0.391	>.05	>.05	12	18.8	3	3.0
				E	15 (14)	258	0.154	>.05	>.05	11	30.7	4	2.7

[a]Homing within 2 h of final release.

flect such disagreement. In theory, the relative importance of the two information sources should vary together with the ratio of the release (or detour) distance to the radius of the familiar territory. The outward journey of routine pigeon releases from most lofts necessarily involves minor driving detours on a scale of a few kilometers, but these result in no widespread disorientation, presumably because they are restricted to within the radius of familiar territory. On a larger scale, Keeton's (1974) negative results might still be consistent with Papi's hypothesis if it were postulated that, for reasons of wind or odor patterns, the radius of familiar territory - of the 'olfactory map' - around Ithaca exceeded the detour release distance. On that note, however, it should be recalled that, for Frankfurt pigeons, a detour effect was detectable at a range of 70 km, but it was in no way enhanced - and, indeed, failed to persist - when the release range was doubled in Series III.

To the preceding complexities must be added the possible effects upon orientation of various nonolfactory stimuli. Specific biases inherent to release sites have been associated with magnetic influences (Keeton et al., 1974) as well as with local landmarks (Wagner, 1972). In addition, one of us has investigated a possible role for magnetic information collected during the outward journey (Kiepenheuer, see this volume). Recent experiments by Papi (see this volume) likewise suggest that magnetism may play an ancillary role in the detour effect.

II. Nasal Pouch Releases

The most direct approach to a hypothesized orientation mechanism dependent on a single sensory modality would seem to suggest interference with that modality in the expectation of a breakdown in orientation. Unfortunately, complex behaviors can be readily disrupted by many treatments that do not disturb directly the underlying causal mechanisms. Thus, only with great difficulty can this alternate explanation - i.e., indirect effects from a generalized behavioral disturbance - be exorcized from sensory deprivation experiments. Likewise, in the technique of masking guidance odors with a strong foreign odor, this uncertainty complicates the interpretation of scattered vanishing bearings or longer homing times. Only if it could be shown that birds possessed normal homing ability despite disturbing odors when released at a familiar site where some nonolfactory mechanism (e.g., landmark orientation) came into play, could the probability of indirect effects be minimized. Such a control technique has been employed in some sensory deprivation experiments (Benvenuti et al., 1973b; Snyder and Cheney, 1975; Fiaschi and Wagner, 1976; Hartwick et al., 1977) but not as yet in nasal pouch experiments.

With negative results, interpretation is still fraught with uncertainty since sensory habituation to the foreign odor could result in a brief period of disrupted orientation. Thus, our Series II technique of placing vials of α-pinene in the cartons before leaving the loft [compare with Keeton and Brown's (1976) Series IV technique] might conceivably have been counterproductive. In any case, an ephemeral disruption would tend to make itself more evident in the initial phases of the homing response - in particular the vanishing bearings - than in the overall homing performance - homing velocity - especially with long-range releases. Our relatively short releasing distance, 25 km, was expected to accentuate any initial effects, but the possibility, mentioned above, of auxiliary homing mechanisms in familiar regions near the loft could have thwarted our intentions. Finally, it may be that the effectiveness of a particular masking odor could vary greatly according to the nature of the unknown guidance stimuli.

In summary, it is evident that, given the variability in our data, we have contributed only modestly toward clarifying the role for olfaction in the homing of German pigeons. The incidence of a detour effect is, at best, sporadic, although it appears to be consistent and unequivocal at one of our release sites. At the other sites it is not obvious whether an effect is lacking or is masked by site-specific disturbances. In the nasal pouch experiments overall effects are marginal but suggest a real disturbance in at least initial bearings.

It seems that techniques of olfactory deprivation or disturbance are less well suited for confirming details of the homing mechanism than are those methods which manipulate olfactory cues and, thereby, homing behavior in precise and predictable fashion (e.g., Baldaccini et al., 1975b; Papi et al., 1974). Our initial efforts with such experiments (Kiepenheuer, 1978) give clearer confirmation than that which we could here provide.

References

Baldaccini, N., Benvenuti, S., Fiaschi, V., Papi, F.: New data on the influence of olfactory deprivation on the homing behavior of pigeons. In: Denton, D. and Coghlan, J. (Eds.): Olfaction and Taste V. Academic Press New York: 1975a, pp. 351-353

Baldaccini, N., Benvenuti, S., Fiaschi, V., Papi, F.: Pigeon navigation: effects of wind deflection at home cage on homing behavior. J. Comp. Physiol. 99, 177-186 (1975b)

Benvenuti, S., Fiaschi, V., Fiore, L., Papi, F.: Disturbances of homing behavior in pigeons experimentally induced by olfactory stimuli. Monit. Zool. Ital. (N.S.) 7, 117-128 (1973a)

Benvenuti, S., Fiaschi, V., Fiore, L., Papi, F.: Homing performance of inexperienced and directionally trained pigeons subjected to olfactory nerve section. J. Comp. Physiol. 83, 81-92 (1973b)

Fiaschi, V., Wagner, G.: Pigeon homing: some experiments for testing the olfactory hypothesis. Experientia 32, 991-993 (1976)

Hartwick, R.: Beach orientation in talitrid amphipods: capacities and strategies. Behav. Ecol. Sociobiol. 1, 447-458 (1976)

Hartwick, R., Foà, A., Papi, F.: The effect of olfactory deprivation by nasal tubes upon homing behavior in pigeons. Behav. Ecol. Sociobiol. 2, 81-89 (1977)

Keeton, W.: Pigeon homing: no influence of outward-journey detours on initial orientation. Monit. Zool. Ital. (N.S.) 8, 227-234 (1974)

Keeton, W., Brown, A.: Homing behavior of pigeons not disturbed by application of an olfactory stimulus. J. Comp. Physiol. 105, 259-266 (1976)

Keeton, W., Kreithen, M., Hermayer, K.: Orientation by pigeons deprived of olfaction by nasal tubes. J. Comp. Physiol.(1977)

Keeton, W., Larkin, T., Windsor, D.: Normal fluctuations in the earth's magnetic field influence pigeon orientation. J. Comp. Physiol. 95, 95-103 (1974)

Kiepenheuer, J.: Pigeon homing: A repetition of the deflector loft experiment. submitted to: Beh. Ecol. and Sociobiol. (1978)

Papi, F.: The olfactory navigation system of the homing pigeon. Verh. Dtsch. Zool. Ges. 1976, 184-205 (1976)

Papi, F., Fiaschi, V., Benvenuti, S., Baldaccini, N.: Pigeon homing: outward journey detours influence the initial orientation. Monit. Zool. Ital. (N.S.) 7, 129-133 (1973)

Papi, F., Fiore, L., Fiaschi, V., Benvenuti, S.: The influence of olfactory nerve section on the homing capacity of carrier pigeons. Monit. Zool. Ital. (N.S.) 5: 265-267 (1971)

Papi, F., Fiore, L., Fiaschi, V., Benvenuti, S.: Olfaction and homing in pigeons. Monit. Zool. Ital. (N.S.) 6, 85-95 (1972)

Papi, F., Ioalê, P., Fiaschi, V., Benvenuti, S., Baldaccini, N.: Olfactory navigation of pigeons: the effect of treatment with odorous air currents. J. Comp. Physiol. 94, 187-193 (1974)

Snyder, R., Cheney, C.: Homing performance of anosmic pigeons. Bull. Psychonomic Soc. 6, 592-594 (1975)

Wagner, G.: Topography and pigeon orientation. In: Galler, S., Schmidt-Koenig, K., Jacobs, G., Belleville, R. (Eds.): Animal orientation and navigation. N.A.S.A. Washington, D.C.: 1972, pp. 259-273

Local Anesthesia of the Olfactory Membrane and Homing in Pigeons*

KLAUS SCHMIDT-KOENIG[1] and JOHN B. PHILLIPS[2], [1]Abteilung Verhaltensphysiologie der Universität Tübingen, 7400 Tübingen, FRG. [2]Section of Neurobiology and Behavior, Langmuir Lab., Ithaca, NY 14853, USA

Abstract

Xylocain spray, tested in heart-rate conditioning experiments, was used to eliminate olfaction. In homing experiments only slight and inconsistent effects of eliminated olfaction on initial orientation and homing performance could be found.

A. Introduction

Various experiments by Papi and co-workers support the view that pigeons use olfactory cues for homing (e.g., Papi et al., 1972). We have attempted to test the use of olfaction in homing by local anesthesia. This approach has the advantage of being very effective in eliminating olfactory information while at the same time being completely reversible. The procedure is only mildly traumatic to the bird and does not impair breathing in contrast to surgically severing the olfactory nerve (Papi et al., 1971; Benvenuti et al., 1973; Hermayer and Keeton, ms. in prep.) or inserting nasal tubes into the air passage (Keeton et al., 1977; Hartwick et al., 1977).

We first tested the effect and duration of xylocain spray in laboratory conditioning experiments and then its effect in homing experiments.

B. Laboratory Experiments

Commercially available xylocain spray was used. The dose was reduced to about one-third of the dose for human oto-laryngeal application. There was no noticeable systemic effect on the bird. No-xylocain spray lacking only the xylocain component was kindly prepared and supplied by the manufacturer. (We thank Astra Chemicals, Wedel, W.-Germany for their helpful cooperation.) Using standard cardiac conditioning techniques, each bird was initially given several training sessions in which air samples containing 4.5% amylacetate at a flow rate of 5 ℓ/min and followed by a mild electric shock, were randomly interspersed at 3 + 6 min intervals with samples of filtered air. A response criterion of a 20 beat/min rate increase during the 10-s sample presentation over that of the 10-s period immediately preceding the trial was used. To test the effectiveness of the two preparations in eliminating olfactory discrimination, an individual that had previously demonstrated

*Supported by the Deutsche Forschungsgemeinschaft.

an ability to distinguish 4.5% amylacetate from filtered air was sprayed through the choanal opening with either xylocain or no-xylocain immediately before a session of sample presentations in which the shock was eliminated for most of the time. Figures 1 and 2 show the results obtained with bird ZT 570. No-xylocain was found not to impair recognition of amylacetate at this concentration, while xylocain eliminated the reaction to amylacetate for 60 - 90 min. In most cases the cumulative response curve for amylacetate began to diverge in the expected manner from that for filtered air after about 70 min although the shock was not turned on until much later. As a control for nonspecific effects, xylocain was administered under the tongue. This did not affect the ability to discriminate filtered air from 4.5% amylacetate. It can be assumed that administration of xylocain through the choanal opening prevented discrimination by eliminating the response to olfactory information and not to nonolfactory artifacts, e.g., valve or relay noise. We should mention that in another series of experiments using cardiac conditioning techniques, we have not yet succeeded in obtaining responses to air samples from different locations around Tübingen containing those odors that the pigeons are suggested to utilize in homing.

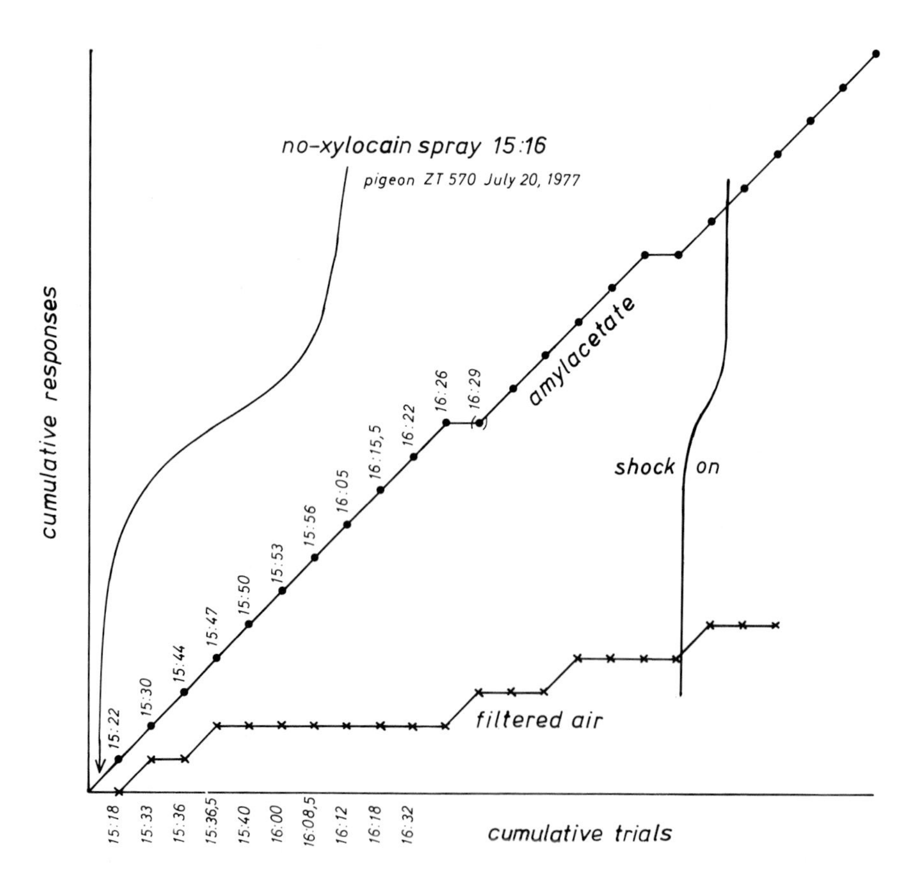

Fig. 1. Response of pigeon ZT 570 in the cardiac conditioning experiment to 4.5% amylacetate as compared to filtered air after a no-xylocain spray into the choanal opening. The time of each trial is given in CET

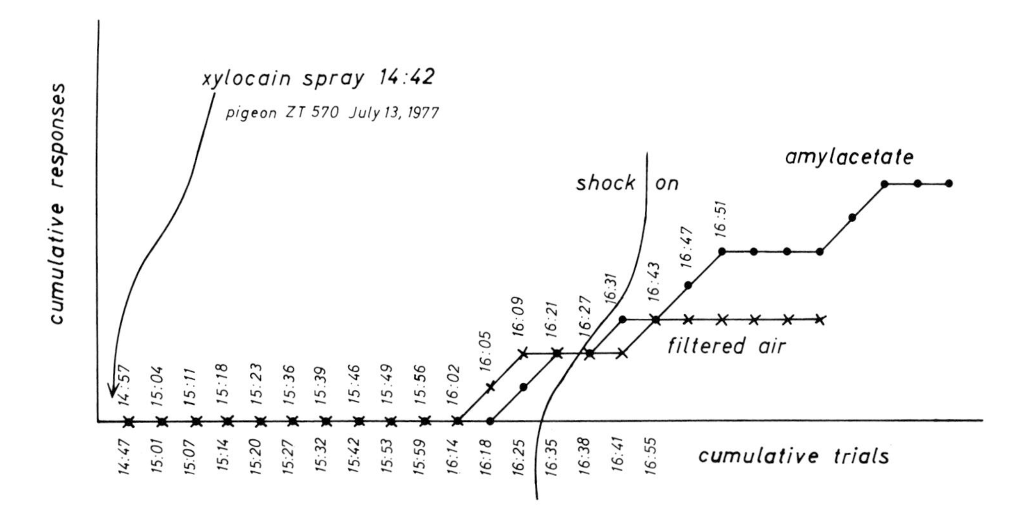

Fig. 2. Response of pigeon ZT 570 in the cardiac conditioning experiment to 4.5% amylacetate as compared to filtered air after a xylocain spray into the choanal opening. Symbols as in Figure 1

C. Homing Experiments

Two series (A and B) of homing experiments were carried out under sun. Initial orientation was recorded with field glasses, vanishing time with stopwatches, and homing performance was timed and calculated in km/h.

The following statistical tests have been applied to initial orientation: the Rayleigh-test for randomness; the V test for homeward directedness; the Watson test for differences between pairs of samples (cf. Schmidt-Koenig, 1975, Appendix). The Mann-Whitney-U test has been applied to homing performance (Siegel, 1956).

Sprays reduced from human dosage were applied as in the laboratory experiments through the choanal opening.

Series A. Two rounds of releases from 40 km E, S, W, and N were carried out with 3 groups of experienced pigeons. The experimental groups was sprayed with xylocain immediately before displacement (XBD), one control group with no-xylocain immediately before displacement (NXBD), the other control group with xylocain upon arrival at the release site (XAR). Treatment was rotated among the groups. We arrived at the release sites within 60 - 80 min after the first experimental had been sprayed. The birds were transported in crates in the open on a rack or inside a car with windows rolled down. Figure 3 presents the summarized initial orientation and homing performance of all 8 releases. Initial orientation was nonrandom ($p < 0.01$) and homeward directed ($p < 0.01$). There was no difference between any of the groups ($p > 0.05$). Likewise, homing performance was at perfectly normal levels with no difference among the groups ($p > 0.05$).

One may, however, select the results of the first three releases in which the groups were receiving their first exposure to each treat-

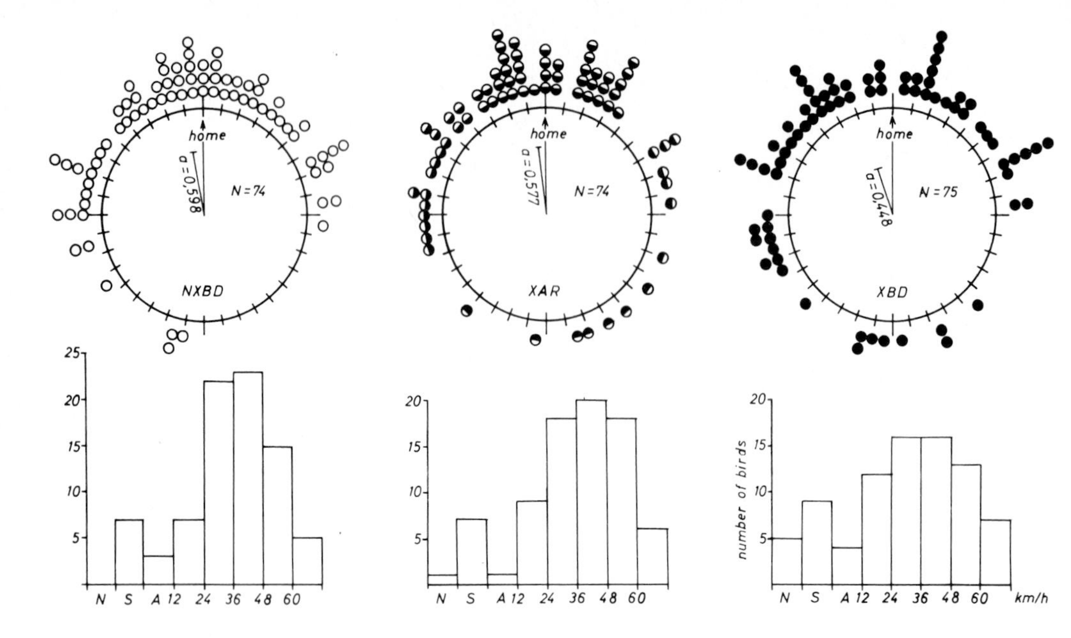

Fig. 3. Summarized initial orientation (*circular diagrams*) and homing performance (*histograms*) of eight releases from 40 km (series A). *NXBD*, no-xylocain before displacement; *XBD*, xylocain before displacement; *XAR*, xylocain at the release site. For more details see text. The mean vector, its length (a), and sample size (N) are given in the circular diagrams. Homing performance has been summarized in eight classes of 12 km/h and below 12 km/h in class *A*, homed on the day of release; *S*, homed after the day of release; *N*, never home

ment. Figure 4 presents initial orientation and homing performance of these releases (also contained in Fig. 3). Initial orientation of experimentals and the no-xylocain controls was different at $p \sim 0.01$ and of experimentals and the xylocain-at-the-release-site controls at $0.05 > p > 0.01$. Only homing performance of experimentals and no-xylocain controls was different at $p \sim 0.02$.

Series B. One round of releases from 70 km E, S, W, and N was performed with the birds of series A, now divided into two groups. The experimental (xylocain) and control (no-xylocain) groups were sprayed immediately before displacement in the loft and at 75-min intervals until the last bird had been released. Figure 5 gives the summarized initial orientation and homing performance of series B. Initial orientation of both groups was nonrandom ($p < 0.01$) and homeward directed ($p < 0.01$) in only the third of the four releases; the summaries (Fig. 5) were random ($p > 0.05$) and not homeward directed ($p > 0.05$); there was no difference between the two groups ($p > 0.05$). Likewise, homing performance was not different ($p > 0.05$).

D. Discussion and Conclusions

The laboratory experiments clearly show that the preparation used temporarily eliminated the birds' ability to discriminate air containing 4,5% amylacetate from filtered air. Amylacetate is a strong odor or

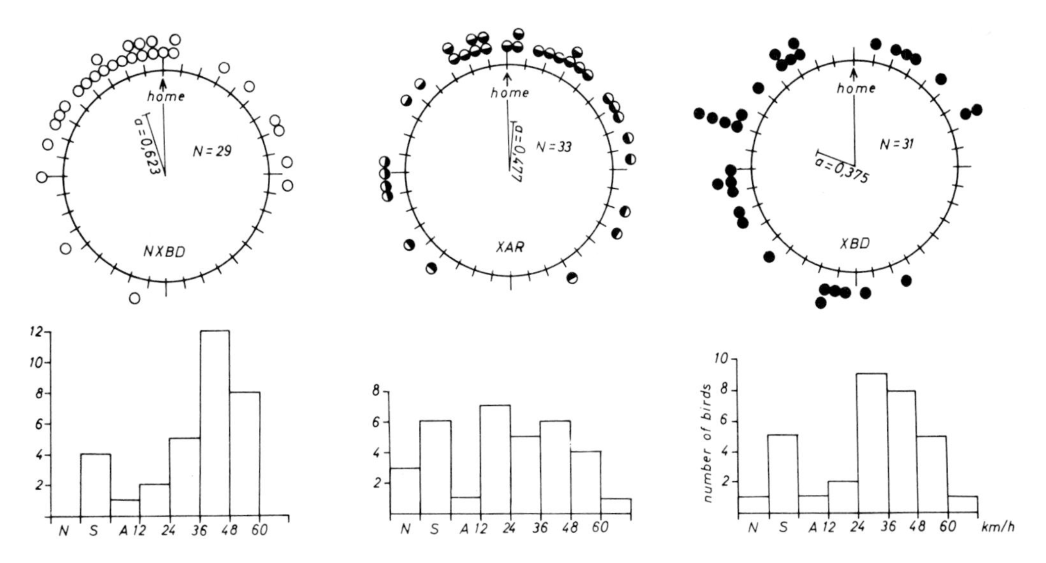

Fig. 4. Summarized initial orientation of the first three releases of series *A*. Symbols as in Figure 3

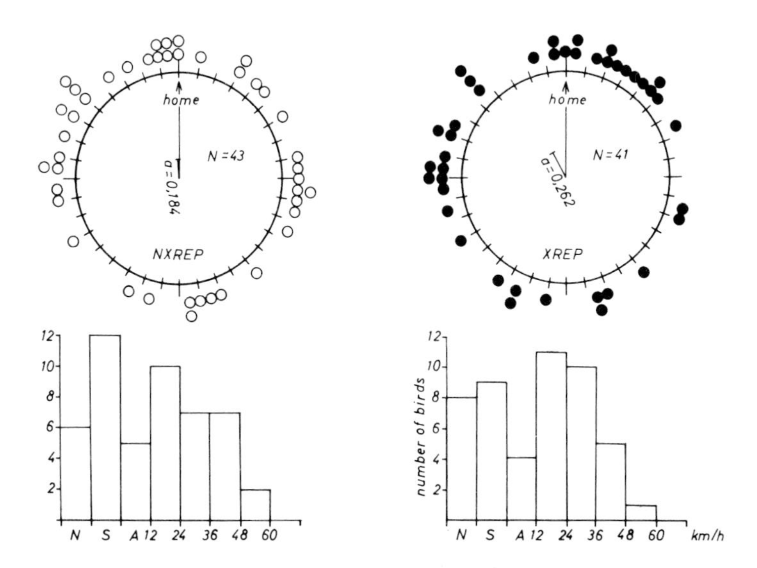

Fig. 5. Initial orientation and homing performance of four releases from 70 km (series B). *NXREP*, noxylocain repeatedly; *XREP*, xylocain repeatedly. Fore more details see text. Symbols as in Figure 3

even a trigeminus irritant at high concentrations but it may be safely concluded that perception of weak odors was eliminated as well. We are concerned, however, that we have not - and to our knowledge nobody else has - (possibly not yet) been able to demonstrate the pigeon's ability to discriminate different environmental air samples. Thus, direct evidence for the recognition of naturally occurring subtle odors is still not available.

Series A of the homing experiments was designed to eliminate perception of odors by the experimental birds during the outward journey. Our summarized results (Fig. 3) are clearly negative as to an effect on initial orientation and homing performance.

There was, however, some significance in the first three releases, accumulated from a highly significant difference in the second release and differences with suggestive significance in the first and third. One may argue that the birds quickly turned to some other system once the olfactory information during the outward journey was eliminated. In series B, olfaction was eliminated for the outward journey *and* beyond the time of release. Unfortunately, nondirected initial orientation of experimental and control birds is a poor basis on which to find an experimental effect. However, nonrandomness and homeward-directedness recorded in the third of the four releases speaks against this randomness being due to experimental or control treatment. The effect of eliminating olfaction by local anesthesia was, thus, slight and inconsistent. The birds do home without olfactory information during the outward journey. The effect was by far not as clear-cut as in similar experiments by Papi and co-workers nor as strong and persistent as, e.g., that of clock shifts. To date, our results can be interpreted to support the view that olfactory information during displacement is not an essential element of initial orientation or homing performance. The diverging results obtained by Papi and co-workers in Italy and by Keeton and co-workes in the United States, and our own intermediate results suggest that pigeons at the various locations may emphasize different cues for homing or they may have been selected with different emphasis to use one of a number of different cues for homing. Of course, modified methologic application of local anesthesia can and must be considered for further experiments.

References

Benvenuti, S., Fiaschi, V., Fiore, L., Papi, F.: Homing performance of inexperienced and directionally trained pigeons subjekt to olfactory nerve section. J. Comp. Physiol. 83, 81-92 (1973)

Hartwick, R.F., Foà, A., Papi, F.: The effect of olfactory deprivation by nasal tubes upon homing behavior in pigeons. Behav. Ecol. Sociobiol. 2, 81-89 (1977)

Keeton, W.T., Kreithen, M.L., Hermayer, K.L.: Orientation by pigeons deprived of olfaction by nasal tubes. J. Comp. Physiol. 114, 289-299 (1977)

Papi, F., Fiore, L., Fiaschi, V., Benvenuti, S.: The influence of olfactory nerve section on the homing capacity of carrier pigeons. Monitore Zool. Ital. (N.S.) 5, 265-267 (1971)

Papi, F., Fiore, L., Fiaschi, N., Benvenuti, S.: Olfaction and homing in pigeons. Monitore Zool. Ital. (N.S.) 6, 85-95 (1972)

Schmidt-Koenig, K.: Migration and homing in animals. In: Zoophysiology and Ecology. Berlin-Heidelberg-New York: Springer Vol. VI (1975)

Siegel, S.: Nonparametric Statistics. New York, Toronto, London McGraw-Hill (1956)

Session III

Chairman:

ARTHUR D. HASLER, Laboratory of Limnology, University of Wisconsin, Madison, WI 53706, USA

Sensitivity of the Homing Pigeon to an Earth-Strength Magnetic Field

MICHAEL A. BOOKMAN, Department of Psychology, Massachusetts Institute of Technology, Cambridge, MA 02139, USA

Abstract

Although the Earth's magnetic field appears to be involved in orientation, previous laboratory experiments have been unable to confirm the magnetic sensitivity of homing pigeons. In the experiments described here, a flight tunnel was constructed within a shielded chamber and used to train three pairs of pigeons (*Columba livia*) to discriminate between the presence and absence of a 0.5 gauss (Earth-strength) magnetic field. Successful discrimination was associated with flutter activity of the birds within the tunnel.

A. Introduction

The ability of homing pigeons to return from an unfamilar release site through a variety of experimental and natural conditions may depend upon a system of "map and compass" navigation (Kramer, 1953; Schmidt-Koenig, 1965; Keeton, 1974) that utilizes several environmental cues. Celestial sources of compass information have been demonstrated with pigeons (Walcott, 1972; Alexander and Keeton, 1974) and migratory birds (Kramer, 1950; Emlen, 1970, 1972). Sources of map information, however, have remained obscure (Schmidt-Koenig, 1965; Keeton, 1973, 1974).

The total strength and vertical component (angle of inclination) of the Earth's magnetic field are maximum at the poles and minimum at the magnetic equator. The angle of declination (between the geographic and magnetic north meridians) varies with longitude. Detection of these parameters either together or individually could contribute to a navigational grid.

Early tests of magnetic orientation compared the homing performance of birds with attached bar magnets to birds with attached nonmagnetic equivalent weights (Yeagley, 1947, 1951). Conflicting results were obtained under sun, and no conclusions could be drawn. A new series of magnet experiments was prompted by the demonstration of Keeton (1969) that pigeons could orient and home from an unfamilar location under complete overcast by using sun-independent cues. The vanishing bearings of experienced birds released with magnets under overcast were then found to be randomized (Keeton, 1971, 1972). Under sun, only experienced birds released at greater distances or first-flight youngsters were disoriented. With battery-powered Helmholtz coils, Walcott and Green (1974) were able to change the inclination of the field experienced by pigeons in flight and obtain predictable and reversible orientation responses.

Natural fluctuations of less than 70 gamma in the Earth's field have been shown to correlate with the orientation performance of pigeons under sun (Keeton et al., 1974), Ring-billed Gulls (Southern, 1971, 1972), and free-flying nocturnal migrants (Moore, 1977). Free-flying migrants also appear to respond to small-scale, artificially induced fluctuations (Larkin and Sutherland, 1977). In pigeons, attached magnets have been used to suppress the correlation (Larkin and Keeton, 1976). Recently, Walcott (pers. comm.) has found a randomization of bearings when pigeons were released under sun in the vicinity of natural magnetic anomalies. Together, these observations suggest that the Earth's magnetic field has a role in the orientation of both homing pigeons and migratory birds.

With specially designed cages, it has been possible to record the orientation preferences of captive migrants in response to a variety of cues. W. and R. Wiltschko (1972a, 1972b, 1974) and Emlen et al. (1976) obtained predictable and biologically appropriate responses from several species of migrants to natural and artificial magnetic stimuli. However, attempts thus far to record the meaningful orientation of pigeons within cages have been unsuccessful and other behavioral techniques have been applied.

Using classical conditioning of the pigeon's heart rate, Kreithen and Keeton were successful in demonstrating a sensitivity to polarized light (1974a) and to small changes in barometric pressure (1974b). Yodlowski et al. (1977) demonstrated a sensitivity to infrasound. Reille (1968) reported some positive results in response to magnetic stimuli. However, Kreithen and Keeton (1974c), Beaugrand (1976), and Bookman (unpublished observations) have been unable to repeat or extend his findings. Attempts by several investigators to use operant conditioning with magnetic stimuli have also been unsuccessful (Meyer and Lambe, 1966) but largely unpublished. Assuming that a magnetic sense does exist, the failure for it to be demonstrated with conditioning techniques is unexplained. Either a proper stimulus was not presented or conditions of the experiments interfered with detection. During conditioning, the pigeons were largely restrained from active movement. Fluttering, however, was permitted in successful experiments with caged migratory birds. Perhaps freedom of movement is a "requirement" of the magnetic sense.

B. Flight Tunnel Training

To allow greater freedom of movement during behavioral testing, a 3.5-m flight tunnel was built within the controlled environment of a mu-metal (see Beischer, 1971) shielded chamber. The dimensions of the tunnel, as illustrated in Figure 1, were large enough to permit limited flutter activity. Training was undertaken to determine if pigeons could learn to discriminate between a 0.5 gauss field induced by Helmholtz coils and the 0.02 gauss background field.

Two feeding boxes were attached at the far end of the tunnel to form a "T" maze. Small bins containing food were concealed within each box. The pigeons were trained to choose and enter one feeding box for each of the two magnetic field conditions. To serve as a positive reinforcement, food was available only from the correct box. Entering the wrong box was punished by confining the bird for 30 s before allowing it to enter the correct box. The sequence of magnet coil activation was determined at random and set before the start of each trial.

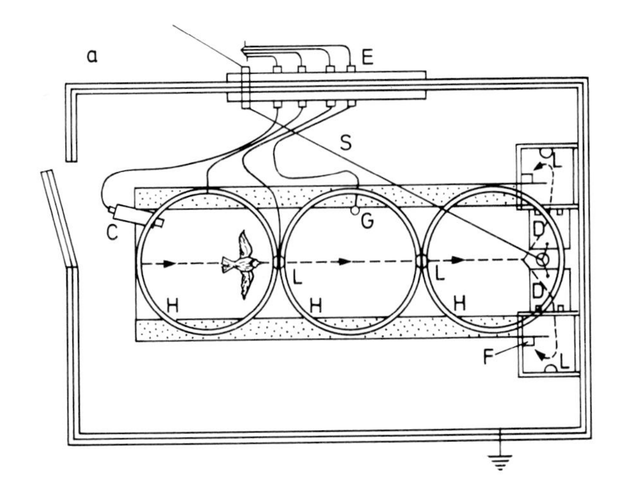

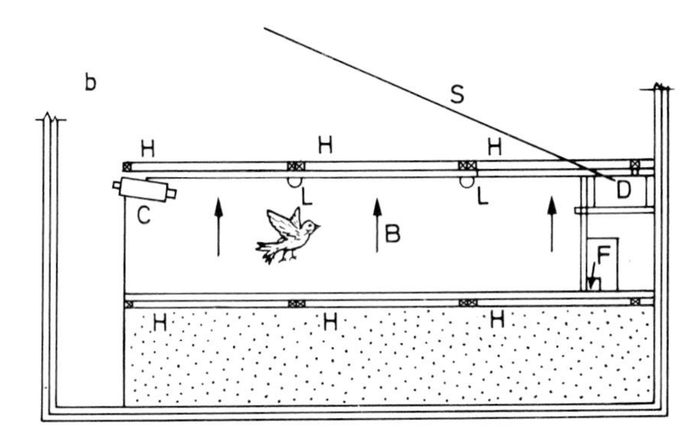

Fig. 1a and b. Pigeon flight tunnel. Top view (a) and side view (b) of flight tunnel as constructed within mu-metal shielded room. Shielding reduced the natural magnetic field to a level of 0.02 ± 0.01 gauss. Three pairs of Helmholtz coils (*H*) were connected in series and arranged to induce a vertical magnetic field (*B*) along the length of the tunnel in as uniform a manner as possible. Each coil was wound with 150 turns of 12-gauge copper wire. Approximately 7.5 V DC at 100 mA was required to induce a 0.5 gauss (Earth-strength) magnetic field, as measured from a central probe (*G*). When the coils were not in use, the current was diverted across an external equivalent resistance. Two feeding boxes with concealed food bins (*F*) were located at the end of the tunnel. For each trial, food in one bin was accessible while food in the opposite bin was covered by an aluminum screen that prevented access. Birds could be confined within the boxes by remote-controlled doors (*D*) which were released by string (*S*). Illumination was provided by 12 V DC lamps (*L*). A closed-circuit television camera (*C*) and electric patchboard (*E*) allowed monitoring and control to procede from a separate room which housed the power supplies, gaussmeter, and television monitor

During a preliminary period in which birds were adapted to the tunnel, it was found that mated pairs of pigeons were more active than single birds. Therefore, three mated pairs were borrowed from several lofts and used as subjects for further testing. The birds were housed in cages within a naturally lighted and ventilated room, and maintained within 10% of their ad lib weight through combined flight tunnel and supplementary feedings.

After two weeks of adaptation, the birds' behavior had been sufficiently shaped to allow the recording of data. Trials began by uniformly releasing a pair of birds at the tunnel entrance, and ended with the recovery of both birds from the correct box. Only the first response of the first bird to enter a box was recorded, as the remaining bird did not respond independently. In this way, the responses from a pair of birds were treated as if from a single bird. Responses were scored by closed-circuit television monitoring; positioning of any part of the body within a feeding box constituted a response.

Frequently, the birds were observed to engage in spontaneous flutter activity within the tunnel before making a choice of boxes. Fluttering consisted of sustained hovering, rapid turning, and short flights. Trials with at least 3 s of continuous flutter activity were recorded as "with flutter," the remainder were designated "no flutter". In the majority of "no flutter" trials, the birds walked the length of the tunnel.

C. Results

Several months were required for the collection of data, as a maximum of only five trials could be run with a given pair of birds each day. A cumulative performance record for each pair of birds is datailed in Figures 2, 3, and 4. A summary of all data is presented in Table 1.

The data indicate that a statistically significant discrimination had occurred. This was even more apparent when the trials were separated according to flutter activity. In all cases, the response records obtained in the presence of flutter were nonrandom, while data collected in the absence of flutter were not significantly different from chance.

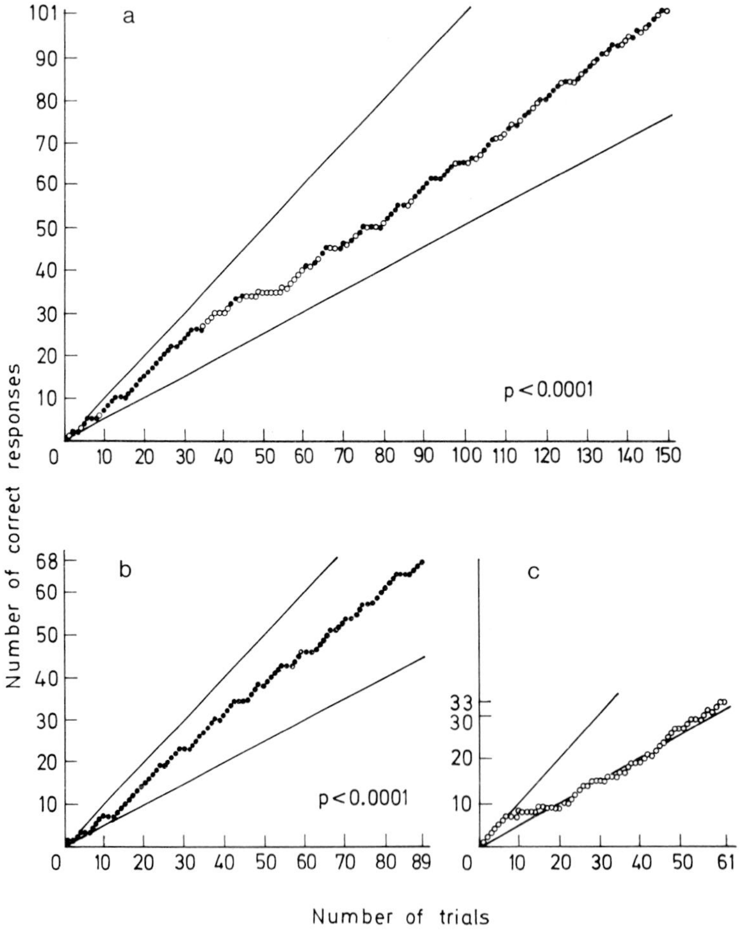

Fig. 2 a - c. Cornell birds: Cumulative record. Graphs showing the number of trials with correct responses as a function of the total number of trials. Only the response of the first bird to make a response was recorded for each trial. *Lines* indicating the 50% (chance) and 100% (perfect; *upper line*) levels of performance have been added. (a) The total trial sequence is shown. Note that a significant discrimination had occurred and that "no flutter" (O) and "with flutter" (●) trials were distributed throughout the testing period. (b) "With flutter" (●) trials graphed separately. (c) "No flutter" (O) trials graphed separately. Significant discrimination occurred only in the "with flutter" trial sequence

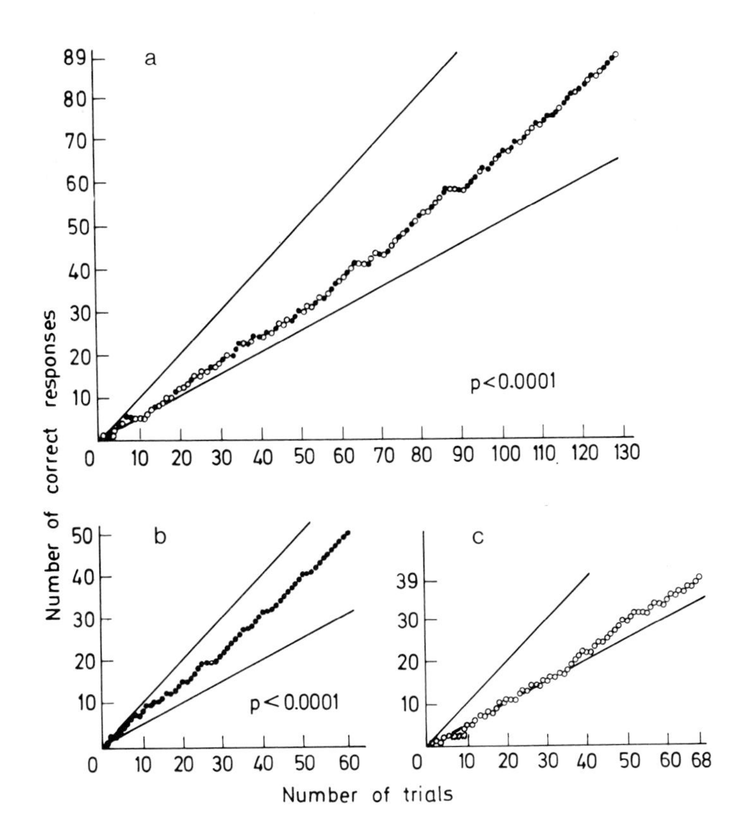

Fig. 3 a - c. Richards birds: Cumulative record. Details same as in Figure 2

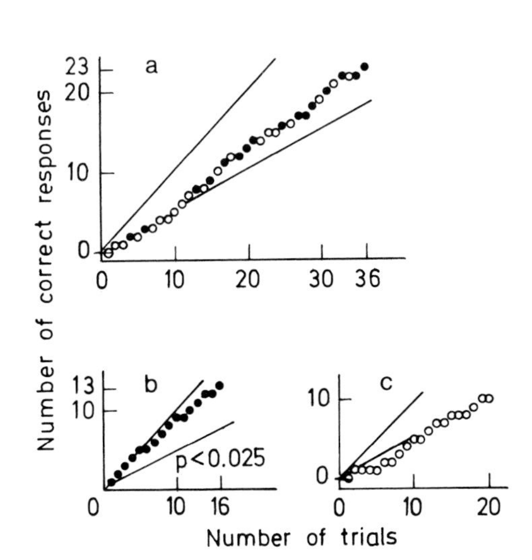

Fig. 4 a - c. Walcott birds: Cumulative record. (a) The total trial sequence is shown. (b) "With flutter" (●) trials graphed separately. Although fewer trials were run with this pair of birds, discrimination is still apparent. (c) "No flutter" (O) trials graphed separately. Again, performance without flutter was random

Performance of the birds was not below the chance level in any category. Trials with flutter were distributed throughout the testing period, during which time the overall levels of performance were not greatly

Table 1. Summary of flight tunnel training results

Pigeons		Number of trials	Number correct	Percent correct	Probability
	Total	150	101	67.3	<0.0001
Cornell pair	With flutter	89	68	76.4	<0.0001
	No flutter	61	33	54.1	NS
	Total	129	89	69.0	<0.0001
Richards pair	With flutter	61	50	82.0	<0.0001
	No flutter	68	39	57.4	NS
	Total	36	23	63.9	<0.15
Walcott pair	With flutter	16	13	81.2	<0.025
	No flutter	20	10	50.0	NS

changed. This suggested that the association between flutter and correct responses was more than coincidence, and that "learning" largely took place during the two-week period of adaptation.

The only factor deliberately altered from one trial to the next was the applied magnetic field. Although attempts were made to prevent inadvertent cues (such as power supply hum), the influence of any such cues on discrimination would probably be independent of flutter activity. Therefore, the finding of significant differences between trials with and without flutter suggests that inadvertent cues or experimenter-induced bias did not influence the results.

D. Conclusions

A major goal of this study has been to obtain a stable behavioral response to magnetic stimuli and to use that response in probing the mechanisms of magnetic perception. It appears from these preliminary experiments that homing pigeons have a sensitivity to magnetic fields that can be demonstrated in the appropriate laboratory setting. However, the results do not speak directly to the ability of pigeons to use magnetic cues for navigation.

Unlike other sensory modalities, there appears to be a requirement for freedom of motion. In these experiments, perhaps the motivation was not strong enough to encourage flutter activity in every trial. This would explain the persistent occurrence of "no flutter" trials with random levels of performance. Although flutter may be directly involved with the mechanism of perception, it may just as likely represent the external manifestation of some internal sensory activation. The mechanism of magnetic sensitivity in birds remains unknown, but has been the subject of recent theories (Leask, 1977).

Although the flight tunnel technique is straightforward, it is hoped that several points can be improved upon. Current studies are directed toward working with single birds and increasing the number of trials per day, using automated techniques of data recording and experiment control.

Acknowledgements. This study was supported by the Sloan Foundation and the Massachusetts Institute of Technology. I thank Dr. W. Richards for continued support and advice. I also thank Dr. W.T. Keeton, Dr. C. Walcott, and L. Richards for supplying pigeons.

Portions of this study have been previously reported (Bookman, 1977).

References

Alexander, J.R., Keeton, W.T.: Clock-shifting effect on initial orientation of pigeons. Auk 91, 370-374 (1974)

Beaugrand, J.P.: An attempt to confirm magnetic sensitivity in the pigeon, Columba livia. J. Comp. Phys. 110, 343-355 (1976)

Beischer, D.E.: The null magnetic field as reference for the study of geomagnetic directional effects in animals and man. Ann. N.Y. Acad. Sci. 188, 324-330 (1971)

Bookman, M.A.: Sensitivity of the homing pigeon to an Earthstrength magnetic field. Nature (London) 267, 340-342 (1977)

Emlen, S.T.: Celestial rotation: its importance in the development of migratory orientation. Science 170, 1198-1201 (1970)

Emlen, S.T.: The ontogenetic development of orientation capabilities. In: Animal Orientation and Navigation. NASA-SP-262, U.S. Govt. Printing Pffice, Wash., D.C., 1972, pp. 191-210

Emlen, S.T., Wiltschko, W., Demong, N.J., Wiltschko, R., Bergman, S.: Magnetic direction finding: evidence for its use in migratory Indigo Buntings. Science 193, 505-508 (1976)

Keeton, W.T.: Orientation by pigeons: is the Sun necessary? Science 165, 922-928 (1969)

Keeton, W.T.: Magnets interfere with pigeon homing. Proc. Nat. Acad. Sci. USA 68, 102-106 (1971)

Keeton, W.T.: Effects of magnets on pigeon homing. In: Animal Orientation and Navigation NASA SP-262, U.S. Govt. Printing Office, Wash., D.C., 1972, pp.579-594

Keeton, W.T.: Release-site bias as a possible quide to the "map" component in pigeon homing. J. Comp. Phys. 86, 1-16 (1973)

Keeton, W.T.: The orientational and Navigational basis of homing in birds. Adv. Study Behav. 5, 47-132 (1974)

Keeton, W.T., Larkin, T.S., Windsor, D.M.: Normal fluctuations in the Earth's magnetic field influence pigeon orientation. J. Comp. Phys. 95, 95-103 (1974)

Kramer, G.: Orientierte Zugaktivität gekäfigter Singvögel. Naturwissenschaften 37, 188 (1950)

Kramer, G.: Wird die Sonnenhöhe bei der Heimfindeorientierung verwertet? J. Orn., Lpz. 94, 201-219 (1953)

Kreithen, M.L., Keeton, W.T.: Detection of polarized light by the homing pigeon, *Columba livia*. J. Comp. Phys. 89, 83-92 (1974a)

Kreithen, M.L., Keeton, W.T.: Detection of changes in atmospheric pressure by the homing pigeon, *Columbia livia*. J. Comp. Phys. 89, 73-82 (1974b)

Kreithen, M.L., Keeton, W.T.: Attempts to condition homing pigeons to magnetic stimuli. J. Comp. Phys. 91, 355-362 (1974c)

Larkin, T.S., Keeton, W.T.: Bar magnets mask the effect of normal magnetic disturbances on pigeon orientation. J. Comp. Phys. 110, 227-231 (1976)

Larkin, T.S., Sutherland, P.J.: Migrating birds respond to Project Seafarer's electromagnetic field. Science 195, 777-779 (1977)

Leask, M.J.M.: A physiochemical mechanism for magnetic field detection by migratory birds and homing pigeons. Nature (London) 267, 144-145 (1977)

Meyer, M.E., Lambe, D.R.: Sensitivity of the pigeon to changes in the magnetic field. Pyschon. Sci. 5, 259-260 (1966)

Moore, F.: Geomagnetic disturbance and the orientation of nocturnally migrating birds. Science 196, 682-684 (1977)

Reille, A.: Essai de mise en évidence d'une sensibilité du pigeon au champ magnétique à l'aide d'un conditionnement nociceptif. J. Phys., Paris 60, 85-92 (1968)

Schmidt-Koenig, K.: Current problems in bird orientation. Adv. Study Behav. 1, 217-278 (1965)

Southern, W.E.: Gull orientation by magnetic cues: a hypothesis revisited. Ann. N.Y. Acad. Sci. 188, 295-311 (1971)
Southern, W.E.: Influence of disturbances in the Earth's magnetic field on Ring-billed Gull orientation. Condor 74, 102-105 (1972)
Walcott, C.: The navigation of homing pigeons: do they use Sun navigation? In: Animal Orientation and Navigation NASA SP-262, U.S. Govt. Printing Office, Wash., D.C., 1972, pp. 283-292
Walcott, C., Green, R.P.: Orientation of homing pigeons altered by a change in the direction of an applied magnetic field. Science 184, 180-182 (1974)
Wiltschko, W.: The influence of magnetic total intensity and inclination on directions preferred by migrating European Robins (Erithacus rubecula). In: Animal Orientation and Navigation NASA SP-262, U.S. Govt. Printing Office, Wash., D.C., 1972, pp. 569-578
Wiltschko, W., Wiltschko, R.: Magnetic compass of European Robins. Science 176, 62-64 (1972b)
Wiltschko, W.: Der Magnetkompass der Gartengrasmücke (Sylvia borin). J. Orn., Lpz. 115, 1-6 (1974)
Yeagley, H.L.: A preliminary study of a physical basis of bird navigation. J. Appl. Phys. 18, 1035-1063 (1947)
Yeagley, H.L.: A preliminary study of a physical basis of bird navigation II. J. Appl. Phys. 22, 746-760 (1951)
Yodlowski, M.L., Kreithen, M.L., Keeton, W.T.: Detection of atmospheric infrasound by homing pigeons. Nature (London) 265, 725-726 (1977)

Inversion of the Magnetic Field During Transport: Its Influence on the Homing Behavior of Pigeons*

JAKOB KIEPENHEUER, Abteilung Verhaltensphysiologie der Universität Tübingen, 7400 Tübingen, FRG

Abstract

Pigeons were subjected to an inversed vertical magnetic field during transport to the release site. This resulted in a significant deviation of the mean vanishing directions by about 30° to the right. Inversion of the field before and after transport did not show any effect on the pigeons.

A. Introduction

It is evident now that birds possess the ability to use the Earth's magnetic field in directional orientation or even in navigational processes. In which way do they use this ability? A compass system functioning during transport to the release site associated with the ability to perceive linear accelerations actually would give the pigeon all the information necessary to determine an approximate home bearing on release. While there is evidence that there is a magnetic sense in the pigeon, it remains unclear whether this provides the bird with information on the outward journey.

In the past, most authors have denied that the pigeon picks up any information on its way to the release site. However, the detour experiments of Papi and his co-workers (1973) and our own experiments (Hartwick et al., 1977) show that in at least some cases and maybe only under certain circumstances, information picked up during transport may provide the information necessary to select the proper homing direction. Is this exclusively to be explained by airborne information? Or are other systems like magnetic and inertial ones involved?

A magnetic compass would deliver the necessary information, which in combination with a sense for time and linear accelerations - sensed quite well by the pigeons as shown by Delius and Emerton 1977 - would enable it to determine the direction of transport and the relative position of the site in respect to the loft. If the compass used does not rely on the movement through the field, an artificial reversal of the field around the pigeon during transport should give the bird the impression of being carried in the opposite direction of actual transport.

Assuming that the Wiltschko model (1972) of the bird's magnetic compass is right, the easiest way to inverse the magnetic field for the bird is to inverse the vertical component of the natural field. This is easily accomplished by a pair of Helmholtz coils generating a

*Supported by the Deutsche Forschungsgemeinschaft.

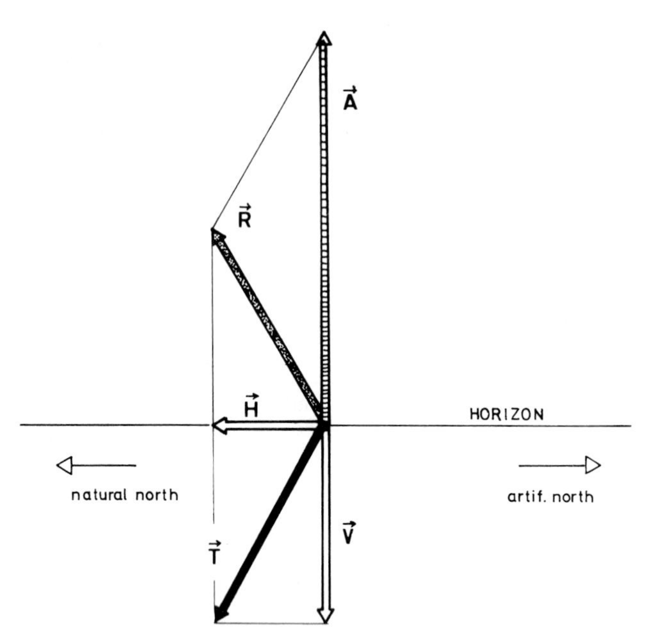

Fig. 1. The magnetic field vector ($\vec{T}$) and the artificial field vector ($\vec{A}$) result in the new total magnetic vector ($\vec{R}$). This results in a reversed direction of the dip

field twice the strength and apposed to the Earth's vertical field. This does not change the polarity of the field but inverses the direction of the smaller angle of inclination, not matter how the car turns (Fig. 1).

The alternative is to change the inclination and polarity of the field by reversing the horizontal component. This requires an elaborate steering system in order to maintain the coil in a north-south direction despite the turns of the road (a publication of the results obtained in this set-up is in preparation).

B. Methods

The Helmholtz coils generating the artificial field were of octogonal form, 90 cm in diameter. The pigeon crate was fitted in the middle of the double coils so that the heads of the pigeons were about in the center plane between the two coils, in the most homogeneous part of the field. The current fed to the coils was delivered by a 12-V battery and regulated by an adjustable resistor. The strength of the resulting magnetic field was equal to that of the natural field and homogeneous within about 15% of the mean value. Irregularities were caused mainly by the car; those caused by passing trucks or steel structures were of about the same order. The Helmholtz coils with the experimental pigeons were carried on the front part of the roof, the controls about 2 m off on the back end of the roof-top carrier, so that the artificial field did not affect them.

Controls and experimentals were transported in cloth-covered crates, allowing any airborne information to be picked up by the pigeons, but hiding landmarks. The inversed field was turned on about 300 m from the loft, and was shut off immediately upon arrival at the release site.

In another series conducted this year, the field was turned on one hour before departure, shut off at departure, and turned on again on arrival, where it stayed on until the end of the release.

In all cases birds were picked randomly for the control and experimental groups. Driving speed to the release site did not exceed the normal flying speed of a pigeon of 70 km/h. The most direct road was taken to the release site. The first-flight pigeons used were 3 - 5 months old and did not have any previous experience except for exercise flights around the loft, while the experienced birds had several flights from varying distances, but usually not from the same site chosen for the experiment. At the site, experimentals and controls were released alternately, vanishing bearings were taken by 7 × 50 binoculars, and vanishing times and homing speeds were recorded.

C. Statistics

Vanishing bearings were compared by the nonparametric Watson U^2 test (Batschelet, 1972) and after certification of approximate circular normal distribution, by the Watson and Williams test (Batschelet, 1965). Differences in vanishing times were tested by the Mann-Whitney-U test and, if normally distributed, by the t-test. Homing speeds were compared by the Mann-Whitney-U test.

D. Results

I. Vertical Magnetic Field, Reversed During Transport

Seventeen releases were carried out, 12 with first-flight birds at distances from 13 - 18 km from varying directions, 5 with the more experienced birds at distances from 14 - 135 km. In all except 2 releases out of 17, the mean vanishing bearing of the experimentals deviated to the right of that of the controls. In 14 out of 17 cases both groups deviated to the right of the home direction. Figure 2 shows the deviation of the mean bearings of the experimentals with respect to those of the controls. Due to the small numbers of birds, the difference was significant at a 4% level in only one experiment. In some, random distribution of bearings is not excluded.

Figure 3 shows the behavior of first-flight birds. Under sunny conditions the difference of the mean bearings of controls and experimentals is 30° and significant by $p < 2\%$ (Watson U^2 test) and $p = 2.5\%$ (Watson and Williams test). Under overcast no difference is detectable ($p \approx 50\%$). Vanishing times of controls and experimentals are about the same, while the homing speeds of controls are somewhat higher than in the experimentals.

Figure 4 shows the data of the more experienced birds. The difference between the mean vanishing bearing of controls and experimentals again is about 28°, significance levels are 7.5% according to the Watson U^2 test and 2.5% applying the Watson and Williams test. Vanishing times of controls again are somewhat shorter than those of the experimentals ($p < 5\%$): It seems that at least some of the experimental birds tend to linger around the release site.

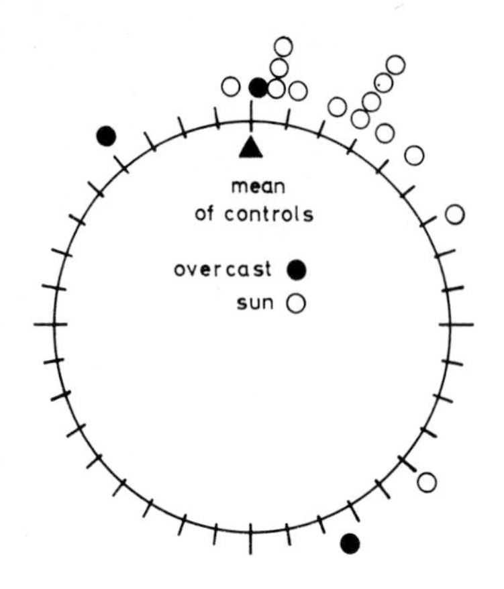

Fig. 2. Each dot represents the mean direction of pigeons transported in a reversed vertical field of one experiment with respect to the respective mean direction of controls (up)

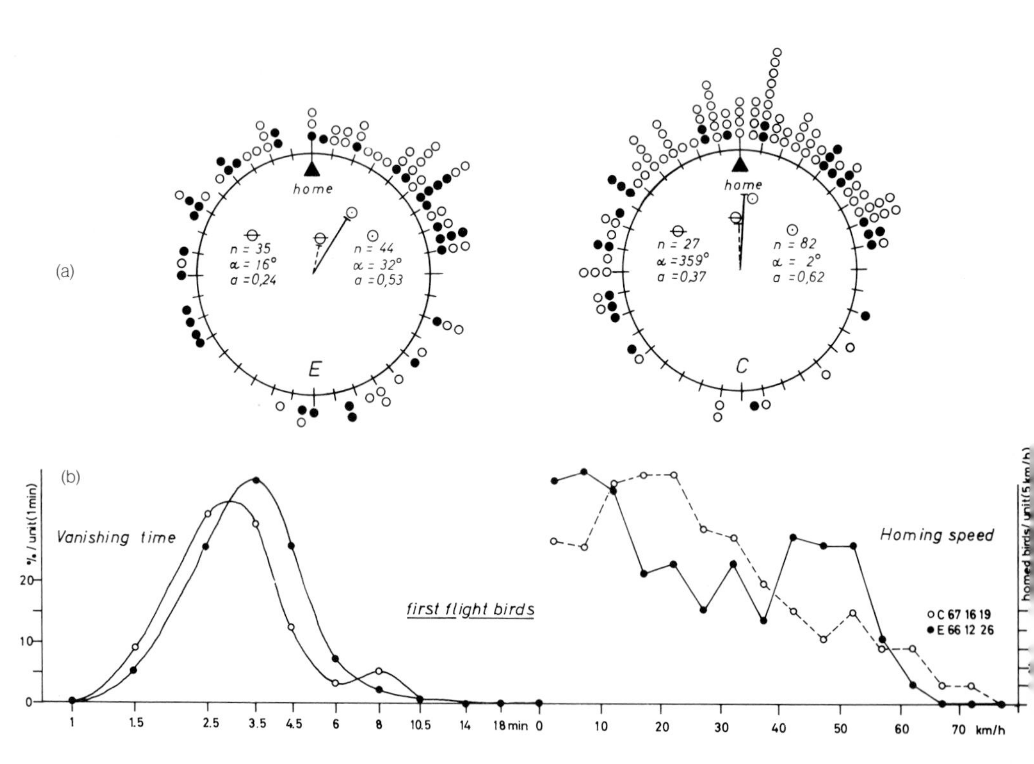

Fig. 3. (a) Vanishing bearings of first-flight birds. Home direction is up. Experimentals *left*, controls on the *right* (• release while sun not visible). (b) *Left:* vanishing times (*log scale*), the ordinate gives the percentage of individuals per 1-min interval. *Right*, homing speeds, the ordinate gives the percentage of homed birds per 5-km/h interval. Numbers on the *right* are of birds returned the first day/later/lost. • Experimentals, o controls

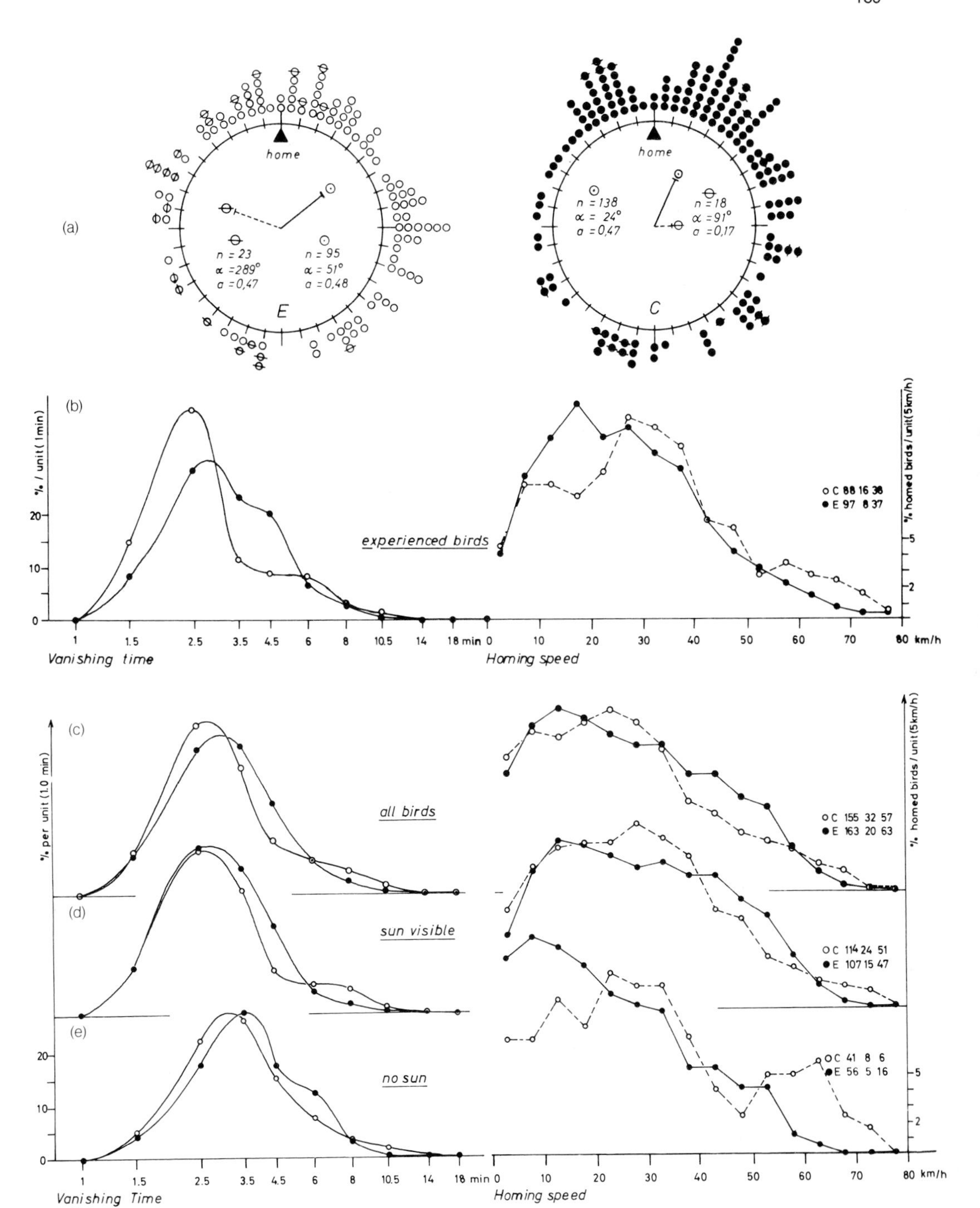

Fig. 4. (a) Vanishing bearings of experienced birds (⊖ sun not visible); (b) vanishing times and homing speeds; (c) vanishing times and homing speeds of first-flight and experienced birds combined; (d) under sunny conditions; (e) under overcast conditions (see Figure 3 for details)

Homing speeds too are higher in controls than in experimentals, but not at a statistically acceptable level.

Combining all data from first-flight and experienced birds under sunny conditions, the difference in the mean vanishing bearing of controls and experimentals of ca. 30° is significant by $p < 0.5\%$ (Watson U^2 test), and since both sets of data are distributed in an approximately circular normal pattern, we may apply the Watson and Williams test, giving a significance smaller than $p = 0.1\%$ for a difference in the mean direction. Under overcast conditions controls and experimentals still show a slight but significant homeward orientation ($p < 1\%$, V test), but no difference in the mean direction.

Figure 4 c shows the combined data of vanishing times and homing speeds. The vanishing times presented in a log scale approximate a normal distribution, making it easier to visualize differences in behavior. Vanishing times in the controls are somewhat shorter than in the experimentals, though the difference is not significant. This seems true for vanishing times under overcast as well as under sunny conditions. The vanishing times under overcast are significantly longer in controls and experimentals than under sunny conditions ($p < 1\%$).

The effect on homing speed is not clear-cut: The differences are suggestive but not significant - the majority of the controls home at higher speeds than the experimentals, especially under overcast conditions.

II. Vertical Magnetic Field Reversed Before and After Transport

Unfortunately controls testing whether application of the reversed field before and after transport interferes with the homing behavior were not carried out until this spring because of too few available birds and bad weather conditions.

Figure 5 shows the pooled data of 7 releases carried out from different sites, 6 - 40 km from the loft. They do not show any significant difference between controls and experimentals. The mean bearing deviates insignificantly by 13° to the left of that of the controls. The data of vanishing times and homing speeds (below) also do not show any significant difference.

E. Discussion

Inversion of the vertical component of the magnetic field during transport, under sunny conditions, has a consistent, though slight effect on the vanishing bearings of the homing pigeon: They are shifted to the right, and the mean vector length is not changed compared to that of the controls. There is no evidence for a bimodal orientation (both distributions of bearings conform well with a circular normal). Under overcast conditions orientation in both groups is bad, though homeward oriented; no difference in vanishing bearings is detectable. Application of the inversed field before and after transport does not seem to have any effect on the behavior of the pigeons.

What conclusions may be drawn from these results?

1. Magnetic information picked up during transport may be of some importance in selecting the appropriate home bearing at the release site.

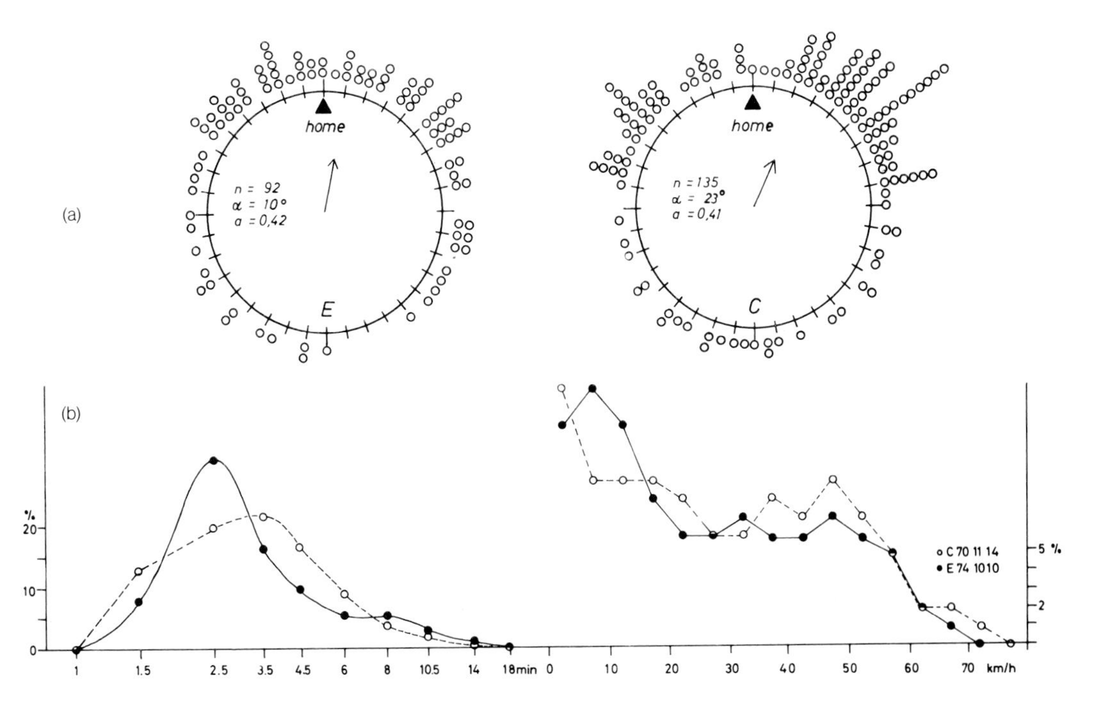

Fig. 5. (a) Vanishing bearings of birds treated with a reversed vertical field before and after transport (*left*) and untreated controls (*right*). (b) Vanishing times and homing speeds (see Figure 3 for details)

2. The navigational process seems to involve a magnetic information system that reacts on stationary fields during transport and does not or not alone depend on a movement in the field involving inductive processes.

3. The magnetic information picked up during transport obviously is not essential to determine the direction of home, since birds deprived of it home almost as well.

4. Let us assume that the magnetic information picked up during transport provides the bird with a vector pointing home. Some other, maybe airborne information system provides the bird with another vector, maybe with a bias to the right (or left) of home. The resultant vector should then be equivalent to the mean vanishing bearing. Reversing, diminishing, or canceling the magnetic vector should lead to a greater bias of the resultant vector. In all 14 releases under sunny conditions the controls were actually deflected to the right of the home direction - as if they had grown up in a deflector loft. I made a preliminary check of the wind directions measured at our loft site (Hohenentringen), 6 m above the ground (500 m above sea level) and compared them to the recordings made on a high tower on a hilltop 6 km away (at 400 m above sea level). Winds from all directions measured at the loft site deviate by roughly 30° clockwise from those measured on the hill 6 km away (Schloß Tübingen). In accordance with the deflector loft experiments of Baldaccini et al. (1975) and my own replication of the experiment (in prep.), the clockwise deflection of control birds at nearly all sites tested so far may be plausible. An evaluation of this effect is in preparation. A reversal (or extinction) of the magnetic vector

information would therefore lead to the deflection of the birds' mean vanishing bearing to the right of the controls. Under overcast conditions the birds have to rely either on very bad compass (sun or magnetic) information leading to almost random departures and/or on a very low level of some map information, so that any difference in the orientation behavior of controls and experimentals will be blurred.

References

Baldaccini, N.E., Benvenuti, S., Fiaschi, V., Papi, F.: Pigeon navigation: Effect of wind deflection at home cage on homing behavior. J. Comp. Physiol. 99, 177-186 (1975)

Batschelet, E.: Statistical methods for the analysis of problems in animal orientation and certain biologigal rhythms. The Am. Inst. Biol. Sci., Wash. (1965)

Batschelet, E.: Recent statistical methods for orientation data. In: Animal Navigation and Orientation. Galler, R.S., Schmidt-Koenig, K., Jacobs, G.J., Belleville, R.E. (eds.) N.A.S.A. Adm. Wash. D.C. 1972

Papi, F., Fiaschi, V., Benvenuti, S., Baldaccini, N.E.: Pigeon Homing: outward journey detours influence the initial orientation. Monit. Zool. Ital. (N.S.) 7, 129-133 (1973)

Wiltschko, W., Wiltschko, R.: The magnetic compass of European Robins. Science 176, 62-64 (1972)

Anomalies in the Earth's Magnetic Field Increase the Scatter of Pigeons' Vanishing Bearings

CHARLES WALCOTT, Department of Biology, State University of New York at Stony Brook, Stony Brook, NY 11794, USA

Abstract

Homing pigeons were released under sunny conditions at places with large variations in the total intensity of the Earth's magnetic field. At five out of six such places, pigeons were more scattered in their homeward orientation than at magnetically normal sites. This suggests that some aspect of the magnetic field affects either the pigeon's sun compass or its navigational system.

A. Introduction

Since Keeton (1971, 1972) first showed that homing pigeons carrying bar magnets were often disoriented under overcast skies, the evidence that magnetic fields play a role in pigeon homing has been increasing. Walcott and Green (1974) reported that reversing the polarity of the magnetic field produced by a pair of Helmholtz coils around a pigeon's head often results in pigeons flying in a direction away from the loft when the sun is not visible. Under sunny conditions, the same coils cause an increase in the scatter of vanishing bearings and a small difference in the angle of the vanishing directions of the two groups (Walcott, 1977). However, Keeton et al. (1974) showed that the vanishing bearings of pigeons also vary with natural variations in the Earth's magnetic field. These fluctuations in the Earth's field are much smaller than those caused by either magnets or Helmholtz coils and are far weaker than local changes in the field at many magnetic anomalies. Graue (1965) and Talkington (1967) report that pigeon orientation was affected by anomalies in the Earth's magnetic field, but the data on which these abstracts were based have never been published. Recently, Wagner (1976, Frei & Wagner, 1976b) has shown that pigeons released near a magnetic anomaly orient to the left of home in roughly the direction of the magnetic gradient. However, these anomalies were rather weak - only about 80 gamma or 0.1% of the normal Earth field. Perhaps some of the stronger local magnetic anomalies in New England would have an even greater effect on pigeon orientation: this paper summarizes the results of releases at six such sites.

B. Materials and Methods

Trained, experienced homing pigeons from lofts in Lincoln, Massachusetts were released under sunny skies and radio tracked at each of the magnetic anomalies. Each bird was released only once at each anomaly and these releases were either preceded or followed by a release at a magnetically normal control site. Anomalies were located using aeromagnetic survey maps published by the U.S. Geological Survey. Figure 1 shows the location of the anomalies, the control sites, and the lofts. The airplane tracking procedure has been described by Michener and Walcott (1967), but the transmitters used were small 148 mHz units (Cochran, 1967). Vanishing bearings were tested for randomness with the Rayleigh test, and the Watson U^2 test was used to compare distributions (Batschelet, 1965, 1972). The scatter of the vanishing bearings was compared with an F test proposed by Watson and explained in simple terms by Emlen and Penney (1964).

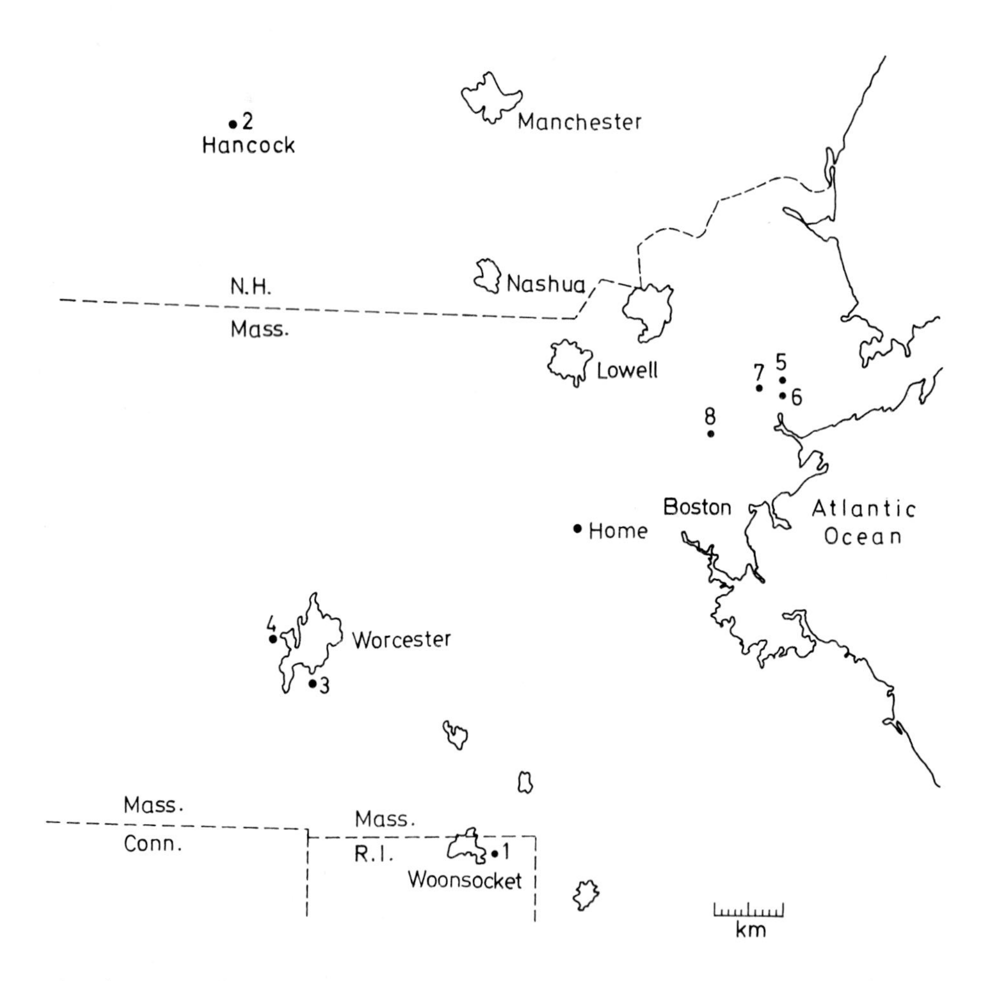

Fig. 1. A map of the release sites; north is up in the diagram and the *bar* at the bottom right corner is 10 km long. The release points are as follows: *1*, Iron Mine; *2*, Hancock, N.H.; *3*, Route 20 in Worcester; *4*, Worcester Airport; *5*, Wenham Swamp; *6*, Beverly Airport; *7*, Danvers; *8*, Lynnfield. The home loft is toward the center of the diagram just west of Boston

C. Results

A group of pigeons was first released at the anomaly at Iron Mine, 4 km east of Woonsocket, Rhode Island (Fig. 2). The vanishing bearings of these birds were random yet their homing performance seemed perfectly normal. These birds were then taken to Hancock, New Hampshire. Hancock was chosen as a control site because it is in almost the opposite direction from Iron Mine and there is little chance that birds released there would have ever flown near Iron Mine. The vanishing bearings of the birds released at Hancock were well oriented toward home. A second group of birds was now released at Hancock, and on their return to the loft, they were then released at Iron Mine. The results for the two releases at Iron Mine and the two releases at Hancock were indistinguishable, so the data for each release site have been combined in Figure 3. Clearly, birds that were well oriented at Hancock were poorly oriented at Iron Mine. The Watson U^2 test shows that the two distributions of vanishing bearings are significantly different; $p = 0.0001$, and the F test shows that the length of the mean vectors is also significantly different, $p = 0.005$.

Since Hancock and Iron Mine are in opposite directions, it could be argued that releasing birds at Hancock is not a suitable control for their release at Iron Mine. To counter this objection, we released birds at an anomaly on Route 20 in Worcester, Massachusetts and for a control point, used the Worcester Airport 13 km northwest of the anomaly. Once again, birds released at the anomaly vanished randomly; the vanishing bearings of the same birds at Worcester Airport were well grouped but show the characteristic release site bias of Worcester Airport. Thus, even though birds released at either site might have flown near the other, the results were essentially identical to those from Hancock and Iron Mine.

Using the aeromagnetic maps, a large, complex anomaly near Beverly, Massachusetts was found. Four release sites in the same general area were selected. The results of releasing pigeons at these sites are interesting: At three of the four sites, Wenham Swamp, Beverly Airport, and Lynnfield, the birds were less well oriented toward home than usual, but their vanishing bearings were not random. At the fourth site, Danvers, the vanishing bearings were completely normal. Thus, even though releases at these anomalies were not alternated with control releases, I think it is fair to say that birds at three of the four sites were less well oriented toward home than they were at Danvers. The obvious question then, is whether there is any relationship between the homeward orientation and the strength of the anomaly. Using the aeromagnetic maps we have tried to devise some way of measuring the strength of an anomaly. One way of considering this problem is to suppose that contours of magnetic intensity are like lines of equal altitude on a topographic chart. From such a chart, one can easily determine the height of a point, but it is more difficult to find a measure of how hilly the terrain in the area may be. Is it the variation in height that is important? The average height? And over how big an area? Somewhat arbitrarily, I have chosen five measures for each release point: (1) the relative strength of the magnetic field at the release point; (2) the difference between the highest and the lowest values of the magnetic field within a circle of a 1-km radius centered on the release point; (3) the same circle but with a radius of 3 km; (4) the largest difference between the field strength at the release point and the field strength anywhere along a 1-km line toward home; and the same measurement but at 3 km. As Table 1 shows, the variation in the field within 1 km toward home seems to show the best correlation with the amount of scatter in the pigeons' homeward orientation.

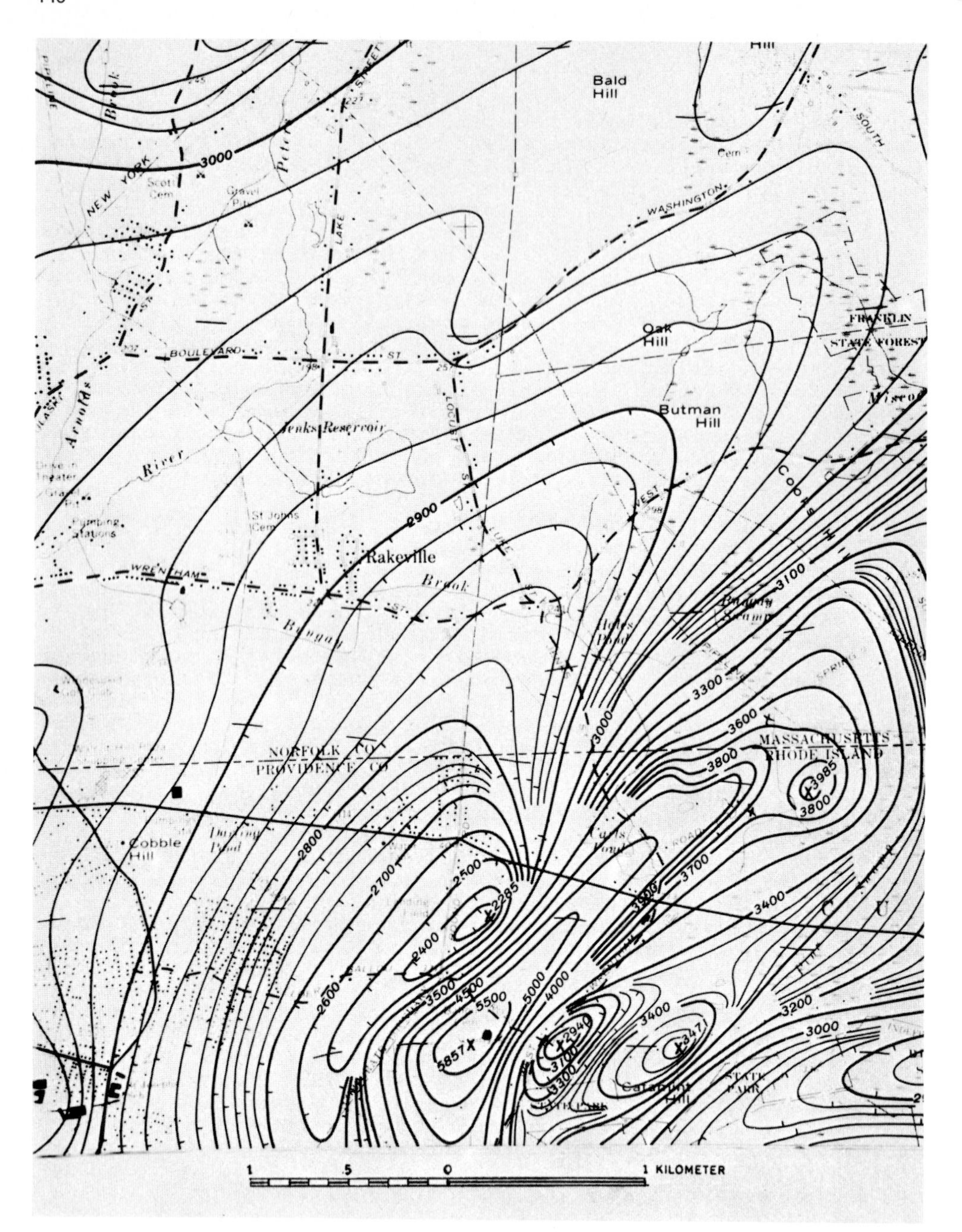

Fig. 2. A portion of the aeromagnetic map showing the magnetic anomaly at the Iron Mine release site. This is indicated by a *dot* at the bottom center of the map just northeast of the spot marked 5857. North on the map is vertical, and the home loft is at a bearing of 17^{0} true from the release site. The numbers on the map represent the total magnetic field strength in gammas relative to some arbitrary and unstated value; probably about 50,000 gamma. This means that a variation in field strength of 1000 gamma is about 2% of the total earth field. This survey was made from an airplane flying at about 500 feet above ground; therefore, the magnetic field at ground level is even more irregular than that shown here

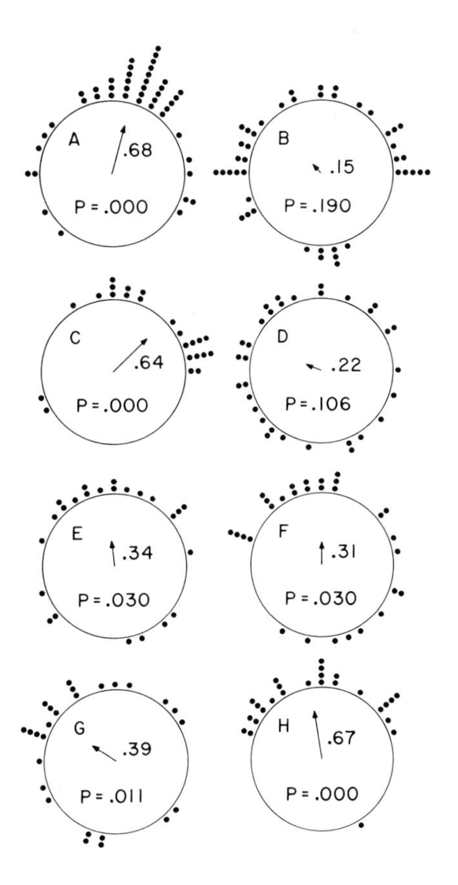

Fig. 3 A - H. The radio-tracked vanishing bearings of pigeons at both experimental and control sites. Each dot around the circle represents the vanishing bearing of a single pigeon at about 18 km from the release point. The *arrow* is the mean vector and its length is indicated beside it. *p* is the Rayleigh probability that the given distribution of points arose by chance. The home direction is at the top of the diagram but since the tracking antenna was aligned with a magnetic compass there may be an error of as much as $\pm 5^{\circ}$ in the alignment of these diagrams. (A) Control releases at Hancock, N.H.; (B) experimental releases at the anomaly at Iron Mine; (C) control releases at Worcester Airport. The mean vanishing bearing is significantly different from the home direction (p = 0.01). (D) Vanishing bearings of the same birds released at the anomaly on Route 20 near Worcester. The Watson U^2 test shows that the vanishing bearings at the two Worcester release sites are significantly different (p = 0.001), and the length of the mean vector is significantly shorter for releases at Route 20 (F test, p = 0.03). (E) Wenham Swamp; (F) Beverly Airport; (G) Lynnfield. In the releases from Wenham Swamp, Beverly Airport and Lynnfield, the birds were significantly more scattered in their vanishing bearings than when they were released at either Worcester Airport or Hancock (F test, p = 0.02). (H) Danvers. These bearings are not significantly different from those at control sites

Plotting the $\log_{10}$ of the length of the mean vector of the vanishing bearings against the $\log_{10}$ of the greatest difference in the total intensity of the magnetic field that occurs anywhere along this 1 km line gives the results shown in Figure 4. This relationship suggests that the more variable the magnetic gradient within 1 km toward home, the more scattered are the pigeons' vanishing directions. However, it is important to bear in mind that this graph is based on relatively few data points and each data point on only a small number of releases. Furthermore, the other four measures of magnetic variability in Table 1 also correlate, although less well, with the scatter of the pigeons'

Table 1. Variability of the Earth's magnetic field at various release points

	Field intensity at release	1 km radius	3 km radius	1 km toward home	3 km toward home	Length of mean vector
Iron Mine	5857	3572	3572	2957	2957	.15
Wenham Swamp	5000	1100	1600	260	600	.34
Beverly	3820	520	1412	140	220	.31
Danvers	4525	825	1177	75	845	.67
Lynnfield	5590	1790	2510	190	2190	.39
Worcester Airp.	3255	35	80	35	35	.64
Worcester Rt 20	4250	930	1426	750	850	.22

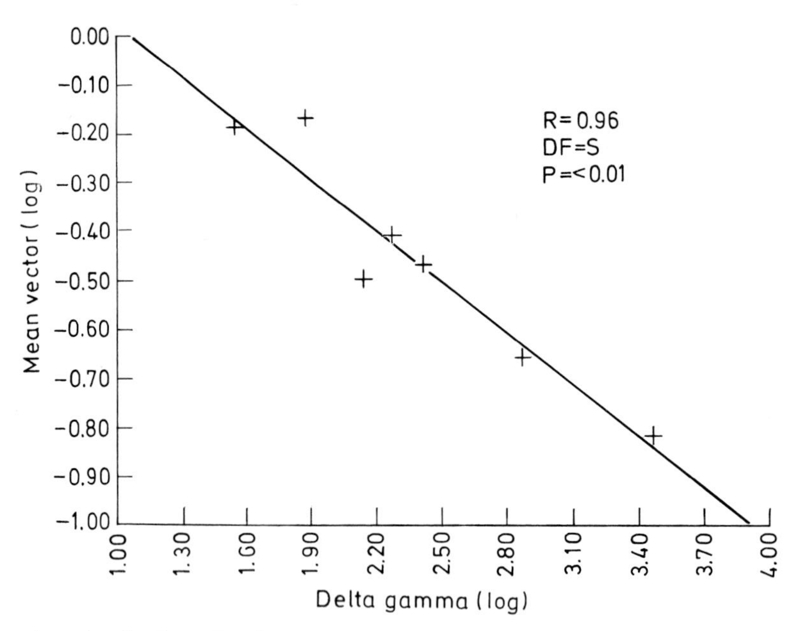

Fig. 4. The length of the mean vector of the vanishing diagram, MEAN VECTOR, is plotted against the maximum difference in the strength of the magnetic field within 1 km toward home from the release site, DELTA GAMMA. In this graph the $\log_{10}$ of each number is used. R is the correlation coefficient, DF the degrees of freedon and p is the probability (two tailed) that the observed distribution arose by chance. The solid line is the regression line plotted by the least squares method

vanishing bearings. There are simply not enough different release points and enough different magnetic field strengths to be sure of which feature of the magnetic field really correlates best with the pigeons' orientation.

The speed with which birds homed and the number of birds which returned home from releases at these magnetic anomalies was indistinguishable from that of birds released at control sites. This implies that pigeons flying incorrect courses must correct them toward home relatively soon after release. To see if this was the case and also to see whether there was any relationship between the birds' tracks and variations in total magnetic intensity as Talkington (1967) reports, we followed some pigeons by airplane. Figure 5 shows the tracks of 17 pigeons released at Iron Mine. Compared to the tracks of pigeons released at other sites (Michener and Walcott, 1967), these tracks were much more erratic at the outset; pigeons tended to twist and turn and to fly back and forth around and over the release site. When pigeons did leave the Iron Mine area, they tended to do so in either of three directions; to the east northeast, the northwest or in two instances, to the south. Once the pigeons had left the release site, they flew directly toward home without the circling and turning that was typical of the first part of their flight. Although we did not track them all the way to the loft, the pigeons' times of arrival at the loft after we left them in the airplane indicated that they must have taken a relatively straight path home. Comparing the pigeons' tracks with the magnetic intensity variations shown on the aeromagnetic maps shows no clear correlation between the two.

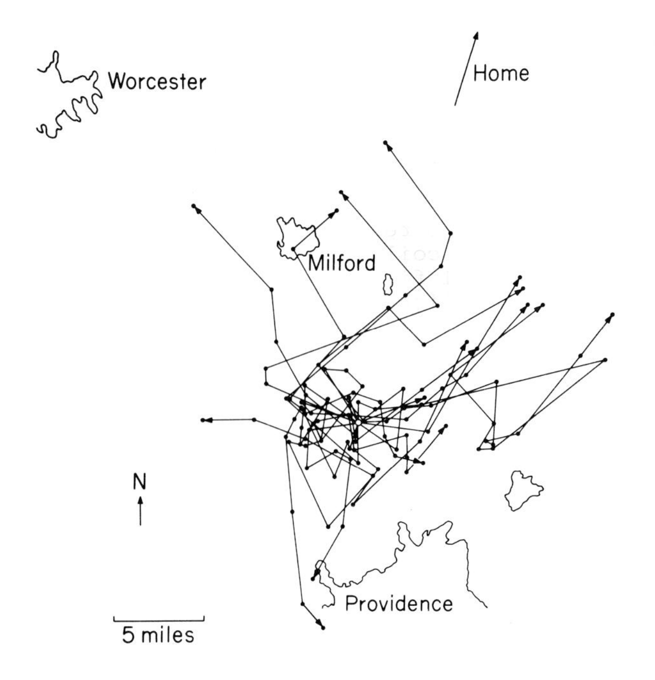

Fig. 5. The tracks of 17 pigeons released at the Iron Mine anomaly. The relase site is indicated by the small open circle. Each dot represents a pigeon's location determined by airplane tracking and the lines connect the points indicating the approximate path the pigeon took. The arrowhead was the pigeon's position when we stopped tracking it. The bar in the lower left corner marked *5 miles*, is 9 km long. It is interesting that no pigeon flew directly toward the home loft from this release site

Does a magnetic anomaly have an effect only on birds released at the anomaly itself or would it disturb birds simply flying over it on the way home? We have not yet done any experiments to answer this question directly, but we have some data from other releases that suggest an answer. Birds released at North Central Airport, 9 km south of Iron Mine, should fly over the Iron Mine release site on their way home. Years ago we released some pigeons at North Central Airport as a control for some other experiments and they oriented accurately toward home (n = 15, length of mean vector = 0.7, direction 19°, home direction 17°, p = 0.0003). Since the range of radio tracking from North Central Airport is about 18 km, we have probably tracked these pigeons released at North Central Airport flying directly over the Iron Mine area. Somewhat more direct evidence comes from following pigeons by airplane. I have examined the tracks of all the pigeons we have followed over the years that have come anywhere close to Iron Mine. These include birds involved in a variety of experiments as well as controls. Several of these tracks pass directly over Iron Mine with no alteration in either direction or speed. Thus, although there is very little data, it appears that once pigeons are flying a course, magnetic anomalies have less effect than they do on pigeons released at the anomaly itself.

D. Discussion

These results suggest that pigeons orient less accurately toward home when released at magnetic anomalies. There is also a hint that anomalies with a greater variation in intensity in the direction toward home result in pigeons being more disoriented than when released at places with a small intensity difference. However, it is not clear from the data whether it is the variation in total magnetic intensity, or whether it is some other aspect of the disturbed magnetic field such as a change in inclination of the field or in differences between true and magnetic north that are important. We are exploring some of these questions at the moment.

Another question is why magnets and Helmholtz coils had so much less effect on the pigeons' orientation under sun than did the anomalies. After all, the field of the magnets and coils was 10 - 100 times stronger than that of the anomalies. This result might be a consequence of the field of the magnet or Helmholtz coil being fixed in relation to the pigeon. This means that the total field would vary with the pigeon's orientation and direction of flight as the field of the coil or the magnet adds to or cancels out the Earth's field. However, a pigeon flying in a straight line might still be able to detect changes in the Earth's magnetic field even in the presence of a strong field from a magnet or Helmholtz coil, and thus be able to use the changes as a cue in its orientation. Perhaps this could explain why Helmholtz coils and magnets have a smaller effect on pigeon orientation under sun than do the anomalies in the Earth's magnetic field.

The other point of interest is that disorientation at magnetic anomalies occurs under sunny skies when the pigeons should presumably be able to use their sun compass. The experiments with Helmholtz coils under sun (Walcott, 1977) suggest that an applied field has a small effect on the pigeons' average vanishing direction; a field of 0.6 gauss often results in a deviation of about 20^{0} in the vanishing bearings. That magnetic fields have an effect when the sun is visible suggests that either there is some relationship between the sun compass and magnetic fields or, that the Earth's magnetic field plays some role in the pigeon's navigation or map. At the moment, I can see no easy way to distinguish unambiguously between these two alternatives. It is becoming increasingly clear that pigeons use many cues in their homeward orientation; perhaps we should not disregard the possibility that the Earth's magnetic field may play a role both as an auxilliary compass for use when the sun is obscured and also have some role in the navigation or map itself.

Acknowledgements. I thank Philip R. Pearson, Jr. for telling me about the anomaly at Iron Mine and R. Charif, M. Corrall, J. Crawford, M. Hyatt, and J. Taylor for tracking pigeons and helping with all aspects of the research. This research was supported in part by a PHS Research Grant No. NS08708-08 from the National Institute of Neurological and Communicative Disorders and Stroke.

References

Batschelet, E.: Recent statistical methods for orientation data. In: Animal Orientation and Navigation (ed. Galler et al.) NASA SP-262, U.S. Government Printing Office Washington, D.C. 1972, pp. 61-89

Batschelet, E.: Statistical methods for the analysis of problems in animal orientation and certain biological rhythms. Am. Inst. Biol. Sci., Wash., D.C. (1965)

Cochran, W.: 145-160 mHz beacon tag transmitter for small animals. BIAC Information Module M-15. Am. Inst. Biol. Sci., Washington, D.C. (1967)

Emlen, J.T., Penney, R.L.: Distance navigation in the Adelie Penguin. Ibis 106, 417-431 (1964)

Frei, U., Wagner, G.: Die Anfangsorientierung von Brieftauben im erdmagnetisch gestörten Gebiet des Mont Jorat. Rev. Suisse Zool. 83, 891-897 (1976)

Graue, L.C.: Initial orientation in pigeon homing related to magnetic contours. Am. Zool. 5, 704 (abstract) (1965)

Keeton, W.: Magnets interfere with pigeon homing. Proc. Nat. Acad. Aci. 68, 102-106 (1971)

Keeton, W.: Effects of magnets on pigeon homing. In: Animal Orientation and Navigation (eds. Galler et al.) NASA SP-262, U.S. Government Printing Office Washington, 1972, pp. 579-594

Keeton, W. et al.: Normal fluctuations in the earth's magnetic field influence pigeon orientation. J. Comp. Physiol. 95, 95-103 (1974)

Michener, M., Walcott, C.: Homing of single pigeons - an analysis of tracks. J. Exp. Biol. 47, 99-131 (1967)

Talkington, L.: Bird Navigation and Geomagnetism. American Zoologist 7, 199 (abstract; 1967)

Wagner, G.: Das Orientierungsverhalten von Brieftauben im erdmagnetisch gestörten Gebiete des Chasseral. Rev. Suisse Zool. 83, 883-890 (1976)

Walcott, C., Green, R.: Orientation of homing pigeons altered by a change in the direction of an applied magnetic field. Science 184, 180-182 (1974)

Walcott, C.: Magnetic fields and the orientation of homing pigeons under sun. J. Exp. Biol. 70, 105-124 (1977)

Effect of Outward Journey in an Altered Magnetic Field on the Orientation of Young Homing Pigeons

ROSWITHA WILTSCHKO[1], WOLFGANG WILTSCHKO[1], and WILLIAM T. KEETON[2],
[1] Fachbereich Biologie der Universität, 6000 Frankfurt, FRG. [2] Section of Neurobiology and Behavior, Langmuir Laboratory, Cornell University, Ithaca, NY 14853, USA

Abstract

To test whether magnetic directional information gained during the outward journey is incorporated in determining the home direction after displacement, young homing pigeons were transported to the release site in an artificially altered magnetic field. Controls were transported in a separate vehicle; a third group of birds was transported together with the controls, but after arriving at the release site, they were placed in the altered field for a period equal to the duration of the journey.

The control birds and the birds that had experienced the altered field while sitting at the release site normally were homeward oriented. The birds transported in the altered magnetic field generally showed a larger scatter, and in many cases a significant directional preference was not found, although this effect was much greater in some releases than in others. These findings indicate that directional information of the magnetic field gained en route does probably play some role in the navigational process.

A. Introduction

Despite the considerable attention given during the last decades to the problem of pigeon navigation, it is not yet clear when or where the birds obtain the information from which they derive the home direction. Is this information gathered during the outward journey to the release site, or while sitting in baskets at the release site, or only after free flight has begun at the release site, or some combination of these? Kramer's (1957) way of describing the navigation process as consisting of a "map" and a "compass" component already implied the use of site-specific information, and the two early hypotheses explaining the navigation mechanisms (Yeagley, 1947; Matthews, 1953) assumed that navigation was based on site-specific factors. After numerous attempts to establish a role of route-specific information had failed (e.g., Wallraff, 1965; Wallcott and Schmidt-Koenig, 1973; cf. Keeton, 1974), the use of site-specific factors in the sense of a "bi-coordinate navigation" became generally accepted (Wallraff, 1974). Recently, however, Papi and his colleques (1973) reported evidence which indicated that olfactory information collected during the displacement played some role in the navigation process, often supplemented by site information.

Earlier models of navigation that emphasized outward route information (e.g., Barlow, 1964) were based almost exclusively on inertial mechanisms. Thus they usually hypothesized detection by the birds of all the angular accelerations of the outward journey, followed by a double integration to determine the homeward direction. This type of route-reversal hypothesis implies that the homeward direction is determined without reference to external cues. However, a much simpler system of route reversal has been demonstrated in insects, especially in ants by Jander (1957) and Wehner (1972) and in honey bees by v. Frisch (1965) and his colleagues. These insects integrate the outward journey relative to an external stimulus, namely, the sun; they memorize their course relative to the sun (time-compensated for changes in azimuth) and reverse direction to return home. The olfactory outward-journey effect described by Papi et al. (1973) represents a route-reversal mechanism of a similar type in birds, because an external stimulus indicating the route of displacement is involved (comp. Papi, 1976).

The possibility that the compass systems normally used by birds may also serve as a reference for a route reversal mechanism, as was discussed for the magnetic compass by Wiltschko and Wiltschko (1976), has not been investigated thus far. As long as the magnetic field is not too badly distorted by the metallic structure of the transportation vehicle, it can provide directional information available to the birds during displacement, which together with measuring the accelerations may be integrated to the direction of movement relative to the magnetic field. So it seemed worthwhile to us to investigate whether magnetic information gained during the outward journey might play a role in pigeon navigation.

Aside from these theoretical condiderations we had been stimulated to conduct these experiments by a chance observation in 1974, the "VW-effect," whereby very young pigeons transported in the back of a Volkswagen Variant ("square back") right above the engine, where, among other factors, the magnetic field was badly distorted, often vanished randomly and had extremely poor homing success, even from a site as near as 8 km from home.

We report here some preliminary results of an investigation on the effect of altering the magninetic field during the outward journey. Instead of distorting the magnetic information we tried to replace it by meaningful false information that would also be in contradiction to the local site-specific "map" factors at the release site.

B. Methods

The experiements reported here were conducted in the years 1974 to 1977; in 15 releases, pigeons from the Cornell pigeon loft near Ithaca, New York, USA and in 5 releases pigeons from a pigeon loft of the University in Frankfurt, Germany, were used.

To keep the applied magnetic field relatively constant during the outward journey, it had to remain in a constant relationship to the natural geomagnetic field. Hence we had to find release sites which could be reached by driving straight in a single direction. In New York, transport was along highway N.Y. 34, which goes almost straight north and passes near our three release sites: Indian Fields Road, 33 km from the home loft; Auburn drumlin, 66 km; and Weedsport drumlin, 73 km. The artificial field was turned on during the parts of the journey when the road proceeded northward or nearly northward, but was turned off (i.e., the birds were transported in the natural field) on two stretches of less than 3 km where the road deviated from northward. In Germany, transport was southward along Autobahn A5 and A67 to a release site near Ludwigshafen, 65 km from the Frankfurt loft (comp. Fig. 1).

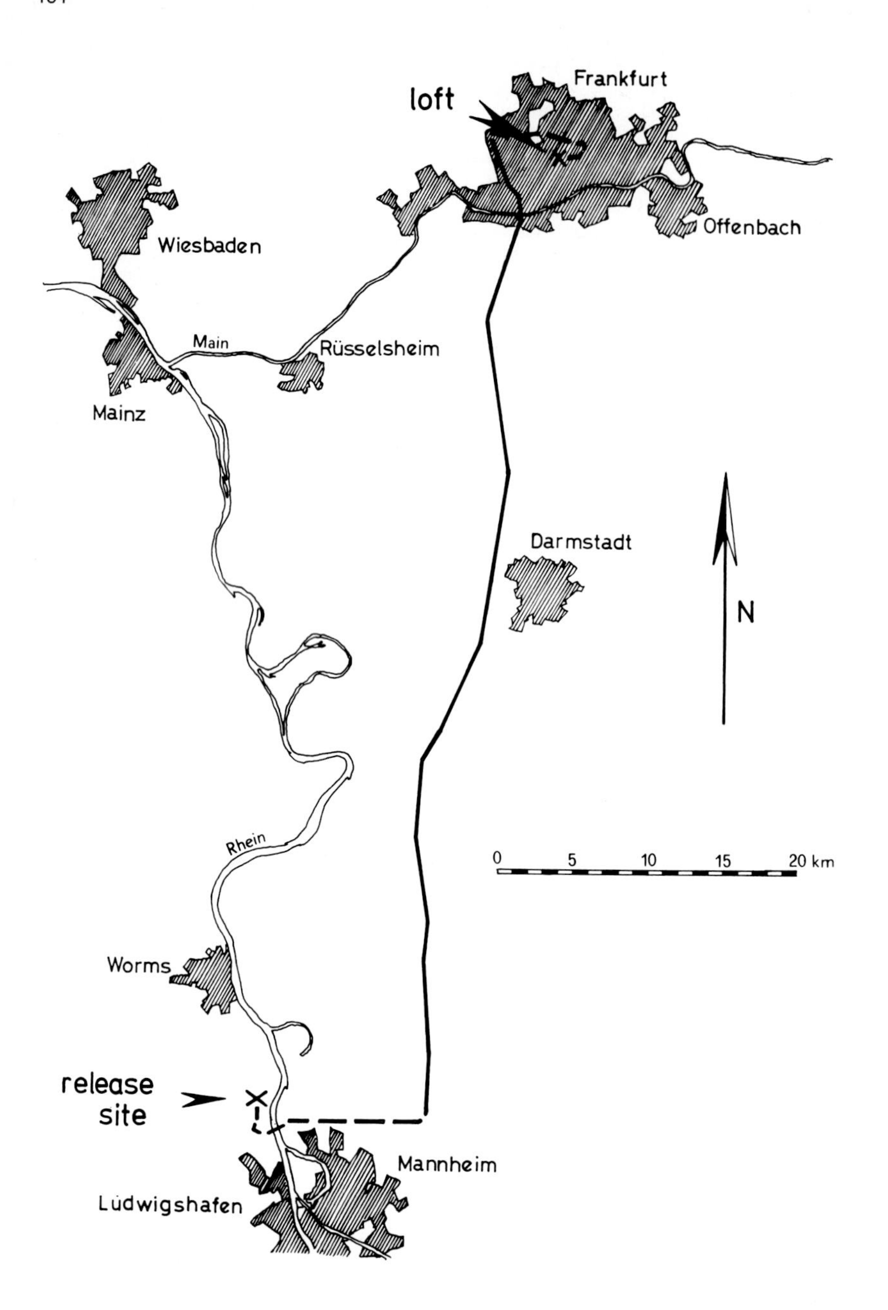

Fig. 1. Map showing the route from the Frankfurt loft to the release site near Ludwigshafen. The portion of the outward journey where the experimental birds were transported in the altered magnetic field is given as a solid line

The experimental pigeons (E birds) were transported in a closed vehicle in which a pair of Helmholtz coils was mounted around the wooden basket in which the birds sat. The coils were energized by an automobile battery in such a manner that the resultant of the field they produced and that of the natural geomagnetic field were of approximately the same total intensity and inclination angle as the natural geomagnetic field but pointed southward when the vehicle was heading northward in New York and southward in Germany. Control birds (C birds) were transported along the same route in a second vehicle, without artificial magnetic field. A third group of pigeons (S birds) was transported with the controls and, after arrival at the release site, was placed in the altered magnetic field for a period equal to the time the E birds had been in the altered field during the outward journey. The intent was to enable us to determine whether any observed effects were caused by the birds' being transported in the altered magnetic field or merely by their sitting in the altered field. This third group was not included in all releases.

In the test releases, each bird was tossed singly from the hand, the direction of the tosses being randomized. Birds of the three groups were released alternately, so that all treatments were exposed equally to any temporally changing conditions such as wind or weather. Each bird was watched by two observers using 10 × 50 or 10 × 40 binoculars until it vanished from sight, and the departure bearing was recorded to the nearest 5°.

The mean vector for each set of bearings was calculated by vector analysis. The Rayleigh test was used to test for significant directional preferences, the V test to compare a sample with the home direction and the Mardia-Watson-Wheeler test to compare the distribution of two samples (Batschelet, 1972).

C. Results

The data of the 20 individual releases are presented in Table 1.

Table 1

Date	Treatment	Number released and (bearings)	Mean bearing (α)	Mean vector length (r)	p (Rayleigh)	Median vanishing interval (min)	Returns same day and (later)
Weedsport, home direction: 173°, distance: 73.5 km							
Aug. 7, '74	C	12 (9)	133°	0.68	$p < 0.05$	5:56	3 (2)
	E	11 (10)	139°	0.21	$p > 0.05$	6:30	2 (4)
Sept.24, '74	C	8 (6)	225°	0.59	$p > 0.05$	17:45	--
	S	5 (3)	120°	0.36	$p > 0.05$	6:57	--
	E	11 (6)	74°	0.17	$p > 0.05$	4:34	--
Auburn, home direction: 170°, distance: 66.2 km							
Aug. 1, '75	C	11 (7)	158°	0.34	$p > 0.05$	5:37	--
	S	23 (20)	174°	0.31	$p > 0.05$	5:15	--
	E	12 (12)	141°	0.46	$p > 0.05$	4:02	--
July 28, '77	C	15 (12)	172°	0.53	$p < 0.05$	6:56	- (1)
	S	15 (12)	164°	0.50	$p < 0.05$	7:45	- (1)
	E	15 (11)	124°	0.39	$p > 0.05$	6:15	- (1)

Table 1 (continued)

Date	Treatment	Number released and (bearings)	Mean bearing (α)	Mean vector length (r)	p (Rayleigh)	Median vanishing interval (min)	Returns same day and (later)
Indian Fields Road, home direction: 163°, distance: 32.6 km							
July 5, '75	C	10 (8)	179°	0.73	$p < 0.01$	9:25	2 (2)
	S	13 (9)	318°	0.47	$p > 0.05$	11:01	- (3)
	E	11 (8)	294°	0.57	$p < 0.05$	7:30	1 (2)
July 17, '75	C	12 (9)	122°	0.74	$p < 0.01$	5:15	2 (6)
	S	15 (12)	140°	0.31	$p > 0.05$	5:26	- (7)
	E	17 (14)	299°	0.23	$p > 0.05$	5:29	1 (7)
July 18, '75	C	10 (8)	136°	0.44	$p > 0.05$	4:55	4 (3)
	S	19 (19)	185°	0.26	$p > 0.05$	4:45	5 (7)
	E	11 (11)	124°	0.45	$p > 0.05$	4:35	2 (5)
June 24, '76	C	11 (9)	110°	0.70	$p < 0.01$	6:33	2 (4)
	E	11 (9)	124°	0.60	$p < 0.05$	4:88	2 (2)
June 18, '76	C	11 (10)	103°	0.83	$p < 0.001$	4:32	5 (4)
	S	23 (20)	110°	0.81	$p < 0.001$	4:00	5 (3)
	E	11 (10)	123°	0.78	$p < 0.001$	3:21	5 (4)
Aug. 2, '76	C	11 (9)	176°	0.74	$p < 0.01$	4:54	7 (1)
	E	10 (8)	209°	0.32	$p > 0.05$	5:52	8 -
Aug. 3, '76	C	10 (10)	153°	0.29	$p > 0.05$	6:46	5 (1)
	E	12 (9)	177°	0.74	$p < 0.01$	4:54	5 (3)
Aug. 4, '76	C	10 (9)	273°	0.71	$p < 0.01$	7:21	- (4)
	E	10 (8)	185°	0.33	$p > 0.05$	8:16	1 (2)
Aug. 5, '76	C	9 (8)	169°	0.36	$p > 0.05$	6:07	6 -
	E	9 (8)	109°	0.17	$p > 0.05$	7:33	3 (2)
Oct. 30, '76	C	11 (8)	223°	0.43	$p > 0.05$	5:40	2 (2)
	S	9 (7)	168°	0.60	$p < 0.05$	4:45	--
	E	7 (7)	185°	0.76	$p < 0.05$	5:40	3 -
Nov. 2, '76	C	9 (8)	166°	0.49	$p > 0.05$	5:16	4 (1)
	S	7 (6)	207°	0.55	$p > 0.05$	6:31	4 -
	E	10 (7)	224°	0.94	$p < 0.001$	5:08	5 -
Ludwigshafen, home direction: 17°, distance: 65.2 km							
July 28, '76	C	12 (11)	319°	0.68	$p < 0.01$	5:22	- (2)
	S	11 (10)	357°	0.45	$p > 0.05$	6:06	--
	E	11 (10)	297°	0.31	$p > 0.05$	5:12	- (1)
Oct. 25, '76	C	15 (12)	27°	0.92	$p < 0.001$	5:00	--
	S	15 (13)	16°	0.84	$p < 0.001$	4:55	--
	E	15 (12)	181°	0.32	$p > 0.05$	5:42	--
Oct. 27, '76	C	15 (14)	15°	0.54	$p < 0.05$	5:25	--
	S	15 (11)	34°	0.62	$p < 0.05$	6:16	--
	E	15 (9)	179°	0.34	$p > 0.05$	3:55	--
May 3, '77	C	13 (10)	303°	0.12	$p > 0.05$	2:59	1 (4)
	S	7 (6)	34°	0.33	$p > 0.05$	3:22	1 (2)
	E	13 (13)	333°	0.39	$p > 0.05$	2:30	1 (3)
Aug. 23, '77	C	14 (12)	341°	0.70	$p < 0.001$	4:48	- (2)
	S	16 (11)	316°	0.85	$p < 0.001$	4:47	--
	E	14 (12)	281°	0.16	$p > 0.05$	4:22	- (1)

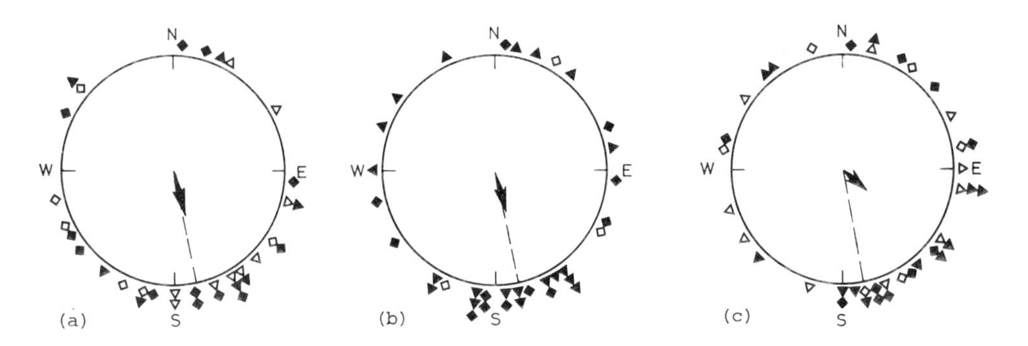

Fig. 2 a – c. Pooled vanishing bearings from the four test releases in Weedsport (home direction 173°, 73.5 km distance) and Auburn (170°, 66.2 km). The home direction is marked as a *dashed line*. The mean vectors are shown as *arrows* whose length is drawn proportional to the radius of the circle = 1. *Open symbols*, data from Weedsport; *solid symbols*, data from Auburn. *Diamonds*, tests with first-flight pigeons. *Triangles*, tests with birds that had been given a few training releases from distances of less than 15 km. (a) Controls: n = 34, α = 165°, r = 0.40 ($p < 0.01$, Rayleigh Test); (b) pigeons that experienced the altered magnetic field while sitting at the release site (S birds): n = 35, α = 166, r = 0.37 ($p < 0.01$); (c) Pigeons that were transported in the altered magnetic field (E birds): n = 39, α = 130°, r = 0.29 ($p < 0.05$)

Figure 2 shows the pooled vanishing bearings from 2 releases at Weedsport and 2 at Auburn, all with relatively inexperienced young birds. Both the controls and the S birds that experienced the altered magnetic field while sitting at the release site showed significant orientation in a homeward direction ($p < 0.001$, V test). The vanishing bearings of the E birds that had been transported in the altered magnetic field were more scattered; their vector length was shorter (although just significant at the 5% level) and their mean bearing was farther away from the home direction than in the other two groups. However, the distribution of their bearings is not significantly different from that of the controls or the S birds under the Mardia-Watson-Wheeler test.

Figure 3 gives the pooled vanishing bearings for 11 releases at Indian Fields Road, all with first-flight youngsters, i.e., pigeons that had

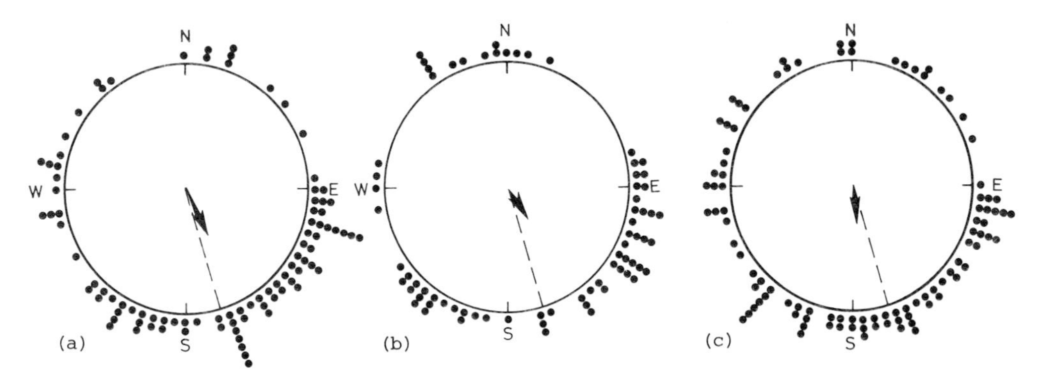

Fig. 3 a – c. Pooled vanishing bearings of the 11 releases from Indian Fields Road (163°, 32.6 km). (a) Controls: n = 93, α = 154°, r = 0.40 ($p < 0.01$); (b) S birds: n = 72, α = 142°, r = 0.32 ($p < 0.01$); (c) E birds: n = 99, α = 172°, r = 0.29 ($p < 0.01$)

never previously been taken away from home. All three treatments were badly scattered, but all three (having very large sample sizes) were significantly nonrandom ($p < 0.01$). Both the E birds and the S birds had shorter mean vectors than the controls, though this difference is not significant. We note, however, that the marginal results obtained when all 11 releases are pooled tend to obscure the fact that the results varied greatly from year to year, and that sometimes there was a dramatic difference between the control and the experimental birds. For example, the 3 releases of 1975 showed good orientation by the control birds but complete randomness by the E birds and by the S birds (Fig. 4). The 1976 releases showed no consistent difference between the treatments (comp. Table 1).

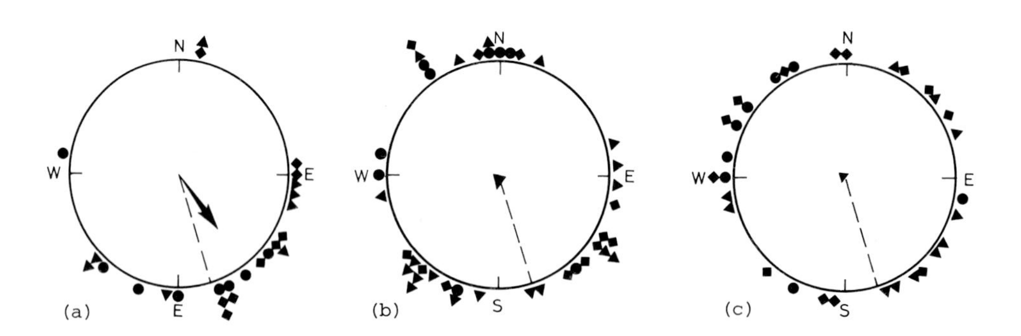

Fig. 4 a - c. Pooled vanishing bearings from the three releases from Indian Fields Road conducted in the year 1975. The different symbols indicate the bearings of the individual releases. (a) Controls: n = 25, $\alpha = 142^{0}$, r = 0.60 ($p < 0.001$); (b) S birds: n = 40, $\alpha = 192^{0}$, r = 0.12 ($p > 0.05$); (c) E birds: n = 32, $\alpha = 313^{0}$, r = 0.07 ($p > 0.05$)

Five test releases with first-flight birds were conducted from the release site at Ludwigshafen in Germany; the pooled vanishing bearings are given in Figure 5. Both the controls and the S birds were significantly homeward oriented ($p < 0.001$, V test), whereas the E birds do not show any significant directional preference. The distribution of their vanishing bearings differs significantly from that of the C

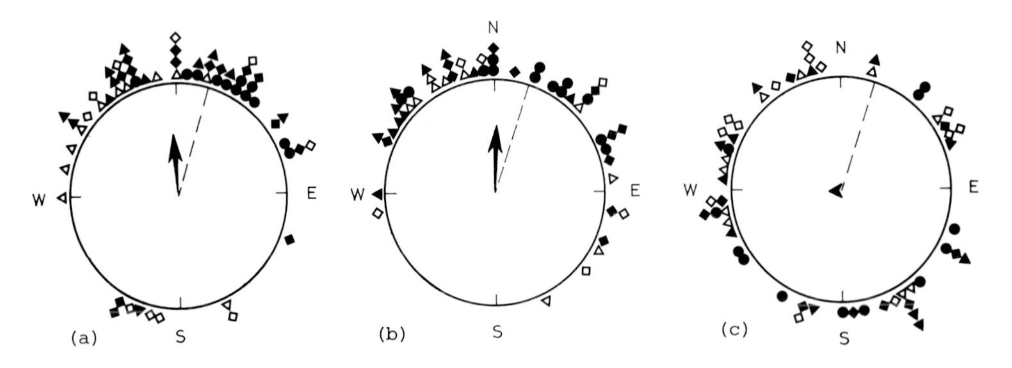

Fig. 5 a - c. Pooled vanishing bearings of the five test releases from Ludwigshafen (17^{0}, 65.2 km); the different symbols indicate the bearings of the individual releases. (a) Controls: n = 59, $\alpha = 357^{0}$, r = 0.53 ($p < 0.001$); (b) S birds: n = 51, $\alpha = 2^{0}$, r = 0.57 ($p < 0.001$); (c) E birds: n = 56, $\alpha = 265^{0}$, r = 0.12 ($p > 0.05$)

birds and from that of the S birds under the Mardia-Watson-Wheeler test ($p < 0.001$).

Figure 6 gives a second-order summary of the data from all 20 experimental releases; the means for each of the three treatments in the individual releases were plotted with 360° set equal to the home direction. Both the distributions of the controls and of the S birds that experienced the altered magnetic field sitting at the site are significantly oriented in a homeward direction ($p < 0.001$, V test), whereas the means of the E birds that were transported in the altered field scatter much wider and are statistically random.

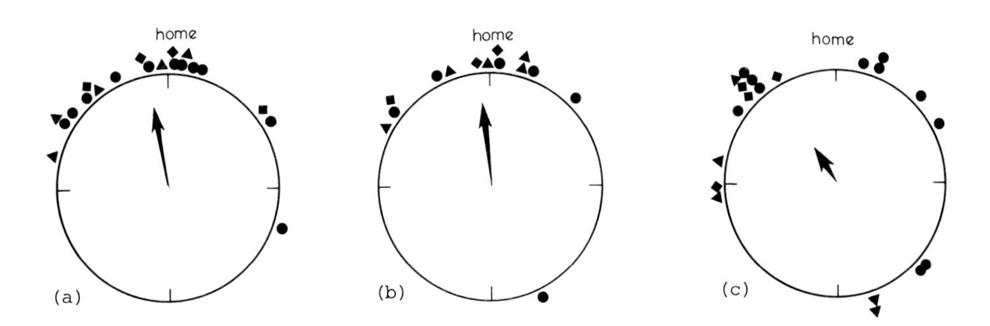

Fig. 6 a - c. The means of all 20 releases plotted with 360° set equal to the home direction. *Diamonds:* means from Weedsport and Auburn, New York; *circles:* means from Indian Fields Road, New York; *triangles:* means from Ludwigshafen, Germany. (a) Controls: $n = 20$, $\alpha = 349°$, $r = 0.75$ ($p < 0.001$); (b) S birds: $n = 14$, $\alpha = 354°$, $r = 0.73$ ($p < 0.001$); (c) E birds: $n = 20$, $\alpha = 332°$, $r = 0.38$ ($p > 0.05$)

D. Discussion

In all three experimental series reported here, the vanishing bearings of the birds transported in the altered magnetic field were more scattered than those of the control birds, although the differences were not always statistically significant. However, when the data are viewed as a whole, especially reinforced by the second-order analysis, a clear pattern of difference seems to emerge, indicating that incorrect magnetic information during the outward journey does have an effect on the initial orientation of pigeons.

In our experimental design, we had tried to replace the normal magnetic information during the outward journey by meaningful false information. We do not know to what extent we succeeded in doing so and what the magnetic field the pigeons experienced during transportation really meant to them. The mechanisms of magnetic field detection in birds are still not clear so that we cannot estimate how birds experience a magnetic field, part of which is transported with them. The data of Wallcott and Green (1974) might be regarded as an indication that such a field may provide meaningful information to birds, but the fact that the route was never ideally straight so that the resultant field varied in total intensity, inclination, and declination because the natural and the artificial magnetic field added differently as a function of the traveling direction, may have added to make our experimental field a distorted field rather than a source of false information.

We are perplexed by the fact that the effect of being transported in this altered magnetic field was much greater in some releases than in others. Thus in 2 releases of the 1975 Indian Fields Road series (comp. Table 1) the difference between the treatments was dramatic. Similarly, in 2 of the Ludwigshafen releases the mean of the E birds pointed almost straight away from home, in contrast to the homeward-directed means of the controls and the S birds in the same releases. Yet in other releases at these same sites the differences between treatments were negligible. Likewise, this variation in the effect of the experimental treatment does not seem to be correlated in any obvious way with meteorologic conditions.

It appears to us that the effect is usually greater when we use very young first-flight pigeons (preferably less than 12 weeks old), but this criterion is certainly not entirely reliable. Another variable we have considered is whether the birds have "ranged" prior to testing. By "ranging", we mean behavior sometimes seen in Cornell pigeons when, during exercise flights around the loft, the birds spontaneously fly out of sight of the loft and remain away for half an hour or more, sometimes even two hours or more. The effect of an altered magnetic field during the outward journey was never very large in any of our experiments in which birds that had ranged were used, but sometimes other first-flight youngsters that had definitely not ranged also showed little effect of the treatment. In short, though the age and flying experience of the birds might influence their repsonse to transport in an altered field, that influence is not entirely consistent.

Though the vanishing bearings of the birds transported in the altered magnetic field were usually more scattered than those of the controls, their means usually lay on the homeward semicircle, despite contradictory or missing magnetic information en route. Thus it appears that magnetic information gained during the outward journey is normally outweighed by other information. Perhaps correct magnetic information during the outward journey has a stabilizing effect on the birds' directional selections, as the better orientation of the control birds indicates.

In summary, magnetic information during the outward journey to a release site does probably play some role in navigation, but the extent of its influence and the factors governing its importance remain to be determined.

Acknowledgments. This research was supported by National Science Foundation Grants GB-35199X and BMS 75-18905 A02 to W.T. Keeton and by grants from the Deutsche Forschungsgemeinschaft to W. Wiltschko. Part of the computer work was carried out by the Hochschulrechenzentrum der Universität Frankfurt a.M. We especially thank Irene Brown, K. Canteres, Elsa Franke, A. Gobert, H. Golle, J. Herold, T. Larkin, and V. Reichard for their help in conducting these experiments and K. Kleine-Borgmann and the pigeon breeders of the RV Groß-Frankfurt for donating young pigeons for the Ludwigshafen releases.

References

Barlow, J.S.: Inertial navigation as a basis for animal navigation. J. Theor. Biol. 6, 76-117 (1964)

Batschelet, E.: Recent statistical methods for orientation data. In: Animal Orientation and Navication. S.R. Galler et al. (eds.): NASA SP-262, U.S. Gov. Print. Off., Washington, D.C., 61-93, 1972

Frisch, K. v.: Die Tanzsprache und Orientierung der Bienen. Berlin-Heidelberg-New York: Springer, 1965

Jander, R.: Die optische Richtungsorientierung der Roten Waldameise (*Formica rufa* L.) Z. Vergl. Physiol. 40, 162-238 (1957)

Keeton, W.T.: The orientational and navigational basis of homing in birds. Adv. Study Behav. 5, 47-132 (1974)

Kramer, G.: Experiments on bird orientation and their interpretation. Ibis 99, 196-227 (1957)

Matthews, G.V.T.: Sun navigation in homing pigeons. J. Exp. Biol. 30, 243-267 (1953)

Papi, F.: The olfactory navigation system of the homing pigeon. Verh. Deut. Zool. Ges. 1976, 184-205 (1976)

Papi, F., Fiaschi, V., Benvenuti, S., Baldaccini, N.W.: Pigeon homing: outward journey detours influence the initial orientation. Monit. Zool. Ital. 7, 129-133 (1973)

Walcott, C., Green, R.P.: Orientation of homing pigeons altered by a change in the direction of an applied magnetic field. Science 184, 180-182 (1974)

Walcott, C., Schmidt-Koenig, K.: The effect on pigeon homing of anesthesia during displacement. Auk 90, 281-286 (1973)

Wallraff, H.G.: Über das Heimfindevermögen von Brieftauben mit durchtrennten Bogengängen. Z. Vergl. Physiol. 50, 313-330 (1965)

Wallraff, H.G.: Das Navigationssystem der Vögel. Ein theoretischer Beitrag zur Analyse ungeklärter Orientierungsleistungen. Schriftenreihe Kybernetik, R. Oldenbourg Verlag, München, Wien (1974)

Wehner, R.: Visual orientation performances of desert ants (*Cataglyphis bicolor*) toward astromenotactic directions and horizon landmarks. In: Animal Orientation and Navigation. Galler, S.R. et al. (eds.): NASA SP-262, U.S. Gov. Print. Off., Washington D.C., 421-436 1972

Wiltschko, W., Wiltschko, R.: Die Bedeutung des Magnetkompasses für die Orientierung der Vögel. J. Ornithol 117, 362-387 (1976)

Yeagley, H.L.: A preliminary study of a physical basis of bird navigation. J. Appl. Phys. 18, 1035-1063 (1947)

Homing Pigeons' Flight Over and Under Low Stratus*

GERHART WAGNER, 3045 Meikirch, Switzerland

Abstract

The Homing pigeons were released and followed by helicopter from mountains emerging like islands from a layer of low stratus clouds in Switzerland, with the loft being under the stratus. Different kinds of behavior were observed: (1) A heading toward a visible mountain-island, flying over the stratus in approximate home direction; (2) flying very low with the ground in sight under the fog; (3) plunging into haze of increasing density, but always with the ground in sight and later coming under the low stratus; (4) perching after a long series of searching flights around the "island." It can be noted that the pigeons never plunged into the closed stratus.

A. Introduction

The initial orientation of homing pigeons seems to be determined by both optical and nonoptical factors. Optical factors can be of celestial (sun) or of terrestrial nature (topography). In our experimental series of the last ten years, we have tried to specify the role of topographical structures. The method of tracking pigeons by helicopter allows us to study these questions not only at the releasing point, but also over longer distances, sometimes over the entire way home (Wagner, 1968, 1970, 1972, 1974).

A special topographical situation is present when the territory in the loft range is covered by fog or by low stratus and the pigeons are released at a point situated on a hill or mountain that emerges from the fog or stratus and is under full sunshine. Low stratus typically occurs in Switzerland between the Alps and the Jura mountains, mainly in autumn and winter, and may remain stable for days or weeks (cf. Figs. 1 and 2).

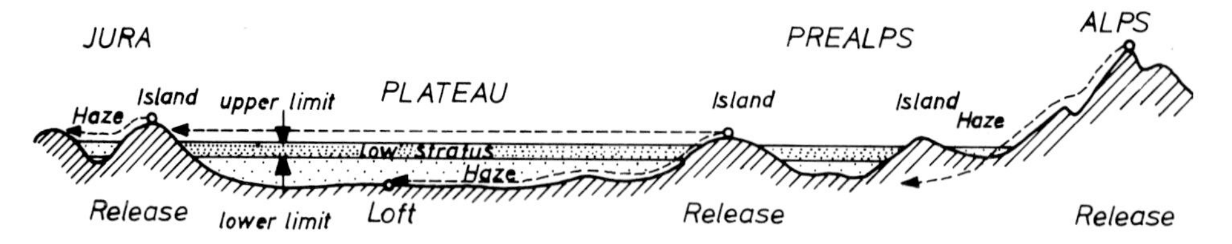

Fig. 1. Low stratus situation between Alps and Jura. Model of a longitudinal section

*Supported by the Swiss National Science Foundation.

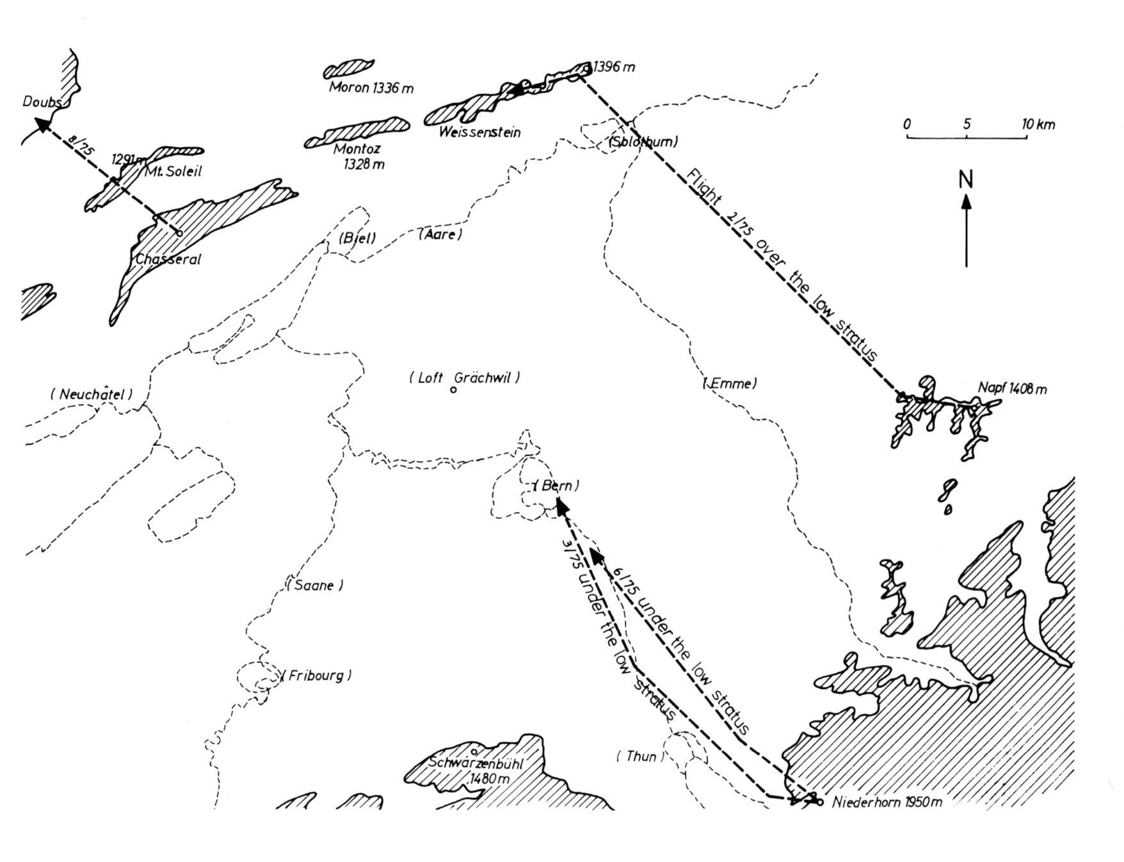

Fig. 2. Low stratus situation between Alps and Jura with the upper limit of the low stratus at 1200 m above sea level. The emerging parts of Prealps and Jura are shown by *hatching*. The cities, rivers, and lakes under the low stratus are *dashed* (names in parenthesis). Flight routes Nos. 2, 3, and 6/75 are indicated

The meteorologic condition is that of a thermic inversion. Depending on the upper level of the stratus, either the Alps alone, or the Alps together with the Jura mountains, or also the highest hill peaks of the plateau emerge from the stratus. As a result of low humidity, the air is very clear over the fog. Pigeons released on a hilltop have a perfect view of the sky and the horizon, but cannot see the surroundings of the loft. The consequence of this experimental situation is a concurrence between normal celestial and abnormal terrestrial optical factors of orientation. The question is whether or not the pigeons will be able to find the loft below the stratus. Will they penetrate the stratus, and if so, where and how will they do it?

B. Materials and Methods

During the fall of 1975 several helicopter flights, using the type "Alouette III" of the Swiss Army, were conducted under the meteorologic conditions just described, with pigeons from my loft at Grächwil, 10 km NW from Bern. The experimental birds had been well trained from all compass directions. As in previous experiments, groups of five birds were released, individually recognizable by different plumage colors. Thus the question of whether there are leading and led individuals could be studied again (cf. Wagner, 1975).

The technique of starting and following the pigeons is described in Wagner, 1970 and 1974. During the experimental days the low stratus over the Swiss plateau had an upper limit at 1000 - 1200 m and a lower limit at 700 - 800 m above sea level. This means that in the loft region (680 m) there was no fog but full overcast with very poor visibility (maximum visibility 2 km). We call that situation in Switzerland "high fog" (*Hochnebel*).

Under such conditions the helicopter pilot faces special problems. Not being equipped for blind flight, he cannot reach the release site by penetrating the stratus, but has to find a gap in the fog. Such gaps were rare over the Plateau, but more frequent near the Alps or in the alpine volleys, where the fog gradually changed into haze.

C. Results

I. The Experimental Flights

Table 1 contains a survey of all experimental flights.

Table 1. Experimental flights: a survey

Flight number	Date	Releasing place (height above sea level)	Distance	Home azimuth	Flight characteristics
2/75	23.10.75	Napf 1408 m	43 km	272°	Flight over the stratus until Jura
3/75	23.10.75	Niederhorn 1950 m	46 km	320°	Flight under the stratus with view to the ground
5/75	12.11.75	Rigi Kulm 1800 m	84 km	268°	Flight round the "island", then plunging under the stratus with view to the ground
6/75	12.11.75	Niederhorn 1950 m	46 km	320°	Flight under the stratus with view to the ground
7/75	9.12.75	Schwarzenbühl 1480 m	30 km	360°	Flight around the "island", finally perching on a tree
8/75	9.12.75	Chasseral 1607 m	28 km	118°	Departure in direction opposite to home away from the stratus

II. Description of the Individual Flights

1. Flight No. 2/75: October 23, 1975, 1147 - 1230, Napf (1408 m) - Grächwil (680 m), 43 km, Az. 272° (Fig. 2).

Pigeons: Five males, one (red) painted artificially.

Weather: Closed low stratus between Alps and Jura with upper limit at about 1200 m. The massif of the Napf (Emmental) was the only island within the fog bank, the length of the woody island being 7 km, the maximum width about 6 km. Sky entirely clear, Temperature under the stratus 7°C, on top of the Napf 15°C (inversion!), calm (Fig. 3).

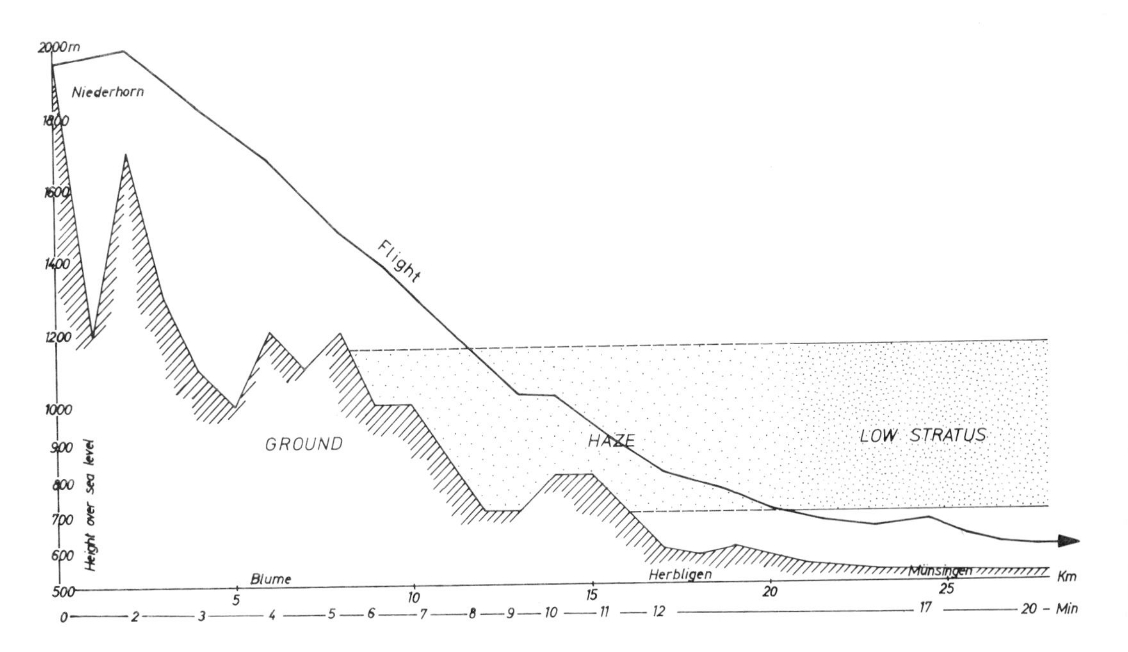

Fig. 3. Flight No. 6/75 from the Niederhorn, 1950 m: model of a longitudinal section of the flight's first part

Flight: The pigeons, after circling twice, flew westward along the ridge of the Napf island, which coincides approximately with the home direction. Ten minutes later they hesitated, drew a loop over the last peak, and finally headed north-westward to the Jura chain.

The most visible part of the Jura was the region of the Weissenstein, whereas the more distant regions in the SW and in the NE hardly emerged from the low stratus. The pigeons, once over the stratus, headed directly toward the Weissenstein, flying as a single flock at a height of approximately 1300 m above sea level. They never changed their course nor did they try to plunge into the stratus. Thus they passed their home loft at a distance of 25 km to their left. The distance of 37 km over the low stratus was flown within 29 min: at 1227, the pigeons reached the Weissenstein. Arriving there, they immediately changed their course and followed the axis of the Jura chain. We followed them for only three more kilometers and then broke off our flight.

At 1345, Black and Red were present in the loft. Which route had they chosen after we had left them? The low stratus did not disappear during the whole day. We suppose that they succeeded in dipping under the stratus with some sight to the ground at the edge of the Jura, enabling them to home under the stratus. Experiments No. 3, 5, and 6/75 (Sect. 2, 3, 4) will show that this hypothesis is not improbable. The other three pigeons of the group did not return the same day.

2. Flight No. 3/75: October 23, 1975, 1527 - 1606, Niederhorn (1950 m) - Grächwil (680 m), 46 km, Az. 320° (Figs. 2 and 3).

Pigeons: Five females, one (Red) painted artificially.

Weather: See flight No. 2/75 (morning of the same day): The low stratus over the Swiss Plateau persisted the whole day, but did not extend into the main alpine valleys: The region of the lake of Thun was hazy, but not covered by low stratus.

Flight: Immediately after the departure from the top of the Niederhorn, the pigeons headed in the home direction, descending in a tremendous downward flight at a speed of 100 km/h and more from a height of 1950 m above sea level, thereby coming from the bright atmosphere at the top into more and more hazy air, but always with the ground in sight. In the region of Thun (600 m), they succeeded in dipping under the low stratus and continued their homeward flight with continuous sight of the ground. The pigeons flew very low (within about 100 m) over the ground, the lower level of the stratus being at about 150 m above ground. Near the town of Berne, we broke off the flight, since the pigeons were going to cross the city very close to the rooftops. At 1606, four of them arrived at the loft, the fifth followed at 1612.

In summary, the pigeons had managed the distance of 46 km within 39 min under very poor optical conditions, flying at a visibility of 3 - 4 km and the sun completely overcast.

3. Flight No. 5/75: November 12, 1975, 1043 - 1104, Rigi Kulm (1800 m) - Grächwil (680 m), 84 km, Az. 268°. (The Rigi massif lies 40 km east of the Napf near Luzern.)

Pigeons: A flock of five, (males and females), three of them artificially painted.

Weather: Homogeneous low stratus over the whole Plateau with upper limit at 1150 m. The massif of the Rigi emerged as a large island from the fog. The Alps were quite clear under a cloudless sky, whereas the Jura was scarcely visible.

Flight: After the departure, the pigeons flew downward from Rigi Kulm (1800 m) to the village of Rigi Staffel (1600 m) in an approximate home direction. After arriving at the "shore" of the low stratus, they returned to Rigi Kulm and again to Rigi Staffel. Then for 20 min, they flew many loops and circles along the "shore" of the "island" with an obvious desire to leave the island in the home direction, but with an apparent hesitation to fly over the stratus offering no visible landmark. Finally, they plunged under the fog, flying barely above the tops of the forest trees, and thus disappeared from our view. This time, the helicopter could not follow them for safety reasons.

Only two pigeons of that flock returned the same day.

4. Flight No. 6/75: November 12, 1975, 1537 - 1555, Niederhorn (1950 m) - Grächwil (680 m), 46 km, Az. 320° (Figs. 2 and 3).

Pigeons: The same flock of five females as in the flight No. 3/75 (see sect. 2.).

Weather: See flight No. 5/75 (morning of the same day): upper limit of the low stratus at 1150 m, but over the lake of Thun, the fog turned to haze as in flight No. 3/75. Consequently, experiment No. 6/75 was a repetition of experiment No. 3/75 with the same animals and under the same meteorologic conditions 20 days later.

Flight: There was a high degree of correlation with flight No. 3/75: Again the pigeons flew immediately downward into the haze and arrived below the stratus. They followed nearly the same route as in the preceding experiment. After 18 min, we broke off the helicopter flight. By 1654, all the pigeons were home (exact time of arrival unknown). Thus, they had repeated their remerkable performance of October 23. Figure 3 shows a longitudinal section of the flight.

5. *Flight No. 7/75:* December 9, 1975, 1002 - 1041, Schwarzenbühl (1480 m) - Grächwil (680 m), 30 km, Az. 360° (Fig. 4).

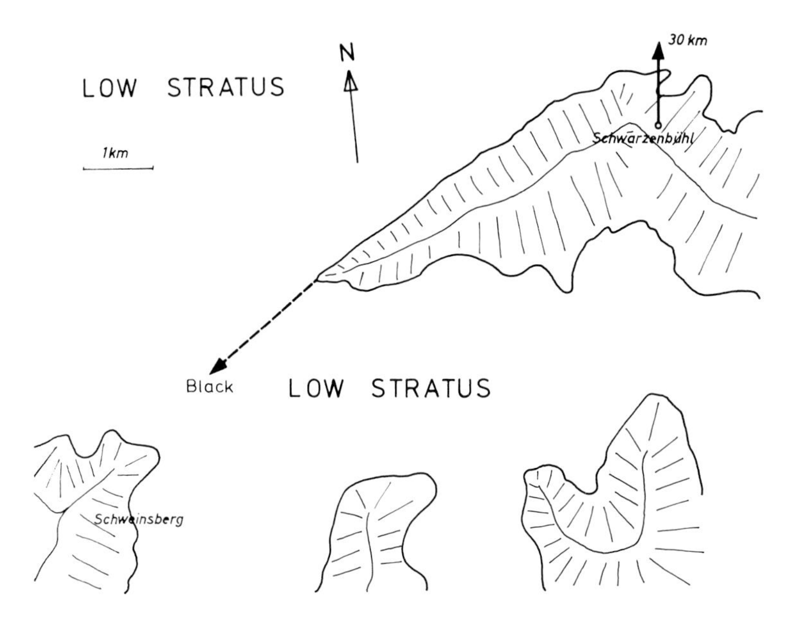

Fig. 4. Flight No. 7/75 from Schwarzenbühl. See explanations in the text

Pigeons: New group of five males.

Weather: Homogenous low stratus between the Alps and the Jura with an upper limit at 1350 - 1440 m. From the releasing point in the Prealps only the highest part of the Jura (Chasseral 1600 m) was visible as a narrow white strip at a distance of 50 km. The temperature at the releasing point was + 6°, at the loft 0° (inversion!). The region of the Schwarzenbühl emerged as a peninsula out of the fog.

Flight: The pigeons drew many circles and loops between the releasing point and the western end of the peninsula. Several times, they intended to leave the peninsula in the home direction and flew some 100 meters over the fog, but returned every time to the visible land. After 14 min, Black departed alone in a SW direction and headed toward the Schweinsberg, a neighbouring "island" at a distance of 4 km. The remaining flock of four pigeons continued to fly circles and loops around the peninsula of Schwarzenbühl and finally, 39 min after release, perched on the top of a tree. Not a single pigeon returned the same day.

6. Flight No. 8/75: December 9, 1975, 1220 - 1235. Chasseral (1607 m) - Grächwil (680 m), 28 km, Az. 118° (Fig. 2).

Pigeons: The flock of five females used in experiments No. 3/75 and 6/75.

Weather: See flight No. 7/75 (morning of the same day). Upper limit of the low stratus at 1450 m. While the stratus was very homogeneous on the Plateau side of the Chasseral (i.e., on the home side), it became broken and turned more and more into haze on the Jura side. As a result, the villages in the valley of St. Imier behind the Chasseral chain were partly visible through the gaps in the fog. Behind the second chain of the Jura (Mt. Soleil), there was no more fog.

Flight: The pigeons started immediately in the direction toward the Mt. Soleil, opposite to the home direction. Flying between fog swathes, they crossed the valley of St. Imier and the Mt. Soleil, where they left the fog region entirely and continued their course with good view of the ground, directly toward the Swiss frontier. Before crossing the frontier, we tried to influence their flight direction: The pilot flew directly in front of them in order to push them back to Switzerland. But it was hopeless: The pigeons made way for the helicopter and continued their flight with a course toward France, Az. 300°. At least, this test confirmed that it is by no means the helicopter that determines the pigeons' flight direction.

III. Leadership Within the Homing Flocks

In most of the flights, the question has been examined whether in the experimental group of five (or four) individually recognizable pigeons, the flight order in terms of leadership varies at random, or whether there are leading and led animals. For this purpose, flight orders have been noted every minute as a rule. The rank sums were statistically examined by the Friedman test as described in Wagner (1975). It could be shown in all flights that flight orders are not random. The results of these investigations will be published separately.

D. Discussion

I. Behavior of Pigeons Released Over High Stratus

All the described experiments show that a homogeneous layer of low stratus as seen from above (in Switzerland called *Nebelmeer* - "fog sea") has a highly hampering or even repelling effect on pigeons. The birds behave exactly as if they were released at the shore of a lake or sea when the home direction goes straight over the surface of the water (Kramer, 1956; Wagner, 1971). Kramer writes: "Die Insel (Helgoland) schien demnach die Tauben gefangenzuhalten. ... Die Tauben klebten bis zu einem gewissen Grad an der Insel. ... Nach einiger Zeit siegte der Heimkehrimpuls über die Abneigung vor dem Flug über offenes Meer" (loc. cit., p. 117). [The island (Helgoland) accordingly appeared to hold the pigeons prisoners. ... Up to a certain point, the pigeons clung to the island. ... After a while the homing impulse won out over the reluctance to fly over the open sea.] It is reasonable to assume that the sight of the fog from above influences the pigeons in the same way as the sight of a large water surface. Never did a pigeon try to plunge into the fog. Only once (Napf, Flight No. 2/75) has the low stratus between the Alps and the Jura been crossed with the pigeons flying 50 - 100 m over the fog, a height similar to that used by the pigeons to cross a lake. The flight direction was obviously determined by the only visible landmark in approximate home direction (the Jura) beyond the fogbank, and not by the loft. Once (Rigi Kulm, Flight No. 5/75) the pigeons, after a long series of searching flights and no landmark being vesible in approximate home direction, followed the mountain slope by flying downward very low over the ground, which permitted them to remain below the stratus.

Twice (Niederhorn, 2 and 4) the pigeons, immediately after the release, flew downward into haze of increasing density, but always with a view of the ground, and homed directly. Once (Schwarzenbuhl, Flight No. 7/75) the pigeons perched after 39 min of searching flights. Only one of them headed toward an "island," with an incorrect direction of 90°. Once (Chasseral, Flight No. 8/75) the pigeons escaped from the fog bank by flying in the opposite direction.

In summary, the experiments show that optical terrestrial factors can play an important role in the orientational system of homing pigeons in a positive or in a negative sense.

Gehring (1963) and Steidinger (1968) found that birds can track during low stratus situations in the Swiss Plateau, over as well as under the stratus, but no or very little within the stratus. However, whereas the track over the stratus is well oriented, it is less-well oriented under the stratus.

For the homing behavior of pigeons the contrary is true: Whereas pigeons flying under the stratus home easily and directly, they cannot find the loft site when flying over the stratus. This difference is probably a consequence of the fact that directional orientation as it occurs in tracking birds is determined by factors other than the homing orientation of pigeons.

Acknowledgments. The Federal Defense Ministry and especially the chief of military communications, Divisionär Guisolan, permitted us to use a helicopter of the Swiss Army. Godi Amsler piloted the aircraft with perfect skill. Hans Senn, Niklaus Loeffel, and my son Andreas noted the flight orders of the pigeon flocks during the experimental flights. My other son Thomas cared for the pigeons in my loft. My wife helped me to prepare the manuscript, Zuzana Hladky drew the figures. Eric Schallenberg and Christopher Dickson checked the English translation.

References

Gehring, W.: Radar- und Feldbeobachtungen über den Verlauf des Vogelzuges im Schweizerischen Mittelland: Der Tagzug im Herbst (1957-1961). Ornithol. Beob. 60, 35-68 (1963)

Griffin, D.R.: Nocturnal Bird Migration in Opaque Clouds. Wallops Station Symposium 1970, NASA SP-262: 169-189 (1972)

Kramer, G.: Über Flüge von Brieftauben über See. In: Steiniger: Natur und Jagd in Niedersachsen, Hannover 1956, 113-118, 1956

Steidinger, P.: Radarbeobachtungen über die Richtung und deren Streuung beim nächtlichen Vogelzug im Schweizerischen Mittelland. Ornithol. Beob. 65, 197-226 (1968)

Wagner, G.: Topographisch bedingte zweigipflige und schiefe Kreisverteilungen bei der Anfangsorientierung verfrachteter Brieftauben. Rev. Suisse Zool. 75, 682-690 (1968)

Wagner, G.: Verfolgung von Brieftauben im Helikopter. Rev. Suisse Zool. 77, 39-60 (1970)

Wagner, G.: Topography and Pigeon Orientation. Wallops Station Symposium 1970, NASA SP-262, 259-273 (1972)

Wagner, G.: Verfolgung von Brieftauben im Helikopter II. Rev. Suisse Zool. 80, 727-750 (1974)

Wagner, G.: Zur Frage des Flugführens in heimkehrenden Brieftaubengruppen. Z. Tierpsychol. 39, 61-74 (1975)

Preferred Compass Directions in Initial Orientation of Homing Pigeons

HANS G. WALLRAFF, Max-Planck-Institut für Verhaltensphysiologie, 8131 Seewiesen, FRG

Abstract

The initial bearings of pigeons are commonly not only homeward oriented, but also polarized toward a preferred compass direction (PCD). The PCD is loft-specific; at ten out of eleven loft sites investigated it is in the western semicircle. It can become shifted in the course of general homing experience. The PCD appeared stable against several experimental treatments (long-term clock shifts, olfactory deprivation, exposure to different winds at the home site), but could be reduced or reversed, respectively, by certain screenings of the aviary in which the birds were raised. It is concluded that "release-site biases" are not necessarily caused by distortions of the physical environment.

A. Introduction

Homing pigeons released at distant sites symmetrically distributed around their loft, depart preferably in a direction that points toward home. However, they do so only in the total mean, i.e., when the vanishing bearings of all the release sites are taken together. At the single sites the pigeons often deviate considerably from the true homeward direction, and these deviations are commonly quite similar in repeated releases at the same place. The site-specific orientation of bearings may in part be due to local influences, but there are obviously more general reasons, too. At all home sites that were investigated more thoroughly, it turned out that the initial bearings are polarized toward a certain compass direction. This direction may be called the "preferred compass direction" (PCD).

In a theoretical study (Wallraff, 1974) the possible significance of PCDs within the functional system of bird navigation has been emphasized. In the present paper I will consider only the empirical aspect. Some additional material on the subject will be presented, and a survey on the PCDs in various areas will be given. It will further be asked whether experimental manipulations of different kinds are able to influence the expression of PCDs, and finally some implications will briefly be stressed.

B. New Data with Inexperienced Pigeons

The existence of PCDs in inexperienced pigeons at four different loft sites in northern Germany has already been demonstrated (Wallraff, 1970a). Additional material from three sites in southern Germany has been collected in the meantime:

(1) Seewiesen. - With first-flight pigeons from Seewiesen (30 km southwest of Munich) two series of releases at distances of about 30 km have been conducted. - Series A: During each of three days in different years (1972 - 1974) four pigeons were released at each of eight regularly distributed places, synchronously at each two opposing sites. By this procedure temporal variations should have been equalized as much as possible. - Series B: At each of the four places in the cardinal directions 30 additional pigeons were released in 1968 - 1971. The releases were not synchronized with each other, and ten birds per release were used. - In addition to these short-distance releases about 100 pigeons were displaced to symmetrically distributed sites on the east-west axis, distances varying between 98 and 216 km. - The vanishing bearings of all these experiments are shown in Figure 1.

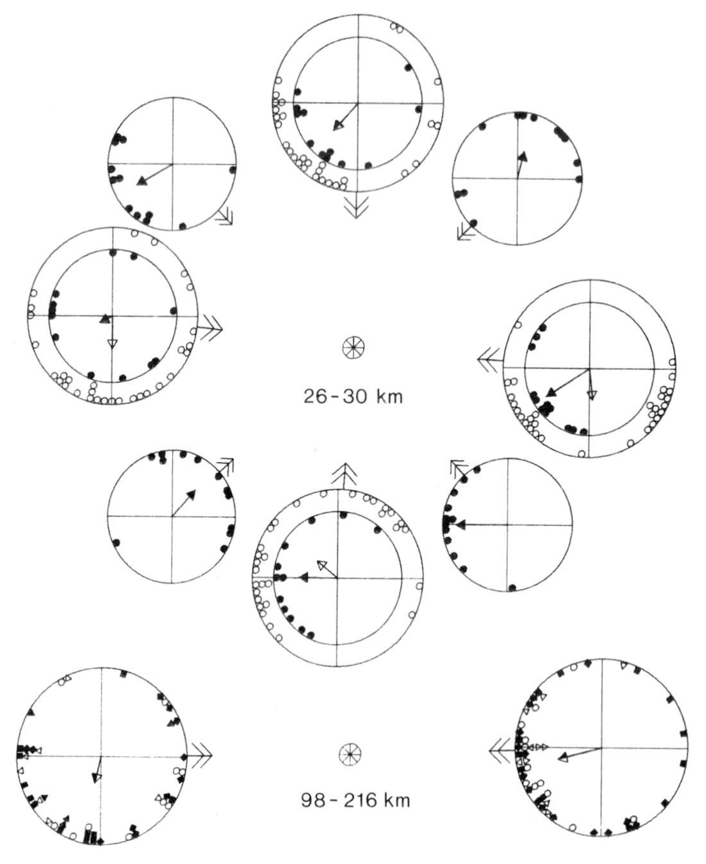

Fig. 1. Vanishing bearings of first-flight pigeons settled at Seewiesen. Eight upper diagrams are arranged according to real direction and distance (distance of eastern site is 27.5 km). *Black dots*, series A; *white*, series B. Radius of the *smaller* circle equals vector length one. Two lower diagrams combine bearings at distances 98 - 112 km (*white symbols*) and at 191 - 216 km (*black*). Different symbols refer to different release days and/or sites. Radii of *these* circles equal unity for the mean vectors

(2) Zorneding. - In 1968 first-flight pigeons of a big loft of a pigeon fancier at Zorneding (45 km ENE of Seewiesen) were used. Two releases were conducted at each of four places about 30 km from home, and one release at two places 70 km east and west (Fig. 2).

(3) Würzburg. - In 1975 a new loft near Würzburg was established (220 km NNW of Seewiesen, 305 km SSE of the formerly used loft site Osnabrück). First-flight pigeons were released from the cardinal directions at distances of about 30, 60, 120, and 180 km. Since these experiments are still in progress, the preliminary results will not be shown in detail.

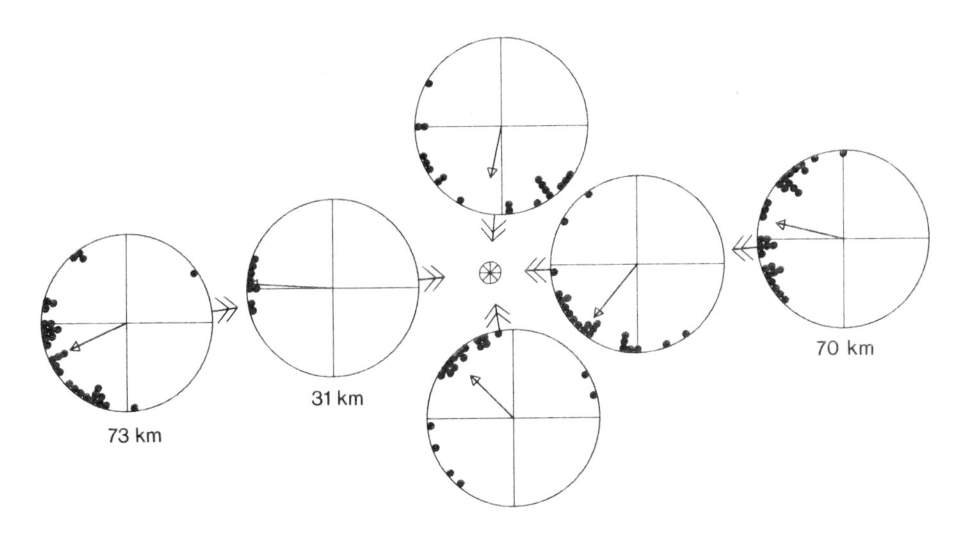

Fig. 2. Vanishing bearings of first-flight pigeons settled at Zorneding. Diagrams are arranged according to real direction and distance. Radii of circles equal vector length one

The Würzburg results are, however, included in Figure 3, which presents the data of all three loft sites in an adequately condensed form. Besides some homeward tendency at least at two lofts (A and C, left), an obvious PCD becomes apparent at all three lofts (right diagrams). It points toward southwesterly directions at Seewiesen and Zorneding, and toward NW at Würzburg.

C. PCDs at Different Loft Sites

If PCDs provide a key for an understanding of the navigational system, it is important to know their orientation in different parts of the earth, large-scale as well as small-scale. A first summary of data published thus far is given in Table 1.

At ten out of eleven loft sites investigated sufficiently (8 in Germany, 3 in USA), the PCD is placed in a quadrant between NNW and WSW (331^{0} and 242^{0}). Even within this quadrant, however, there are significant differences between different lofts. Wilhelmshaven, with preference of ESE in inexperienced birds, remains an exception.

Preferences of more or less westerly directions are also indicated in first-flight pigeons of the following loft sites: Osnabrück II (Wallraff, 1970a, Fig. 5), Tübingen (loc. cit., Fig. 2), and Pisa, Italy (Papi et al., 1971). Birds with little experience preferred WSW at Sacramento, California (Graue and Pratt, 1959). There are, however, other indications, too: Inexperienced pigeons of Bowling Green, Ohio, strongly preferred NNE (Wallraff and Graue, 1973, Fig. 5), and pigeons settled at Freiburg, SW Germany, inexperienced as well as experienced, deviated to the east from their homeward direction (Pratt and Wallraff, 1958; Wallraff, 1970a, Fig. 2). Thus, although preference of westerly directions predominates, it is not a rule without exceptions.

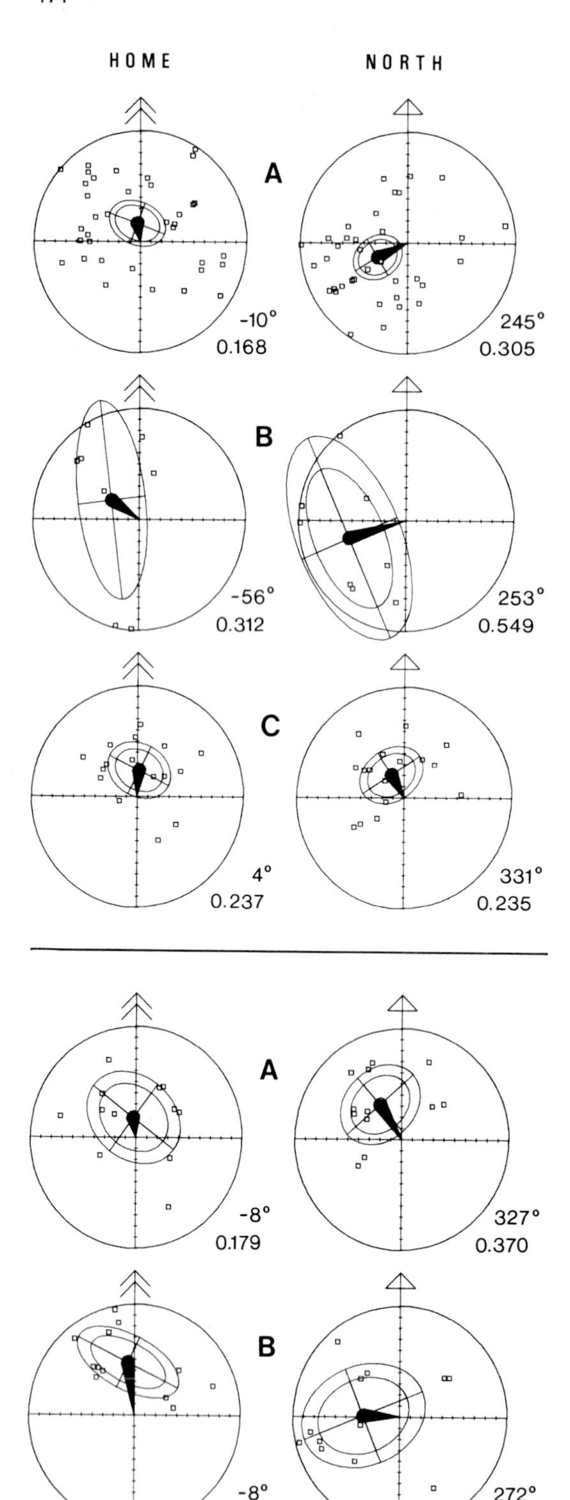

Fig. 3 A - C. Means of vanishing bearings of first-flight pigeons settled at Seewiesen (A), Zorneding (B), and Würzburg (C). *Dots* represent end points of mean vectors of symmetric single releases at 26 - 31 km (A and B) or release sites at distances ca. 30 - 180 km (C), respectively. The *dots* are to be thought of as being connected with the center. At the *left*, the results are combined with homeward direction pointing upward, at the *right* with north direction upward. Club-shaped *vectors* represent means of means, and their numerical values (*left:* $\bar{\bar{\alpha}}_H$, $\bar{\bar{a}}_H$; *right:* $\bar{\bar{\alpha}}_N$, $\bar{\bar{a}}_N$) are indicated. Also included are the confidence ellipses of these vectors at the 0.05 and 0.01 levels of significance (or 95% and 99%) according to Hotelling's T^2 test (Batschelet, 1978). If only one ellipse is drawn, it is for 95%. Radii of circles equal vector length one

Fig. 4 A and B. Means of vanishing bearings of 12 simultaneous releases of first-flight (A) and experienced (B) pigeons settled at Würzburg. Each dot represents the mean vector of 9 (A) and 5 - 7 (B) bearings. Release sites were in cardinal directions at distances about 30, 60, and 120 km. For further explanation see Figure 3

Table 1. PCDs in vanishing bearings of pigeons settled at different home sites. All the data included are collected from releases symmetrically distributed in at least four directions around the loft. If sample sizes differed, means of mean vectors of single releases or release sites were calculated and n reduced to the smallest sample multiplied by the number of samples. All the mean vectors shown represent distributions with random probability $p < 0.001$ (Rayleigh test). Homeward component $\bar{\bar{c}}_H = \cos \bar{\bar{\alpha}}_H \cdot \bar{\bar{a}}_H$ (cf. Fig. 3)

	Loft Site	Release Sites		n (bear.)	$\bar{\bar{c}}_H$	PCD		Reference
		number	distances (km, ca.)			$\bar{\bar{\alpha}}_N$	$\bar{\bar{a}}_N$	
inexperienced birds	Hohenkirchen	4	160	92	0.180	328°	0.548	Wallraff 1970a
	Nordenham	4	160	116	0.254	329°	0.310	Wallraff 1970a
	Wilhelmshaven	4	160	156	0.171	104°	0.398	Wallraff 1970a
	Osnabrück I	4	90	120	0.364	300°	0.365	Wallraff 1970a
	Würzburg	16	30 - 180	144	0.236	331°	0.235	Fig. 3 C
	Zorneding	4	30	32	0.174	253°	0.549	Fig. 3 B
	Seewiesen	8	30	144	0.165	245°	0.305	Fig. 3 A
experienced birds	Frankfurt	32	5 - 300	1270	0.41	242°	0.21	Schmidt-Koenig 1970
	Würzburg	12	30 - 120	60	0.472	272°	0.339	Fig. 4 B
	Ithaca, N.Y.	28	15 - 190	168	0.44[a]	326°	0.536[a]	Windsor 1975
	Bowling Green, Ohio	32	8 - 320	1280	0.25	288°	0.17	Graue 1970
	Durham, N.C.	32	5 - 300	1274	0.42	304°	0.13	Schmidt-Koenig 1966, 1970

[a]Radio-tracked. (All other bearings were recorded by visual observation).

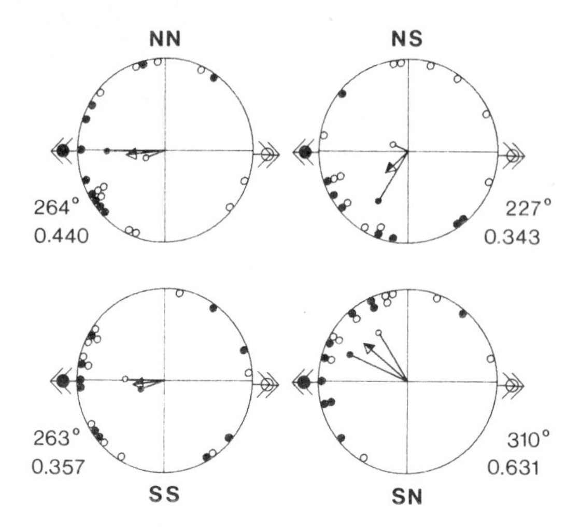

Fig. 5. Initial bearings after long-term clock shifts in young pigeons at Seewiesen. Releases from 112 km east (Sept. 13 and 16; *black dots*) and 98 km west (Sept. 12 and 17; *white dots*). Numbers refer to the combined mean vector $\bar{\bar{\alpha}}_N$, $\bar{\bar{a}}_N$ (*arrow*). For abbreviations, see text. The difference between NS and SN is significant with $p < 0.05$

D. Influences and Noninfluences of Experimental Treatments

I. General Experience

Figure 4 shows the results of simultaneous releases of first-flight pigeons and experienced birds whose training pattern (12 to 17 preceding flights) was symmetric around the loft at Würzburg. The shift of the PCD from NW to W is significant with $p < 0.01$ (calculated from original data). Even larger differences are indicated at Durham, N.C. (Schmidt-Koenig, 1966), and at Bowling Green, Ohio (Graue, 1970; Wallraff and Graue, 1973). On this background it seems doubtful whether the respective shift from ESE to NNE at Wilhelmshaven (Wallraff, 1959a, b) is due only to asymmetric training patterns.

In general, experience changes the relation between PCD and homeward orientation in favor of the latter (cf. Table 1).

II. Long-term Clock Shifts

It has been hypothesized (Wallraff, 1974) that the PCD is in some way correlated with the direction of a physical gradient serving as a navigational coordinate. The bird should determine this direction during its long stay at home and adjust it by help of its sun compass. If this holds true, wrong adjustment should occur when the circadian clock is phase-shifted during exposure time.

From May to September of 1974 a group of young pigeons was kept, from fledging age onward, in a LD (light/dark) schedule advanced by 4 h. During most of the overlapping daytime the loft was open and birds could fly around. By appropriate feeding they were accustomed to return to the loft before the entrance was shut and the lights were switched off. A control group was kept under natural LD conditions.

A week before release the birds were divided into four groups and kept in two rooms with artificial LD (daylength as outside). The groups are named by two letters N (natural LD phase) and/or S (LD shifted 4 h ahead), the first of which designates the 3 - 4 months exposing time and the second the 6 - 7 days preparing time. Room N (normal LD) contained groups NN and SN, room S (shifted LD) groups NS and SS. The birds were released on successive days at sites about 100 km east and west.

The results (Fig. 5) are summarized for both places, since the birds headed westward irrespectively of their relation to home. If the exposure time from May to September would have had no effect, the bearings of the SN birds should coincide with those of the NN group (controls), and SS as well as NS should deviate by roughly 60° to the left. However, the SS birds behaved "normally", and only the NS birds deviated to the left. The SN pigeons deviated to the right as if their clock had been shifted backward, although it was in phase with the natural day. Thus, the findings are in agreement with the assumption that the PCD is determined in relation to sun-compass directions *as they are available* for the bird while it gets acquainted with its home site. (That this also holds true for homeward orientation has been shown in the meantime by Wiltschko and Keeton, 1976, who discuss the related problems in more detail. Since the results coincide, no more material is required despite the insufficient statistical evidence of my data.)

III. Olfactory Deprivation

As olfaction has recently been shown to be an important aspect of pigeon homing (see Papi, 1976), it may be asked whether the expression of PCDs and/or their development depend on an intact olfactory system.

Three types of inexperienced pigeons were produced: "early operated" (EO), "late operated" (LO) birds, and controls. In the EO pigeons both olfactory nerves were bisected 3 - 4 months and in the LO birds 8 - 15 days (exceptionally 55 - 57 days) prior to their release. (The nerve stumps were tilted backward leaving a gap of 3 - 5 mm between them.) Controls were either sham-operated or remained untreated. A few days before release the olfactory sensitivity was proved by cardiac response tests. Results were only positive in the controls, showing that even in the EO pigeons olfactory functions had not recovered (cf. Oley et al., 1975; Bedini et al., 1976).

Three series of releases were conducted: (1) 112 km east and 98 km west of Seewiesen (July, 1974); (2) 112 km east and 199 km west of Seewiesen (November, 1974); (3) about 180 km N, E, S, and W of Würzburg (September, 1975). All birds of series (1) and a part of series (2) were several months old and were allowed to fly around the loft before and after the operations. All birds of series (3) and a part of series (2) had been confined at Seewiesen in closed rooms or at least between high walls up to an age of 2 - 3 months when the early operations were conducted. Afterward birds of all three types received free flight. In this case EO birds had no opportunity to experience odors and winds simultaneously (cf. Papi, 1976), and the EO birds of series (3) did not experience the odors of their home site (Würzburg) at all.

Results are summarized in Figure 6. Some reduction of homeward orientation is indicated, though not proved. However, PCDs appear unaffected by the experimental treatments. The findings suggest that not only the choice of a certain PCD at the release site, but also the development of a PCD at the home site is independent of olfactory stimuli.

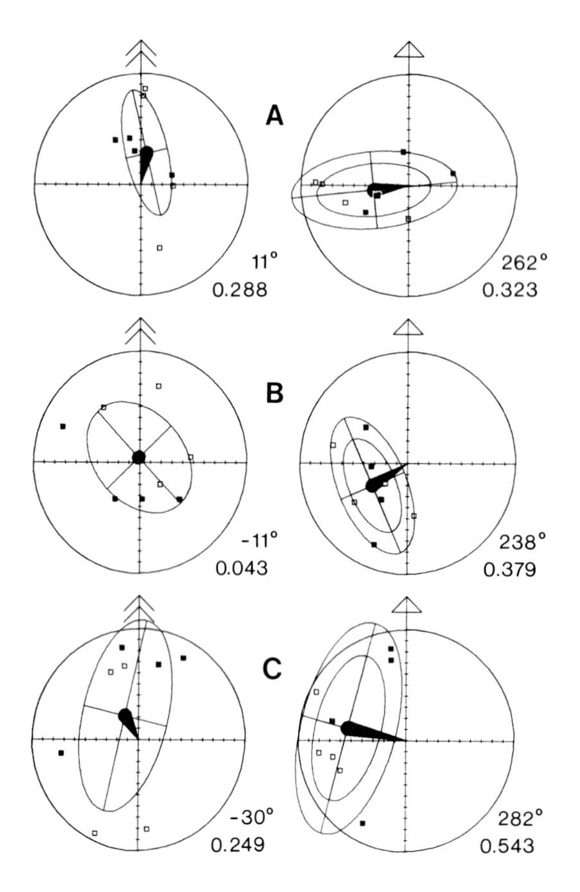

Fig. 6 A - C. Means of vanishing bearings of 8 simultaneous releases of controls (A) and of pigeons with bisected olfactory nerves (B, early, EO; C, late, LO). Each *dot* represents the mean vector of 2 - 14 bearings. *White dots*, Seewiesen; *black*, Würzburg. For further explanation see Figure 3

IV. Screening of Air Flows at the Home Site

It has been shown that homeward orientation of pigeons that were raised in aviaries surrounded by solid walls is impaired as compared with birds that were confined between walls pervious to air flows (Kramer, 1959; Wallraff, 1966, 1970b, 1978). Figure 7 summarizes means of vanishing bearings of simultaneous releases of pigeons from "mechanically open" aviaries (partially consisting of only wire mesh, but in most cases optically opaque) and pigeons that lived, up to their release, between walls made of glass. Besides the fact that homeward orientation is reduced in those birds (Fig. 7 B, left), there is also a reduction in the length of the mean vector with respect to the compass directions (right). This reduction is mainly a consequence of a significantly increased scatter at each of the four release sites (for more details see Wallraff, 1978).

It should be mentioned that the PCD of aviary pigeons of the Seewiesen area, as calculated from the data of the four release sites used, is SSW (Fig. 7 A) and thus deviates from that shown in Figure 3 A. It is not yet clear whether this difference depends on the aviary conditions or on the different sites of release.

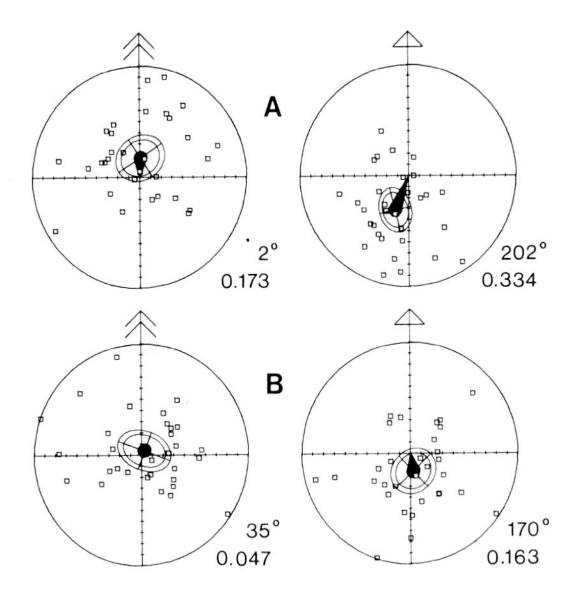

Fig. 7 A and B. Means of vanishing bearings of 31 simultaneous releases of pigeons confined in aviaries with walls pervious to air flows (A) and with impervious walls made of glass (B). About equal numbers of releases were made at sites ca. 100 km N, E, W, and 26 km S of the aviary near Seewiesen. Each *dot* represents the mean vector of 2 - 11 bearings. For further explanation see Figure 3

V. Exposure to Different Weather Conditions at the Home Site

The development of PCDs might be considered to be connected with prevailing weather conditions. For instance, the general westward tendency (Table 1) could be seen in context with a predominance of winds coming from westerly directions.

Young pigeons at fledging age were confined in a building connected with a small aviary surrounded by a stone wall exceeding the wire-meshed roof by 70 cm. From time to time half of the pigeons were placed individually in small wire cages (35 × 35 cm, 21 cm high) arranged in four-storied shelfs on the roof of the institute at Seewiesen. They remained there, in each case, for either two days (1972) or one day (1973). Days for exposure were chosen according to expected weather conditions: "W" birds were exposed with winds coming predominantly from westerly directions, "E" birds were exposed with predominantly easterly (or northeasterly) winds. Four groups of both types were used and exposed at different times. The birds were kept to an age of at least four months during which time they were sitting on the roof for 10 - 19 ("W") and 9 - 16 ("E") days, respectively.

The mean bearings of these birds at sites about 100 km east and west are shown in Figure 8. There is no obvious difference between the two types of pigeons. If there is anything different at all, it is the relation between the strength of the homeward component and that of the compass component rather than between the angular values of the PCD.

It may be worth mentioning that all bearings together (n = 120) were significantly homeward oriented ($p < 0.01$, V test).

VI. Orientation of Corridor-shaped Screens at the Home Site

If the aviary in which young pigeons are confined does not have solid walls at all sides but at only two opposing sides, the direction the birds prefer after release depends on the orientation of these walls (Wallraff, 1970b, 1978). It does not affect the degree of homeward

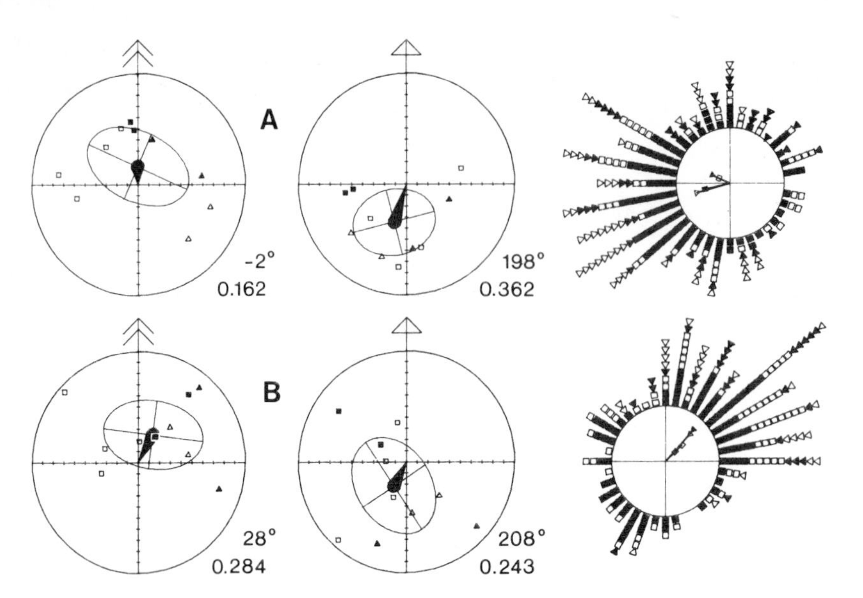

Fig. 8 A and B. Means of vanishing bearings of 10 simultaneous releases of caged pigeons that were exposed to different weather conditions (A, "W" birds; B, "E" birds). Five releases were from 112 km east and 5 from 98 km west. Each *dot* represents the mean vector of 4 - 11 bearings. Different symbols refer to different groups of birds exposed during different days (*squares* 1972, *triangles* 1973). The diagrams at the *right* show the directions *from which* the wind (minimum speed 2 knots) blew during exposure time of the respective groups. Measurements were made at Munich airport at 3-h intervals. The mean vector per sample is indicated (radius equals length one)

orientation (Fig. 9, left), but it reverses the sign of the PCD (Fig. 9, right). If the walls constitute a corridor that is open to air flows in the north-south axis, the PCD remains normal (Fig. 9 B, cf. Fig. 7 A).

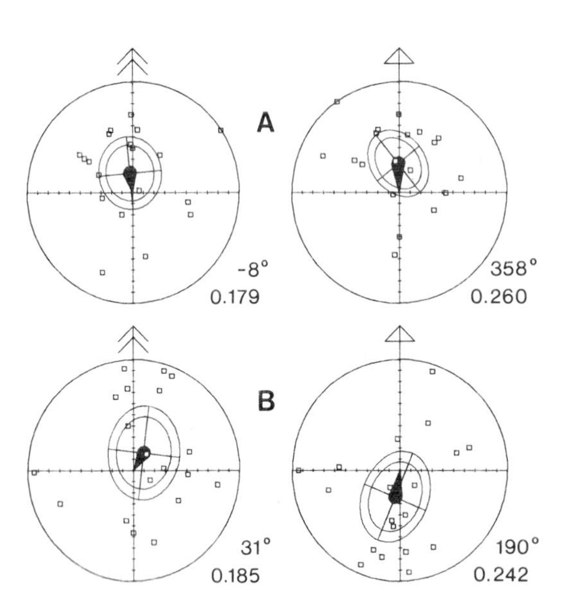

Fig. 9 A and B. Means of vanishing bearings of 19 simultaneous releases of pigeons confined within corridors open at east and west (A) or at north and south (B). About equal numbers of releases were made at sites ca. 100 km N, E, W, and 26 km S of the aviary near Seewiesen. Each *dot* represents the mean vector of 2 - 7 bearings. For further explanation see Figure 3

If the corridor is open in the east-west axis, however, the PCD points in the opposite direction (Fig. 9 A). The effect is statistically significant and independent of the visual conditions in the aviary (for details see Wallraff, 1978).

I cannot give an explanation of this effect, but I feel that the results do not fit into what might be expected on the basis of the hypothesis of olfactory navigation as put forward by Papi et al. (1972) and Papi (1976). On this basis one should expect that "corridor birds", when, for instance, released at a southerly point, are better oriented homeward if they lived within a corridor that was open in the north-south direction, than if they lived in a corridor open at east and west. However, just the opposite is true (see Wallraff, 1978).

E. Conclusions

I conclude that PCDs represent a widespread and probably general peculiarity in pigeon homing. It has been demonstrated at each of the eleven home sites at which systematic investigations have been conducted. The degree of its expression is variable within a considerable range (length of mean vector from 0.13 to 0.55; see Table 1). No correlation becomes evident so far between the strength of the PCD and the level of the homeward component. However, the quantitative relation between compass tendency and homeward tendency may depend on a variety of circumstances such as, e.g., home site, distance from home, and experience, and the possible interrelations have not yet been investigated sufficiently.

PCDs seem to develop dependent upon some unknown environmental conditions at the home site. The pigeons seem to be exposed to these conditions even if they have no opportunity to make exercise flights (aviary pigeons) and even if they are deprived of olfactory cues. Some reduction in the strength of the PCD is indicated only in pigeons confined between solid walls (Fig. 7). It may be noteworthy that this reduction is not accompanied by an "automatic" increase of the homeward tendency. In contrary, also homeward orientation is maximally reduced by this treatment. I take this as an argument for the assumption that PCDs are not only curious effects superimposed on the homing process, but that they are an unavoidable expression of its functional structure.

The only treatment at the home site that changed the preferred direction itself, was the confinement within an E-W corridor (Fig. 9 A). It is unclear why this change occurred, but it may not be by chance that the direction switched to its opposite, i.e., that it remained on the same axis.

Commonly the birds' flight path at a certain release site deviates more or less from the air-line direction toward home. As long as this deviation agrees with the general pattern determined by the loft-specific PCD, it makes little sense to search for physical peculiarities that characterize the local environment of that particular release site. For instance, I would not expect environmental peculiarities at the Castor Hill Fire Tower that was extensively used for pigeon releases by Keeton (1973). The preferred direction at this place fits perfectly into the general pattern of directional tendencies shown by pigeons of the Cornell lofts (Windsor, 1975). Thus, within this pattern the birds behaved quite normally. I do not see a reason why we should start a new at a stage we had reached some fifteen or twenty years ago (cf. Kramer, 1959; Pratt and Wallraff, 1958; Wallraff,

1959b) and look for an external factor that might, for example, be rotated either clockwise or counterclockwise at a certain place. Pigeons from different lofts released at the same site do not necessarily deviate from the homeward direction with the same sense of rotation (see, e.g., Wallraff, 1967, Fig. 3, 1970a, Fig. 2).

We should look for local features that might be responsible for a "release-site bias" only in those cases in which this bias does not correspond to the general patterns of preferred directions. Such cases do exist (cf. Fig. 1), and certainly many small deviations from an idealized geometric scheme are due to local influences. However, larger deviations seem to be exceptions rather than the rule. These considerations should be taken into account if one decides to search for the possible effect of local distortions of particular physical parameters on the directional choices of pigeons (e.g., Wagner, 1976; Frei and Wagner, 1976).

More practical conclusions such as those above do not depend on the position of PCDs within the functional system of pigeon navigation. Some ideas about this position have been communicated earlier ("null-axis hypothesis"; Wallraff, 1974), and they will be revisited at a later occassion.

Acknowledgements. I gratefully acknowledge the possibility to establish a pigeon loft in a building that belongs to the Zoological Institute (Prof. M. Lindauer) at the University of Würzburg. I thank Dr. E. Batschelet for providing his unpublished paper on Hotelling's T^2 test, Dr. O.G. Gelderloos for a critical reading of the manuscript, and K. Wielander for conducting most of the pigeon releases.

References

Batschelet, E.: Second-order statistical analysis of directions. This volume pp. 3-24

Bedini, C., Fiaschi, V., Lanfranchi, A.: Degenerative and regenerative processes in the olfactory system of homing pigeons. Arch. Ital. Biol. 114, 376-388 (1976)

Frei, U., Wagner, G.: Die Anfangsorientierung von Brieftauben im erdmagnetisch gestörten Gebiet des Mont Jorat. Rev. Suisse Zool. 83, 891-897 (1976)

Graue, L.C.: Orientation and distance in pigeon homing (Columba livia). Anim. Behav. 18, 36-40 (1970)

Graue, L.C., Pratt, J.G.: Directional differences in pigeon homing in Sacramento, California and Cedar Rapids, Iowa. Anim. Behav. 7, 201-208 (1959)

Keeton, W.T.: Release-site bias as a possible guide to the "map" component in pigeon homing. J. Comp. Physiol. 86, 1-16 (1973)

Kramer, G.: Recent experiments on bird orientation. Ibis 101, 399-416 (1959)

Oley, N., Dehan, R.S., Tucker, D., Smith, J.C., Graziadei, P.P.C.: Recovery of structure and function following transection of the primary olfactory nerves in pigeons. J. Comp. Physiol. Psychol. 88, 477-495 (1975)

Papi, F.: The olfactory navigation system of the homing pigeon. Verh. Deut. Zool. Ges. 69, 184-205 (1976)

Papi, F., Fiore, L., Fiaschi, V., Baldaccini, N.E.: Orientation of homing pigeons released over the sea. Z. Vergl. Physiol. 73, 317-338 (1971)

Papi, F., Fiore, L., Fiaschi, V., Benvenuti, S.: Olfaction and homing in pigeons. Monitore Zool. Ital. (N.S.) 6, 85-95 (1972)

Pratt, J.G., Wallraff, H.G.: Zwei-Richtungs-Versuche mit Brieftauben: Langstreckenflüge auf der Nord-Süd-Achse in Westdeutschland. Z. Tierpsychol. 15, 332-339 (1958)

Schmidt-Koenig, K.: Über die Entfernung als Parameter bei der Anfangsorientierung der Brieftaube. Z. Vergl. Physiol. 52, 33-55 (1966)
Schmidt-Koenig, K.: Entfernung und Heimkehrverhalten der Brieftaube. Z. Vergl. Physiol. 68, 39-48 (1970)
Wagner, G.: Das Orientierungsverhalten von Brieftauben im erdmagnetisch gestörten Gebiete des Chasseral. Rev. Suisse Zool. 83, 883-890 (1976)
Wallraff, H.G.: Über den Einfluß der Erfahrung auf das Heimfindevermögen von Brieftauben. Z. Tierpsychol. 16, 424-444 (1959a)
Wallraff, H.G.: Örtlich und zeitlich bedingte Variabilität des Heimkehrverhaltens von Brieftauben. Z. Tierpsychol. 16, 513-544 (1959b)
Wallraff, H.G.: Über die Heimfindeleistungen von Brieftauben nach Haltung in verschiedenartig abgeschirmten Volieren. Z. Vergl. Physiol. 52, 215-259 (1966)
Wallraff, H.G.: The present status of our knowledge about pigeon homing. Proc. 14th Int. Ornithol. Congr. (Oxford), 331-358 (1967)
Wallraff, H.G.: Über die Flugrichtungen verfrachteter Brieftauben in Abhängigkeit vom Heimatort und vom Ort der Freilassung. Z. Tierpsychol. 27, 303-351 (1970a)
Wallraff, H.G.: Weitere Volierenversuche mit Brieftauben: Wahrscheinlicher Einfluß dynamischer Faktoren der Atmosphäre auf die Orientierung. Z. Vergl. Physiol. 68, 182-201 (1970b)
Wallraff, H.G.: Das Navigationssystem der Vögel. München and Wien; Oldenbourg, 1974
Wallraff, H.G.: Goal-oriented and compass-oriented movements of displaced homing pigeons after confinement in differentially shielded aviaries. In preparation (1978)
Wallraff, H.G., Graue, L.C.: Orientation of pigeons after transatlantic displacement. Behaviour 44, 1-35 (1973)
Wiltschko, W., Keeton, W.T.: Effects of a "permanent" clock-shift on the orientation of young homing pigeons. Behav. Ecol. Sociobiol. 1, 229-243 (1976)
Windsor, D.M.: Regional expression of directional preferences by experienced homing pigeons. Anim. Behav. 23, 335-343 (1975)

Further Investigations of the Effect of "Flight During Clock Shift" on Pigeon Orientation

WOLFGANG EDRICH[1] and WILLIAM T. KEETON[2], [1] Universität Freiburg, Biologie III, 7800 Freiburg, FRG. [2] Section of Neurobiology and Behavior, Cornell University, Ithaca, NY 14853, USA

Abstract

Keeton and Alexander have shown that use of the sun compass by homing pigeons can be altered by giving the birds exercise flights at the home loft while they are undergoing 6-h phase shifting; such birds, when tested at a release site, usually choose bearings deflected much less than 90° from those of control birds. As a follow-up to these experiments, we used exercise in flight cages (1.8 × 7.3 × 1.8 m) to try to determine how much flight is necessary to produce such a result. All 16 of our cage-exercised groups had mean departure bearings intermediate between those of control birds and those of a group of normal clock-shifted birds. This is strong evidence that the very restricted flight in the cages was sufficient to alter the pigeons' response to being clock-shifted. Less definitive, but nonetheless strongly suggestive, evidence indicates that exercise in cages aligned E-W was more effective than exercise in cages aligned N-S, probably because of the differing relationships to the home direction. In a second series of experiments, pigeons given repeated unidirectional training flights while living under clock-shifted conditions were found to choose bearings only slightly deflected from those of controls. When such birds were put back under a normal photoperiod and retested, their bearings were substantially deflected, suggesting that they had previously recalibrated their sun compass, though alteration of cue utilization may also have been involved.

A. Introduction

Although the important role of the sun compass for homing pigeons has been demonstrated many times, it is also now well established that there must be other factors (some of them probably not yet specified) that enable pigeons to find their way home from unfamiliar release sites (for review, see Keeton, 1974). The role of these other factors is most easily examined if the sun is not visible to the birds or if its usually dominating influence on the birds' orientation has been downgraded experimentally.

Keeton and Alexander (in press) have recently shown that one way to alter the birds' use of the sun compass is to release them for exercise flights at the home loft during the 5 - 6 days while their internal clocks are being phase-shifted 6-h. Although the same authors (Alexander and Keeton, 1974) have shown that allowing the birds to sit in an aviary from which they can see the sun during the overlap period between the true day and the shifted day does not diminish the effect of the phase-shift on the birds' initial orientation at a release site (Fig. 1 A), the addition of flight under sun during the phase-shifting period does often have an effect; in many cases (though not always)

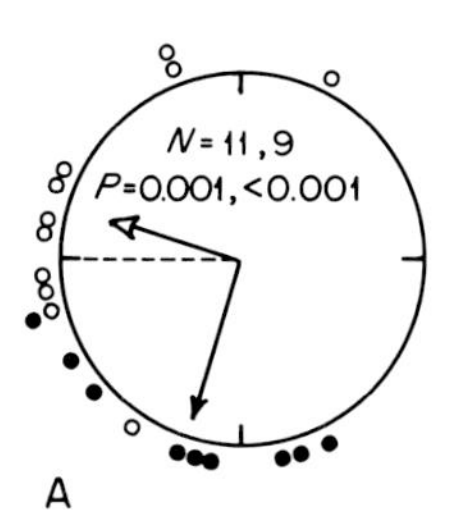

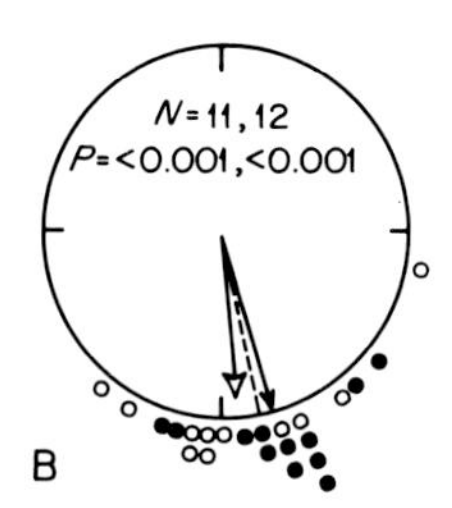

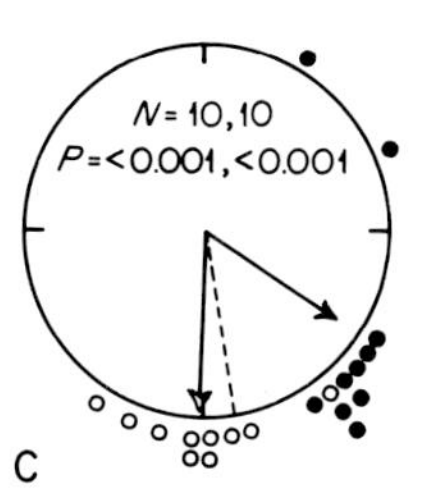

Fig. 1. (A) An example of bearings obtained from pigeons allowed to see the sun while sitting in an aviary during the phase-shifting period (from Alexander and Keeton, 1974). (B, C) Two examples of bearings of pigeons given free exercise flights during phase-shifting, showing the variability of the extent of compensation (from Keeton and Alexander, in press). (In this and all later figures, *small lines* on the inside of the circle indicate N, E, S, and W; a *dashed line* indicates the home direction. Mean bearings are shown as *arrows*, with *open* and *solid heads* corresponding to open and solid samples of bearings; the length of the vectors is proportional to the tightness of clumping of the bearings. Within each *circle*, the sample sizes, *N*, and uniform probability, *p*, under the Rayleigh test are shown; the first-given values are for the sample shown as open symbols, and the second for the solid symbols.) In this figure, *open symbols* are for control bearings, *solid symbols* for experimentals. Clock shifts were 6-h fast

the magnitude of deflection of the birds' bearings is much less than the 90° expected from an ordinary 6-h clock shift (Fig. 1 B, C) (Keeton and Alexander, in press). In the present paper, we report two kinds of preliminary experiments designed to follow up the "flight during clock shift" results: (1) experiments to determine how much flight (and what kind) is necessary to alter the birds' orientational response to being phase-shifted, and (2) experiments to explore the basis of the birds' changed response to the sun compass.

B. Methods

The subjects were homing pigeons bred at the Cornell lofts and given training flights from all directions. Most were less than 6 months old; the birds in two tests were one year old. Clock shifts were performed in artificially ventilated rooms with lighting controlled automatically by astronomic timers set for the proper latitude and date. In one room (the shift room) the L/D cycle was shifted 6-h out of phase from true time (i.e., from eastern standard time), and in the other room (the control room) the L/D cycle was in phase with true time. In experimental Series I, three groups of experimental birds were housed in the shift room - one for each of two exercise cages and one (designed SR) to be shifted but not exercised. A fourth group (designed C) was housed in the control room. Birds were allocated randomly to the four groups. For the 1974 experiments, the clock-shift rooms were located at the main loft; for 1975, they were located at the Liddell Laboratory.

Each day during the 5 - 6 day phase-shifting period, Keeton and Alexander (in press) released their pigeons during the overlap between the true day and the shifted day, and forced the birds to fly around the loft for a minimum of half an hour. We followed the same procedure in several early tests to determine whether such treatment would have the reported effect on our pigeons. However, in our experimental Series I, the pigeons were not released, but were exercised in two flight cages consisting of light wooden frames covered with plastic netting. The cages were 1.8 m wide, 7.3 m long, and 1.8 m high. Three long perches were provided at each end. For the 1974 tests and two 1975 tests at Orwell, the cages were set up in an open field within sight of the main loft buildings; for the 1975 experiments, tests at Auburn and one test at Orwell, the cages were in a field near the Liddell Laboratory, 2 km north northwest of the lofts. For most experiments, the long axis of one cage (E_1) was oriented north-south, while the long axis of the other cage (E_2) was oriented east-west; for two 1975 experiments with young birds at Orwell, one cage (E_3) was oriented northwest-southeast and the other (E_4) northeast-southwest.

For our experimental Series II (1975 only), birds that returned fast enough from Series I releases (after cage treatment) to be still phase-shifted were immediately put back into the light-controlled rooms. These birds were then given "training under shift," i.e., they were given many training flights while living under clock-shifted conditions. The training flights for one group of pigeons were from release sites south of the loft, including 2 flock tosses and 2 single tosses from the same site (Orwell, Pa., 65.5 km S) where they had been tested in the Series I (cage) experiments and where they were going to be tested again in the Series II experiments. The training flights for a second group of pigeons were from release sites north of the loft, including one single toss from near Auburn, N.Y., and one flock toss from Auburn (66.2 km N), the location of their Series I test and of their forthcoming Series II tests. Each group of birds was then tested twice - first without further manipulation to determine the effect of their having been repeatedly trained while phase-shifted, and second after being normalized (i.e., exposed to the natural L/D cycle for 5 days).

All test releases were conducted under sunny skies. Vanishing bearings (to the nearest 5°), vanishing intervals, and homing times were recorded. The mean vector of each distribution of bearings was calculated, and its length was used in computing the uniform probability under the Rayleigh test. Homing speeds were compared under the matched-pairs sign test (all birds that returned with another were omitted from the calculations; lost birds were assigned a speed of O).

C. Results

I. Preliminary Experiments (Free Flight During Clock Shift)

Before beginning our experiments with flight cages, we conducted two tests using the shift plus free flight technique of Keeton and Alexander, to make sure the results previously reported would hold for this experimental season (1974). As Figure 2 shows, the results were of two Kinds. In one release, there appeared to have been some compensation, the deflection of the mean bearing of the experimental birds from that of the controls being only 46° (Fig. 2 A). In the other release, there appeared to have been no compensation at all, the deflection of the mean bearing of the experimentals being 133° (Fig. 2 B). These releases, which were from a single site with birds of similar background, illustrate the variability of results obtained from the shift plus free flight technique. The very large deflection of the experimentals in the release of Figure 2 B is not unusual in normal clock-shift tests at this release site, where deflections of considerably more than 90° are the rule rather than the exception (Keeton and Brown, unpublished data), but such a large deflection *is* unusual in a shift plus free flight test.

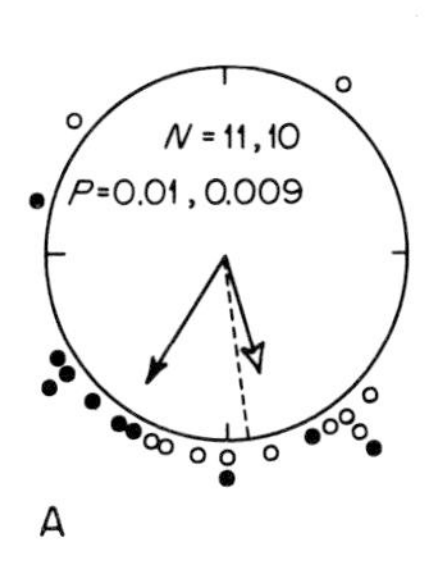

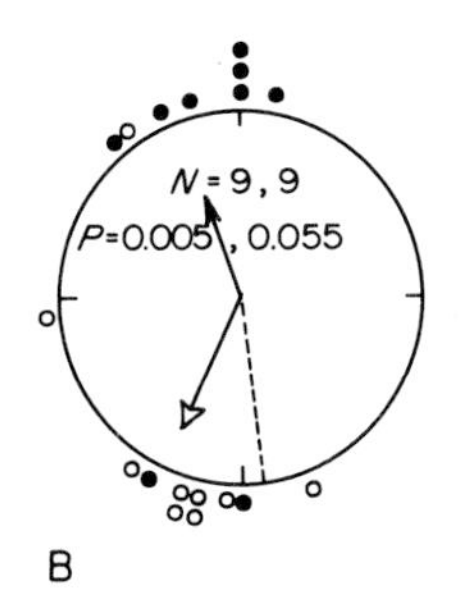

Fig. 2 A and B. Bearings from two experimental releases at Weedsport, N. Y., 73.6 km N, utilizing pigeons given free flight experience during clock-shifting. (A) 23/24 Aug. 1974. (b) 5 Sept. 1974. *Open symbols* indicate bearings of control birds, *solid symbols* bearings of experimentals. Clock shifts were 6-h slow

The control birds had significantly faster homing speeds than the experimentals in both releases ($p = 0.011$ and $p = 0.004$).

II. Series I Experiments (Flight in Cages)

Experiments with birds exercised in N-S and E-W cages. In 1974 we conducted three test releases of young birds that had been exercised in the flight cages during the period of phase-shifting (some other releases, using birds that had returned from a release the previous day and were retested, will not be discussed in this paper). All three were from very short distances - two from 1.4 km W and one from 5.8 km E (Fig. 3). In all three the control birds (C) were well oriented homeward. In two of the three the *un*exercised clock-shifted birds (SR) had mean bearings deflected more than 90° from the control means (Figs. 3 A, C), but in the other the SR birds were random (Fig. 3 B). In all three releases, the E_1 and E_2 birds were significantly oriented but with deflections from the controls of less than 90°; moreover, in all three the deflection of the mean of the E_2 birds (average 40°) was less than that of the E_1 birds (average 61°). Accurate homing speeds could not be determined for these very short flights.

In 1975 we conducted three tests of cage-exercised birds at Auburn, New York, 66.2 km N (Fig. 4). In all three the C birds were well oriented homeward, and in all three the SR birds had mean bearings deflected more than 90° (average 127°) from the control means. As in the 1974 experiments, the mean bearings of the E_1 and E_2 birds were deflected less than the means of the SR birds in all three tests. And in all three, the mean of the E_2 birds was deflected less than that of the E_1 birds (average 41° for E_2 versus 80° for E_1), though in one release the E_1 birds were not significantly oriented (Fig. 4 A_2) and in another the E_2 birds were not (Fig. 4 B_2).

We performed one experiment with old birds (more than one year old) from 65.5 km S, a site familiar to most of the birds. As Figure 5 shows, the mean bearings of both the E1 and the E2 groups were intermediate between those of the C and SR groups, as predicted from the experiments with young birds. However the E_1 and E_2means differed from each other by only 2°.

In summary, the bearings from the seven test releases of this type (N-S, E-W) yielded consistent results - in every case the means of the exercised groups (two groups per test, or a total of 14 groups) were intermediate between the mean of the controls and the mean of the SR birds (or between the mean of the controls and 90° to the left of the controls in the one release where the SR birds were random). Such consistency is highly significant ($p < 0.01$). In addition in six of seven

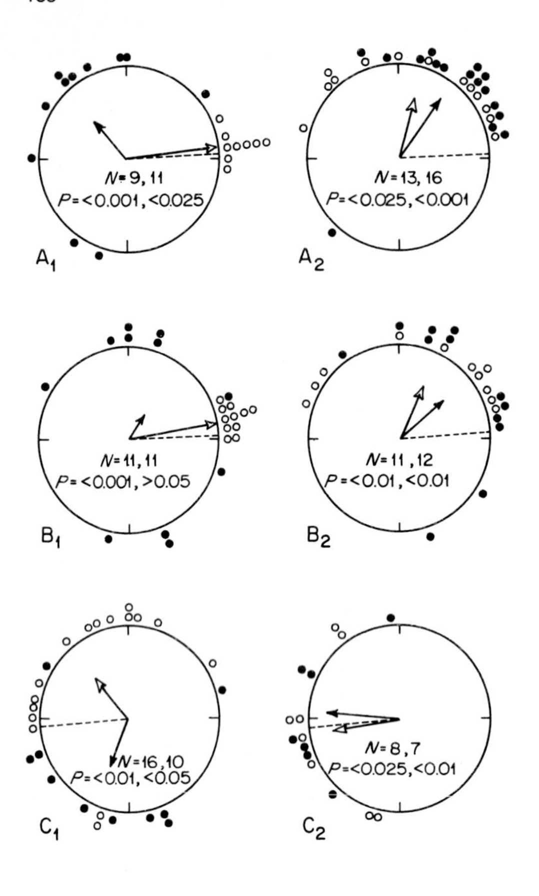

Fig. 3 A - C. Bearings from three short-distance releases utilizing young pigeons given flight in NS and EW cages during clock-shifting. (A) 8 - 11 Oct. 1974, from 1.4 km W. (B) 22/24 Oct. 1974, from 1.4 km W. (C) 15/16 Nov. 1974, from 5.8 km E. In this Figure (and in Figs. 4, 5, and 6), the *open* and *solid symbols* on the subscript-1 circles indicate control and SR bearings, respectively; the *open* and *solid symbols* on the subscript-2 circles indicate E_1 and E_2 bearings, respectively

cases the deflection of the E_1 bearings from the controls was greater than the deflection of the E_2 bearings and in the one case where this was not true (Fig. 5) the two means differed by only 2°. However, the fact that some sets of bearings were not statistically oriented under the Rayleigh test requires that the results be regarded as tentative only.

The homing speeds from the Auburn releases reinforce the picture of E_1 and E_2 birds being intermediate in their behavior between the C and SR birds. Pooling the three releases, we found that the speeds of all three clock-shifted groups were significantly slower than those of the controls, but that the speeds of both the E_1 and the E_2 birds were significantly faster ($p = 0.008$ and < 0.001, respectively) than the speeds of the SR pigeons. There was a suggestion that the E_1 birds were a bit slower than the E_2 birds, which would be consistent with the results from the bearings, but the difference was not quite significant ($p = 0.09$). The speeds of the C birds at Orwell were significantly faster than those of the E_1, E_2, and SR birds; there were no significant differences between the latter three groups.

The numbers of birds homing on the day of the release again support the picture of cage-exercised birds as intermediate in their behavior. In the five tests of 1975, 36 of 54 (66.7%) controls, 21 of 51 (41.2%) E_2 birds, 17 of 52 (32.7%) E_1 birds, and 9 of 52 (17.3%) SR birds homed on the release day (birds returning together with others are omitted from this analysis).

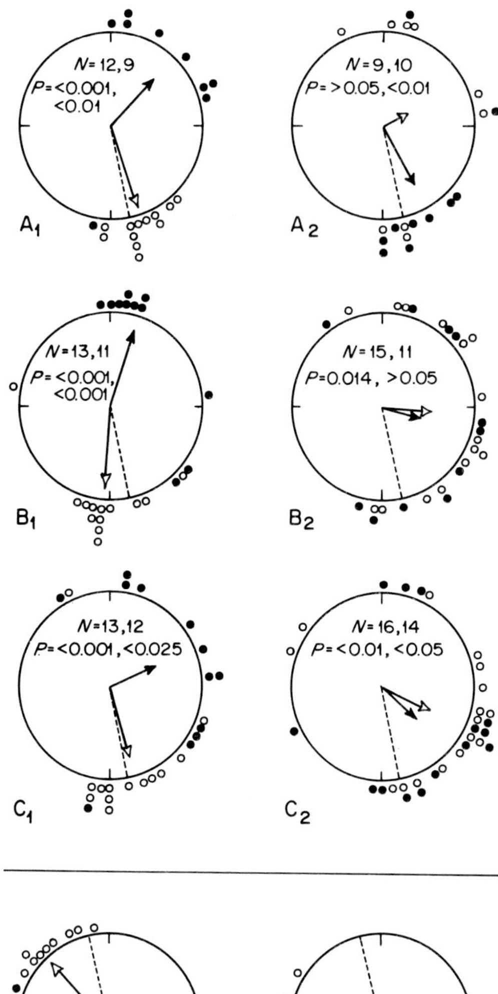

Fig. 4 A - C. Bearings from three releases from Auburn, N.Y. (66.2 km N), utilizing pigeons given flight in NS and EW cages during clock-shifting. (A) 15 Aug. 1975. (B) 20 Aug. 1975. (C) 25 Aug. 1975

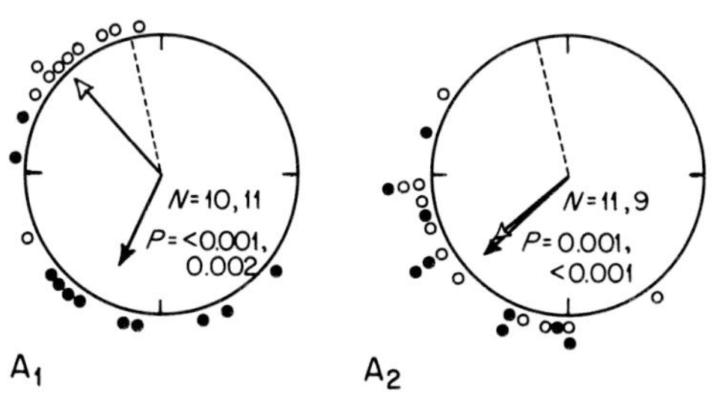

Fig. 5. Bearings from a release from Orwell, Pa. (8 Aug. 1975), utilizing old birds given flight in NS and EW cages during clock-shifting

An experiment with young birds exercised in NW-SE and SW-NE cages. Having found suggestive evidence that N-S and E-W cages affect pigeons differently, we tried one preliminary experiment (two tests) with different cage alignment - E_3 oriented NW-SE and E_4 NE-SW. The releases were from Orwell, Pennsylvania, 65.5 km S, a site not familiar to the birds. Figure 6 of the pooled results shows that, as in all the other tests, both of the exercised groups chose bearings intermediate between the means of the C and SR birds. In this case, however, the means of the two experimental groups differed from each other by only 2°.

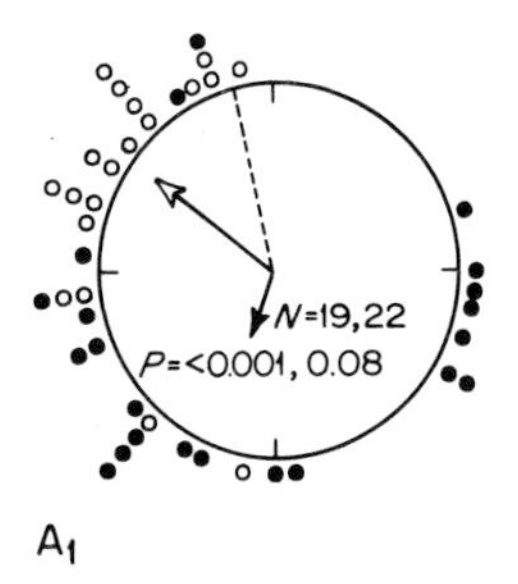

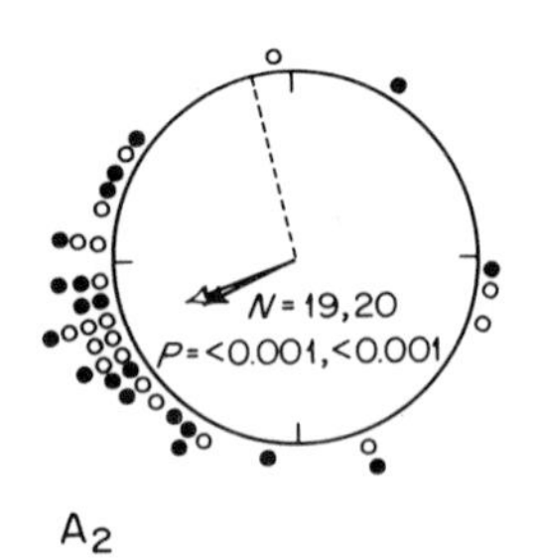

Fig. 6. Pooled bearings from two releases from Orwell. Pa. (65.5 km S; 22/26 July 1975), utilizing young pigeons given flight in NW-SE and NE-SW cages during clock-shifting (the results of the two tests were very similar, hence pooling makes no change in the overall outcome)

The control birds had significantly faster homing speeds than the three clock-shifted groups. There were no significant differences between the shifted groups.

III. Series II Experiments (Training Under Shift)

Two groups of pigeons were used in these experiments. One group was composed of birds that returned rapidly from the Auburn releases (Fig. 4) and the other of birds that returned rapidly from the Orwell releases (Figs. 5 and 6). The pigeons were immediately put back into the 6-h-fast-shifted rooms and held there while being given training flights almost daily - the Auburn group from sites north of the loft and the Orwell group from sites south. At the end of the training period, the groups were tested at Auburn and Orwell, respectively. The resulting bearings are shown in Figure 7 A, B; the means of the clock-shifted birds are deflected counterclockwise from the means of the controls in one case only 9° and 45° in the other.

Next the birds were exposed for 5 - 6 days to the true photoperiod and were then tested again at Auburn and Orwell. Figure 7 C, D show the resulting bearings. Now the means of the experimental birds were clockwise from those of the controls in both cases. For these birds, being

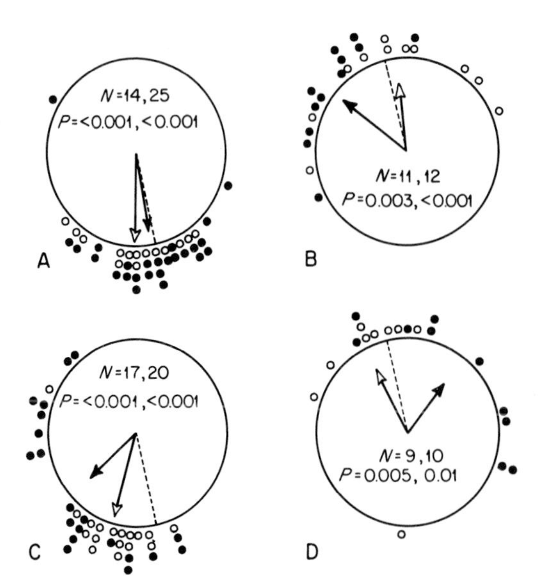

Fig. 7 A - D. Bearings from releases from young pigeons given training flights while living under clock-shifted conditions. (A and B) Releases before normalization at Auburn (4 Sept. 1975) and Orwell (14 Aug. 1975), respectively. (C and D) Releases of the same birds after normalization at Auburn (9/10 Sept. 1975) and Orwell (19 Aug. 1975), respectively

moved to normal time had apparently had some of the effect of being clock-shifted 6 h slow.

C. Discussion

We performed a total of 8 test releases of cage-exercised pigeons, from 4 release sites. All 16 cage-exercised groups had mean departure bearings intermediate between the C and SR means (or intermediate between C and 90° left of C in the one instance when the SR birds were random - the second of the short-distance releases). We consider these results to be a strong indication that the very restricted flight in our cages is sufficient to alter significantly the pigeons' response to a 6-hour clock shift. We think it likely that flight was the thing that made this treatment have an effect not seen in the experiments of Alexander and Keeton (1974), where pigeons were allowed to sit in a sun-lit aviary during the days of the phase-shifting process; we do point out, however, that our experimental set-up differed in one other respect also, namely, the exercise cages were detached from the birds' home loft, whereas the aviaries used by Alexander and Keeton were attached directly to the loft.

The rationale for the "flight during clock shift" experiments of Keeton and Alexander (in press) was that, since clock-shifted pigeons released less than 1.5 km from home choose deflected bearings (Graue, 1963; Alexander, 1975; Keeton, 1974), the birds must utilize their navigation system at very short distances. Pigeons flying for exercise around the loft, and ranging away 1 - 2 km, might therefore "turn on" their navigation system. If so, then exercise flights during phase-shifting might be confusing to pigeons, because the sun compass component of their navigation system would be giving them information at odds with their visual contact with the loft from which they have just flown up. Moreover the conflict would be different each day of the 5 - 6 day phase-shifting period, because the setting of the birds' internal clock would be different each day. Regular exposure to this conflict of cues might cause the pigeons either to reduce their dependence on the sun compass or to recalibrate their sun-compass, the consequence being diminished deflection of their bearings in test releases.

The amount of flight possible in our cages was, of course, far less than the free flight given the birds by Keeton and Alexander. Nonetheless, our birds regularly showed diminished deflection of bearings. The combination of two conditions may have contributed to our results: (1) The detachment of our cages from the loft may have brought the birds into a motivational state necessary for homing; (2) the possibility to fly in the cages may have helped them to attend or helped them to better perceive additional orientation cues, which enabled the pigeons to determine the approximate home direction even under shift and with the loft out of sight - better than when they were simply sitting in baskets in the field. Such a formulation would be consistent with Bookman's (1977) evidence that flight may be a prerequisite for response to magnetic stimuli.

If we are correct in our very tentative conclusion that flight in E-W cages was more effective in reducing the orientational effect of phase-shifting than flight in N-S cages, then we must ask what might be the basis for this difference. The two cages were equally exposed to topographic, celestial, and atmospheric factors. They differed most obviously with respect to their alignment relative to the magnetic field and to the home direction (because both locations where cages were used were roughly on a line running N-S through the loft, the N-S cage was aligned toward home and the E-W cage at right angles to the home direction). Note that in the test with NW-SE and NE-SW cages, the alignments did not differ with regard to either the magnetic field or the home direction, and in this case the bearings of the E_1 and E_2 birds were nearly the same. Our experiments should be repeated, using cage locations east or west of the loft.

One piece of informal observational data suggests that the alignment of the cages relative to the home direction influences the birds' behavior. Birds in the E-W cages, but not in the N-S cages, were sometimes observed flying to and fro along the long axis of the cage in a peculiar sideways weaving manner, with their heads oriented toward the long side of the cage in the homeward direction. If this type of attentiveness to the home direction is necessary for cage flight to produce maximal compensation of the orientational effects of clock shifts, then perhaps the motivational condition of the birds will prove to be an important variable in these experiments also.

Since both the flight during clock-shift treatment used by Keeton and Alexander and the cage-flight procedure reported here are effective in reducing the deflection of bearings produced by clock shifts, and since Wiltschko et al. (1976) found that training flights reduced the deflecting effect of normalization on their permanently clock-shifted pigeons, we must ask whether the pigeons are downgrading their dependence on the sun compass while elevating the importance of alternative cues, or whether they are recalibrating their sun compass. Wiltschko et al. (1976) addressed this question, but their results did not lead to a clear-cut answer; it appeared that their birds might have been using both procedures.

Our Series II experiments, though only preliminary and in need of further replication, give evidence of recalibration by our pigeons. However, in only one of the experiments was the deflection produced by normalization 90° or more (from 45° left of C to 69° right of C, equals 114° total; see Fig. 7 B); in the other experiment the deflection was only 43° (from 9° left of C to 34° right of C; see Fig. 7 A). Hence these preliminary tests are not in disagreement with the results of Wiltschko et al. (1976); both recalibration and alteration of cue weighting may be involved.

Wiltschko et al. (pers. comm.) have carried out some experiments similar to ours, but with training flights from all directions rather than on a line to a familiar release site. It will be interesting to compare their results with those reported here.

We conclude by saying that, in our opinion, techniques similar to those reported here have considerable potential for aiding in the task of determining how pigeons integrate diverse orientational cues into the remarkable navigational system that enables them to home from unfamiliar release sites. It is clearly possible to modify the ratings of cues within that system experimentally in ways that expose details difficult to see in the more usual operation of the system.

Acknowledgments. We extend our tanks to Irene Brown, Timothy Larkin, and Michael Weiler for their help in conducting the experiments, and to Wolfgang Wiltschko, Roswita Wiltschko, and Andre Gobert for valuable discussions. This research was supported by grants from the Deutsche Forschungsgemeinschaft and the Deutsche Akademische Austausch Dienst to W. Edrich, and by NSF grant BMS 75-18905 AO2 to W. T. Keeton.

References

Alexander, J.R.: The effect of various phase-shifting experiments on homing in pigeons. Thesis, Cornell Univ. Ithaca, N.Y. (1975)

Alexander, J.R., Keeton, W.T.: Clock-shifting effect on initial orientation of pigeons. Auk 91, 370-374 (1974)

Bookman, M.: The sensitivity of the homing pigeon, *Columba livia*, to an earth-strength magnetic field. Nature (London) 267, 340-342 (1977)

Graue, L.C.: The effect of phase shifts in the day-night cycle on pigeon homing at distances of less than one mile. Ohio J. Sci. 63, 214-217 (1963)

Keeton, W.T.: The orientational and navigational basis of homing in birds. Adv. Study Behav. 5, 47-132 (1974)

Keeton, W.T., Alexander, J.R.: The effect of exercise flights during phaseshifting on the orientation of homing pigeons (in press)

Wiltschko, W., Wiltschko, R., Keeton W.T.: Effects of a "permanent" clock-shift on the orientation of young homing pigeons. Behav. Ecol. Sociobiol. 1, 229-243 (1976)

Accurate Measurements of the Initial Track (Radius 1500 m) of Homing Pigeons

BURKHARD ELSNER, Universität Göttingen, I. Zoologisches Institut, 3400 Göttingen, FRG

Abstract

The homing of pigeons is a complicated process that depends upon many parameters. The measurement of vanishing bearings instead of the entire flight tracks reduces the obtained information from more than 10 kbit to only 8 bit. Therefore, the entire initial flight tracks should be measured. An apparatus has been assembled for such measurements. It consists of telescopes that are coupled to electronic digitizers. The positions of the pigeons are stored (max. 10 measurements/s) and processed on line with a HP 9815A calculator.

Possible sources of error and some preliminary measurements are discussed.

The navigation of birds and homing of pigeons is a very complicated and difficult problem. Many details are known, but so far there is no good working hypothesis available. To explain the difficulties let us describe the problems involved. The main issue appears to be finding the mechanisms that underlie the pigeons' homing ability.

To obtain a better idea of the basic mechanisms involved, we have to look for variables $y_1 - y_m$, which are subject to measurement, such as homing, speed of homing, energy consumption, physiologic parameters (blood pressure, pulse rate, respiration rate) and flight tracks, including initial choice of direction and vanishing bearings.

These variables are related to independent variables, $x_1 - x_n$, such as electromagnetic fields (from static fields to high frequency), olfaction, air pressure, noise, gravitation, optical cues (landmarks, high voltage lines), geographic and geophysical attributes of the landscape, condition, training, inheritance, experience, weather, wind, cosmic radiation.

Very often even the fact of the dependence is not known; in other cases the relations between the variables are far from being understood.

A couple of years ago there was the notion that all x_i except one would be zero, which means that the birds would navigate only by optical *or* olfactory *or* other mechanisms. Apparently this notion turned out to be wrong. Navigation seems to work through many channels, and one mechanism can replace the other.

With so many parameters involved exact measurements of the behavior and of physiologic data are necessary to obtain more insight into navigational mechanisms. Following correlation analysis, theories can be developed on the basis of these obtained data. The theories in turn should be helpful in the planning of improved experiments.

When a working hypothesis is not available, one should consider which of the possible measurements could give the most useful information. A lot of information is given by the flight tracks, especially by the initial flight tracks of homing pigeons. To measure these tracks I have designed and built an apparatus as follows.

After release, the pigeons fly many curves, which possibly are correlated with certain landmarks, roads, rivers, or trees. To find this correlation the measurement must be exact to ± 10 m. The maximum speed of the birds is 20 m/s. The minimum radius of their flight curves is about 10 m, which means that to get all information one has to make 2 measurements/s or more.

The accuracy of 10 m means 1% at 1 km. This, together with 2 measurements/s over many minutes or hours, requires electronic data analysis of some kind.

We use an optical method and calculate the location of the birds from measured angles (Fig. 1). Two telescopes, positioned at a known distance from each other (about 100 m) are used with electronic digitizers. They measure the horizontal angles α and β. A third digitizer measures the zenith angle γ.

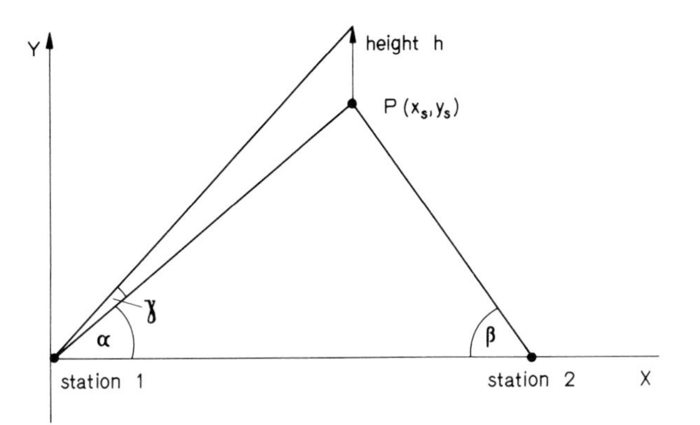

Fig. 1. Localization of the birds from measured angles

From the measured angles and the known distance between the measuring stations c, the rectangular coordinates of the bird (projection on the surface of the earth) can be calculated:

$$x_s = \frac{c \cdot tg\beta}{tg\alpha + tg\beta}, \qquad y_s = x_s \cdot tg\alpha$$

The height of the pigeon over this point is:

$$h = tg\gamma \cdot \sqrt{x_s^2 + y_s^2}$$

and the speed of the birds between two points is:

$$v = \frac{\sqrt{(x_2 - x_1)^2 + (y_2 - y_1)^2}}{t_2 - t_1}$$

($t_2 - t_1$ is the time of flight between the points)

If we postulate a certain accuracy for the coordinates, we have to use the law of error propagation of Gauss or other numerical methods to calculate the influence of errors in the measurement of the angles on the calculated coordinates. The most appropriate method involved a small calculator program for the HP 25. It calculated the coordinates from the measured angles, added "typical" errors of measurement to the angles (e.g., 0.2°), calculated again the coordinates, and finally the relative deviations of the two results.

The result of these calculations, which could also qualitatively be seen from Figure 2, is a strong dependence of the final accuracy on the measured angles.

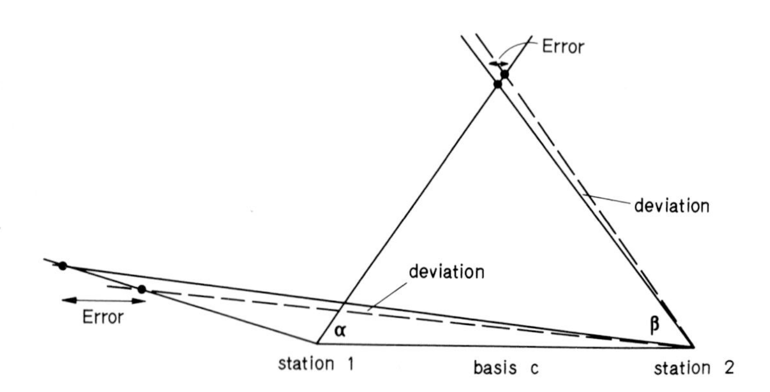

Fig. 2. Dependence of localization errors on the measured angles

The measurements are accurate if the pigeons are perpendicular to the line between the measuring stations, but the x-coordinate cannot be measured at all if the pigeons are between the stations or on the extension of this line. The final results could be made more accurate by using a third station, but the expenditure (personal and technical) would be quite high.

The digitizers give 1000 pulses for an angle of 2π. This number is electronically multiplied by a factor of 4. So we get 1000 pulses of a right angle or, if we use decimal grads ($2\pi \approx 400$ grads), 1 pulse for 0.1 grad.

How much information have we to store and to analyze?

Digitizer for α and β, each	4 digits	32 digits BCD
Digitizer for γ	3 digits	12 digits BCD
Quartz clock	5 digits	20 digits BCD
		64 digits BCD

If we see a pigeon for 3 min, we have to store and to process 25,000 bit/pigeon.

Release experiments with pigeons need much time and long preparations. Therefore it is essential to get fast feedback from experimental results in order to optimize the experimental parameters, if possible during the experiment.

We use a small computer HP 9815A for storing the data during the experiment and processing the data either on line during the experiment or later in the laboratory.

Figure 3 shows a block diagram of our electronic apparatus. The pulses from the digitizers (Stegmann DG-S-1000) pass to a central electronic

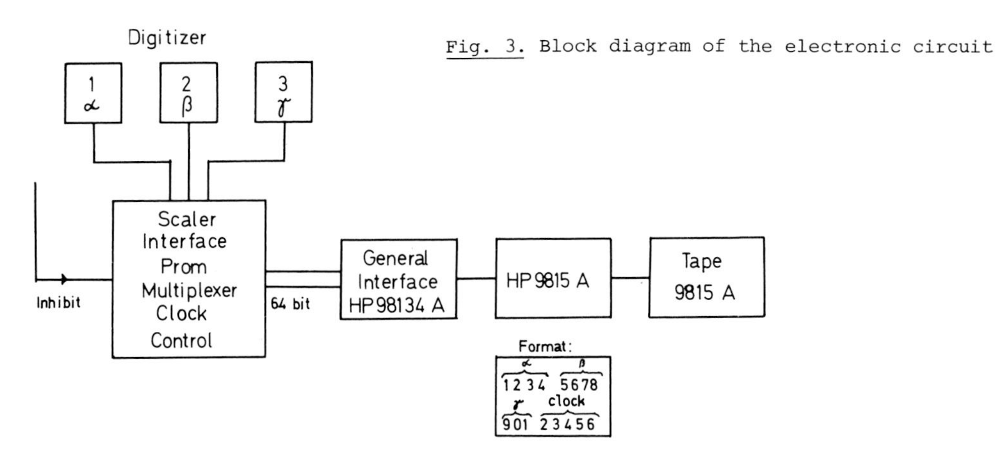

Fig. 3. Block diagram of the electronic circuit

unit with forward/backward discriminators, scalers, latches, multiplexers, and a PROM. Here the information is serialized and passes through a general input/output-interface (HP 98134A) to the computer HP 9815A. Under program control the data are stored in 200 storage registers. If these registers are full, the data are transferred to the magnetic tape in the computer. This transfer requires 3 s, during which time no information can be stored. Following this procedure again, the storage registers in the computer are used. One measurement requires two registers, the first for the angles α and β, the second for the angle γ and for the time (accuracy 0.1 s).

The power supply comes from two 12-V batteries; the computer runs from a 12 VDC/220 VAC-converter.

Data are only taken when the observers on both telescopes press a switch in order to indicate that they see the bird in the cross wires of the telescope. As long as both switches are pressed, data are automatically taken at a rate between 10/s and 0.1/s. All the observers have to do is to hold the pigeon in the telescope near the center of the cross wires.

So far we have measured approximately 150 tracks. The whole experimental set-up seems to cope satisfactorily with the input data. Data loss due to misfunction of the apparatus is less than 10% (Fig. 4).

The analysis of the data is very difficult, because there is so much information in the flight tracks. Furthermore it is completely unknown which information is relevant for the problem. For that reason simple digital analysis cannot be applied. First one has to look at the flight tracks and find out whether there are typical flight patterns, curves, or speeds correlated with experimental parameters. In case some correlation can easily be made out, the tracks will be analyzed in detail using Fourier-analysis and statistical methods.

The flight track can be plotted by hand from the computer print-out of the coordinates, although this appears to be a tedious job. We therefore use a normal x-y-plotter with a handmade interface to the HP 9815A, thus enabling us to draw a track in less than a minute.

The arrangement of the position of the telescopes is not symmetric with regard to rotation. Even if one looks with only one telescope for the vanishing bearings, there is always a bias due to the different background of the landscape, different illumination, and other irregularities. This bias is even greater when two observers have

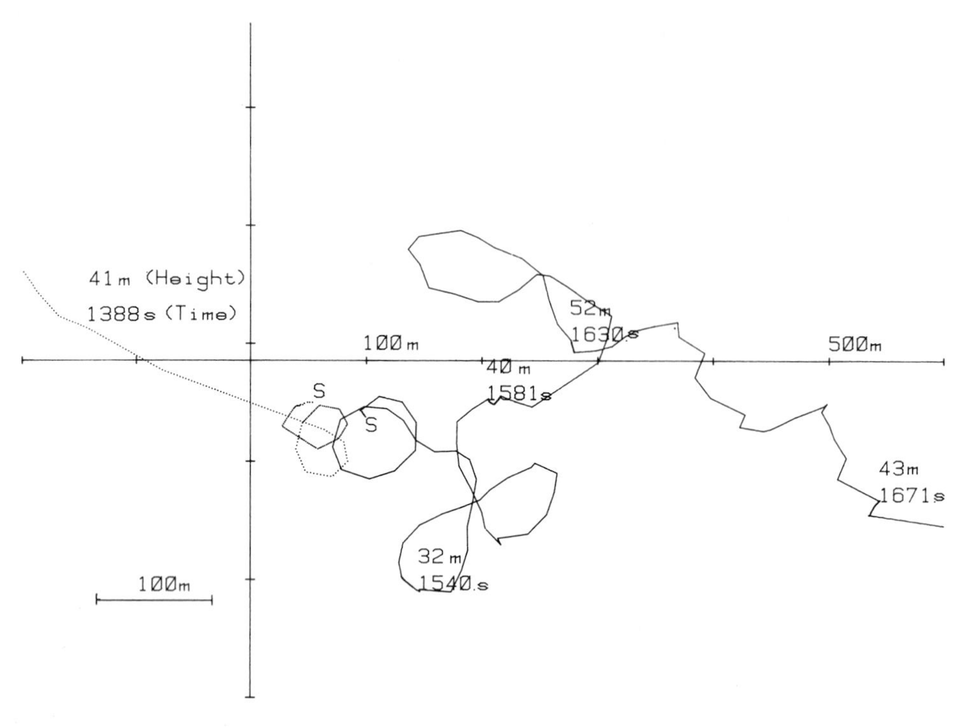

Fig. 4. Two typical flight curves of released pigeons, plotted directly from the obtained data with HP 9815A calculator and 9872A plotter. (*S* start point)

to watch the bird at the same time. This effect adds to the dependence of the measuring errors on the measured angles and could lead to apparent anisotropies of the vanishing bearings that are in fact not related to experimental parameters.

Until now tracks were measured to see differences in the behavior between young and experienced pigeons, and to compare the initial tracks of pigeons whose internal clocks were shifted with those of control pigeons. Particularly in the case of clock-shifted pigeons, we expected clear differences between shifted and unshifted pigeons.

Surprisingly, at first examination no definite differences in the flight tracks were found. In both cases the tacks were very complicated, with many curves and circles. The birds were lost at different positions, and most of them could be followed over a longer distance with hand-held binoculars.

From these tracks it became very clear to me that the mean vanishing bearing is a very critical parameter.

It is clear from our measurements that often the vanishing point is not in the direction of the radius vector from the releasing point. The bird could be lost flying on a circle around the releasing point, and the vanishing bearing is not correlated with the direction of flight. It is possible that the results of experiments could be changed by a bias as described.

Anyhow, these results are preliminary and it is too early to draw definite conclusions from them. A future theory of bird navigation will surely explain all the data so far collected.

Session IV

Chairman:

WILLIAM T. KEETON, Sect. of Neurobiology and Behavior, Langmuir Lab., Cornell University, Ithaca, NY 14853, USA

Echolocation of Extended Surfaces

DONALD R. GRIFFIN and EDWARD R. BUCHLER*, The Rockefeller University, New York, NY 10021, USA

Abstract

Relatively large surfaces return echo complexes not only to flying bats but to birds that emit flight calls. The spreading loss during the return of a ground echo to a flying bird is only 6 dB, because the total path length has doubled. Over the range from 1 - 6 kHz the ground reflection loss (GRL), or ratio of echo intensity from an actual surface below that to be expected from a perfect reflector, varies even for relatively smooth ground with grass and low vegetation from less than 5 to more than 30 dB, tending to increase with frequency. Due to complicated interference patterns between reflected sound waves from different parts of a large and usually irregular surface, GRL may differ by 10 - 20 dB from the average value. By utilizing only the peaks, a flying animal could hear components of ground echoes at much greater altitudes than would be predicted from average values. Since these interference patterns are frequency dependent, broad-band signals such as frequency sweeps should be audible to higher altitudes than constant frequencies, as well as containing potentially useful information about the ground or water surface.

Echolocation is best known through extensive studies of bats and cetaceans, which are highly specialized for this mode of gathering information about objects at some distance via echoes that they return from sounds emitted by an animal (Simmons, 1977). Many other animals also employ echolocation to a limited extent, including shrews (Buchler, 1976), cave dwelling birds, and blind people (Griffin, 1973). Flight calls are not emitted by all migrants, but are common enough to warrant consideration as one potential source of information, not only in the form of communication signals from other birds, but as echoes from the ground, as suggested by D'Arms and Griffin (1972). When a direct view of the surface is not possible because of fog or low clouds, acoustic information might not only indicate a bird's altitude but tell it something about the ground or water below.

To estimate the likelihood that a bird emitting a given flight call could hear its echo from the surface the following information is desirable: (1) the bird's altitude, (2) the emitted intensity and frequency pattern of the flight call, (3) the fraction of the incident acoustic energy that is reflected back in the direction of the bird, (4) the reduction in sound intensity due to atmospheric absorption, and (5) the auditory threshold of the bird *in flight*, which may be higher than under quiet laboratory conditions due to wind noise caused by motion through the air. Birds can be treated as point sources gen-

* Present address: Department of Zoology, University of Maryland, College Park, Maryland 20742 U.S.A.

erating spherical waves, and under most conditions the speading loss (the inverse square relationship between distance and sound intensity) can be assumed to be much more important than atmospheric absorption. These two assumptions allow us to make a surprising approximation with relatively minor error: If the surface of the earth were a perfect reflector, the sound level in the echo returning to a bird would be only 6 dB less than that incident at the surface. The flight calls of migrants, audible on quiet nights, are relatively faint, very roughly of the order of 40 dB SPL at ground level. Were the ground a perfect reflector and if atmospheric absorption was negligible, the ground echo of such a flight call might reach a bird at a level of the order of 30 - 34 dB SPL.

Rates of atmospheric absorption are reasonably well known, and the best available values for the frequency and humidity range likely to be significant for the flight calls of migrating birds are shown in Table 1. Note that the effect of relative humidity is complex and frequency

Table 1. Atmospheric absorption of sound as a function of frequency and relative humidity (from Evans and Bass, 1972). These values are for still air at 20°C, but temperature effects are relatively small

Absorption in dB/100 m at various relative humidities					
Freq. (kHz)	20%	40%	60%	80%	100%
1.0	0.513	0.334	0.339	0.371	0.405
1.25	0.747	0.429	0.401	0.424	0.459
1.6	1.16	0.596	0.507	0.507	0.534
2.0	1.17	0.834	0.655	0.619	0.629
2.5	2.63	1.21	0.885	0.790	0.769
3.15	4.04	1.81	1.26	1.07	0.993
4.0	6.23	2.80	1.87	1.52	1.36
5.0	9.21	4.24	2.78	2.19	1.90
6.3	13.5	6.59	4.25	3.27	2.77
8.0	19.4	10.3	6.66	5.06	4.22
10.0	26.1	15.5	10.2	7.70	6.35

dependent (Griffin, 1971). Except at the higher frequencies, or at very considerable altitudes, atmospheric absorption is likely to reduce by only a few decibels the estimated intensity of a ground echo. While altitudes of migrants vary widely, a large fraction are often below 1000 m. This means that it is not as important as it might seem to know the bird's altitude. Provided one assumes that the major factor determining echo intensity is the geometric spreading loss, one can simply reduce the measured level incident at the surface by 6 dB and make whatever further subtraction is indicated by estimates of the reflectivity of the ground surface.

These considerations indicate that the acoustic reflectivity of natural ground or water surfaces is the major variable affecting the audibility of ground echoes from a given altitude. Since scarcely any published data are available, we have made some preliminary measurements. It is convenient to express this in terms of "ground reflection loss," that is, the ratio of the actual echo level to what would be expected from a perfectly reflecting surface. Very few natural land surfaces, especially when covered with vegetation, are perfect reflectors of sound, so that ground reflection losses might be quite large. Smooth water, however, is an excellent reflector of incident sound.

In a preliminary experiment, a 1.57 kHz pulse generator was suspended below a tethered kite balloon that also carried a sensitive radio microphone of the type used to measure ambient sound levels available to migrating birds (Griffin, 1976). Pulse duration was about 125 ms and repetition rate roughly one Hz. The ground being echolocated was a mowed field that varied in altitude by one or two meters over the area below the apparatus. A series of tape-recorded signals obtained with this radio microphone is shown in Figure 1. Each horizontal sweep represents a time interval of 360 ms. In most cases the oscilloscope sweep was triggered by the intense outgoing pulse (which overloaded the radio microphone), but the oscilloscope triggered spontaneously for the 4th pulse so that the entire outgoing pulse and part of the ground echo are clearly displayed. The interval between the two shows that the altitude was approximately 33 m.

The most striking feature of Figure 1 is the enormous variability from one ground echo to the next. While a moderately turbulent wind caused some motion of the apparatus suspended below the kite balloon, most of the variability seems to have resulted from complex interference between echo components from different areas of ground. In an attempt to

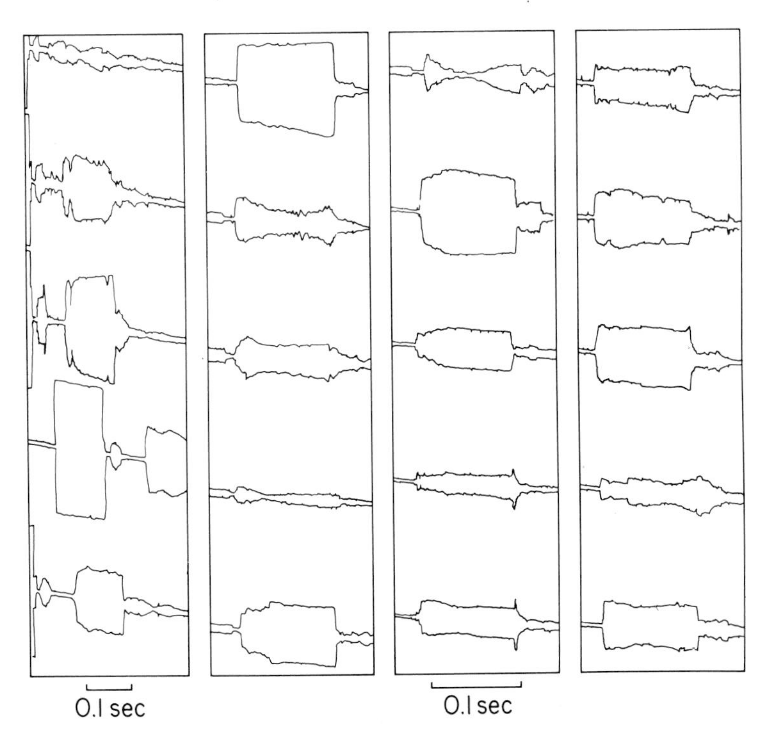

Fig. 1. Echoes of 1.57 kHz sound pulses from a mowed field. Pulse duration about 125 ms, repetition rate about 1/s. Only envelopes of pulse and ground echo are shown, and in all but the 4th pulse of the left-hand column, the oscilloscope sweep was triggered by the outgoing pulse of which only the end is shown. The great variability of echo amplitude resulted largely from interference between echo components from areas of ground at varying path lengths

reduce this variability we turned to direct measurements of ground reflection losses with loudspeaker and microphone held in fixed positions. The most effective arrangement employed to date has been a moderately directional loudspeaker and the calibrated microphone of a sound level meter mounted near the end of a light weight crane at an altitude of 10 - 11 m. The base of the crane is attached by a set of hinges to a trailer, so that the entire apparatus can be moved to locations with various types of vegetation. Either a pulse generator or a tape recorder provides brief signals to the loudspeaker via an appropriate power amplifier. Signal duration is short enough (10 - 20 ms) so that the downward directed signal has ended well before the first reflected waves reach the microphone. The calibrated microphone of a second sound level meter is used to measure the incident intensity at the ground. An initial frustration was that measurements of incident intensity were even more variable than the ground echoes measured by the microphone at 10 m. This was found to result from a mixture of direct waves and reflections from the ground near the microphone. To alleviate this problem the microphone was mounted on the upper end of a vertical pole approximately 4.2 m in length located directly under the loudspeaker and the microphone at the top of the crane. This microphone on the pole measured the signal traveling downward from the loudspeaker en route to the ground, and reflections from the ground reached it only after a long enough interval that they could be easily distinguished from the downward-traveling pulse.

For convenience the microphone mounted on the crane will be called the "echo microphone", and the microphone on the 4.2-m pole the "en route microphone." In no case discussed below could any difference be detected in the signal at the echo microphone when the pole and en route microphone were removed, presumably because the ground surface was large enough that the pole, and its supporting tripod, contributed very little energy compared to the large surface of ground below the crane. The ground echo recorded at the echo microphone was of course lower in intensity than the outgoing signal; and, when gain of the system was adjusted to optimize ground echoes, the outgoing signal was usually distorted. It did, however, provide a timing signal so that the acoustic path length to the en route microphone and to the reflecting ground surface could be measured quite accurately. The two microphones were compared by placing them both on 4.2-m poles within a few centimeters of each other and all measurements discussed below have been corrected for any small differences that existed between the sensitivities of the two sound level meters (Bruel and Kjaer 2209 and General Radio 1561). Varying the length of the pole confirmed that levels at the en route microphone varied inversely as the square of the distance, i.e., the loudspeaker generated spherical wave fronts.

To measure ground reflection loss, the signal level recorded by the en route microphone was corrected to allow for the round trip distance to the echo microphone. This corrected value corresponds to the signal expected at the echo microphone if the ground were a perfect reflector. Measurements with this apparatus from relatively level surfaces covered with coarse grass or a mixture of grass, weeds, and low bushes showed a surprising amount of variability. Over the frequency range from 1 - 6 kHz the ground reflection loss varied between approximately 0 and 20 dB with a pronounced tendency to increase with frequency. Table 2 shows some typical measurements from three areas described as "rough weeds," "ca. 30 cm grass," and "small leafy trees." The first area was a highly irregular mixture of many types of vegetation characteristic of recently abandoned gardens. The low trees were locust trees of mixed height ranging up to about 3 m, with a wide variety of grasses and low shrubs also in the area below the loudspeaker. The variability indicated by Table 2 is typical of many such measurements, and consideration of

Table 2. Measurements of ground reflection loss (dB) using directional loudspeaker and calibrated microphone at about 10 m

Frequency (kHz)	Rough weeds	Mean	ca. 30 cm grass	Mean	Small leafy trees
1.0	7.1, 6.2	6.7	4.0, 5.2	4.6	5.4
1.5	4.5, 10.7	7.6	1.1, 8.0	4.6	8.8
2.0	1.3, 10.1	6.7	7.5, 7.5	7.5	12.6
2.5	-1.2, -1.8, 3.0	0.	3.0, 4.8	3.9	18.4
3.0	10.3, 7.3 7.3	8.4	3.0, 3.3, 8.3, 8.1	5.5	14.3
3.5	8.5, 12.8	10.7	8.4, 8.7	8.6	16.1
4.0	10.7, 8.8, 11.0	10.2	7.9, 4.2	6.1	16.5
4.5	14.8 14.6	14.7	8.5 3.8	6.2	15.8
5.0	13.7, 10.6	12.2	11.7, 6.1	8.9	17.1
5.5	8.1, 12.9, 10.8	10.6	19.3, 10.8	15.1	13.8
6.0	16.3, 12.8, 15.0, 12.2	14.1	14.1, 11.9	13.0	16.8

these unsatisfying data led to a realization that simple physical phenomena involving interference between different components of the ground echo were responsible. Furthermore it become clear that such interference patterns must be of great importance whenever sounds emitted by flying animals return to their ears audible echoes from large surfaces such as the ground. One simple indication that complex mixtures of constructive and destructive interference were responsible for this variability was to make small changes in either the frequency or the position of the echo microphone. In either case large fluctuations (10 dB or more) were often observed, including occasional negative values of ground reflection loss - as at 2.5 kHz in Table 2. These presumably resulted from unusually favorable interference patterns. On windy days movements of the vegetation or of the light crane caused similar fluctuations. Such variability is likely to be even greater for birds flying at altitudes of 100 m or more, because there will be even greater differences in effective path lengths for returning echoes and even greater opportunities for wave interference to produce strong peaks or nulls.

It is well known that a smooth surface of water is an almost perfect reflector of sound waves. The apparatus described above has been used at the edge of a small fresh-water pond, and when there was no wind and the water surface was glassy smooth the echoes were, as expected, strong and nearly constant from moment to moment. Echo levels agreed very well with predictions based upon measurements with the en route microphone as discussed above. In the course of recording echoes from the surface of this pond it became apparent that even relatively small waves sometimes produce large fluctuations in echo level. When small waves (not more than 3 - 5 cm in peak-to-peak amplitude) were set up by means of a canoe paddle moved vigorously through the water about 10 m from the point directly below the echo microphone, fluctuations of 10 dB or more often occurred, and some frequencies showed greater fluctuations than others.

It seems clear that water waves produce interference patterns which fluctuate from moment to moment. Furthermore these are quite frequency dependent. Under the circumstances of these particular measurements the fluctuations of frequencies below 3 kHz were small compared with those observed at 3.5, 4, or 5 kHz. Larger waves would presumably have similar effects at lower frequencies.

Partly to study the frequency-dependent character of fluctuations in ground echoes and echoes from water surfaces with waves, similar measurements were made using chirps instead of constant frequencies. Both the flight calls of some migrating birds, and the orientation sounds of bats under many different conditions, involve just such rapid frequency sweeps. Despite the complexities of the echoes of chirps, they may offer certain advantages to animals concerned with echoes from large surfaces. For example some frequency within the range included in the chirp will encounter favorable interference patterns. In almost no case when chirps with an octave of frequency sweep were employed did all parts to the echo drop to very low levels, although this did often happen with constant frequency signals. Signals sweeping from approximately 4 - 2 kHz show much more complicated interference patterns than 2 - 1 kHz chirps. This is the case both for echoes from ground and water. A typical example is shown in Figure 2 in which 4 - 2 kHz chirps were recorded after reflection from a water surface disturbed by small waves. When the surface was smooth all frequencies gave strong echoes, and the echoes were constant within ± 1 dB for every chirp. However, it is clear from Figure 2 that within a few seconds, the echoes underwent strong and frequency dependent fluctuations.

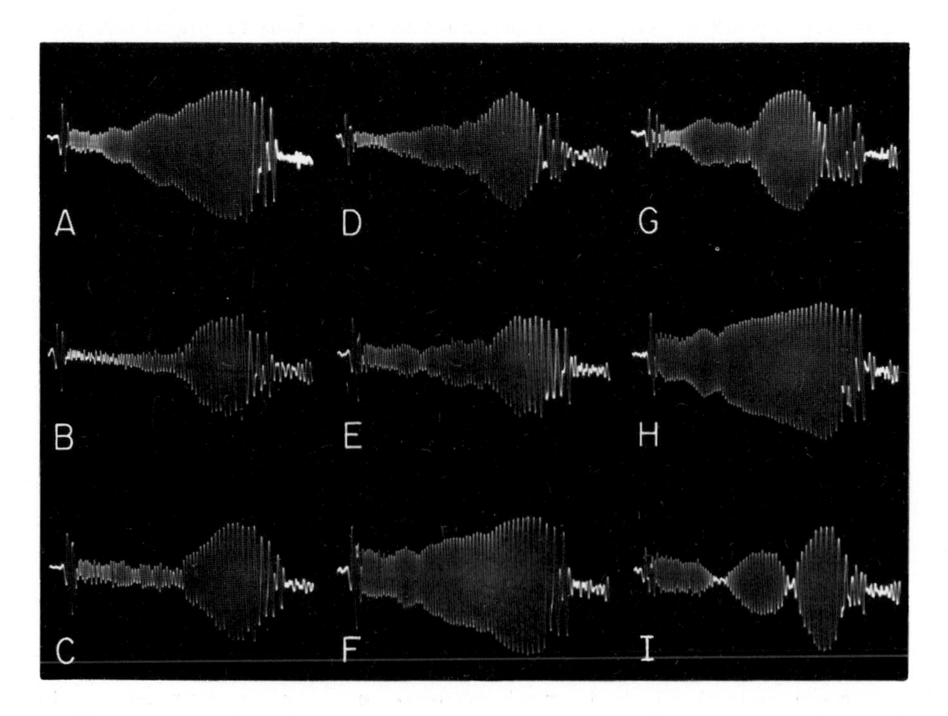

Fig. 2. Echoes of 4 - 2 kHz chirps from a water surface disturbed by small waves. A to F and G to I are two continuous series at a repetition rate of about 1/s

Discussion

Despite the exasperating variability of measured echo levels from natural ground surface and from water disturbed by even small waves, an important conclusion may be drawn from the data discussed above: The interference patterns that produce this variability may serve to increase the altitude at which a bird or bat can hear some fraction of the echoes reaching it from the ground or other large surfaces. The erratic fluctuations in echo level due to interference effects must result in many echoes being far less intense than the average values encountered at a particular point. However, it may well be that migrating birds do not need to hear echoes from *every* flight call and that hearing such echoes from one in ten or even a hundred flight calls would suffice to inform them of their altitude. The same consideration applies to different frequencies in a call containing a broad band of frequencies. While interference effects will reduce some components below the average levels, others will be correspondingly enhanced. Thus if a bird can obtain the information it needs from some fraction of the total number of echoes, or of frequency components within these echoes, it might be able to hear some ground echoes at altitudes very much greater than one would expect from estimates based on average values. This consideration raises the question of the magnitude of such interference-induced fluctuations. None of our measurements to date provide adequate answers to this question, because our instruments were at relatively low altitudes. However, the data shown in Figure 1, with the echo microphone at about 33 m altitude, and some measurements with the crane from 10 - 11 m, have shown fluctuations as great as 20 dB over periods of several seconds or frequency changes of 100 - 200 Hz or less.

If we take a figure of 10 dB as a conservative rough estimate of common magnitudes of such fluctuations in ground echo levels, this can be converted to an increase in altitude at which a given ground echo might be audible to a migrating bird. In such calculations the actual altitude of the bird and the frequencies employed become quite important. This is because atmospheric absorption varies both with distance and with frequency, as shown in Table 1. If only the speading loss is considered, a variation of 10 dB in ground reflection loss can produce a threefold increase in altitude of audibility. This will be reduced, however, to the extent that atmospheric absorption becomes larger relative to the geometric spreading loss. However, it does seem possible that twofold variations in altitude of audibility may occur as a result of variations in the kinds of interference patterns discussed above.

Fluctuations in the intensity of echoes from large surfaces may also be important to bats and cetaceans. Although most considerations of echolocation have involved targets small enough that each echo is a relatively simple train of sound waves, large surfaces are also commonly encountered by animals moving through the air or under water. Under many circumstances echoes from such large surfaces doubtless constitute interference or clutter, but it may on some occassions be important for the animal to determine the distance and nature of large surfaces such as the ground, the bottom of the ocean, or the air-water interface above an aquatic animal. The same basic considerations outlined above for the case of a bird listening for ground echoes of its flight calls are applicable to any situation in which an echolocating animal is obtaining information from the complex of echoes from a large surface.

References

Buchler, E.R.: The use of echolocation by the wandering shrew (*Sorex vagrans*). Anim. Behav. 24, 858-873 (1976)

D'Arms, E., Griffin, D.R.: Balloonists' reports of sounds audible to migrating birds. Auk 89, 269-279 (1972)

Evans, J.B., Bass, H.E.: Report WR72-2, Wyle Laboratories, Huntsville, Alabama (1972)

Griffin, D.R.: The importance of atmospheric attenuation for the echolocation of bats (Chiroptera). Anim. Behav. 19, 55-61 (1971)

Griffin, D.R.: Echolocation. In: Basic Mechanisms in Hearing Moller, A.R. (ed.) New York: Acadmic Press, 1973

Griffin, D.R.: The audibility of frog choruses to migrating birds. Anim. Behav. 24, 421-427 (1976)

Simmons, J.A.: Localization and identification of acoustic signals, with reference to echolocation. In: Recognition of Complex Acoustic Signals Bullock, T.H. (ed.), Berlin, Dahlem Konferenzen, 1977

Radar Observations of Behavior of Migrating Birds in Response to Sounds Broadcast from the Ground

RONALD P. LARKIN, The Rockefeller University, New York, NY 10021, USA

Abstract

Field experiments with a tracking radar and loudspeaker system have been performed to investigate birds' behavior in response to sounds (thunder, bird calls) heard during migration. Some, but not all, sounds elicited changes in flight paths. Most reactions occurred quickly and were turns away from the source of the sound. Thunder was more effective on totally overcast than on clear or partly cloudy nights.

The experiments constitute evidence that birds migrating at night attend to and, in a least some cases, respond appropriately to sounds heard at moderate levels of intensity.

Birds on the breeding or wintering grounds are mainly diurnal animals (sometimes nocturnal predators or frugivores); one takes it for granted that sounds and hearing are important to them in many ways. However, the role of sounds has often been ignored in discussions of the largely nocturnal migrations of birds in favor of the visual, vestibular, and proposed magnetic sensitivity of birds. Partly this is because sounds at the conventional audio frequencies attenuate in air over distances that are short compared to the distances with which birds must contend while migrating or homing. Therefore, audio sounds cannot directly furnish long-distance orientational cues. Nevertheless sounds may provide information on altitude, landmarks, habitats, winds, and weather systems, or may furnish a means for social communication during nocturnal migration.

At least two kinds of sounds are audible to birds during migrations. Some birds vocalize while migrating at some times of the year; although the functions of such "nocturnal flight calls" are almost completely open to speculation, they are an important feature of the birds' acoustic environment (Ball, 1952; Graber and Cochran, 1960; Hamilton, 1962; Larkin, unpublished observations). In addition, natural and human-produced sounds from the ground are audible even at altitudes of 1 - 2 km (Griffin and Hopkins, 1974; Griffin, 1976).

To investigate the possible use of sounds by migrating birds, several colleagues and I have broadcast sounds from a site on the ground and, using a tracking radar, have studied the reactions of nocturnally migrating birds. I report here a partial analysis of reactions to several sounds that were chosen for their potential biologic meaningfulness (or lack thereof) across species, since it is usually not possible to select or identify species using radar.

A. Methods

Experiments were conducted at the Rockefeller University Center for Field Research near Millbrook, New York. The radar has been described in several previous publications (Griffin, 1972; Larkin et al., in press). It is a conical-scan nutating tracker operating in X-band with a 4-kHz nominal pulse repetition frequency. It can track single birds at close range (under 100 m) and it has been modified to be quiet while tracking. The radar controls a separate "slave" unit, located 10 m distant. During the entire time a track is in progress, the slave unit accurately and quietly aims an audio loudspeaker at the bird which is being tracked by the radar. The speaker mounted on the slave unit is surrounded by a large (4 m across) wooden cuff which attenuates horizontal sound transmission but provides a free sound path to the bird. The experiments took place at night, starting between 1800 and 2130, depending on the season, and the use of lights possibly visible to the birds was avoided during experiments. Sessions ended at 2400 regardless of whether reactions to sounds occurred. Data were taken from October, 1975 to June, 1977.

Sounds were broadcast from a conventional high-fidelity system consisting of a Uher 4000 Report-L tape recorder, one channel of a Dynaco SCA-80 amplifier with all settings approcimately neutral so that the signals were not filtered, and an Electrovoice Musicaster IIa outdoor loudspeaker system with a Model 30R reentrant paging projector. Each sound was calibrated for intensity by connecting an oscilloscope across the speaker connections and adjusting the amplifier gain until the voltage of the strongest part of the signal was almost saturated. The gain setting that produced this voltage was then recorded and used, with periodic readjustments, throughout the experiment. The quality of the sound was also monitored by ear each night; on a few nights the sound system audibly malfunctioned and the data from these nights were discarded. The broadcast sounds were measured and recorded using a calibrated microphone suspended on-axis about 25 m from the speaker at an elevation of 45^{0} and the sound levels at the distance of the birds were calculated considering attenuation and spreading loss.

Sounds that were used as stimuli are shown in Figures 1 and 2. The call note of a Veery (*Catharus fuscescens = Hylocichla f.*) (1A) is commonly heard during nocturnal migration. Calls of a Golden-crowned Kinglet (*Regulus satrapa*) (1B) are not heard during nocturnal migration. A single thunderclap (1C) was followed by 8 s of rumbling sounds of diminishing intensity, then by a second clap of lower intensity than the first. Calls given by a Sharp-shinned Hawk (*Accipiter striatus*) (2A) were originally recorded from an adult while its nest was being disturbed. Distress calls of a Mockingbird (*Mimus polyglottos*) (2B) were originally recorded by holding the bird in the hand and recording the ensuing vocalizations. An artificial pulsed sound (2C) was recorded from an electric alarm clock. The particular selections of these stimuli were: Veery, 2 notes separated by 4 s; Kinglet, 7 - 9 s of call notes; thunder, two claps making a sound 10 s long; Hawk, 3 bouts total duration 8 - 10 s, the intensity decreasing within a bout but increasing over the 3 bouts; Mockingbird, a continuous series 9 - 11 s long; and artificial pulsed sound, 8 - 11 s. The bird sounds were obtained from the Cornell Laboratory of Ornithology Library of Natural Sounds (cuts 1, 4, 20, and 50), the thunder from a sound-effects record.

The procedure was standardized after pilot trials. A radar operator searched for bird targets in the azimuths from which most migrants were approaching the radar, using a fixed search elevation of 25 - 30^{0} and tracking any bird in the range interval of 275 - 450 m. After

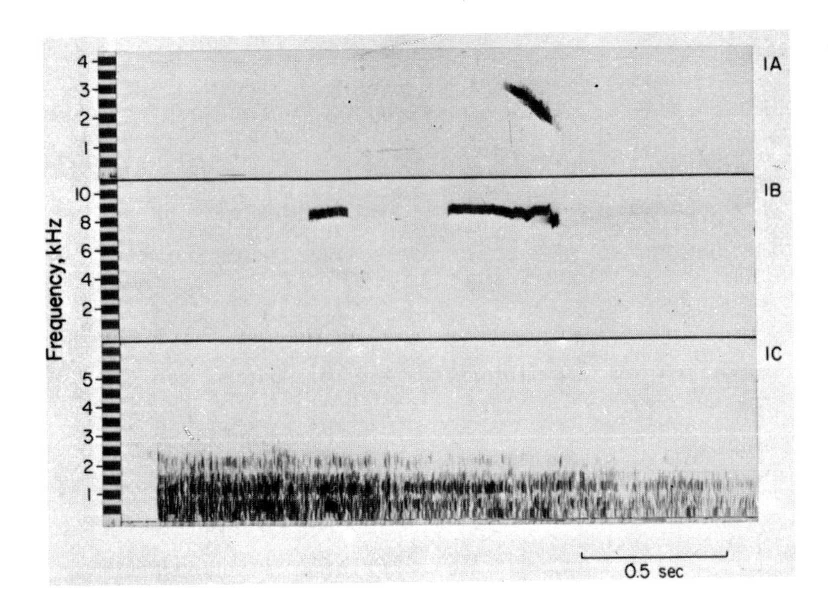

Fig. 1 A - C. Sound spectrograms of three stimuli. (A) Veery call note. (B) Golden-crowned Kinglet call notes. (C) Thunder. Spectrograms were made on the wide-band setting from recordings taken 25 m from the speaker

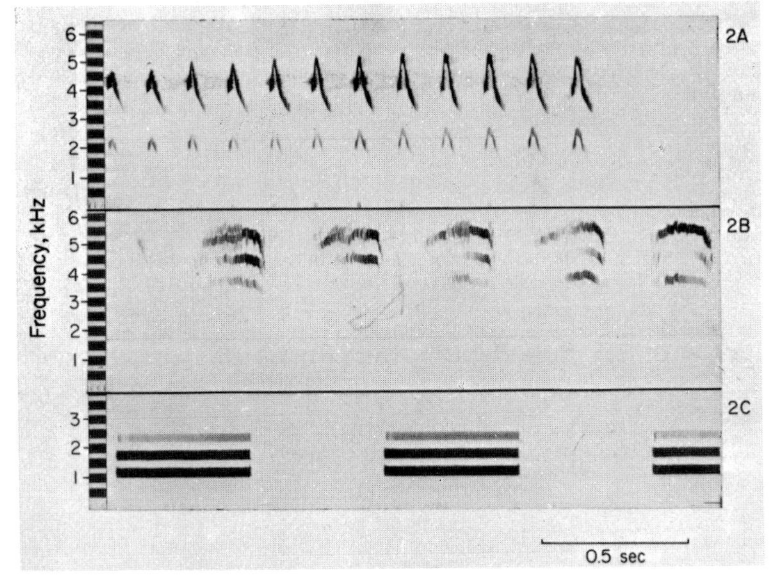

Fig. 2 A - C. Sound spectrograms of three stimuli. (A) Call of a Sharp-Shinned Hawk. (B) Distress calls of a Mockingbird. (C) Artificial pulsed sound (alarm tone from an electric clock-radio)

automatic tracking began, the operator kept the radar gain properly adjusted and watched an A-scope display for signs of switching between targets. When the bird approached to a range of 250 m, a second experimenter started the tape recorder, the specific stimulus having been chosen from a predetermined series of recorded sounds and controls (no sound) that was unknown to the operator. The bird was thus acquired and tracked by an operator working under blind conditions. The radar continued tracking until the bird target was lost for some reason (usually intervening trees) or until at least 20 - 30 s had elapsed after the end of the sound.

The period during and after a sound broadcast or a control will be called a "trial." Trials were aborted during an experiment or discarded later due to: (1) the bird curving or beginning to climb or drop during the period before the onset of a trial; (2) less than 5 s of straight track before onset; (3) the radar switching between one target and another; or (4) failure to track the bird during the entire trial.

A minicomputer system at the radar site displayed each bird's position while the bird was being tracked and automatically recorded it once per second in XYZ coordinates with a resolution of 1 m. In addition, the computer record indicated time of occurrence of the trial to the nearest second.

From the birds' position the sounds occurred at an angle from 35° to an extreme of about 75° down from the horizontal, from somewhere ahead of the bird, and always at the same intensity. Only one sound was played to each bird.

To estimate the frequency of reactions to the sounds, large plots of the birds' XY tracks and altitudes were examined by eye. Tracks were classified according to whether they turned toward the source of the sound, away from it, or not at all in the XY plane during the first 10 s following the onset of the trials. Similarly, changes in altitude were classified as ascents or descents according to whether the bird changed its angle of climb relative to the initial section of track before the stimulus; that is, no reaction would be scored on a bird which had been flying straight but ascending before the stimulus onset and which continued to ascend at the same rate. Most birds, in fact, flew level before stimulus onset.

B. Results

Peak sound levels for each of the stimuli, as calculated for the position of the bird at 250 m away, are shown in Table 1. The frequency-dependent atmospheric absorption of sound was an important variable determining the sound levels present at the position of the bird. The thunder, Veery call, and artificial pulsed sound were approximately

Table 1. Estimated peak sound levels for broadcast sounds at 250 m from the source.

Sound	Approximate peak frequency, kHz	Calculated attenuation dB/km[1]	Sound level 250 m, dB
Thunder	0.5	1.5	50
Sharp-Shinned Hawk	4	46.5	39
Golden-Crowned Kinglet	8	144.5	11
Mockingbird	5	68.7	33
Veery	2.5	19.7	52
Alarm clock	1.2	5.6	53

[1] At 0 C and 80% R.H. From Evans, 1974.

equal in intensity, the Sharp-Shinned Hawk and Mockingbird less intense, and the Kinglet call almost certainly inaudible to the birds. Ambient sound level readings at the appropriate altitudes were not made on nights when sounds were broadcast.

The fraction of birds that reacted to sounds varied both among the different sounds used and from night to night. Reactions included turns, altitude changes, and speed changes, or sometimes a combination of these. Sample responses to one of the stimuli, the distress call of the Mockingbird, are illustrated in Figure 3. The top four show the first four times the distress calls were used on one particular night. They illustrate a pattern of response to sound that is typical for several stimuli and many nights. The birds turn in the XY plane, and the turns are away from the sound. Most reactions occur quickly, within a few seconds after the onset of the sound. After the initial turn, some birds continue on their new course, but most birds, even those making sharp turns, regain their former course within a short time after the end of the sound. Altitude changes, primarily

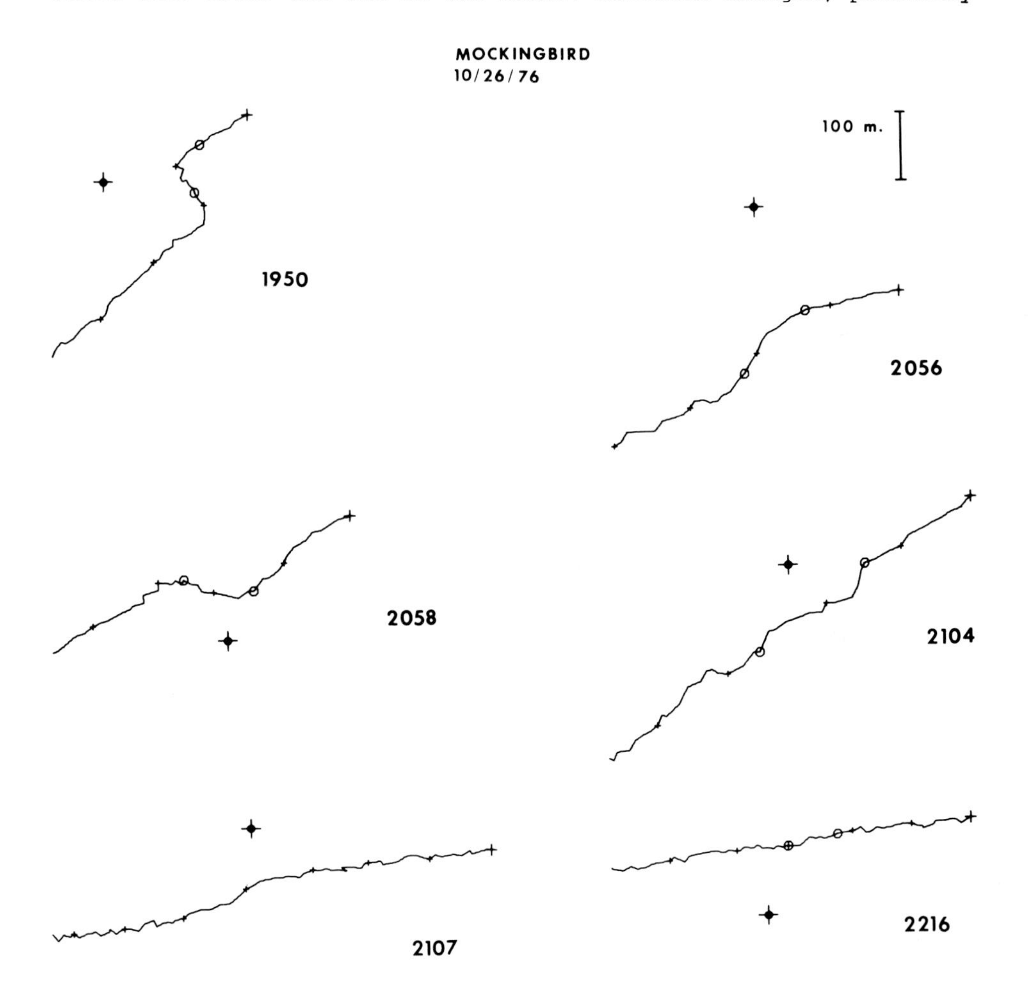

Fig. 3. Examples of reactions to distress call of a Mockingbird. Tracks proceed from right to left and are records of XY position taken once per second. *Small + marks* appear each 10 s and a *large + mark* indicates the position of the radar and sound source for each track. A *small circle* denotes the onset of the trial and another *circle* the end of the trial

descents (not illustrated in Fig. 3) sometimes accompanied XY turns and sometimes occurred without any noticeable XY deviations.

The bottom two tracks are at variance with the turning pattern of the top four birds in Figure 3. Some birds do not visibly react to the distress call (2216). And some birds in the control condition turn even when no sound is broadcast (2107). "Spontaneous" turns such as this might occur at any time, such as before a trial (resulting in an aborted trial) or many seconds after the end of a trial when distant from the radar (not affecting the data).

Estimates of whether a reaction occurred showed reasonably good agreement between investigators. In addition, an objective, quantitative analysis (to be reported elsewhere) agreed with the subjective estimates.

Differences between several bird calls in the percent of reactions elicited from migrants are shown in Figures 4 and 5. Few or no reac-

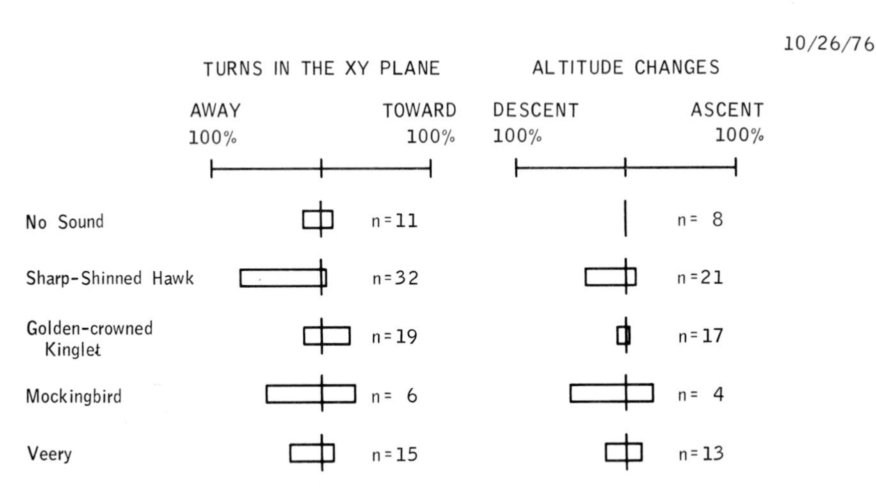

Fig. 4. Percent of birds reacting to control trials (no sound) and each of four bird calls during a night of fall migration. *Left*, percent of birds turning toward and away from the radar in the XY plane; *Right*, percent of birds ascending or descending relative to their previous flight. The data for XY turns and altitude changes were tabulated separately. Some tracks containing altitude artifacts were discarded prior to the analysis

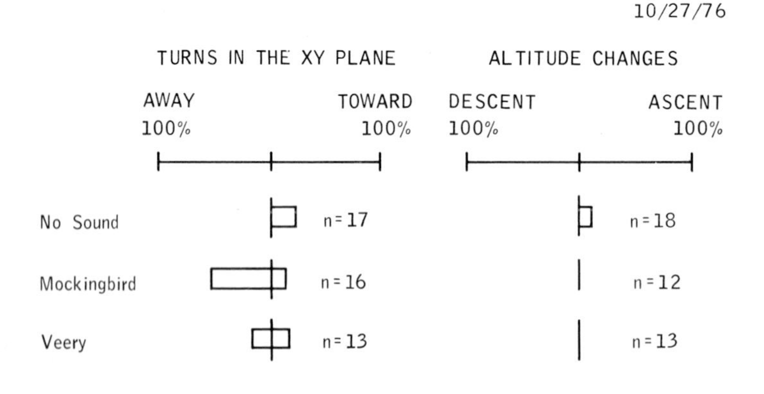

Fig. 5. Percent of birds reacting to control and each of two bird calls during a night of fall migration. For explanation see Figure 4

tions are elicited by the control (no sound) condition and by the recordings of the Veery and Golden-crowned Kinglet call notes. The call note recordings have elicited no more reactions than the control trials on any night. The calls of the Sharp-Shinned Hawk and the distress calls of the Mockingbird generally elicited many more reactions in the XY plane than did controls, and sometimes more descents as well (Fig. 4). When responses to sounds occurred, turns away from the radar predominated over turns toward the radar and descents were more frequent than ascents after the sound onset.

The artificial pulsed sound elicited no more reactions than the control trials on two nights in the fall of 1976. Experiments conducted on five nights in spring, 1977, showed no more responses to any of the bird call stimuli than to control trials. A seasonal effect might be indicated, although a tape recorder malfunction could have decreased the sound intensity on some trials during spring, 1977.

On two nights when thunder was used as a stimulus, nearly half the control population and the same fraction of the stimulated population "reacted" during a trial. Each of these nights was characterized by strong, gusty winds. The data from these two nights will not be further considered. The sound of thunder was effective in eliciting reactions from migrants on some nights but not on others. With the aforementioned exceptions, the number of birds which reacted in control trials was low, averaging 11%. Reactions to thunder occurred in spring and fall. On nights when reactions to thunder occurred, 90% were turns away from the sound in the XY plane, sometimes preceded by brief "hovers," or periods of a few seconds when the XY ground position remained virtually stationary. Since migrating birds usually fly in a generally downwind direction on a night of heavy migration, those which "hover" are usually facing upwind, away from the sound, during the "hover."

As shown in Figure 6, barometric pressure was not an important variable influencing the frequency of reactions on a given night. Cloud cover, however, had a strong effect on whether birds reacted (Fig. 7). In clear, partly cloudy, or clearing conditions, fewer than about half of the birds reacted; in total overcast, more than about half of the birds reacted.

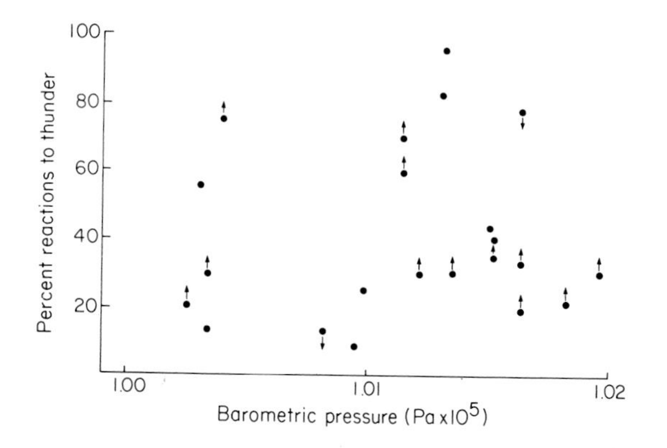

Fig. 6. Percent of birds reacting to broadcast sounds of thunder plotted against barometric pressure. Barometric pressure was measured at a site about 30 km away at the time of the first track in the experimental session. *Small arrows* affixed to points indicated rising or falling barometric pressure

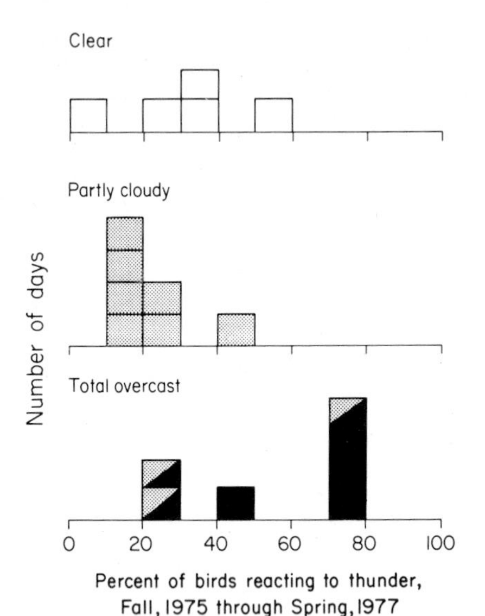

Fig. 7. Percent of birds reacting to broadcast sounds of thunder in different cloud conditions. *Partly cloudy* conditions are 10 - 80% cloud cover estimated by an observer at the radar site. In the *Total overcast* condition, *half-shaded boxes* indicate nights when total cloud cover cleared during an experimental session (2 boxes in the 25% interval) or was high altostratus cloud (1 box in the 75% interval). All nights during which the recording of thunder was played at least four times and during which controls flew straight are included (n = 4-32, mean n = 15 thunder trials per night)

C. Discussion

The data presented here indicate that on some nights migrating birds attend to and discriminate between certain sounds, and that in at least one instance they react to the sound of thunder in a biologically meaningful fashion.

The broadcast sounds were adjusted to be intense enough so that most birds probably heard the sounds, but not so intense that a startle reaction was likely because of loudness alone. Most sound levels approximated ambient levels of natural sound levels recorded at the same site during warm nights in May (Griffin, 1976). The fidelity of sound reproduction appeared to be good, both to the human ear and on the sound spectrograms. A visitor to the radar site on one clear evening returned to obtain his raincoat when he heard the tape recorded thunder from a distance.

On most nights, a small proportion of birds "reacted" during control trials. As suspected during previous work showing reactions to extremely low-frequency radiation (Larkin and Sutherland, 1977), abrupt changes in wind velocity near the ground are important causes for small turns and other maneuvers made by undisturbed migrants. On several nights, subjective impressions of observers on the ground, anemometer readings at tree-top height, and radar tracks of free-floating balloon targets all indicated that gusty wind or turbulence was highly correlated with a high proportion of "spontaneous" reactions. Error in measuring and recording the XY position of the bird was not a significant source of artifact in these experiments, although some altitude artifacts occurred.

When actual reactions to sound occurred, they often occurred soon after the onset of the stimulus. Since the sound began when the bird was 250 m distant, it took approximately 0.7 s for the sound to reach the bird. In addition, an average delay of 0.5 s was introduced during the recording procedure. One of the recordings, the Golden-crowned Kinglet call, was attenuated to an extremely low level (Table 1); it elicited no more responses than the control trials. The Sharp-Shinned

Hawk call recording increased in intensity over a period of 8 - 10 s. Responses to this stimulus sometimes occurred near or immediately after the last and most intense calls; at this time the bird was also approaching the sound and usually at a range of less than 250 m. Especially in the case of birds in this small category of "delayed" reactions, but also for birds which failed to react to any one of the sounds, the inevitable increase in stimulus intensity as the bird approached the sound source must be considered. The Sharp-Shinned Hawk call (which can probably be considered both a predator call and an alarm call) and the Mockingbird distress call were chosen because of their salience for many species of migrating birds. They elicited many more reactions than the call notes of the Veery, even though they were probably not as loud as the Veery call, indicating that different bird sounds have different effects upon migrating birds. Although differences in numbers of responses between the Veery call and the Sharp-Shinned Hawk and Mockingbird calls cannot be attributed to differences in sound intensity, any of several acoustic or biological factors could account for the difference. The spectral distributions of the calls, their repetition rate, their familiarity to nocturnal migrants, or their specific function in communication may be important.

An effect of weather upon responses to thunder was suspected during the first season of experiments; consequently, much more effort has been devoted to thunder than to other stimuli. No correlation with barometric pressure is present. The relationship to cloud cover, necessarily a loose one when cloud conditions change during a night's work, is unambiguously positive: Birds reacted to the sound of thunder only on nights with low, total cloud cover. Unfortunately, conditions with thunderclouds in the distance and relatively clear sky overhead did not occur during these experiments. Without such a fortunate circumstance, it is impossible to discriminate clearly between two hypotheses that can explain the reactions to thunder. One hypothesis is that the orientation mechanisms used by the birds are different under conditions of total overcast, as has been repeatedly demonstrated for homing pigeons (Keeton, 1974). If so, then perhaps the orientation mechanism used under total overcast is more susceptible to being perturbed by the occurrence of a sound ahead of the bird. An alternate and perhaps more attractive hypothesis is that many migrating birds associate total overcast with the possibility of rain and turbulence, and that the reactions are part of a normally occurring pattern of avoidance of storms.

Although no amplitude spectra on the thunder recordings have been made, it is certain that the natural infrasonic portions of this sound below 20 Hz were lacking from the broadcast stimulus. Infrasonic sounds may be an important part of birds' acoustic environment (Yodlowski, et al., 1977).

Except for occurrence of altitude changes following stimulus onset on some nights (Fig. 3) we have thus far not been able to relate any specific kind of turn or maneuver to the stimulus being used. Since a wide variety of unknown species of birds take part in these experiments, this is not too surprising. Future experiments will employ a variety of different stimuli, seeking to find one that has an effect other than causing birds to turn away from the source of the sound.

Previous experiments have raised the possibility that lights mounted on aircraft may be able to help in preventing collisions between aircraft and birds (Larkin et al., 1975). The present experiments suggest that distress calls or other sounds might also prove effective in scaring birds away from the path of an aircraft in flight, provided that the aircraft is moving more slowly than the speed of sound.

Acknowledgments. National Science Foundation Grant BNS 74-07693 supported this work. The data were collected and analyzed with the collaboration of Donald R. Griffin, Pamela J. Sutherland, and David Thompson. The recordings of bird calls were supplied by the Cornell Laboratory of Ornithology Library of Natural Sounds through the generous cooperation of Dr. James Gulledge.

References

Ball, S.C.: Fall migration on the Gaspe Peninsula. Peabody Mus. Nat. Hist. Bull. 7, 1-211 (1952)

Evans, L.B.: Atmospheric absorption of sound: temperature dependence. Report no. D3-9190, September 23, 1974, Boeing Corp., Wichita, Kansas, 1974, 265 pp.

Graber, R.R., Cochran, W.W.: Evaluation of an aural record of nocturnal migration. Wilson Bull. 72, 253-273 (1960)

Griffin, D.R.: Nocturnal bird migration in opaque clouds. In: *Animal Orientation and Navigation.* Galler, S.R., Schmidt-Koenig, K., Jacobs, G.J., Belleville, R.E. (eds.), Washington, D.C.: U.S. Government Printing Office, 1972, pp. 169-188

Griffin, D.R.: The audibility of frog choruses to migrating birds. Anim. Behav. 24, 421-427 (1976)

Griffin, D.R., Hopkins, C.D.: Sounds audible to migrating birds. Anim. Behav. 22, 672-678 (1974)

Hamilton, W.J. III: Evidence concerning the function of nocturnal call notes of migratory birds. Condor 64, 390-401 (1962)

Keeton, W.T.: The orientational and navigational basis of homing in birds. In: Advances in the Study of Behavior 5, 47-132, New York: Academic Press, 1974

Larkin, R.P., Griffin, D.R., Torre-Bueno, J.R., Teal, J.: Radar observations of bird migration over the Western North Atlantic Ocean. Behavioral Ecology and Sociobiology (in press)

Larkin, R.P., Sutherland, P.J.: Migrating birds react to Project Seafarer's electromagnetic field. Science 195, 777-779 (1977)

Larkin, R.P., Torre-Bueno, J.R., Griffin, D.R., Walcott, C.: Reactions of migrating birds to lights and aircraft. Proc. Nat. Acad. Sci. USA 72, 1994-1996 (1975)

Yodlowski, M.L., Kreithen, M.L., Keeton, W.T.: Detection of atmospheric infrasound by homing pigeons. Nature (London) 265, 725-726 (1977)

Importance of the Daytime Flights of Nocturnal Migrants: Redetermined Migration Following Displacement

SIDNEY A. GAUTHREAUX, JR., Department of Zoology, Clemson University, Clemson, SC 29631, USA

Abstract

The daytime flights of nocturnal migrants are examined in relation to the principal axes of migration (P.A.M.) based on calm nights during spring and fall in northwest South Carolina. When nocturnal flights were to the east or west of the calm night directions (30° spring, 232° fall), flights occurred the next morning basically perpendicular to these directions. Because most migrants were displaced east of their P.A.M. in spring and fall, most daytime flights were directed toward the NW, but when nocturnal flights were west of the P.A.M., flights the next morning were to the SE. The daytime redetermined migrations permit the displaced migrants to regain their migratory route by flying the shortest distance.

A. Introduction

Ornithologists in North America and in Europe have frequently observed, in the spring and fall, flights of typical nocturnal migrants during the morning hours after sunrise. Often these flights are in directions that are "reversed" with reference to the normal seasonal direction of nocturnal migration, and because the movements are often nearly into the wind, some investigators have called the morning flights "headwind migrations" (Rabøl, 1964, 1967, 1969, 1974; Rabøl and Hindsbo, 1972; Christensen and Jacobsen, 1969; Gruys-Casimir, 1965). Although these movements have been extensively described in the literature, their function and importance have been points of conjecture. Along the Atlantic coast of the United States in the fall, flights of nocturnal migrants in northwesterly and northerly directions in the hours following sunrise are often spectacular (Baird and Nisbet, 1960; Bagg, 1950; Murray, 1976; Able, 1977). Similar flights occur along the Pacific coast of California (DeSante, 1973). It is thought that these flights enable the birds to regain land after being drifted offshore by wind during the previous night's migration. In this paper I report on my studies of a similar phenomenon, but the flights take place *inland*. The occurrence of these morning flights at several inland localities in the United States during the spring and fall suggests that the movements are adaptive and function to correct for drift sustained during nocturnal migration. I refer to the morning flights as redetermined migrations as defined by Lack and Williamson (1959).

B. Methods

I gathered data on nocturnal migration from the fall of 1970 through the spring of 1977. Three methods were used in acquiring the data: (1) direct visual observations using the ceilometer technique (Gauthreaux, 1969; Able and Gauthreaux, 1975); (2) direct visual observations using the moon-watching technique (Lowery, 1951); and (3) radar observations using the ASR-4 radar operated by the Federal Aviation Administration (Gauthreaux 1973, 1974, 1975, 1976 a, b). During the fall of 1975 and the spring and fall of 1976, I used a 7x image intensifier (AN/TVS-2) directed up the vertical beam of a 300 W very narrow spotlight to view nocturnal migration. In the spring of 1977 similar observations were made with a newer model 7x image intensifier (AN/TVS-5). The image intensifiers markedly enhanced the detection and quantification of nocturnal migration aloft (Gauthreaux, in prep.). Radar observations were made at the Greenville-Spartanburg Airport, and direct visual observations were made at the Airport and in the Clemson-Seneca areas of South Carolina.

During the fall of 1976 and the spring of 1977, I made detailed observations of the early morning (daytime) flights of nocturnal migrants on the shore of Lake Hartwell at Clemson. All birds detected overhead or taking off from surrounding woodlands were followed with 10 × 50 binoculars until they disappeared on the horizon, and the disappearing azimuth of each was recorded using a surveying instrument. These watches usually lasted for several hours, beginning shortly after sunrise and ending about 10:00 E.S.T. The statistical analyses of directional data in this paper follow Batschelet (1965), Mardia (1972), and Zar (1974).

C. Results

I present a diagrammatic model in Figure 1 that summarizes the hypothesis set forth in this paper - that after sunrise the morning flights

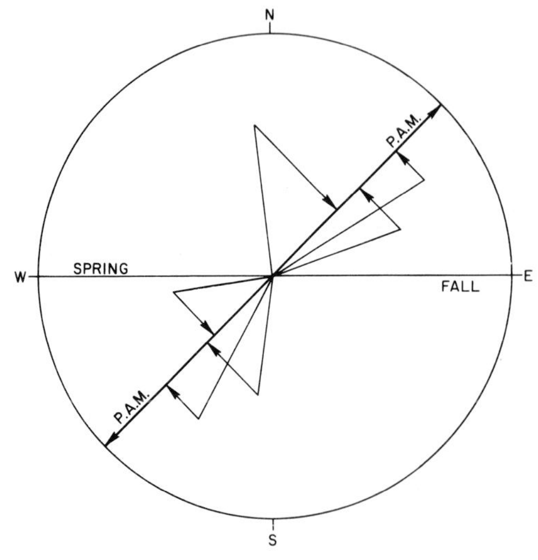

Fig. 1. A diagrammatic model of redetermined migration. The vectors on either side of the principal axis of migration (*P.A.M.*) represent the tracks of drifted migrants during nocturnal migration. The *arrows* directed perpendicular to the P.A.M. represent the morning redetermined migrations. By flying perpendicular to the P.A.M. a migrant can correct for its displacement by the shortest route. Note that no matter where the birds land on one side of the P.A.M. after being drifted, if all had the same P.A.M., the direction of their redetermined migrations will be the same

of nocturnal migrants are redetermined migrations that enable the birds to correct for the displacement sustained during nocturnal migration. According to the model the direction flown by a displaced migrant after sunrise is perpendicular to its normal direction of migration on a calm night. This direction will require the shortest flight to correct for the displacement and permit the migrant to return to its principal axis of migration (P.A.M.).

To establish the preferred directions of nocturnal migrations for northwestern South Carolina, I examined the distribution of the directions of nocturnal migration on calm nights in spring and in fall. For direct visual observations in the spring, only 10 of 98 nights with migration from 1971 through 1977 met the criterion for calm nights (winds less than 2 mps at the surface and at the 950 mb level aloft), and these data are given in Table 1. When the weighted mean directions

Table 1. Directions of nocturnal migration on calm nights in spring[a]

Date	Type obs.[b]	Migration				Wind (mps)[d]		
		θ	r	s	n[c]	surface 1900	surface 2100	950 mb 1900
28 Mar. 72	MW	14°	.857	32°	8925	300° (2)	calm	282° (2)
26 Apr. 72	MW	7°	.943	20°	3047	350° (1.5)	calm	336° (2)
14 Apr. 73	MW	32°	.958	17°	10,277	280° (1.5)	calm	259° (2)
14 May 73	MW	360°	.603	58°	6712	calm	150° (2)	106° (2)
23 Apr. 76	I^2	44°	.773	41°	12,800	350° (2)	calm	346° (1)
9 May 76	I^2	27°	.854	32°	9600	350° (2)	calm	286° (1)
12 Apr. 77	I^2	32°	.898	27°	12,000	360° (2)	60° (1.5)	333° (2)
16 Apr. 77	I^2	20°	.609	57°	7400	340° (1.5)	calm	calm
17 Apr. 77	I^2	346°	.773	41°	6200	calm	320° (1.5)	calm
11 May 77	I^2	45°	.954	18°	32,400	60° (2)	calm	104° (1.5)

[a]Calm nights when wind was 2 mps or less.
[b]Type of observation: MW = moon watch; I^2 = image intensifier.
[c]Migration Traffic Rate = number of birds crossing a mile of front per hour.
[d]Wind direction is direction to which wind is blowing; time is E.S.T.

of migration on these nights are pooled and a resultant vector computed, the mean preferred direction of spring migration is 29.5° (r = .956; s = 17°; migration traffic rate = 109,200 birds). For direct visual observations in the fall, again only 10 of 93 nights with migration from 1970 through 1976 met the criterion for calm nights (Table 2). When these nightly means were pooled and a resultant vector calculated, the mean preferred direction of fall migration is 231.5° (r = .951, s = 18°; migration traffic rate = 132,500 birds). Although the mean azimuth directions of spring and fall migrations are not back azimuths - they differ by 22° - the nightly directions of the heaviest migrations in spring and fall are much closer to being back azimuths (cf. Tables 1 and 2).

Data from radar observations lend additional support to the computed preferred directions of migration derived from direct visual observations. On nine of 70 nights when radar showed migrations underway in spring and winds were calm at the surface and at the 950 mb level aloft, the mean direction of migration was 18° (r = .972, s = 14°; migration traffic rate = 114,100 birds). By analyzing track and heading directions of nocturnal migrations displayed on the ASR-4 radar, it was possible to compute the average preferred direction by regressing

Table 2. Directions of nocturnal migration on calm nights in fall[a]

Date	Type obs.[b]	Migration				Wind (mps)[d]		
		Θ	r	s	n[c]	Surface 1900	Surface 2100	950 mb 1900
12 Sept. 73	MW	229°	.724	46°	13,230	calm	calm	283° (2)
3 Oct. 74	MW	209°	.851	33°	18,383	270° (1.5)	190° (2)	285° (2)
5 Sept. 76	MW	233°	.621	56°	5731	90° (2)	90° (2)	19° (1)
16 Sept. 76	I^2	243°	.661	52°	12,600	170° (2)	240° (2)	29° (2)
18 Sept. 76	I^2	255°	.714	47°	17,600	calm	calm	90° (1)
22 Sept. 76	I^2	235°	.776	41°	13,800	calm	calm	55° (1)
23 Sept. 76	I^2	273°	.540	64°	9300	calm	calm	292° (2)
10 Oct. 76	MW	219°	.843	33°	16,858	330° (1.5)	170° (2)	310° (2)
11 Oct. 76	MW	217°	.653	53°	19,371	calm	190° (2)	250° (2)
16 Oct. 76	I^2	226°	.955	17°	5600	calm	180° (2)	41° (2)

[a]Calm nights when wind was 2 mps or less.
[b]Type of observation: MW = moon watch; I^2 = image intensifier.
[c]Migration Traffic Rate = number of birds crossing a mile of front per hour.
[d]Wind direction is direction to which wind is blowing; time is E.S.T.

track directions on α, the track direction minus the heading direction (Alerstam, 1976). Following this procedure I found the average direction of spring migration to be 25° when α equalled zero. In fall, on only one of eight nights did the radar show a migration underway when winds were calm, and on this night the mean direction of migration was 225°. When the track directions of the remaining seven nights were regressed on the track directions minus their respective heading directions, I found the average preferred direction of fall migration to be 217° when α equalled zero. Thus the radar data when analyzed by two different methods yield similar results, and these results are in general agreement with those obtained from direct visual observations.

The radar and direct visual data also provide evidence that passerine migrants flying in the night sky are often drifted by the wind. On the basis of the analysis of track and heading directions of nocturnal migrants (Alerstam, 1976), the linear regression coefficient of track directions on α is 0.648 for spring radar data (61 nights) and 0.834 for fall radar data (seven nights). By making direct visual observations it is often possible to accurately record the heading and track directions of birds as they pass before the disc of the moon or through the ceilometer beam. The linear regression coefficient of track direction on α is 0.728 for spring direct visual data (n = 38) and 0.630 for fall direct visual data (n = 45). The coefficients derived from radar and direct visual data strongly suggest that nocturnal migrants can only partially compensate for wind drift while flying or that some migrants are drifted by the wind while others compensate for wind drift almost completely while aloft.

In Figure 2 I have attempted to derive the expected directions of morning redetermined migratory flights based on the preferred directions of nocturnal spring and fall migrations under calm wind conditions. The lighter gray shaded arcs in Figure 2 represent the mean preferred directions of nightly spring and fall migrations and their angular deviations computed from direct visual observations. The unshaded arcs are perpendicular to those for the preferred nighttime directions of migration. Thus if nocturnal migrants in the morning fly perpendicular to their preferred directions at night, they would be expected to fly in directions indicated by the unshaded arcs. The darker shaded arcs

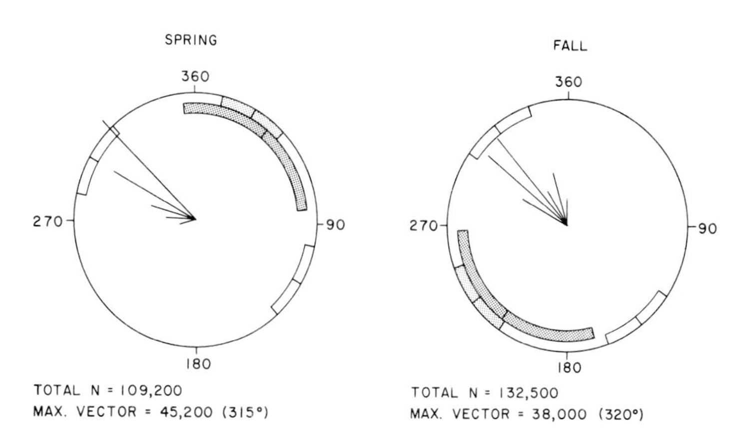

Fig. 2. The expected directions of morning redetermined migrations based on the preferred directions of nocturnal migration. The *lighter shaded arcs* represent the means and angular deviations of the preferred directions of spring and fall migrations: The *darker shaded arcs* represent the means and angular deviations of the actual directions of nocturnal migrations in spring and fall. The *unshaded arcs* are perpendicular to the arcs of preferred directions of nocturnal migration. The *vectors* are perpendiculars to the specific preferred directions of nocturnal migration and weighted the same as the specific preferred directions. The vectors are drawn toward the northwest to reflect the fact that most birds are drifted east rather than west of their preferred routes, but some movements to the southeast are expected. *N* equals the number of birds crossing a mile of front per hour, i.e., the migration traffic rate

represent the mean and angular deviation of the track directions of nightly spring and fall migrations gathered by direct visual observations in 1977 and 1976, respectively (see Fig. 3 for plots of these data). As can be seen in Figure 2 most of the track directions at night

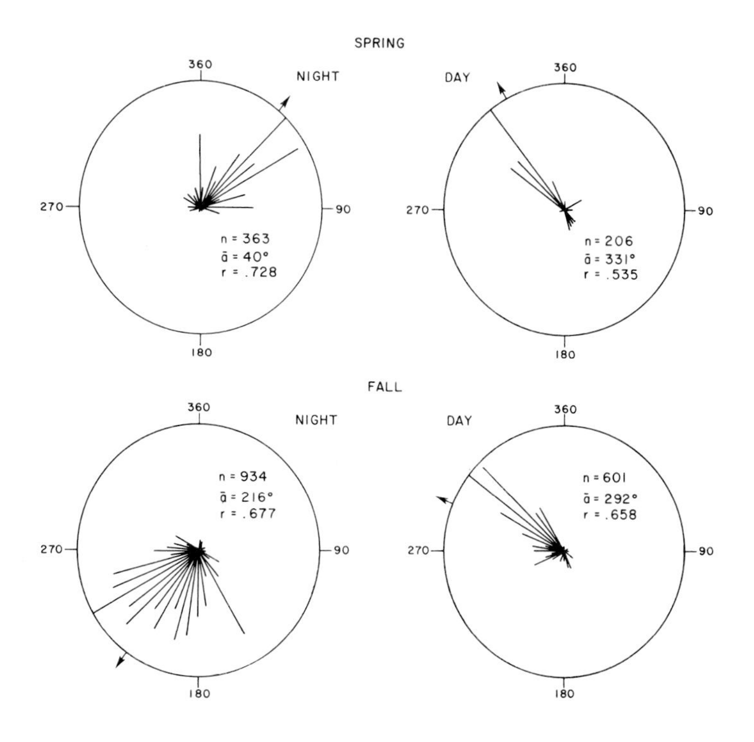

Fig. 3. The flight directions of migrants at night and the directions of daytime redetermined migrations in the fall of 1976 and in the spring of 1977. The data are from direct visual observations at night (image intensifier) and during the early daylight hours (10 × 50 binoculars). *N* equals the number of birds observed. The *arrows* on the circles indicate the mean azimuth direction ($\bar{a}$); *r* is a coefficient of directionality

are shifted east of the preferred directions in both spring and fall, indicating that far more birds are drifted east of their preferred headings than are drifted west of their preferred headings. In light of this, one would expect that far more migrants would fly toward the northwest in the morning than toward the southeast. Accordingly, I have plotted in Figure 2 the radii of expected directions of the morning redetermined migratory flights toward the northwest and omitted the expected directions for the considerably smaller morning movements toward the southeast. The direction of each radius is a perpendicular to a specific preferred direction at night, and the length of each radius is identical to the amount of nocturnal migration recorded in that specific preferred direction. Even though the unshaded arcs suggest that there might be a slight difference in the major directions of the morning flights in spring and fall, the radii indicate that most birds in the morning should fly in about the same direction irrespective of season. This is in fact the case and is clearly demonstrated in Figure 3.

The tracks of nocturnal migrants viewed through the image intensifier and by moon-watching in the spring of 1977 and in the fall of 1976 are plotted in Figure 3. The means and angular deviations of these movements are also plotted in Figure 2 as darkly shaded arcs. The angular deviation in the directions of movement is slightly smaller in the spring (46°) than in the fall (51°), and about a third fewer birds were observed in the spring than in the fall. The plots of the daytime movements of nocturnal migrants recorded in the fall of 1976 and in the spring of 1977 are basically toward the northwest in both spring and fall and are in close agreement with the directions expected if migrants in the morning fly perpendicular to their preferred direction of nocturnal migration. The morning movements toward the southeast are considerably smaller than the northwest movements as expected, and the density of the southeast movements is in keeping with the proportion of nocturnal migrants drifted west of their preferred headings at night (cf. Fig. 2).

Although a detailed analysis of the morning redetermined migratory flights of nocturnal migrants is not possible in this paper, there is evidence to suggest that the flights are better oriented (less confused flight behavior and a smaller angular deviation) and of greater magnitude when the sun is visible. On overcast mornings migrants frequently circled, called, changed directions abruptly, and often landed after a brief period of flight. The magnitude of the redetermined movements was also greatly reduced under solid cloud cover. Even though migrants attempted to fly on overcast mornings they usually alighted before they disappeared on the horizon, and under such circumstances it was meaningless to take disappearing azimuths. On those mornings when solid cloud cover was present for the first hour or two of a watch, the number of migrants flying was greatly reduced, but when the cloud cover dissipated and the sun was visible, the number of migrants showing redetermined flights increased dramatically. A more detailed discussion of the redetermined flights of nocturnal migrants with regard to species composition, age, and general behavior is currently in preparation.

D. Discussion

The daytime redetermined migrations of nocturnal migrants I report on in this paper appear to be quite similar to the offshore daytime movements of drifted migrants along the Atlantic coast in fall (Baird and Nisbet, 1960). It is important to note that in the studies of Baird and Nisbet many of the migrants set out in northwestern directions even though in some cases another direction could have involved a shorter water crossing. Similarly DeSante (1973) found that nocturnal migrant warblers that have been drifted off the California coast in the vicinity of the Farallon Islands fly toward the northeast after dawn despite the fact that flights toward more easterly and southeasterly directions would permit a water crossing of the same distance. These instances and the occurrence of the morning redetermined migrations inland argue against the notion that the birds are simply trying to regain land. Baird and Nisbet (1960) have said that the northwestward flights are in fact the simplest reaction that would counter the effects of offshore drift and return the migrants, whether over land or over water, to their preferred migration route. Rabøl und Hindsbo (1972) pointed out that in spring the north-northwest tendency of several migrants during the day in Denmark might be interpreted as "navigation" toward goal areas on the migratory route following displacement toward the east by the prevailing southwest-west winds in the area. There is unfortunately a paucity of exact data on the preferred directions of nocturnal migration and the precise azimuth directions of the redetermined morning flights in previous studies, and exact comparisons between my findings and those from other studies are not possible.

When nocturnal migrants are displaced from their principal axis of migration or from their preferred migration route several possible strategies exist for countering the displacement: (1) The birds could attempt to regain the location on their principal axis of migration where they would have landed if no displacement occurred; (2) the birds could fly perpendicular to their principal axis of migration and regain their preferred migration route in the shortest distance; (3) the birds could choose a direction intermediate between that perpendicular to their principal axis of migration and their ultimate migratory goal; and (4) they could change their preferred migratory route and fly directly toward their ultimate goal from the point of displacement. The evidence presented in this paper supports the second strategy, if the redetermined migrations take place during the daylight hours, but other possibilities (e.g., the fourth strategy at night) cannot be ruled out. It is important to realize that all of the above strategies require that the migrant be able to assess its displacement. Its manner of doing so, however, is beyond the scope of this paper.

The prevalence of the northwestward direction in the morning redetermined flights of nocturnal migrants is clearly a function of the mean orientation of the principal axes of spring and fall migrations in northwestern South Carolina and the prevailing westerly winds aloft in the area. However, it is interesting to note that a northwestward tendency in the homing orientation of many birds has been recorded and has been termed "nonsense orientation" by Matthews (1968). Keeton (1973) and Windsor (1975) both showed that there is a pattern to the release-site biases of homing pigeons from the Cornell loft at many different release sites within 193 km of Ithaca and northwest is often the preferred direction of flight. Windsor (1975) suggested that the directional preferences of homing pigeons seemed to represent a varying compromise between flying to the north-northwest and flying toward home, and he found that homeward directedness of vanishing bearings, time to leave the release point, and time to reach home were all superior for releases conducted to the south and southeast of the loft. My findings on the daytime redetermined flights to the northwest and the northwestward tendency in many homing studies may not be related, but the possibility of there being some common denominator for the two phenomena is indeed thought provoking.

Acknowledgments. I benefited greatly from lengthy discussions with K.P. Able, F.R. Moore, and H.E. LeGrand, Jr. about this research, and I thank all who offered suggestions at the minisymposium on migration, orientation, and navigation at the State University of New York at Albany in December, 1976. This work and most of my work on bird migration would not have been possible without the continued grant support from the Air Force Office of Scientific Research.

References

Able, K.P.: The orientation of passerine nocturnal migrants following offshore drift. Auk 94, 320-330 (1977)

Able, K.P., Gauthreaux, S.A.: Quantification of nocturnal passerine migration with a portable ceilometer. Condor 77, 92-95 (1975)

Alerstam, T.: Bird migration in relation to wind and topography. Ph. D. Thesis, Lund, Sweden, Univ. Lund, Dept. Anim. Ecol. 51 pp. (1976)

Bagg, A.M.: Reverse warbler migration in the Connecticut valley. Auk 67, 224-245 (1950)

Baird, J., Nisbet, I.C.T.: Northward fall migration on the Atlantic coast and its relation to offshore drift. Auk 77, 119-149 (1960)

Batschelet, E.: Statistical methods for the analysis of problems in animal orientation and certain biological rhythms. Washington, D.C., Amer. Inst. Biol. Sci. 57 pp. (1965)

Christensen, N.H., Jacobsen, J.R.: Headwind migration. Dansk Ornith. Foren. Tidsskr. 62, 153-159 (1969)

DeSante, D.F.: An analysis of the fall occurrences and nocturnal orientations of vagrant wood warblers (*Parulidae*) in California. Unpubl. Ph. D. Thesis, Palo Alto, California, Stanford Univ. (1973)

Gauthreaux, S.A.: A portable ceilometer technique for studying low-level nocturnal migration. Bird-Banding 40, 309-320 (1969)

Gauthreaux, S.A.: Quantification of bird echoes on airport survaillance radars. Proc. Conf. Transparent Aircraft Enclosures, Las Vegas, Nevada. Wright-Patterson AFB, Ohio, AFML-TR-73-126. pp. 515-529 (1973)

Gauthreaux, S.A.: The detection, quantification, and monitoring of bird movements aloft with airport surveillance radar (ASR). Proc. Conf. Biol. Aspects Bird/ Aircraft Collision Problem, Clemson, South Carolina. Bolling AFB, D.C. AFOSR-TR-74-1240. pp. 289-307 (1974)

Gauthreaux, S.A.: Radar ornithology - bird echoes on weather and airport surveillance radars. Tech. Report, Clemson, South Carolina, Clemson Univ. Dept. Zool. 47 pp. (1975)

Gauthreaux, S.A.: Radar ornithology and bird/aircraft colisions. Proc. Conf. Aerospace Transparent Materials and Enclosures, Atlanta, Georgia. Wright-Patterson AFB, Ohio, AFML-TR-76-54. pp. 777-790 (1967a)

Gauthreaux, S.A.: The role of radar in the avoidance of bird/aircraft collisions. Proc. Bird Hazards to Aircraft Seminar and Workshop, Atlanta, Georgia, Federal Aviation Admin. pp. 179-190 (1976b)

Gruys-Casimir, E.M.: On the influence of environmental factors in the autumn migration of Chaffinches and Starlings - a field study. Arch. Neerl. Zool. 16, 175-279 (1965)

Keeton, W.T.: Release-site bias as a possible guide to the "map" component in pigeon homing. J. Comp. Physiol. 86, 1-16 (1973)

Lack, D., Williamson, K.: Bird-migration terms. Ibis 101, 255-256 (1959)

Lowery, G.H.: A quantitative study of the nocturnal migration of birds. Univ. Kansas Mus. Nat. Hist. 3, 361-472 (1951)

Mardia, K.V.: Statistics of directional data. New York: Academic Press. 357 pp (1972)

Matthews, G.V.T.: Bird navigation. Cambridge: University Press. 197 pp. 1968

Murray, B.G.: The return to the mainland of some nocturnal passerine migrants over the sea. Bird-Banding 47, 345-358 (1976)

Rabøl, J.: Fugletraekket ved Knudshoved. Dansk Ornith. Foren. Tidsskr. 58, 49-97 (1964)

Rabøl, J.: Visual diurnal migratory movements. Dansk Ornith. Foren. Tidsskr. 61, 73-99 (1967)

Rabøl, J.: Headwind-migration. Answer. Dansk Ornith. Foren. Tidsskr. 62, 160-165 (1969)

Rabøl, J.: Correlation between coastal and inland migratory movements. Dansk Ornith. Foren. Tidsskr. 68, 5-14 (1974)

Rabøl, J., Hindsbo, O.: A comparison of the bird migration recorded by radar and visible field observations in the middle of Sjaelland, Denmark, spring 1971. Dansk Ornith. Foren. Tidsskr. 66, 86-96 (1972)

Windsor, D.M.: Regional expression of directional preferences by experienced homing pigeons. Anim. Behav. 23, 335-343 (1975)

Zar, J.H.: Biostatistical analysis. Englewood Cliffs, New Jersey: Prentice-Hall, 620 pp. 1974

Field Studies of the Orientation Cue Hierarchy of Nocturnal Songbird Migrants

KENNETH P. ABLE, Department of Biological Sciences, State University of New York at Albany, Albany, NY 12222, USA

Abstract

Tracking radar, surveillance radar, and visual observation of free-flying nocturnal migrants in the United States are revealing relationships among several compass cues. Whereas passerine migrants in the southeastern United States selectively flew downwind even under clear skies, downwind flight in seasonally inappropriate directions in the northeastern United States occurred only under solid overcast that prevented the birds from seeing both stars and the sun near the time of sunset. The hypothesis that a view of either stars or the sun near sunset is sufficient to allow the birds to perform "correct" orientation was supported by data from experimentally released birds.

Research during the past decade on the orientation mechanisms of migratory birds has revealed several cue systems that may supply directional information. Among species that migrate principally at night, potential cues include stars, the Earth's magnetic field, the sun, and wind direction (Emlen, 1975, has recently reviewed nocturnal orientation mechanisms).

With the demonstration of multiple cue systems, the problem of investigating migratory orientation has become substantially more complicated. Only in the field can one be certain that the full capabilities of the birds will be expressed. Under natural conditions, a free-flying migrant may experience situations in which some potentially usable cues are absent, give conflicting information, or are changing in time or space. This paper presents tracking radar data on free-flying nocturnal migrants that reveal relationships between stars, sun, and wind direction as orientation cues. Preliminary data obtained from a series of White-throated Sparrows (*Zonotrichia albicollis*) released aloft wearing translucent contact lenses were similar to the pattern shown by the free-flying migrants.

A. Methods

I. Free-Flying Migrants

Flight directions of nocturnal migrants were obtained using a tracking radar and portable ceilometers. During spring, 1972 and 1973, I used the AN/GPG-1 automatic tracking radar described by Griffin (1972, 1973) at Millbrook, Dutchess County, New York. In subsequent seasons, tracks were obtained with an AN/MPQ-10 automatic tracking radar located at Berne, Albany County, New York. In automatic tracking mode, this 10-cm radar is capable of tracking a target with an accuracy of ± 18 m in range and ± 0.08° in azimuth and elevation. Positions of birds were routinely recorded every 10 s. Flight directions were obtained with portable ceilometers as described by Gauthreaux (1969).

For the analyses presented here, radar tracks were selected on the basis of the following criteria: (1) length > 20 s; (2) straightness, $s < 40^\circ$; (3) average change in altitude less than 1 mps over the length of the track.

To qualify as solid overcast, cloud cover must have been 10/10 during the entire observation period. Overcast during daytime must have been 10/10 and sufficiently thick that I could not determine the position of the sun. Radar tracks were accompanied, within at least 2 h, by a balloon track obtained with the same radar equipment. Ceilometer data obtained on nights when no radar tracking was done are compared with winds aloft data obtained at 18:00 at Albany, New York, by the National Weather Service.

II. Experimental Releases

Using techniques similar to those devised by Emlen (1974) I released individual White-throated Sparrows from boxes carried aloft by balloons. Individuals showing consistent, intense nocturnal restlessness were selected for release. Just before dark, the birds were transferred to a closed carrying case that afforded no outside view. They were removed individually from the carrying case and placed in the release boxes just prior to launching.

Contact lenses were made of vacuum formable vinyl (5 mil) using a Mattel Vac-u-Form (Mattel, Inc.) with a 7-mm ball bearing as the form. To make the lenses translucent, I lightly sanded the entire outer surface with fine emery paper until it became uniformly cloudy in appearance. Through these lenses I was unable to resolve even the most contrasting forms at more than 2 cm. Only the grossest patterns of light and dark are visible at distances greater than 2 cm.

B. Results

I. Orientation Under Solid Overcast

In an earlier paper (Able, 1974b) I noted a tendency for downwind flight in seasonably inappropriate directions in New York to occur only under solid overcast skies. Since that time, I have accumulated more data in the rare condition of solid, extensive overcast accompanied by winds opposed or at large angles to the seasonal migration direction in this area. These data permit an examination of free-flight orientation in the absence of input from two potentially important cues, stars and sun.

Table 1 presents data from six nights when solid overcast commenced by at least 15:00 EST (6 May 1975) and usually before 12:00 (all other days). In each case the overcast remained solid in the observation area throughout the remainder of the day, and sunset position was not perceptible to me. On 6 May 1975, and 20 September 1974, very light, scattered rain occurred during tracking. On the other nights birds were flying beneath high clouds and on none of the nights did any of the birds tracked appear to be within clouds.

For each night I have computed the mean direction of all straight and level tracks, the length of the mean vector (r), angular deviation (s), the angle between the mean track and wind directions ($|W - T|$), the downwind component of the mean vector (h), and performed a Rayleigh test (Z) on the distribution using methods described in Batschelet

(1965) (see Table 1). The flight directions of the birds were significantly oriented on each night and the concentrations of directions (measured by r) were not significantly different from tracks (Table 2) obtained in similarly opposed winds but under clear skies (Mann-Whitney-U test).

The flight directions of birds under solid overcast are shown relative to winds in Figure 1. The birds showed a strong tendency to fly downwind under these conditions even though the winds were very opposed to the normal directions of migration in this area (northeastward in spring, south-southeastward in fall).

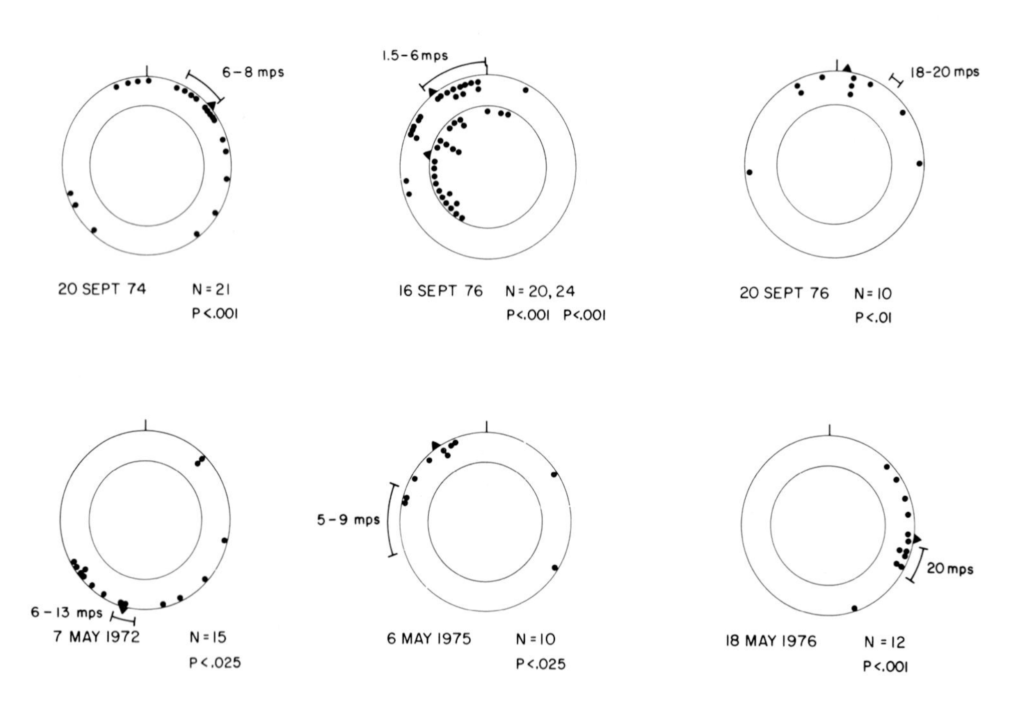

Fig. 1. Flight directions of nocturnal migrants flying in opposed winds under solid overcast which began at least 5 h before sunset. Each *point* on the *outer circle* represents the mean direction of the segments comprising the radar track of a single bird. Each *point* on the *inner circle* represents the flight direction of a single bird observed with the portable ceilometer. The mean track direction for all birds on each circle is represented by the *black arrow* and the Rayleigh test probabilities associated with the radar tracks and ceilometer data, respectively, are given at the *lower right* of each Figure. The range of directions toward which winds were blowing and the range of wind speeds are shown just *outside* each circle

II. Orientation Under Clear Skies

I have obtained flight directions of nocturnal migrants on 25 nights when winds were at least 45° from the normal seasonal directions and skies were essentially clear (3/10 cloud cover or less) (Table 2).

Table 1. Flight directions of nocturnal migrants in opposing winds under overcast beginning before sunset

Spring	N	Wind dir. (°)	Speed (mps)	Mean dir. (°)	r	s (°)	Z	$\|W - T\|$ (°)	h
7 May 1972	15	186-200	6-13	194.3	.536	55.2	4.30^b	1.3	.536
6 May 1975	10	250-290	5-9	325.2	.636	48.9	4.05^b	55.2	.363
18 May 1976	12	112	20	97.1	.876	28.5	9.21^d	14.9	.847
Fall									
20 September 1974	21	26-50	6-8	48.5	.527	55.7	6.10^d	10.5	.518
16 September 1976	20	336-360	1.5-6	281.3	.679	45.9	11.07^d	66.7	.269
[a]16 September 1976	24	350-360	1.5-6	322.3	.855	30.8	14.63^d	32.7	.720
20 September 1976	10	37-45	19-22	6.7	.718	43.0	5.15^c	34.3	.593
			$\bar{X}$ = 9.93		$\bar{X}$ = .690			Θ = 30.63	$\bar{X}$ = 0.549
			S.D. = 7.46		S.D. = .139			s = 21.91	S.D. = .198

[a]Denotes ceilometer data.
[b]= $p < .05$
[c]= $p < .01$
[d]= $p < .001$

Table 2. Flight directions of nocturnal migrants in opposing winds under clear skies

Spring	N	Wind dir. (°)	Speed (mps)	Mean dir. (°)	r	s (°)	Z	|W - T| (°)	h
17 April 1972	28	146-164	10-12	96.4	.853	31.0	18.94[d]	58.6	.444
29 April 1972	10	120-140	5-7	89.7	.683	45.6	4.66[c]	40.3	.521
22 May 1973	27	130-160	3-9	35.0	.860	30.4	19.95[d]	110.0	-.294
2 May 1975	14	112-130	15-19	81.2	.954	17.5	12.73[d]	39.8	.733
9 May 1975	13	125	3	35.2	.975	12.9	12.35[d]	89.8	.003
13 May 1975	29	180-215	6	348.3	.575	52.9	9.57[d]	150.8	-.502
16 May 1975	13	135-150	11-14	88.0	.924	22.3	11.11[d]	54.5	.537
23 May 1976	19	112-146	12-17	64.9	.982	10.7	18.34[d]	64.1	.429
27 May 1976	20	105-140	13-20	62.4	.784	37.7	12.29[d]	60.1	.391
[a]27 May 1976	11	105-140	13-20	50.4	.960	16.2	10.14[d]	72.1	.296
12 June 1976	13	160-240	4-8	317.9	.518	56.3	3.49[b]	117.9	-.242
Fall									
[a]4 September 1973	6	300-315	1-3	205.2	.910	24.8	4.90[c]	102.3	-.193
[a]19 September 1973	18	0-20	5-6	233.7	.624	49.7	7.00[d]	136.3	-.451
[a]21 September 1973	10	50-85	2-4	234.9	.674	46.2	4.50[c]	167.4	-.658
[a]25 September 1973	10	325-345	4-5	286.0	.852	31.2	7.26[d]	49.0	.559
5 September 1974	8	318-338	4-9	227.6	.887	27.3	6.29[d]	100.4	-.160
10 September 1974	41	55-70	10-17	141.5	.682	45.7	19.07[d]	79.0	.130
5 October 1974	10	70-115	15-25	222.7	.700	44.4	4.90[c]	130.2	-.452
6 October 1974	18	25-60	12-19	133.5	.639	48.7	7.35[d]	91.0	-.011
4 October 1976	8	300-330	4-9	185.1	.533	55.4	2.28[b]	129.9	-.343
[a]5 October 1976	12	310-320	3-6	270.7	.893	26.5	9.58[d]	44.3	.639
			$\bar{X}$ = 9.68		$\bar{X}$ = .784			Θ = 88.77	$\bar{X}$ = .066
			S.D. = 5.54		S.D. = .154			s = 36.14	S.D. = .435

[a]Denotes ceilometer data.
[b]= $p < .05$
[c]= $p < .01$
[d]= $p < .001$

Under these conditions, birds did not tend to fly downwind (mean $|W - T| = 88.8^{o}$). Deviations of track directions from downwind under the different overcast conditions are shown in Figure 2. In comparably opposed winds of nearly identical speed, birds under clear skies had a significantly smaller downwind flight component ($\bar{h}_{overcast} = .549$; $\bar{h}_{clear} = .066$; $p < .01$, Mann-Whitney-U test, one-tailed). I conclude that birds migrating under solid overcast skies, which have also prevented them from seeing the sun late in the day, selectively flew downwind even when winds were blowing in seasonally inappropriate directions. Under clear skies other, presumably visual, cues enabled the birds to select a normal migration direction.

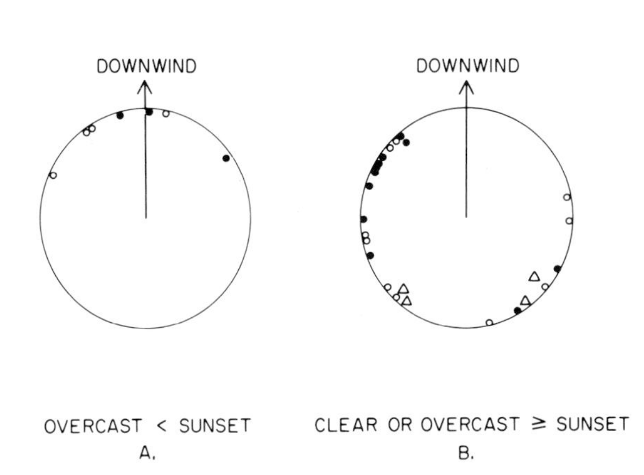

Fig. 2 A and B. The deviatiations of mean track directions from wind direction under clear and overcast skies. Directions have been normalized so that downwind is always at the top. (A) Data obtained in opposed winds under solid overcast which began at least 5 h before sunset. *Solid circles* are spring data, *open circles* refer to fall. (B) Data obtained in opposed winds under clear skies or when overcast began near the time of sunset. *Solid circles* represent spring data, *open circles* refer to fall, and *triangles* represent data from nights when overcast began near sunset (Table 3)

III. Orientation Under Overcast Beginning Near or After Sunset

We may ask what would happen if birds were allowed to see the sun near the time of sunset but incoming overcast precluded their seeing stars after dark. Coupled with opposed winds, these are extremely rare events and more data are needed, but flight directions on three nights are given in Table 3 and illustrated in Figure 3. These birds behaved like those flying under clear skies, flying in the predicted migratory direction, but almost into the wind (mean $|W - T| = 138.1^{o}$; $\bar{h} = -.541$).

IV. Releases of Visually Impaired Migrants

The data on free-flying birds given above yield the hypothesis that a view of either stars or the sun near the time of sunset is sufficient to allow the birds to perform "correct" orientation. In the absence of both of these inputs, downwind orientation should occur. I predicted that birds released after dark wearing translucent contact lenses would behave like the latter group and fly downwind.

Tracks of nine birds released on two nights are summarized in Table 4 and plotted in Figures 4 and 5. There are at present far too few tracks to be more than preliminary, but those obtained so far, especially on 29 May 1977, are in agreement with the free-flight data. Three of four control birds climbed at rates comparable to birds initiating migration from the ground (Able, 1977). The birds were not disoriented and did not make loops or zigzags on a scale resolvable by the radar. Superimposed on the typical flap-pause fluctuations, several released birds showed irregular changes in echo amplitude sug-

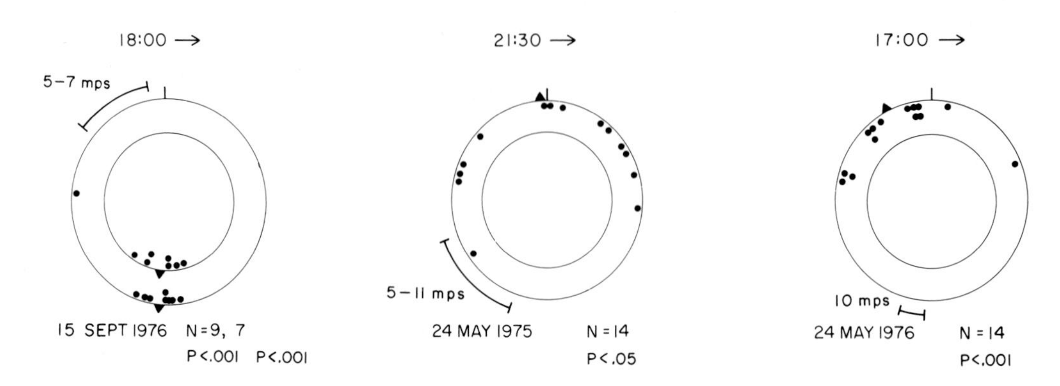

Fig. 3. Flight directions of nocturnal migrants flying in opposed winds under solid overcast which began near the time of sunset. Times above each circle denote the time (EST) solid overcast began. Other notations as in Figure 1

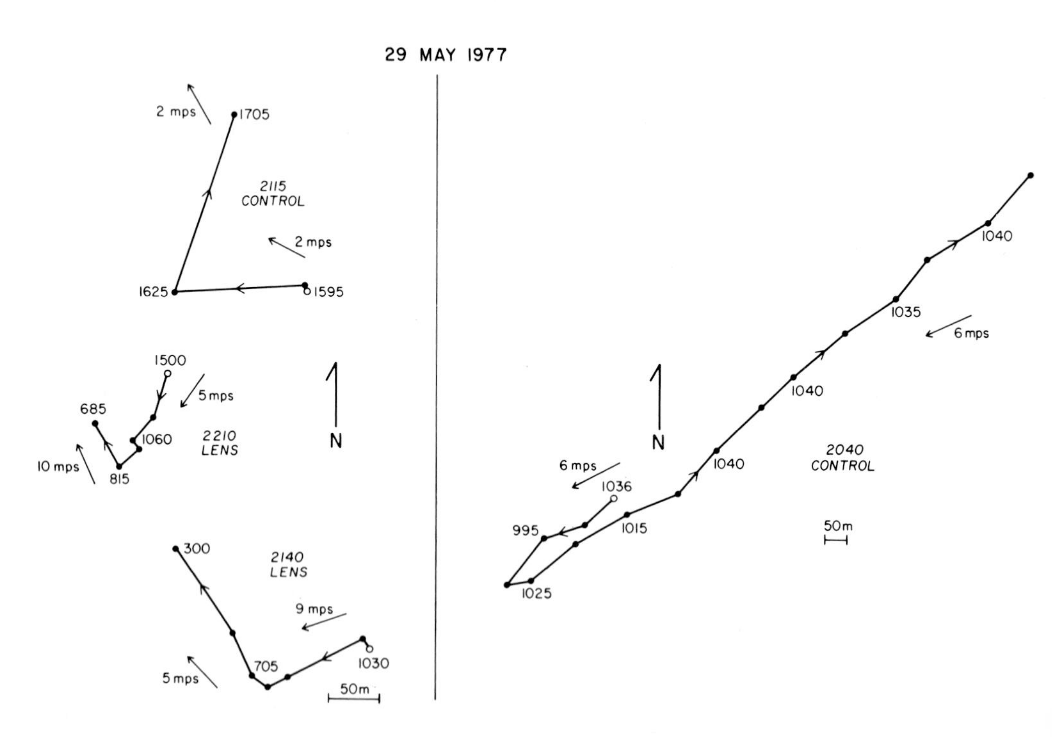

Fig. 4. Tracks of White-throated Sparrows (*Zonotrichia albicollis*) released on 29 May 1977, from boxes carried aloft by balloon. The *open circle* denotes the starting point of the track. The *numbers* at alternate points along the tracks are the altitudes of the birds in meters. Each track is identified by its time and a designation as LENS or CONTROL. LENS birds wore translucent contact lenses, CONTROL birds were released without lenses. *Arrows* give wind direction and speed

Table 3. Flight directions of nocturnal migrants in opposing winds under overcast beginning near or after sunset

Date	N	Wind dir. (°)	and Speed (mps)	Mean dir. (°)	r	s (°)	Z	|W - T| (°)	h
24 May 1975	14	205-250	5-11	357.5	.459	59.6	3.15[b]	131.0	-.319
24 May 1976	14	190	10	333.2	.815	34.9	9.30[c]	143.2	-.653
15 September 1976	9	310-350	5-7	193.2	.885	27.4	7.06[c]	136.8	-.447
[a]15 September 1976	7	310-350	5-7	188.5	.953	17.6	6.36[c]	141.5	-.746
			$\bar{X}$ = 7.5		$\bar{X}$ = .778			Θ = 138.1	$\bar{X}$ = 0.541
			S.D. = 1.91		S.D. = .220			s = 4.73	S.D. = .194

[a]Denotes ceilometer data.
[b]= $p < .05$
[c]= $p < .001$

Table 4. Releases of White-throated Sparrows wearing translucent lenses, spring, 1977

Date	Time	treatment	Track length (s)	Wind dir. at release point (°)	and Speed (mps)	Altitude range (m)	Mean dir. (°)	r	s (°)
29 May	2040	control	150	267	5	995-1060	54.6	.592	51.7
	2115	control	30	298	2	1595-1705	330.1	.708	43.8
	2140	lens	62	251	9	1030-292	301.7	.797	36.5
	2210	lens	60	218	5	1500-685	239.4	.468	59.1
8 June	2120	control	90	57	12	1865-1570	77.7	.974	13.0
	2205	lens	60	80	17	1500-1949	77.3	.981	11.3
	2230	lens	70	100	14	1310-958	81.5	.998	3.6
	2320	lens	70	85	17	1495-1105	76.7	.908	24.6
	2340	control	50	91	15	1295-1520	73.2	.954	17.5

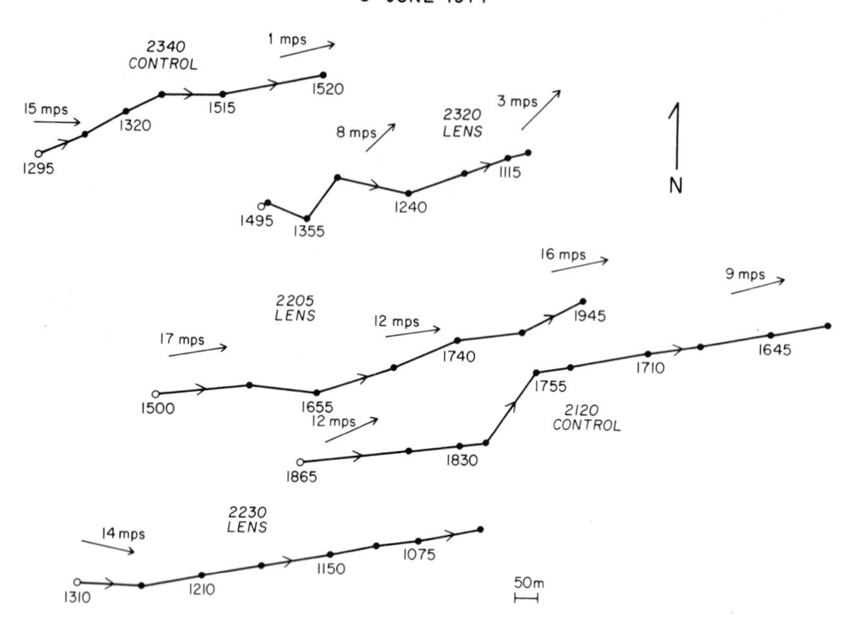

Fig. 5. Tracks of White-throated Sparrows (*Zonotrichia albicollis*) released 8 June 1977, from boxes carried aloft by balloon. Notations as in Figure 4

gesting abrupt changes in aspect. Such fluctuations were very rarely seen in free-flying migrants. All but one of the lens birds descended rather steeply.

On 29 May, the two control birds initially flew downwind, but then turned through large angles and flew northeastward. Each of the lens birds oriented downwind for the duration of its track and made downwind turns if it descended through altitudes where winds shifted. On 8 June, winds aloft were strong and blowing in a more or less appropriate migratory direction. Both lens and control birds flew downwind toward the east-northeast.

C. Discussion

In his recent review of migratory orientation, Emlen (1975) predicted that the weightings of the several usable orientation cues would be found to vary between species, seasons, meteorologic conditions, and regions. Radar and visual observations of nocturnal migrants in the United States are confirming this prediction. In the inland southeastern states wind direction predominates over other cues (Gauthreaux and Able, 1970; Able, 1974a), whereas individuals of the same species migrating through the northeastern states use wind as the primary cue only under the special circumstances described above. Although the adaptive significance of this difference is not clear, it is apparent that in the Northeast birds preferentially use directional information either from the sun or stars when selecting a flight direction. Wind seems to rank above other potential cues (e.g., magnetism). In the absence of both sun and stars, these free-flying migrants appeared unable to determine the correct migration direction. These results appear to agree with Emlen's (1975) unpublished data on White-throated Sparrows released at Wallops Island.

A number of studies have reported migrants maintaining seasonally appropriate tracks beneath, between, or even inside clouds (Bellrose and Graber, 1963; Hebrard, 1971; Williams et al., 1972; Griffin, 1973). In the cases in which the birds were not flying downwind, it is not clear that they initiated migration without seeing both sun and stars. The observed behavior may have reflected maintenance rather than selection of a direction (see Emlen, 1975, p. 182).

As a whole, the migrations initiated under solid overcast were as well oriented as those in comparably opposed winds under clear skies. The absence of celestial cues, at least for 5 - 11 h prior to the initiation of migration, did not result in a marked disruption of orientation. This conclusion may well not apply to individual birds, however. For the analysis presented here I have selected only relatively straight tracks.

To view the flight behavior of birds released from boxes carried aloft by balloons as reflecting migratory motivation requires a considerable act of faith. Based on a large number of similar releases at Wallops Island, Emlen (1974; pers. comm.) believes that appropriately selected and handled birds often initiate migratory flights following release. My radar does not yield such long tracks, but especially my control birds behaved as if migrating. Virtually all of the lens birds descended and are less convincing in this regard.

In the inland northeastern United States input from either the sun or stars appears sufficient for the determination of the migratory direction by nocturnal migrants. The possible role of the sun in nocturnal orientation has been largely ignored although suggestive data exist (Kramer, 1949; Hebrard, 1971; Frank R. Moore, pers. comm.). The apparent primacy of celestial cues and the downwind orientation under overcast shown by my data indicate a hierarchical relationship among usable directional cues, but do not preclude the possibility that under any given set of conditions information is integrated from all available sources. The results are, however, clearly at odds with recent experiments from which the Wiltschkos (Wiltschko and Wiltschko, 1975a, b) concluded that magnetic information overrides even stellar orientation in several European migrants. Whereas some common ground has been found, there remain considerable inconsistencies between studies of orientation by free-flying migrants and birds in orientation cages.

Acknowledgments. During the springs of 1972 and 1973 I enjoyed use of tracking radar facilities at the Rockefeller University Center for Field Research through the courtesy of D.R. Griffin. My tracking radar was obtained with the kind assistance of J. Hughes, Office of Naval Research, and B. Vonnegut, State University of New York at Albany. The electronics expertise of R. Zeh was indispensable in renovating, modifying, and maintaining that radar. I thank S.A. Gauthreaux for his hospitality and assistance during two trips to capture birds at Clemson, South Carolina. V.P. Bingman assisted with bird releases and S.T. Emlen and N.J. Demong kindly shared with me their experiences in releasing birds. This work was funded by the Research Foundation of State University of New York and the National Science Foundation (grants BO41339 and BNS76-18524).

References

Able, K.P.: Environmental influences on the orientation of nocturnal bird migrants. Anim. Behav. 22, 225-239 (1974a)

Able, K.P.: Wind, track, heading and the flight orientation of migrating songbirds. In: Proc. Conf. Biol. Aspects of the Bird/Aircraft Collision Problem (Gauthreaux, S.A., ed.), 1974b, pp. 331-357

Able, K.P.: The flight behaviour of individual passerine nocturnal migrants: a tracking radar study. Anim. Behav. 25, 924-935 (1977)

Batschelet, E.: Statistical methods for the analysis of problems in animal orientation and certain biological rhythms. Am. Inst. Biol. Sci., Washington, D.C. (1965)

Bellrose, F.C., Graber, R.R.: A radar study of the flight directions of nocturnal migrants. Proc. 13th Int. Ornithol. Congr., 362-389 (1963)

Emlen, S.T.: Problems in identifying bird species by radar signature analysis: intraspecific variability. In: Proc. Conf. Biol Aspects of the Bird/Aircraft Collision Problem (Gauthreaux, S.A., ed.), 1974, pp. 509-524

Emlen, S.T.: Migration: orientation and navigation. In: Avian Biology. vol. V, Farner, D.S., King, J.R. (eds.) New York: Academic Press, 1975, pp. 129-219

Gauthreaux, S.A.: A portable ceilometer technique for studying low-level nocturnal migration. Bird-Banding 40, 309-320 (1969)

Gauthreaux, S.A., Able, K.P.: Wind and the direction of nocturnal songbird migration. Nature (Lond.) 228, 476-477 (1970)

Griffin, D.R.: Nocturnal bird migration in opaque clouds. In: Animal Orientation and Navigation. NASA SP-262, U.S. Gov. Printing Office, Washington, D.C., 1972, pp. 169-188

Griffin, D.R.: Oriented bird migration in or between opaque cloud layers. Proc. Am. Phil. Soc. 117, 117-141 (1973)

Hebrard, J.J.: Fall nocturnal migration during two successive overcast days. Condor 74, 106-107 (1971)

Kramer, G.: Über Richtungstendenzen bei der nächtlichen Zugunruhe gekäfigter Vögel. In: Ornithologie als biologische Wissenschaft, Mayr, E., Shüz, E. (eds.) Carl Winter, Heidelberg, 1949, pp. 269-283

Williams, T.C., Williams, J., Teal, J., Kanwisher, J.: Tracking radar studies of bird migration. In: Animal Orientation and Navigation. NASA SP-262, U.S. Gov. Printing Office, Washington, D.C., 1972, pp. 115-128

Wiltschko, W., Wiltschko, R.: The interaction of stars and magnetic field in the orientation system of night migrating birds. I. Autumn experiments with European warblers (Gen. *Sylvia*). II. Spring experiments with European robins (*Erithacus rubecula*). Z. Tierpsychol. 37, 337-355, 39, 265-282 (1975a, b)

Orientation of Transatlantic Migrants

TIMOTHY C. WILLIAMS and JANET M. WILLIAMS, Department of Biology, Swarthmore College, Swarthmore, PA 19081, USA and Marine Biological Laboratory, Woods Hole, MA 02543, USA

Abstract

Observations from a network of radars on ships and on islands in the western North Atlantic Ocean and on the eastern coast of North America suggest that successful autumnal migrants moving between North and South America utilize simple compass orientation and do not require bicoordinate navigation. Successful migrants maintained a southeast heading from the North American coast to the Caribbean. Unsuccessful migrants became disoriented over the Atlantic. Visual observations suggest that successful migratory behavior is restricted to certain taxonomic groups of passerines.

A. Introduction

Each fall large numbers of birds move over the western North Atlantic Ocean from the eastern coast of North America to the Caribbean and South America. Visual observation from ships at sea (Hurdis, 1897) and studies of species distribution indicate that several species of shore birds (Charadriiformes) use this migration route (Lincoln, 1939; Hagar, 1966; Palmer, 1968; McNeil and Burton, 1973). Drury and Nisbet (1964) and Nisbet (1970) conclude that several species of passerines (Passeriformes) including the Blackpoll Warbler (*Dendroica striata*) also use this route, although Murray (1965, 1976) disputes this point.

Radar has been used to observe the nocturnal departures of migrants from the North American coast (Drury and Keith, 1962; Drury and Nisbet, 1964; Swinebroad, 1964; Williams et al., 1972; Richardson, 1972a; Williams et al., 1977a). The size and speed of movement of the radar echoes detected suggested that both passerines and shore birds depart southeast over the Atlantic. Radar observations in the Caribbean reveal large numbers of birds flying over Antigua and Puerto Rico moving both southeast and southwest (Hilditch et al., 1973; Richardson, 1976). These observers concluded that both passerines and shore birds reached the Caribbean by an Atlantic route.

From 1969 to 1974 we used a network of radars on ships, islands, and the North American coast to make simultaneous observations from up to nine sites over a 2 - 4 week period each fall. A general description of these observations can be found in Williams et al. (1974) and Williams et al. (1977b). Observations made at Miami, Florida have been reported in Williams et al. (1977a). A descriptive analysis of the data from Bermuda is reported in Ireland and Williams (1974). Observations from ships are described in Larkin et al. (in press) and McClintock et al. (in prep).

B. Methods

Observations were made for a 2 - 4 week period in late September and early October. Figure 1 gives the location and dates of observations for the years 1971, 1972, and 1973. In 1969 and 1970 we used radars at Cape Cod, Wallops Island, Bermuda, and in 1970 also at Antiqua.

Whenever possible we used narrow beam radars of the type used for weather observations or missile tracking because these instruments indicate the altitude as well as the horizontal position of radar echoes. Data reported in this paper were obtained with weather radars located at Chatham (on Cape Cod), Mass., Miami, Florida (WSR-57), and the Caribbean islands of Antigua, Barbados, and Tobago (Mitsubishi RC-32-B), and with missile tracking radars at Wallops Island, Virginia (SPANDAR and FPS-16), Bermuda (FPS-16 and FPQ-6), and Antigua (FPQ-6). The beamwidth of the radars ranged from 2° to 1.39° at the 3 dB points and their peak power ranged from 500 kW to 3.8 MW. (See Williams et al., 1977b for additional information and dates of operation of these radars.)

The radars were rotated at a constant angle of elevation (between 0.5° and 15°) and the echoes displayed on a plan position indicator (PPI). A time exposure of from 3 - 8 min was made of the PPI with a Polaroid camera. Slowly moving targets such as birds produced streaks or a series of dots on the film that were used to determine the direction, speed of movement, and density of migrants. Altitude of birds was computed from range, angle of elevation with correction factors for curvature of the earth, and index of refraction of the atmosphere.

We scored over 1700 PPI photographs on a four point density scale (no movement, light, moderate, and heavy) established for each radar. Detailed comparisons between different radars are unreliable (see Eastwood, 1967; Gauthreaux, 1970; Richardson, 1972b); thus, neither a more precise scale nor a single scale for all radars was used.

U.S. Weather Service charts of the northern hemisphere were used to obtain surface weather conditions over the Atlantic. Direction and speed of the winds aloft were obtained from radiosonde weather balloons launched within 5 km of all sites. In almost all cases these wind measurements were within 4 h of our radar observations. Velocity of birds relative to a stationary air mass (V_a) was calculated by vector addition from the velocity relative to the ground (V_g) as taken from the PPI photographs and the velocity of the wind (V_w). The scalar components of V_a are airspeed and heading.

Observations from ships in 1971 and 1972 were made with a 40 kW tracking radar described in Griffin (1973) and Larkin et al. (in press). In 1973 and 1974 we used 40 and 75 kW marine radars with their antennas modified by being tilted upward at 7° to 30° as described in McClintock et al. (in prep.). The tracking radar detected few birds above 1000 m and the modified marine radars could not have detected birds above 500 m. Winds aloft for 1971 and 1972 were obtained with radar-tracked balloons launched every 4 h or more often during a change in weather. Winds during 1973 and 1974 were taken from wind sensors on the ships at a height of 30 m above the sea. In almost all cases these wind data were in good agreement with geostrophic winds calculated from surface weather charts.

Dates and times of observation from ships at sea will be found in Larkin et al. (in press) and McClintock et al. (in prep.).

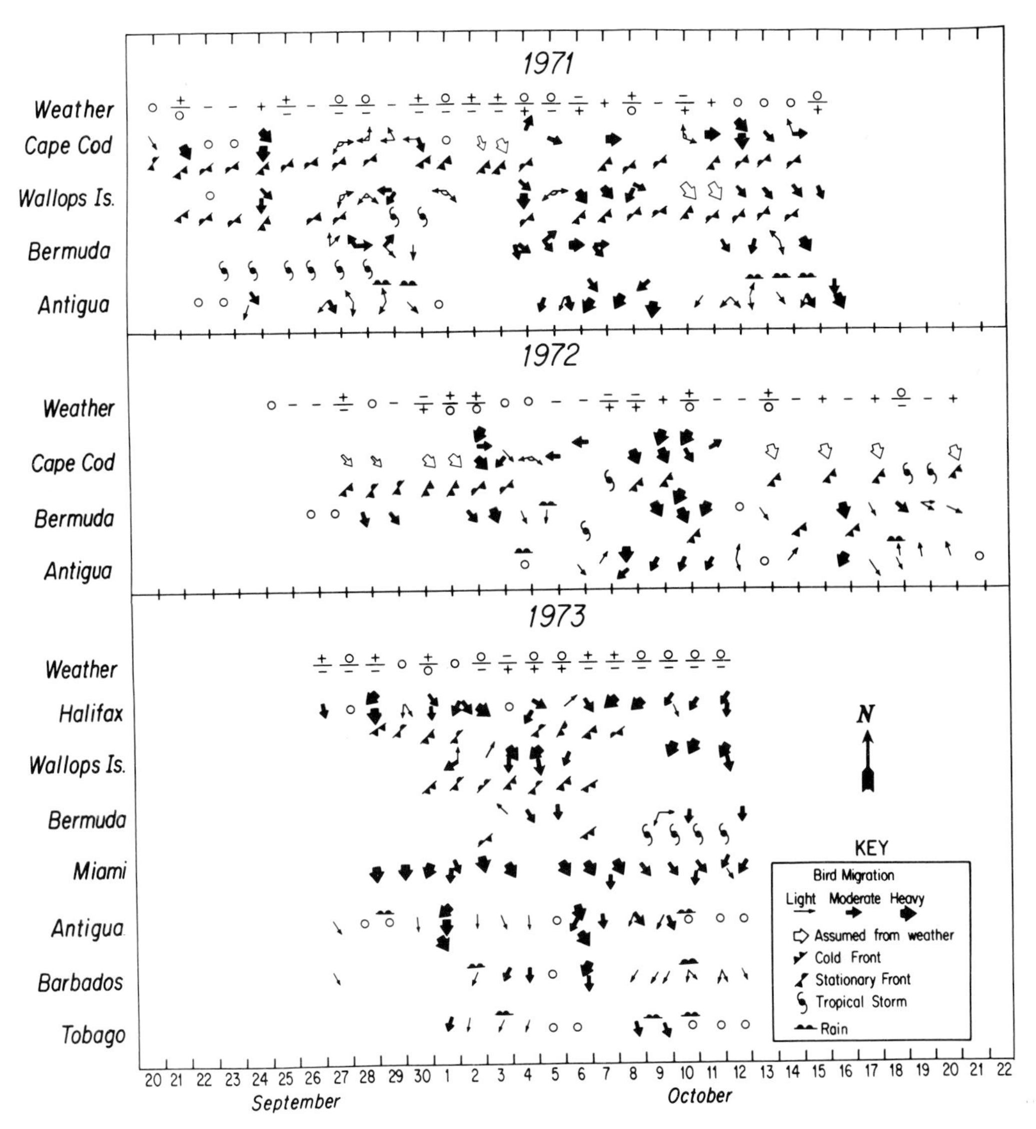

Fig. 1. Average (V_g) direction and density of bird migration over the North Atlantic Ocean as detected by radar. Light, moderate, or heavy migration is indicated by the *width of the arrow*. An *0* indicates no significant migration. *Direction of arrows* shows the average direction of the most dense movement(s) seen during an 8-h observation period. An *arc* between two light arrows indicates targets were spread between these directions. *Weather symbols* indicate prevailing frontal conditions between northern sites, and the presence of widespread rain (tropical depression) at Caribbean sites. Rows labeled *weather* indicate the suitability of weather conditions along the North American coast for departures of migrants: + indicates weather favorable for departure (north winds, recent passage of a cold front), - indicates unfavorable conditions (south winds, cold front approaching, rain), and 0 indicates no clear advantage or disadvantage (stable high pressure system, light and variable winds). The symbol $\frac{+}{-}$ indicates favorable conditions on the northern part of the coast and unfavorable on the southern part

C. Results

I. Observations from Coastal and Island Radars

Figure 1 presents the data from all radars except those on ships for 1971, 1972, and 1973. The Figure shows the complex nature of migratory behavior over the Atlantic. It is not possible to identify clearly the movement of single migration waves over the study area. Departures typically occurred over 1000 km of coastline during a period of two or three nights after a cold front passed through the observation area, as may be seen by the weather symbols in Figure 1. This association has been more fully documented by Nisbet and Drury (1968) and Richardson (1972a). Departures not associated with cold fronts in our observations tended to be moderate movements along the coast occurring under stable weather conditions ("O" in Fig. 1). Observations at Miami revealed almost continuous southeast movements toward the greater Antilles regardless of weather conditions (see Williams et al., 1977a).

Cold fronts, with accompanying northwest winds, which trigger departures of birds from the eastern coast of North America, rarely penetrate the North Atlantic beyond Bermuda. Instead these weather systems become stationary, producing a band of cloudiness with locally turbulent wind conditions stretching more or less parallel to the coast 500 - 1000 km offshore. Beyond this stationary front, the winds are usually light and from the south or east. As shown in Figure 1, many birds penetrate the frontal zone and continue their course to reach Bermuda (7 Oct. 1971, 29 Sept., 2, 9, 10, 16, 18 Oct. 1972, and 4. Oct. 1973; also four cases in 1969 and 1970). Thus, although downwind orientation may be important in departure from the coast, some other orientation system is needed for continuing the flight.

Figure 2 shows the distribution of the modal direction(s) of all moderate and heavy migrations detected at coastal and island radar sites. The great majority of birds over Bermuda were moving to the southeast, but at Antigua we most frequently detected migratory movements to the southwest, indicating that migrants passing over Bermuda later turned toward the southwest in the area of the Sargasso Sea. Figures 1 and 2 show a second route to the Caribbean, which also involves a change in V_g en route (few birds apparently make straight flights between North and South America). Birds on this route first move southwest along the North American coast and then turn in the vicinity of Florida to continue southeast along the Antilles.

II. Observations from Ships at Sea

Bird migration observed at sea both visually and with radars was intermittent. Typically there would be a period of no migratory activity followed, often suddenly, by a period of 4 - 12 h when many birds were seen around the ship or on radar. Within a migratory period, birds usually exhibited consistent migratory behavior, with the next period of activity showing different direction, speed, size, or dispersion of headings. Following the terminology of Larkin (in press) we refer to these consistent periods as blocks of migratory behavior.

Figure 3 shows the location of all blocks of migratory behavior detected by radar during our observations from ships in the western North Atlantic Ocean. In each case we indicate the density of migration and the average V_g and V_a of the block. If the angular deviation (see Batschelet, 1965) of V_a exceeds 60° this is indicated by the letter "S" for scattered headings. Inspection of Figure 3 reveals that migratory behavior over the Atlantic may be grouped into four categories.

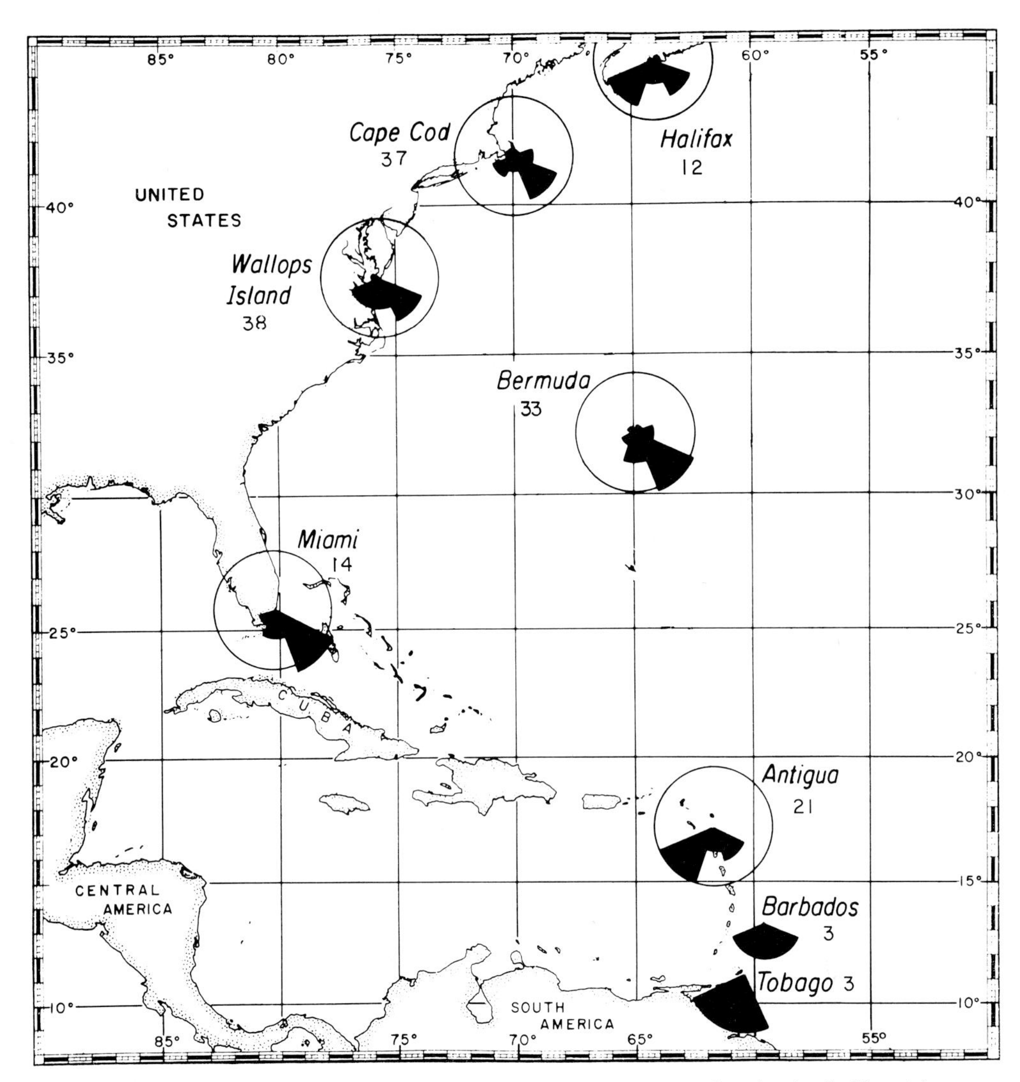

Fig. 2. Map of North Atlantic Ocean showing radar sites and principal directions (V_g) of autumnal bird migration. Circular histograms at each radar site indicate per cent of moderate and heavy migrations with average directions in eight compass sectors (*ring* indicates 50%; *numeral* indicates number of nights with moderate or heavy migration)

First are those periods showing no significant migration. Second are groups of birds that have relatively little dispersion in air headings and are making net progress in a southerly direction. Third are groups of birds with widely scattered headings (V_a), whose airspeed, in most cases, was less than 20 km/h. [As discussed by Larkin et al. (in press) and McClintock et al. (in prep.), the strength of the radar echoes and the range at which the echoes were detected make it improbable that these were insects.] It appears unlikely that any significant proportion of birds in this category would be likely to reach the Caribbean or South America if they persisted in poorly oriented flight at low speed. The fourth type of behavior consisted of birds that appeared to be

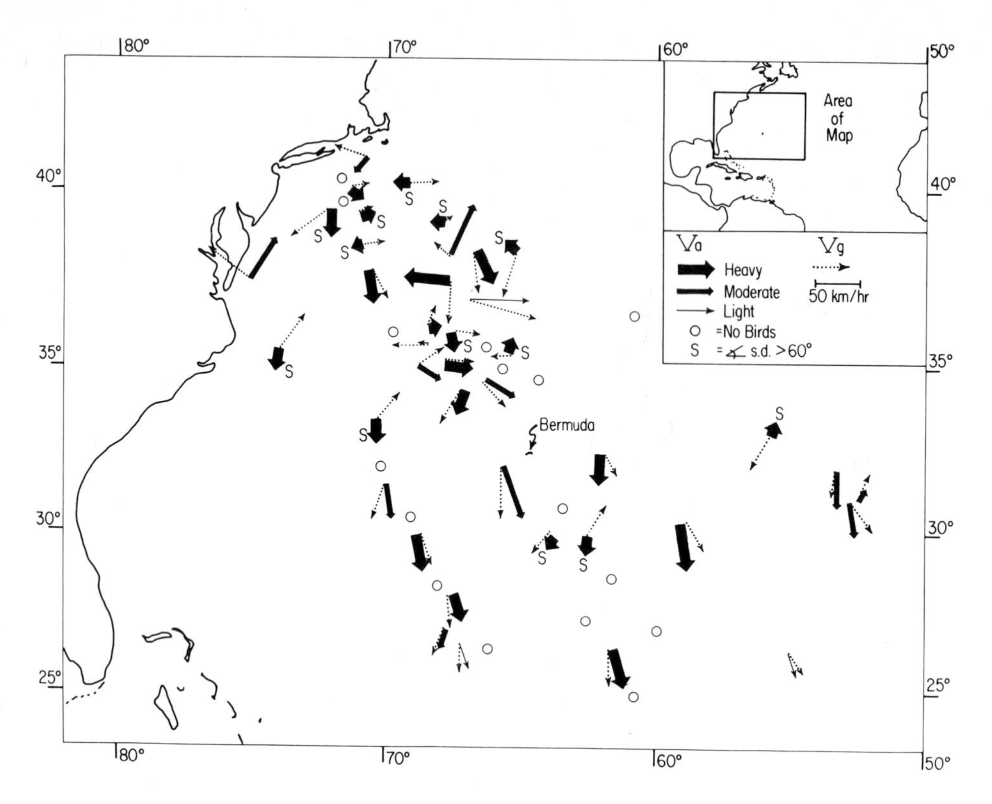

Fig. 3. Location of all blocks of bird migration detected from ships in the western North Atlantic by radar (see text). *Solid arrow* gives average V_a and *dotted arrow* gives average V_g of radar echoes. *S* indicates that angular deviation of V_a is greater than 60°

making an effort to move in a direction other than toward the Caribbean. An example of this type is the most westerly of our observations just off the coast of Virginia at 37° N in Figure 3.

Figure 3 reveals that the behavior of birds over the Atlantic during autumn migration is not easily described geographically; there do not appear to be clearly defined routes of movement. If we divide the Atlantic into two areas by a line running ENE to WSW through Bermuda in an attempt to separate coastal from true transatlantic migrants, and then examine the distribution of the four classes of migratory behavior, we obtain a χ^2 of 6.0, indicating that the distribution does not differ significantly from a uniform distribution. Perhaps most surprising is the observation of birds in apparently disoriented flight to the south and east of Bermuda, up to 2000 km from the North American coast.

Figure 4 shows the same data as Figure 3, but plotted on a schematic diagram of weather patterns rather than on a map of the Atlantic. This technique, adapted from Richardson (1972a), allows one to place any observation in our study area in relation to the predominant weather patterns. The schematic weather diagram (reading upper right to lower left) consists of a low-pressure center or hurricane, an actively moving cold front, and a stationary front. Two high-pressure air masses are shown; the one in the lower right of the Figure represents a subtropical

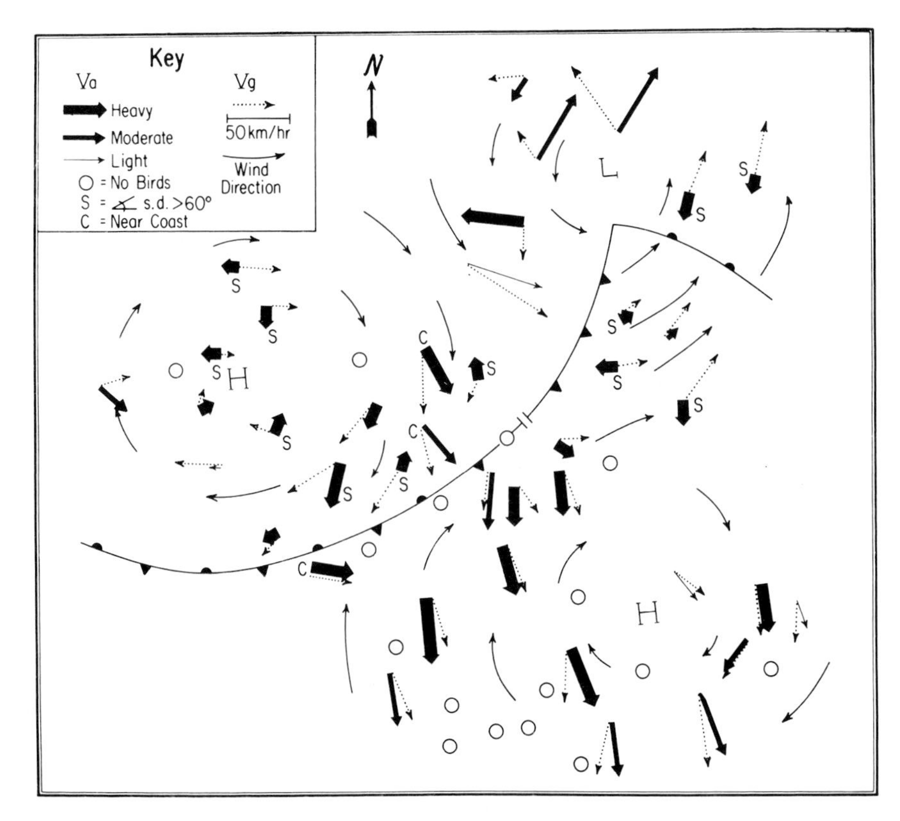

Fig. 4. Data in Figure 1 replotted on a weather diagram. See text for explanation. *C* indicates observations made within 400 km of the North American coast

high-pressure center and the one northwest of the frontal system represents air recently moved off the North American continent. The placement of an observation in Figure 4 was determined first by the observed V_W (indicated by light curving arrows in Fig. 4) and secondly by the relative position of dominant weather patterns at the time of observation. Note that there is no implied distance relationships in Figure 4; an observation in the lower right of Figure 4 may have been physically closer to an advancing cold front than one near the center of the Figure.

Unlike in Figure 3, migratory behavior in Figure 4 is clearly different for several areas of the Figure. In the area around the subtropical high in the lower right of the Figure we find only blocks of migrants with well-grouped headings making effective progress toward the Caribbean. Groups of birds with scattered air headings were found only northwest of frontal systems at sea, or south and east of low-pressure centers. In both cases these birds would have recently experienced strong offshore winds. These areas of Figure 4 also contain almost all the observations of groups of birds with average airspeeds of less than 20 km/h. The three cases of birds moving north or west are in the upper right of Figure 4 and probably represent birds attempting to avoid an advancing tropical storm.

Based upon the analysis outlined above, we can tentatively classify migrants observed over the Atlantic as "true migrants" if they have penetrated the frontal system that triggered their departure, and unsuccessful migrants if they are observed in a high-pressure center northwest of a frontal system or in a low-pressure system. Some true migrants will be included in the latter category but, as shown by the letter "C" in Figure 4, these will usually result from observations made near the North American coast when birds had not yet penetrated the frontal system. With a single exception, the average headings of all blocks of birds classed as true migrants lie between 100° and 200°.

Table 1 and Figure 2 show that a generally southeast heading is typical of Atlantic migrants detected at coastal and island-based radars as

Table 1. Average heading (V_g) of "true transatlantic migrants"[a] detected by radar

Radar site	V_g[b]	Mean vector (V_a)	Angular deviation
Chatham, Mass. (Cape Cod)	000° - 219°	148°	± 52°
Wallops, Is., Va.	000° - 169°	142°	± 39°
Bermuda	000° - 169°	143°	± 28°
Miami, Fla.	000° - 360°	152°	± 24°
Antigua (Atlantic)[c]	170° - 360°	163°	± 35°
Antigua (Caribbean)	000° - 169°	132°	± 32°
Ships at sea[d]	---	163°	± 26°

[a]Birds likely to make a successful crossing of a portion of the western Atlantic, see text.
[b]Range of observed direction of movement used as a basis for selecting migrants.
[c]At Antigua birds were divided into two groups; the first group of birds approached the island from the open Antlantic Ocean, the second group approached from the Caribbean islands westnorthwest of Antigua.
[d]These migrants not selected on the basis of V_g, see text.

well as from ships at sea. Identification of "true migrants" at these sites was complicated by the frequent simultaneous occurrence of two or more patterns of migratory behavior. Thus, at these sites we separated individual tracks of birds on the basis of ground direction and then computed the average headings of each group. Although the average heading of all migrants at Cape Cod, Wallops Island, Miami, and Bermuda was to the southeast, the analysis at Antigua is probably the most important. At Antigua we divided birds into two groups: those that were moving along the Antilles to the southeast and those that were arriving from the Atlantic moving southwest. Despite this arbitrary division, the average heading of both groups was to the southeast; the difference in observed V_g was due to flight in different wind conditions.

The orientation of Atlantic migrants departing Bermuda to the southeast may be compared with that of migrants arriving at Antigua moving southwest. A significant difference would indicate re-orientation by the birds in midocean; a lack of any significant difference would indicate that the entire flight had been made with a single compass heading. Due to the truncated nature of the distributions, none of the standard statistical tests appears applicable. However, the average V_a in Table 1 for Bermuda and Antigua differ by only 20° (less than one angular deviation) despite the fact that the V_g of the two groups are mutually exclusive. Thus, if there had been a change in average V_a en route, it would have been small in comparison to the change in V_g direction.

III. Visual Observations at Sea

Most of the birds seen on or around the ships during the period of our observations were warblers and sparrows (Passeriformes) (see Larkin et al., in press, and McClintock et al., in prep.). Figure 5 shows the distribution of eight species of warblers (Parulidae) that were seen at least twice at sea. These species, which winter in the Caribbean and South America, were all seen south of Bermuda as well as between Bermuda and the coast. Figure 6 shows the occurrence of four species of sparrows (Fringillidae) seen at sea. Of these birds, which do not regularly migrate to the Caribbean and South America, only one junco was seen south of Bermuda. The difference in the distribution of these two groups of birds, which are physically similar, suggests that there are differences in the migratory behavior of at least two groups of passerine species regularly found over the Atlantic in autumn.

D. Discussion

Our observations suggest that compass orientation at a constant heading is a necessary and sufficient explanation of the orientation of autumnal migrants over the Atlantic Ocean. Unlike continental migrants, these birds do not fly in relatively constant wind and weather conditions that might aid in orientation (see Gauthreaux and Able, 1970, and Able, 1974). Although the migratory route over the Atlantic involved a major change in V_g, it appears that this does not require a change in V_a and the entire flight is made at a constant southeast heading. Our observations did not reveal any evidence for bicoordinate navigation by these migrants.

If, as suggested by Murray (1965, 1976), all passerines seen over the Atlantic are drifted by strong offshore winds, one would not expect to find major differences in the distribution of physically similar species. Observations from ships, however, reveal species specific patterns of distribution and, in conjunction with our radar observations, suggest two types of orientation for passerines over the Atlantic. Several species of warblers appear to use the Atlantic as well as a coastal route between North and South America. These birds must be able to penetrate frontal weather systems and maintain their southeast orientation until they reach the Caribbean (see Williams, in prep.) and South America. Birds such as the sparrows, on the other hand, appear unable to penetrate these frontal systems, perhaps because they lack a strong southeast orientation, and thus rarely move beyond Bermuda, where they are recorded as regular winter residents (Wingate, 1973).

Acknowledgments. We are extremely grateful to many government, organizations, government personnel, research assistants, graduate students, and friends who made it possible for us to obtain the data reported here; we regret that space does not permit us to individually acknowledge them. We are especially grateful to C. McClintock, D. Griffin, R. Larkin, and J. Teal for allowing us to use data collected by them from ships at sea. This research was supported by NSF Grants GB-13246 and GB-43252, NASA grant NGR 33-183-003, and the Bache Fund of the National Academy of Sciences.

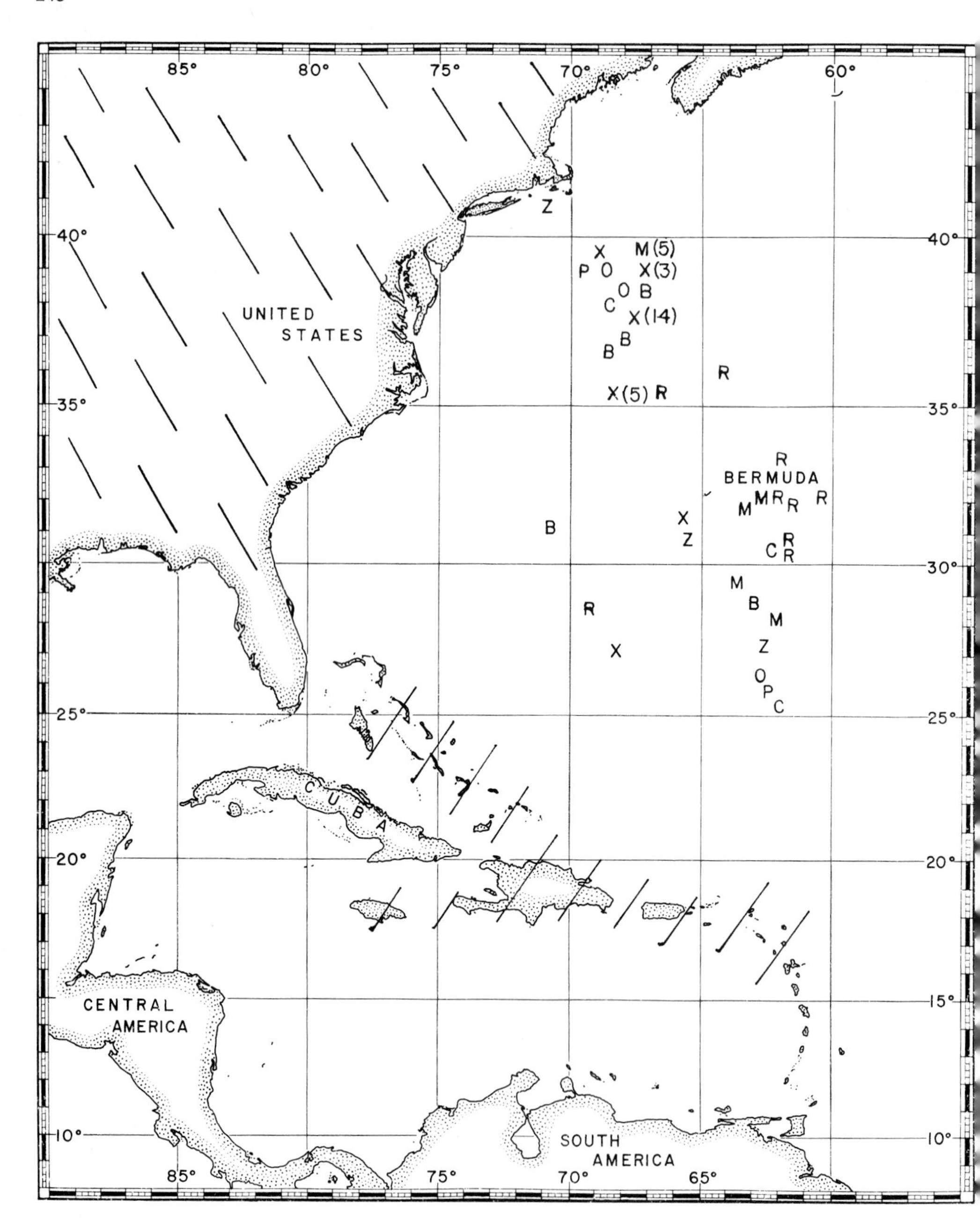

Fig. 5. Map of the western Atlantic showing the distribution of eight species of warblers seen from ships. *Hatching downward from left to right* indicates summer territories and areas in which the birds are seen during fall migration in eastern North America. *Hatching downward from right to left* indicates wintering grounds in the West Indies (Williams, in prep.). Several of these species also winter in Central and South America. Species indicated as follows: *P*, Northern Parula Warbler (*Parula americana*); *C*, Cape May Warbler (*Dendroica tigrina*); *B*, Black-Throated Blue Warbler (*D. caerulescens*); *M*, Myrtle Warbler (*D. coronata*); *Z*, Prairie Warbler (*D. discolor*); *X*, Palm Warbler (*D. palmarum*); *O*, Ovenbird (*Seiurus aurocapillus*); *R*, American Redstart (*Setophaga ruticilla*)

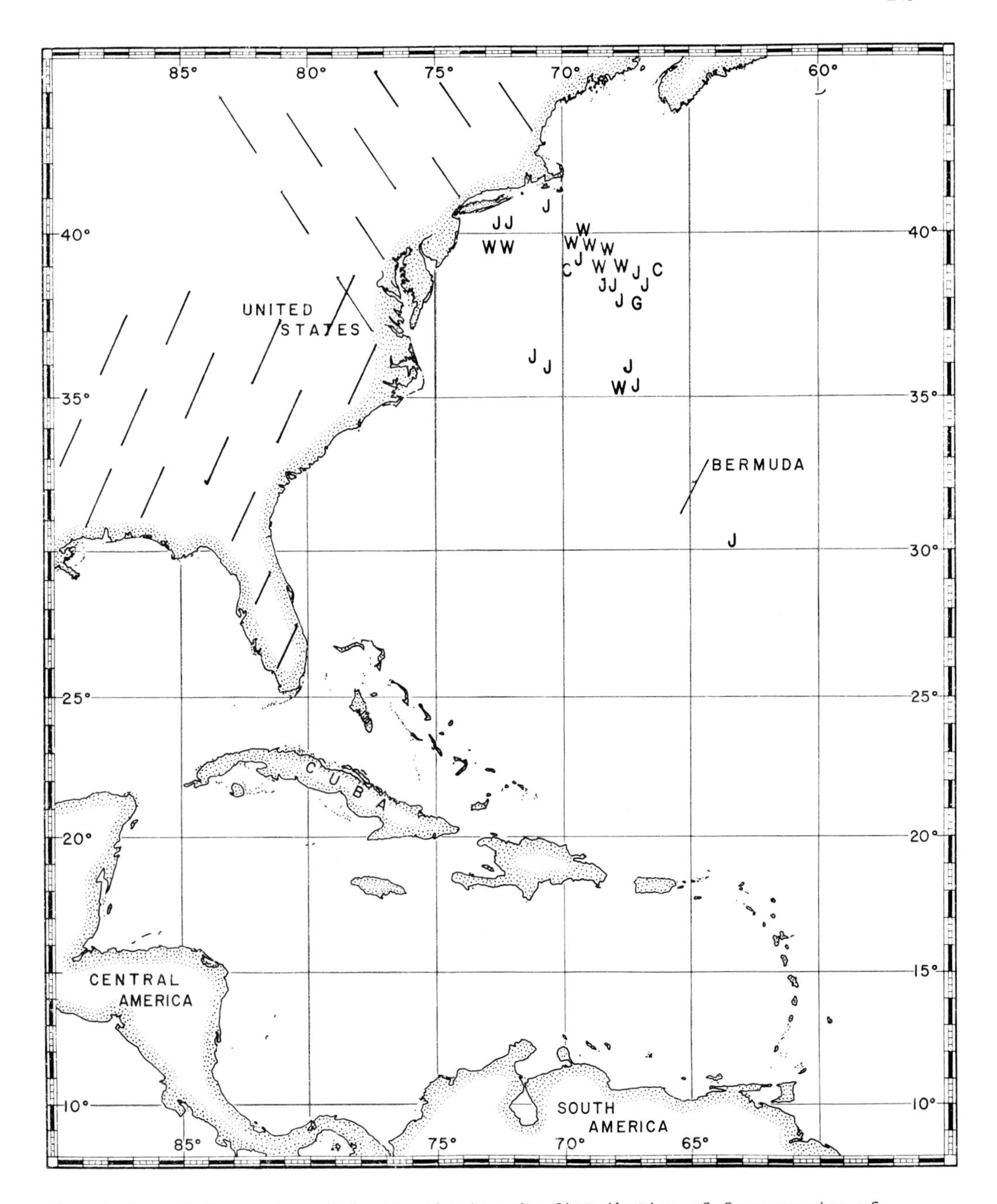

Fig. 6. Map of the western Atlantic showing the distribution of four species of sparrows seen from ships. Symbols as in Figure 5. Species indicated as follows: *G*, Grasshopper Sparrow (*Ammodramus savannarum*); *J*, Slate-Colored Junco (*Junco hyemalis hyemalis*); *W*, White-Throated Sparrow (*Zonotrichia albicollis*); *C*, Chipping Sparrow (*Spizella passerina*)

References

Able, K.P.: Environmental influences on the orientation of free-flying nocturnal bird migrants. Anim. Behav. 22, 224-238 (1974)

Batschelet, E.: Statistical methods of analysis of problems. In: Animal Orientation and Certain Biological Rhythms. AIBS monograph, Washington, D.C, 1965

Drury, W.H., Keith, J.A.: Radar studies of songbird migration in coastal New England. Ibis 104, 449-489 (1962)

Drury, W.H., Nisbet, I.C.T.: Radar studies of orientation of songbird migrants in southeastern New England. Bird Banding 35, 69-119 (1964)

Eastwood, E.: Radar Ornithology. London: Methuen, 1967, 278 pp.

Gauthreaux, S.A., Able, K.P.: The direction of nocturnal songbird migration. Nature (London) 25, 476-477 (1970)

Gauthreaux, S.A.: Weather radar quantification of bird migration. Bioscience 20, 17-20 (1970)

Griffin, D.R.: Oriented bird migration in or between opaque cloud layers. Proc. Am. Phil. Soc. 117, 117-141 (1973)

Hager, J.A.: Nesting of the Hudsonian Godwit at Churchill, Manitoba. Living Bird 5, 1-43 (1966)

Hilditch, C., Williams, T.C., Nisbet, I.C.T.: Autumnal bird migration over Antigua. Bird Banding 44, 171-179 (1973)

Hurdis, J.L.: Rough notes and memoranda relating to the natural history of the Bermudas by the late Hurdis, John L. edited by his daughter, H.J. Hurdis. London: R.H. Porter, 1897

Ireland, L.C., Williams, T.C.: Radar observations of bird migrations over Bermuda. Proc. Conf. on the Biol. Aspects of the Bird/Aircraft Collision Problem. Gauthreaux, S.A. (ed.), U.S.A.F. Office of Scientific Research, Arlington, Va. (1974)

Larkin, R.P., Torre-Bueno, J., Griffin, D.R., Teal, J.M.: Radar observations of bird migration over the western North Atlantic Ocean. Behavioral Ecology and Sociobiology (in press)

Lincoln, F.C.: The Migration of American Birds. New York: Doubleday, Doran and Co. 1939, 189 pp.

McClintock, C., Williams, T.C., Teal, J.M.: Radar observations of bird migration from ships in the western North Atlantic (in prep.)

McNeil, R., Burton, J.: Dispersal of some southbound migrating North American shorebirds away from the Magdalen Islands, Gulf of St. Lawrence, and Sable Island, Nova Scotia. Carib. J. Sci. 13, 257-278 (1973)

Murray, B.G.: On the autumn migration of the Blackpoll Warbler. Wilson Bull. 77, 122-133 (1965)

Murray, B.G.: The return to the mainland of some nocturnal passerine migrants over the sea. Bird-Banding 47, 345-359 (1976)

Nisbet, I.C.T: Autumn migration of the Blackpoll Warbler: evidence for long flight provided by regional survey. Bird-Banding 41, 207-241 (1970)

Nisbet, I.C.T., Drury, W.H.: Short term effects of weather on bird migration: a field study using multivariate statistics. Anim. Behav. 16, 496-530 (1968)

Palmer, R.S.: Species accounts. In: The Shorebirds of North America. Stout, G.C. (ed.) New York: Viking Press, 1968, 270 pp.

Richardson, W.J.: Autumn migration and weather in eastern Canada: a radar study. Am. Birds 26, 10-17 (1972a)

Richardson, W.J.: Temporal variations in the ability of individual radars in detecting birds. Field note #61, Associate Committee on Bird Hazards to Aircraft, National Research Council, Ottawa (1972b)

Richardson, W.J.: Autumn migration over Puerto Rico and the western Atlantic, a radar study. Ibis 118, 309-332 (1976)

Swinebroad, J.: The radar view of bird migration. Living Bird 3, 64-75 (1964)

Williams, J.M.: Atlantic migrants (in prep.)

Williams, J.M., Williams, T.C., Ireland, L.C., Teal, J.M.: Transatlantic bird migration. Proc. Conf. on the Biol. Aspects of the Bird/Aircraft Collision Problem, Gauthreaux, S.A. (ed.), U.S.A.F. Office of Scientific Research, Arlington, Va, 1974

Williams, T.C., Berkeley, P., Harris, V.: Autumnal bird migration over Miami studied by radar: A possible test of the wind drift hypothesis. Bird-Banding 48, 1-10 (1977a)

Williams, T.C., Williams, J.M., Ireland, L.C., Teal., J.M.: Autumnal bird migration over the western North Atlantic Ocean. Am. Birds. In press.(1977b)

Williams, T.C., Williams, J.M., Teal, J.M., Kanwisher, J.W.: Tracking radar studies of bird migration. In: Animal Orientation and Navigation: a symposium. Galler, S. and Schmidt-Koenig, K. (eds.) NASA SP262. Washington, D.C. U.S. Government Printing Office. 1972, pp. 115-128

Wingate, D.B.: A Checklist and Guide to the Birds of Bermuda. Bermuda: Island Press, 1973

Effects of Alpine Topography and Winds on Migrating Birds*

BRUNO BRUDERER, Schweizerische Vogelwarte, 6204 Sempach, Switzerland

Abstract

Radar tracks of about 150 single migrants and 2 to 3 measurements of upper winds per each of 139 nights in the lowlands of Switzerland and 35 nights at an Alpine pass show that the principal direction of migration is parallel to the mean course of the border and to the main ranges of the Alps and that winds against or from the right side of the main vector of migration induce pseudodrift (increased migration of northern and northwestern populations) while winds from the left are usually compensated (this to a lesser extent in poor visibility). The principal direction remains astonishingly constant at every altitude and seems not to depend on actually visible leading-lines, while in single flight paths close adaptations to topographical features are recognicable, most pronounced at lower levels and in unfavorable winds.

A. Methods and Data

The basic methods used for the present study were described by Bruderer (1971) and by Bruderer and Steidinger (1972). Some improvements introduced in the meantime affect only the way of recording (simultaneously on magnetic tape and on a two-channeled XY-plotter) but not the sort and amount of data: About 150 night migrants and two to three pilot balloons have been tracked with the fire-control radar "Superfledermaus" (Radar equipment made available by Contraves AG, Zurich, and by the Swiss Army) to obtain information on the distribution of bird tracks and wind directions in every height-band desired during each night of observation. The data used for this paper originate from 64 spring nights (1969 and 1971) and 75 autumn nights (1968 and 1971) in the lowlands of Switzerland. The radar sites used for these lowland studies were located about 15 km to the N and to the SW of Zurich (cf. Fig. 1), about 400 m ASL. Additional data were collected at the pass of Hahnenmoos, in the Bernese Alps (about 1950 m ASL); we included 30 nights of autumn 1974 and selected nights of autumn 1975.

It is important to bear in mind that each of the diagrams on the spread of directions given in this paper represents the relative quantity of birds migrating toward each direction under the given conditions, but they do not include absolute numbers indicating the intensity of migration during different nights.

---

*Supported by the Swiss National Foundation for Scientific Research, grants Nr. 3.244.69, 3.038.73 and 3.058.76.

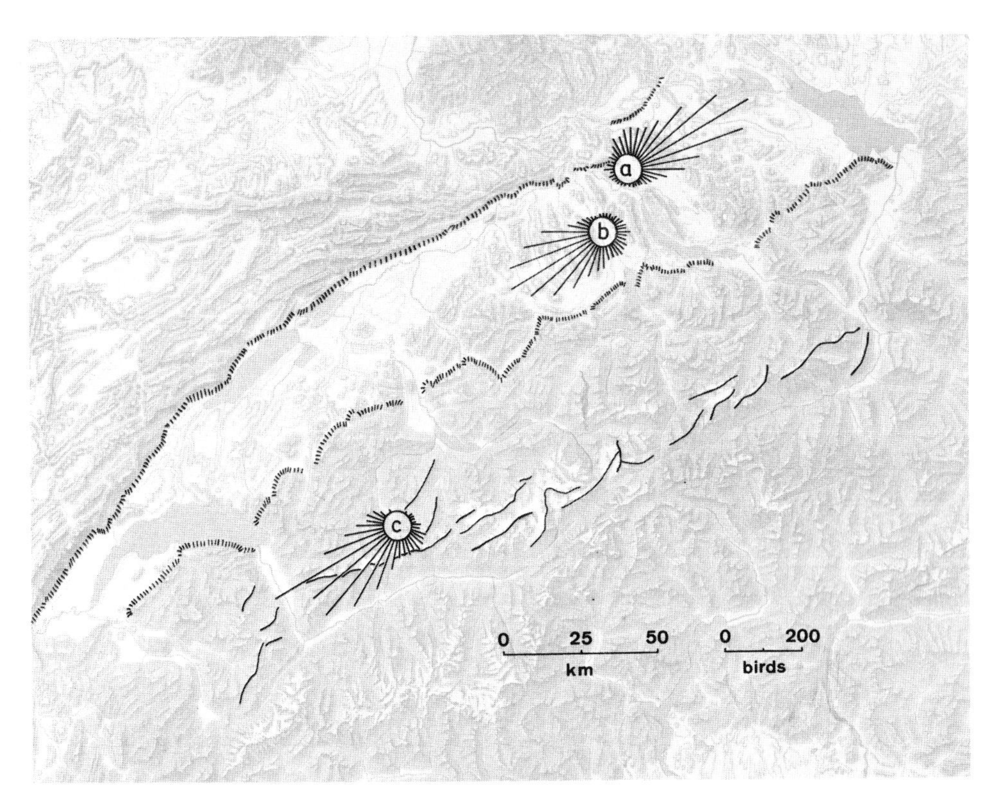

Fig. 1. Map of Switzerland emphasizing the border of the Jura and the Alps as well as the main ridges of the northern Alps. The radar sites are indicated by *circles* at Bachenbülach *a*, Merenschwand *b*, and the pass of Hahnenmoss *c*. The *lines* around the circles show the distribution of track directions during spring migration 1969 and 1971 *a*, during autumn migration 1968 and 1971 *b*, and during autumn 1974 *c*. Copyright of the map: Eidg. Landestopographie

B. Problems

Large-scale alterations of the direction of migratory flights are necessary, if after an initial SW flight, long-distance migrants from eastern or central Europe are to reach their winter quarters in Africa. Many passerines breeding in or passing throught Great Britain fly initially S or SSE toward France, where they change direction to SW to reach Portugal (cf. Zink, 1973 and Zink, 1977). In contrast to the situation over the western Atlantic (cf. Williams and Williams, this volume) these changes in direction are not identical with the prevailing winds (cf. Bruderer, 1977). Berthold (1977) suggested an endogenous and therefore inherited control of such changes in direction; Wiltschko (1977) provided some experimental evidence for this hypothesis.

Yet, the strong dependence of migratory movements on external factors, especially on weather, demands a high amount of flexibility in the suggested endogenous timing, probably combined with an additional measuring system (e.g., for the time of active flight or the distance flown) to achieve necessary changes of direction within the appropriate region. Beyond that, one could imagine that the combined endogenous system is, under natural conditions, primarily responsible for a lowering of the threshold, thereby increasing the birds' readiness to react in a given manner to external stimuli. This possibility rises the question of whether smaller changes in direction (e.g., along the Alps) are purely dependent on the actual influence of winds and visible leading-lines or, wheter again a predetermined amount and sense of shift in direction could be involved in such medium scale changes.

For the Alpine station of Hahnenmoos we could imagine the following principal directions of migration and correspondant interpretations:

1. 255° parallel to the main Alpine chains in that region
2. 240 - 250° parallel to the principal direction in the lowlands, and toward the next visible passes
3. 210° parallel to the Enstligen Valley and its main bordering ridges (visible in Fig. 1), coinciding with the expected direction for central Europe.

The influence of local topography would lead to a larger scatter of directions, while larger proportions of northern and northwestern populations would cause a general shift toward S.

C. Topography, Winds and General Features of Migration in the Swiss Lowlands

Switzerland may be considered as a large valley bordered by the mountain ridges of the Alps and the Jura, respectively. The main summits of the Alps are about 4000 m ASL, the main ridges of the order of 3000 m, and the main passes between 2000 and 2500 m. In the lowlands 90% of the migrants fly lower than 2000 m AGL or 2500 m ASL. Thus, for most of the birds even the crossing of a pass requires additional climbing; this and the lack of suitable feeding places would make it advantageous to avoid the Alps.

If we extract from *Zink's Atlas of Bird Migration* (1973 and 1975) the European recoveries of some night migrants of which the ideal direction crosses a defined area including Switzerland and its immediate surroundings, we get a distribution with a mode pointing toward 210° (Fig. 2). An indication of this direction S of SW in autumn and N of NW in spring can also be deduced from the experiments with caged birds by Sauer (1957), Wiltschko (1968, 1972, 1974), and Wiltschko and Merkel (1971). It is again confirmed by the only valid radar information I could find (within the region of interest, but outside the influence of the Alps); it is 30° for spring migration in the region of Paris (Laty, 1972).

The directions of tracked birds above the Swiss lowlands show for spring and autumn an asymmetry toward the same side as in the recovery diagram, but a much wider scatter and a mode W of SW and E of NE, respectively (cf. Bruderer, 1975 and Fig. 1). The principal direction of migration in spring and autumn (definition in Bruderer, 1971) is parallel to the border as well as to the main ridges of the Jura and the Alps; it is the same at two different observation points despite the differing topography (Fig. 1), and it remains constant at different altitudes (Fig. 3). The wind diagrams for three height bands in spring and autumn (Fig. 4) show that the main air currents all tend to be parallel to the mountain chains. One could argue that the directions of migration are simply reflecting the distribution of winds. If - according to Gauthreaux and Able (1970) and Able (1973) - birds always fly with the wind regardless of its direction, they simply had to wait for favorable winds; and flying with this wind, they would parallel the Alps.

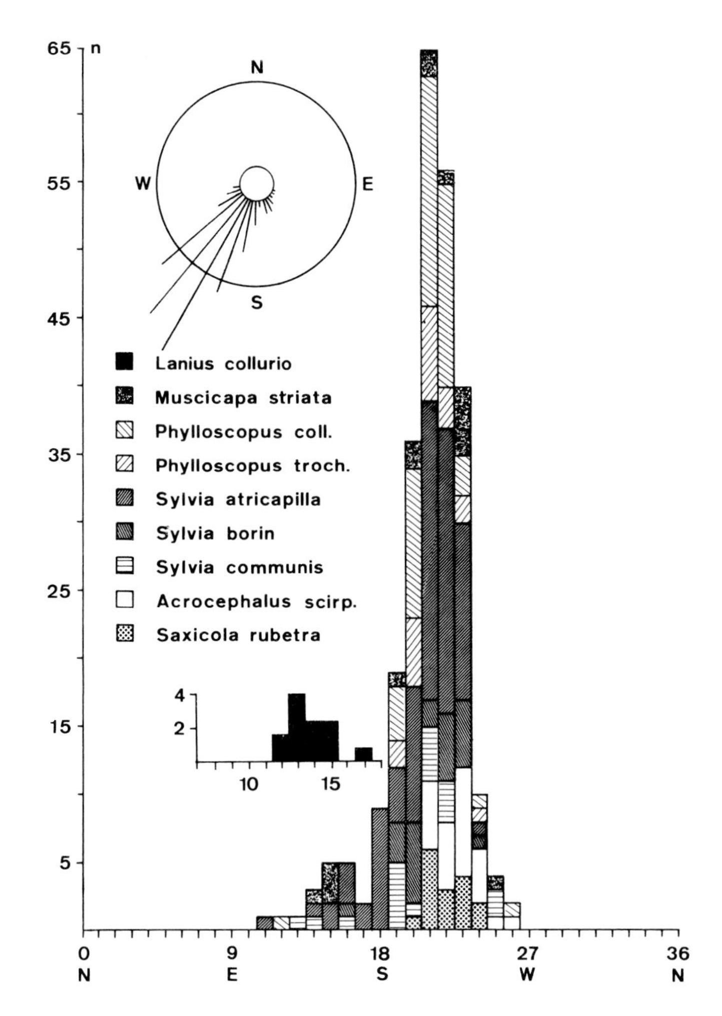

Fig. 2. Distribution of the ideal direction (*straight line*) between ringing sites and European recoveries of different night migrants crossing an area between 46 and 49° N, and 2,5, and 10° E (from Zink, 1973; 1975). *Lanius collurio* is given as an example of a species migrating SE

The diagrams in Figure 5 show that there is a certain tendency to make use of the wind; but at the same time they show that this does not necessarily mean flying with the wind: Winds from a 90°-sector around SW are positive for most spring migrants in central Europe; they lead to a narrow distribution around the principal direction of 60° and to a slight asymmetry toward N, indicating that southwesterly winds have a positive component also for north-bound migrants. Under these conditions strongest spring migration occurs. Winds from sector SE are especially positive for northwestern and northern populations. However, the relative dominance of the northeast-bound migrants is still present despite the cross-wind. NE winds are head winds for the main mass of spring migrants. The proportion of these northeastern birds is reduced, especially when head winds are combined with cloudy weather; whereas the proportion of northwestern populations and reversed migration is increased. NW winds are cross-winds for the main mass of migrants and contain no positive component toward a meaningful goal of spring migration (except perhaps some compensatory movements); under good visibility there is no shift of directions according to winds.

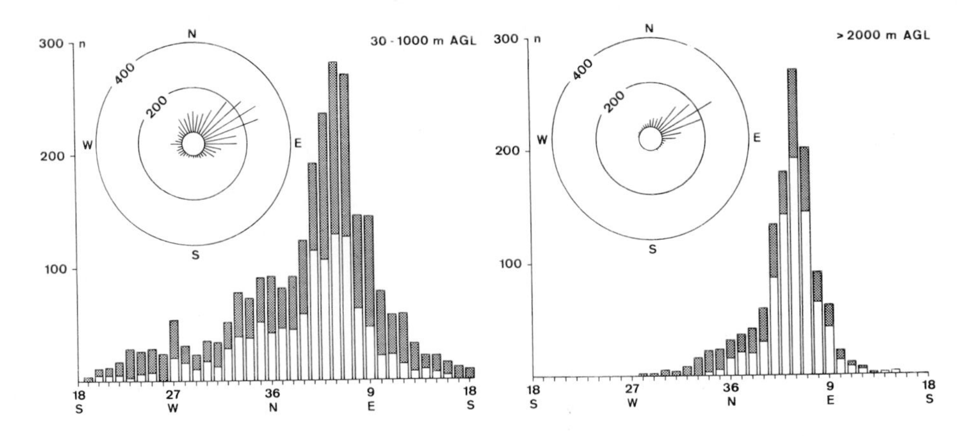

Fig. 3. Distribution of the directions of radar-tracked night migrants during spring above the Swiss lowlands. *Left diagram* below 1000 m AGL, *right diagram* above 2000 m AGL. *White columns* refer to situations with fine weather, *shaded columns* to unfavorable meteorologic conditions. There is a larger scatter in bad weather than in fine weather at both levels and larger scatter at low levels than at high altitudes under both meteorologic conditions

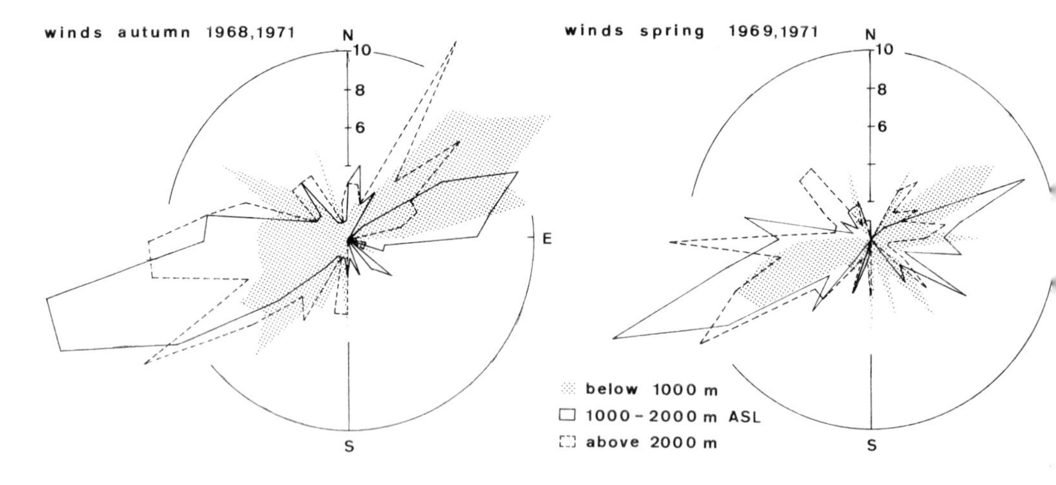

Fig. 4. Wind diagrams for the radar sites in the Swiss lowlands for three height bands, indicating a strong dependence of winds on the course of the mountain ranges and the lowlands between. The *polygons* give the number of measurements with this wind direction (direction of origin) for each ten-degree interval

Generally (for spring and autumn) we can state that winds from the left side of the main vector of migration are usually compensated, whereas winds from the right induce stronger migration of the northern and northwestern populations. Apart from these situations with important movements of northern populations the principal direction is usually recognizable at 60° in spring and at about 245° in autumn.

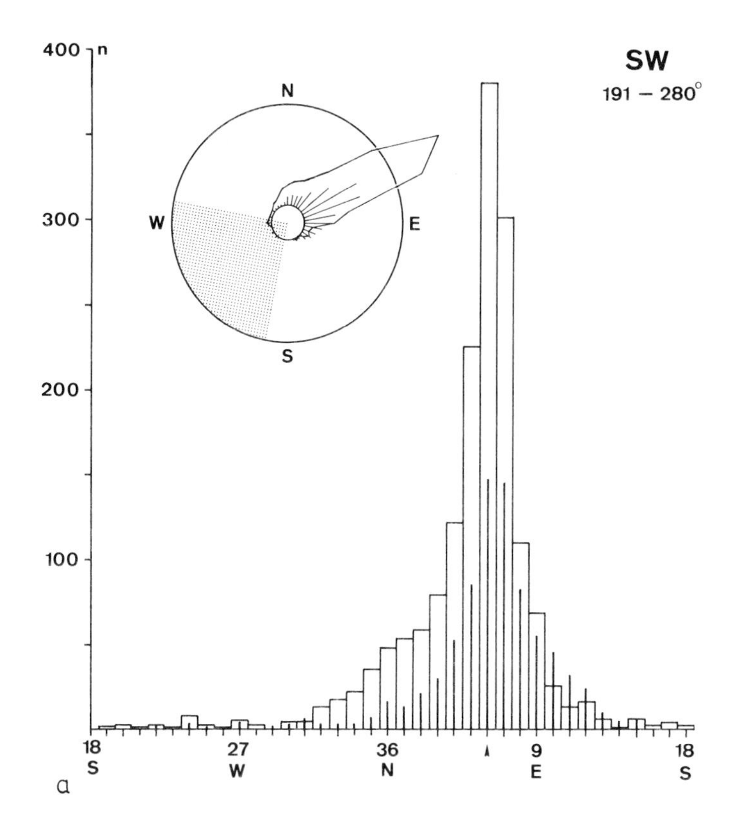

Fig. 5 a–d. The four diagrams contain all the flight directions of birds tracked during 64 spring nights, separated according to the winds prevailing during the nights and at the flight levels considered. The wind sectors are chosen according to the principal direction of migration (parallel or perpendicular to it); they are *shaded* in the circular diagrams. The *columns* and the *polygons* refer to situations without low or medium clouds, the *lines* to situations with reduced visibility to the Alps (by more than 5/8 clouds at low or medium levels).

The *small arrow* below the abscissa marks the principal direction of migration

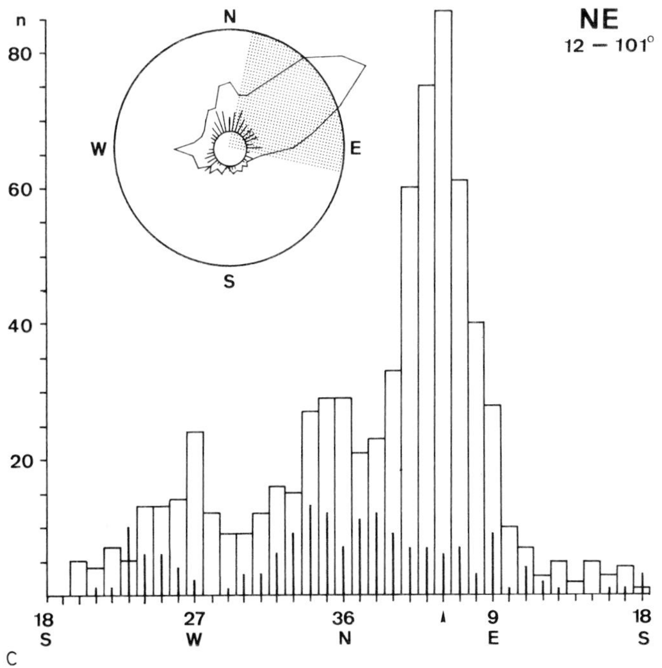

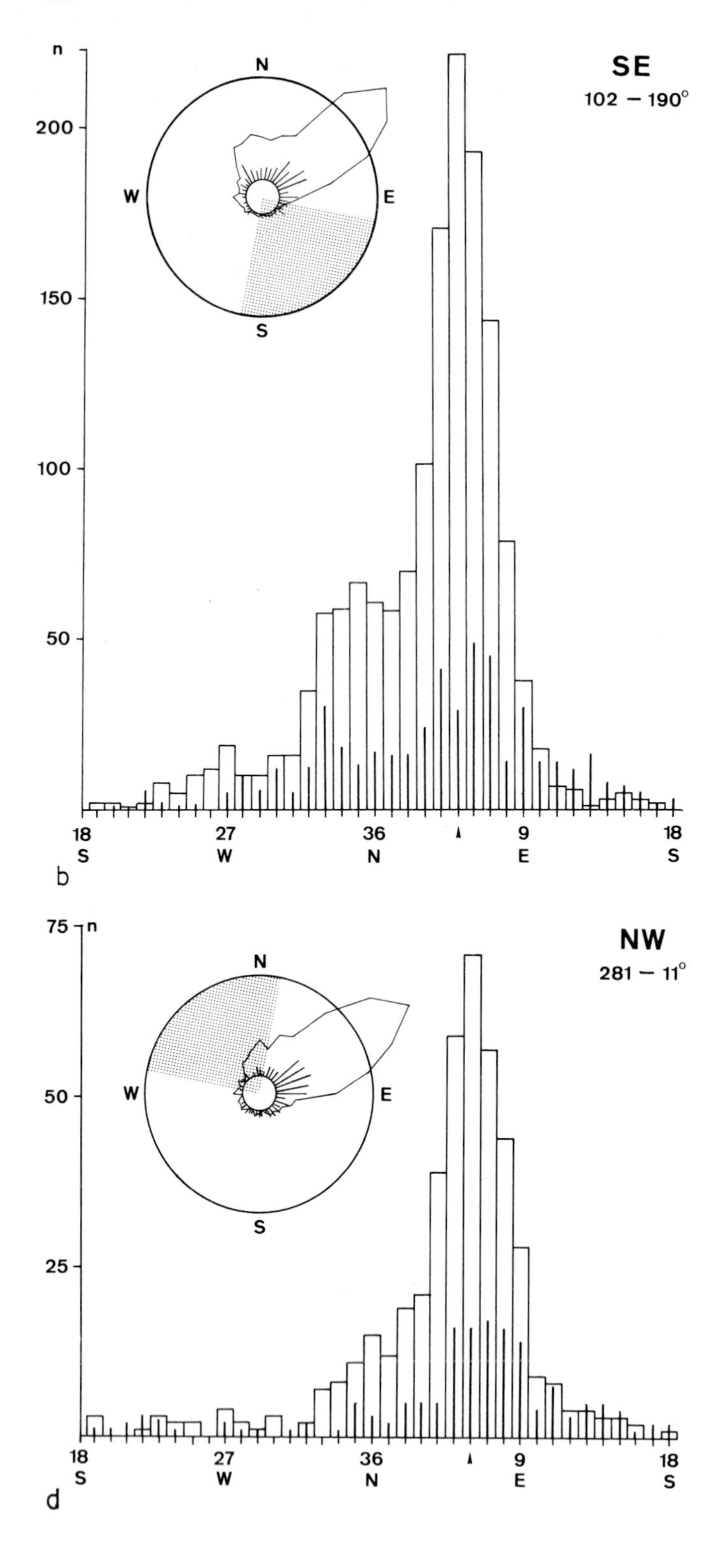

Fig. 5 b and d
Legend see page 257

D. Direction of Migration Under Reduced Visibility

If the mountain barrier of the Alps influences the individual birds as a visual leading-line, its influence should be reduced at night and at altitudes with poor visibility; the direction should shift toward about 30° in spring and toward about 210° in autumn. We have selected conditions with more than 5/8 stratocumulus or altocumulus based between 800 and 2500 m aboveground as nights with poor visibility. The autumn diagram with winds from sector NW (Fig. 6) supports this hypothesis, because there is a distortion of the direction under poor visibility conditions (indicated by lines) against the clear-weather columns. The direction of 210° is fairly pronounced. Yet, an alternative explantation is possible: In cloudy weather the birds' ability to find or keep the preferred direction, and thus to compensate for wind drift, could be reduced.

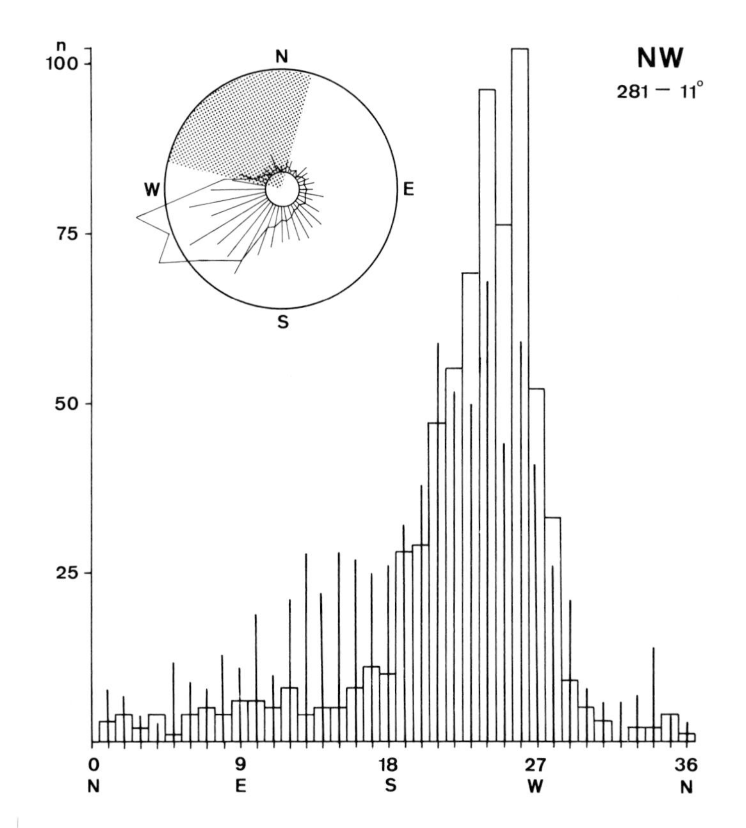

Fig. 6. Same representation as is Figure 5 but for autumn and for winds from sector NW. Notice the large number of birds in poor visibility conditions and their distortion toward S

In the spring migration diagrams (Fig. 5) the situations with SE and NE winds may be interpreted in either way. The decisive situations are those with neutral southwesterly winds, which show no distortion at all, and the NW wind situations, which show a shift in the sense of the second hypothesis (more wind drift in cloudy weather).

Checking the autumn nights with neutral NE winds and SE winds (which induce no distortion under good visibility conditions) led to similar results: There is practically no distortion in cloudy weather with neutral winds or with side winds from the left.

We checked the same question in birds flying below the upper limit of inversions. Within this stable layer of cold air there is usually a dense haze, reducing visibility during the first days to 30 or 40 km, later to 20 or even 10 km. Under these conditions the Alps are not visible from the lowlands, and as confirmed by pilots, one has to climb above the inversion to get sight of the Alps. We selected nights with visibility ranges of less than 20 km at Zurich Airport. The result was the same as under reduced visibility by low clouds: no distortion in neutral winds; slight distortion in the sense of side winds; but no preference for the expected directions of 30 or 210°, respectively.

These results led us to the conclusion that the prevailing directions of migration, while being parallel to the Alps, are not actually induced by the visual impression of the Alpine barrier. We put forward the hypothesis that an evolutive adaptation of the flight paths crossing the region of the Alps and their foreland could induce this astonishingly constant principal direction of migration in the Swiss lowlands and the Swiss Alps (cf. Fig. 1 and next section).

E. Migration, Topography, and Winds in a Central Part of the Alps

A comparison of the directions of migration at the Hahnenmoos pass (Figs. 1 and 7) with the directions in the lowlands reveals a slightly increased scatter in the Alps, but the principal direction tends only some 10° more southward than in the lowlands, and still deviates from the direction expected for central Europe (210°). Neither the direction of the main Alpine chains in that region (about 255°) nor the Engstlinger Valley and its bordering ridges (about 210°) has a dominating leading-line effect. It seems that we are dealing with the same principal direction as in the lowlands, except that the proportion of south-bound migrants seems to be somewhat increased, causing a small southward shift of the mean direction. Observations at other points in the Alps are necessary to confirm this idea.

With respect to the larger scatter of directions, we must take into account that birds at an Alpine pass are closer to the mean ground level in the region considered than the main mass of birds in the lowlands. Comparing only the low-flying birds in Figure 3 with the diagrams of the Alps leads us to the conclusion that there is no basic difference between the birds' behavior in the Alps and in the lowlands; however, the very pronounced topographic features in the Alps permit us to correlate some of the most frequent flight paths with topographic elements.

The influence of winds seems to be very similar to that in the lowlands, except the stronger appearance of leading-line effects (cf. Fig. 7). In SW winds, the scatter of directions and the tendency to follow local guiding-lines is most pronounced. The main mass of migrants fly toward 220 to 240°. In SE winds the southward migration is pratically stopped and the principal direction is very close to the direction in the lowlands; possibly because of the known tendency to undergo some drift in the cloudy weather during the three days available with SE winds. Winds from sector NW induce extreme scatter; the tendency toward S is more pronounced than in the lowlands, probably induced by the strong leading-lines of the Lafey Ridge and the Simmen Valley.

Talking about leading-lines in the present context implies the assumption that flights parallel to long visible objects are real expressions of cause and effect and not only coincidence. The interpretation is

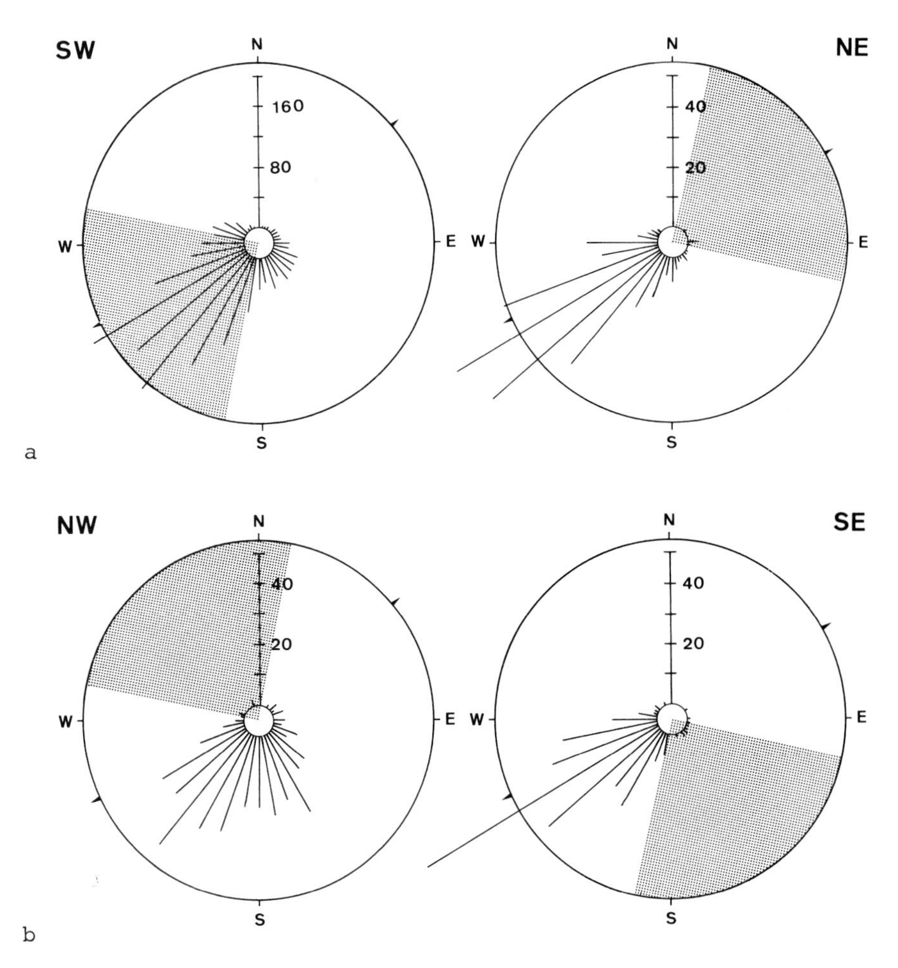

Fig. 7 a and b. Directions of migration at the pass of Hahnenmoos in relation to winds. Same representation as in the circular diagrams of Figure 5. *Small arrows* indicate the principal direction of migration in the lowlands. The *scale* to the north of the circle is a measure for the number of birds per 10-degree interval

complicated by the fact that birds seem, if at all, to use the available leading-lines in very different, perhaps specific, individual or even incidental ways. While some flights show no visible similarities to topographic elements, many others seem to follow a valley, the top or the slope of a ridge, or even artificial elements like a funicular. At the moment we fail to see any preference for either of these elements nor do we know whether birds primarily follow leading-lines to the side of the birds, just below, or in front of them.

Three main local effects of leading-lines can be recognized under appropriate wind and visibility conditions (cf. Fig. 8): (1) the tendency to follow the course of the Simmen Valley; (2) the tendency to maintain the direction flown along the Lafey Ridge, or to turn slightly S after leaving the leading-line; (3) a pronounced funneling effect of the pass, especially in southwesterly winds. In Figure 9 some selected examples of flights paths under different wind conditions are illustrated and commented.

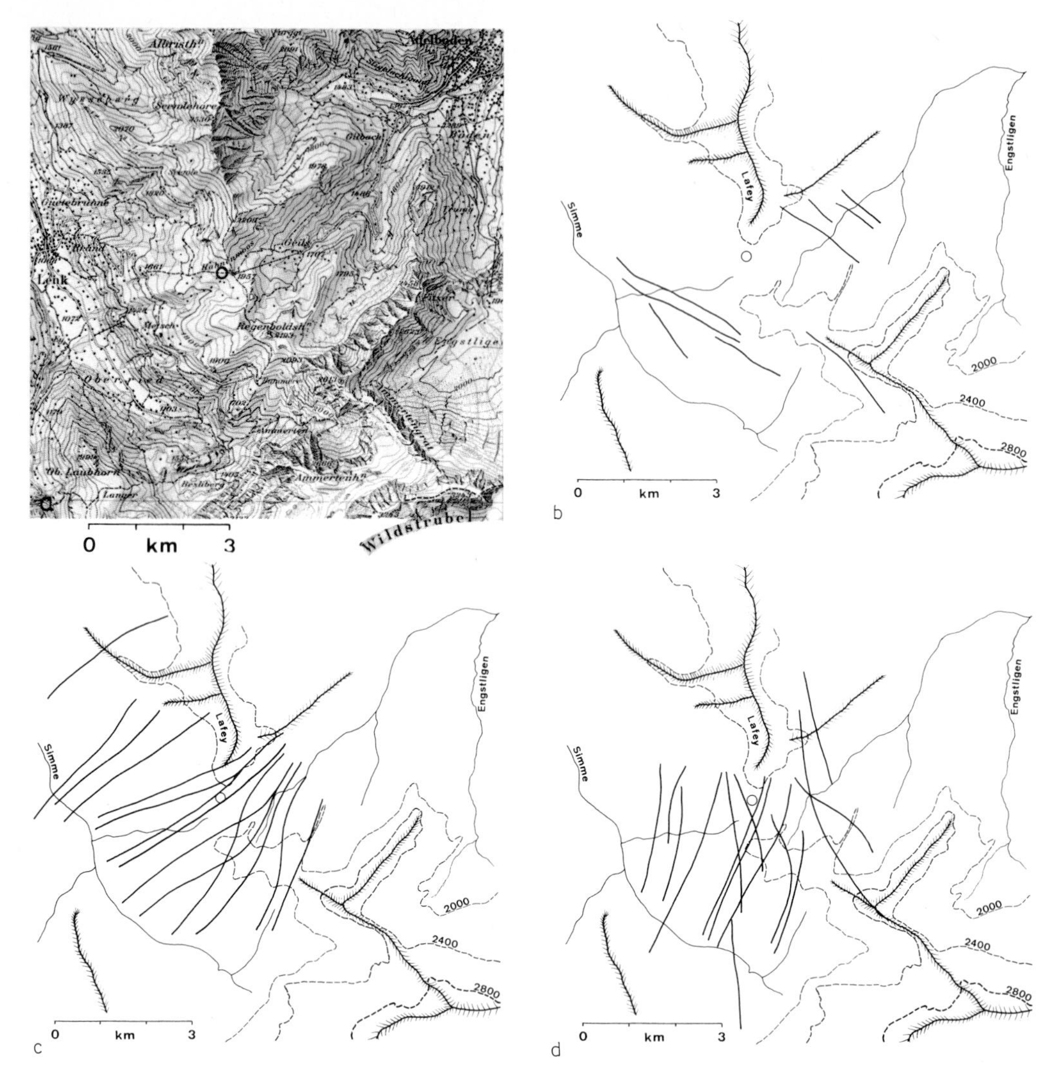

Fig. 8 a - d. Map of the region of the Hahnenmoos Pass. The *circle* left of center indicates the position of the radar. Reproduced with permission of Eidg. Landestopographie (1 September 1977). (b - d) sketches of the same region containing the isohypses of 2000, 2400, and 2800 m, the main ridges, the source of the Simme river, and the sources of the Engstlingen river. Superposed on the sketches of the landscape are lines summarizing the most typical and most frequent flight paths of birds (cf. text)

The sum of the data available makes clear that there is no common strategy of crossing Alpine ridges or passes, except that most of the birds show a tendency to reduce altitude changes and too large horizontal deviations. These tendencies become more pronounced with increasing altitude and in favorable winds. First indications suggest that it is more accentuated in waders than in passerines.

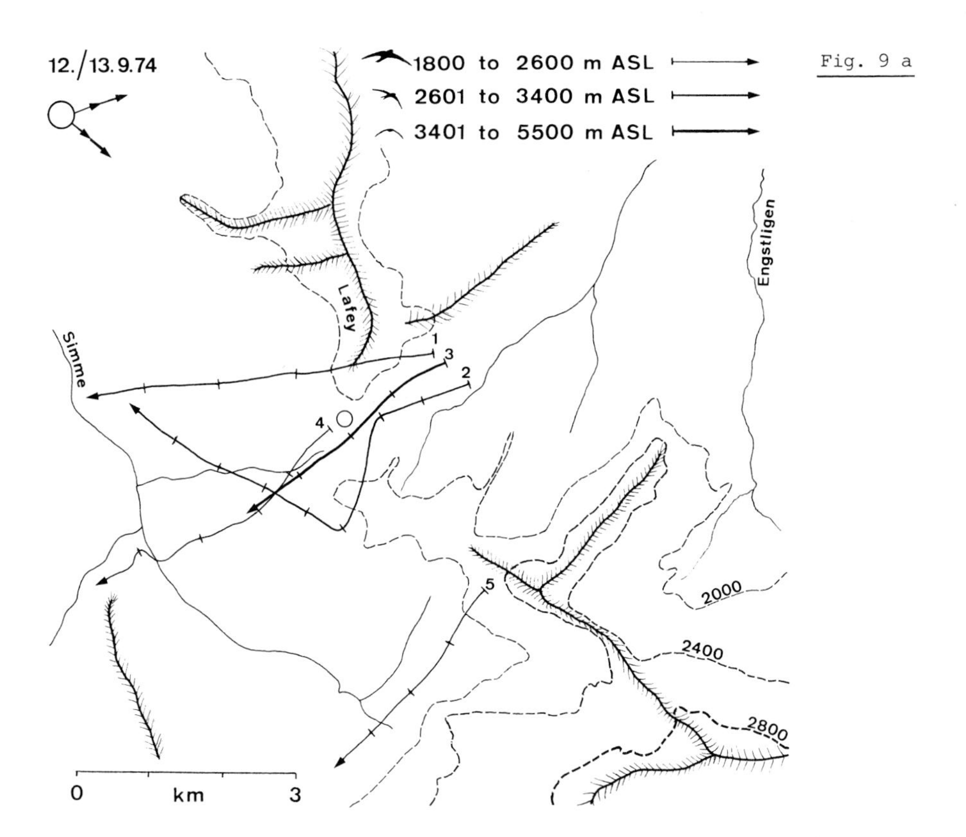

Fig. 9 a

Fig. 9 a - c. Selected flight paths of birds under different wind conditions in the region of the Hahnenmoos Pass showing both striking adaptation of flight paths to topography and straight flights without any similarity to topographic features. The length of the *arrows* indicates the way flown during 20 s (on the *left* side, below date, the correspondent way flown by the pilot balloon).

Catchwords to the most illustrative flight paths: (*1*) straight, but passing edge of Lafey; (*2*) following funicular, turing S at terminal, turning toward Simmen Valley after 1.5 km; (*3*) crossing the pass at 2000 m AGL (~ 4000 m ASL); (*4*) after crossing the pass, slight descent toward Simmen Valley (but remaining at 1000 m above the valley), slight bend indicates a tendency to fly down the main valley, afterward flight toward the small valley, (probably closer to the preferred direction); (*5*) following slope of main ridges.

(*6, 7, 8, 12, 13, 16, 17*) in the direction of nearby topographic elements, simple coincidence cannot be excluded; (*9*) no adaption to landscape; (*10,11*) perhaps maintaining direction along Lafey, both climbing toward high ridges; (*15*) crossing small pass but climbing toward high ridges; (*18*) climbing toward high ridges like the last three birds, but changing direction when arriving at the barrier of the Wildstrubel. (*19-27*) only low flights in headwinds and rainy weather, close adaption to landscape, many birds on reversed migration; important Nr. *20*, coming along Simmen Valley, crossing the Hahnenmoos Pass, but turning again into the former direction on the other side of the pass

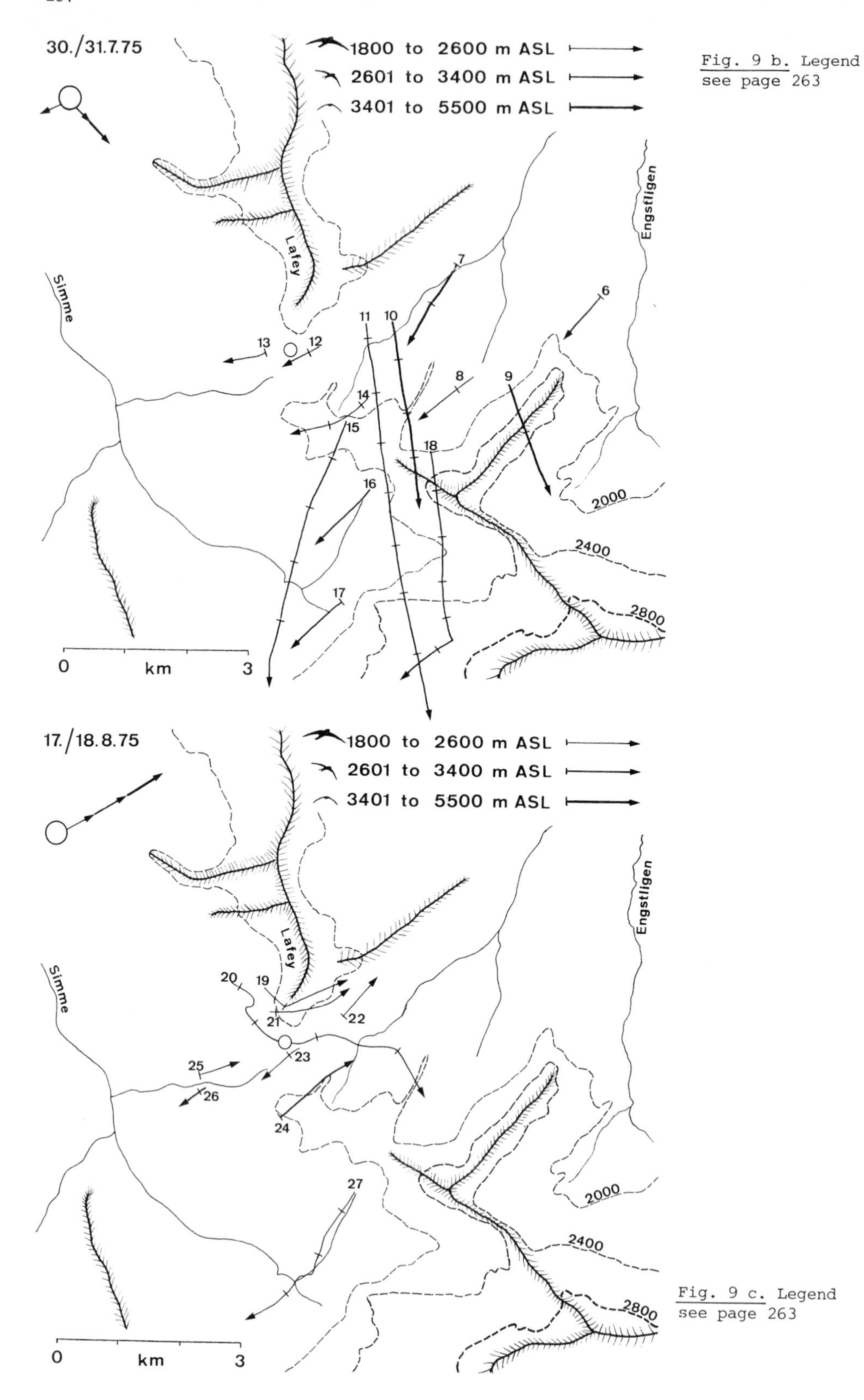

Fig. 9 b. Legend see page 263

Fig. 9 c. Legend see page 263

References

Able, K.P.: The role of weather variables and flight direction in determining the magnitude of nocturnal bird migration. Ecology 54, 1031-1041 (1973)

Berthold, P.: Endogene Steuerung des Vogelzuges. Oekophysiologische Probleme in der Ornithologie. Sonderheft Die Vogelwarte 29, 4-15 (1977)

Bruderer, B.: Radarbeobachtungen über den Frühlingszug im Schweizerischen Mittelland. Orn. Beob. 68, 89-158 (1971)

Bruderer, B.: Zeitliche und räumliche Unterschiede in der Richtung und Richtungsstreuung des Vogelzuges im Schweizerischen Mittelland. Orn. Beob. 72, 169-179 (1975)

Bruderer, B.: Beiträge der Radar-Ornithologie zu Fragen der Orientierung, der Zugphysiologie und der Umweltabhängigkeit des Vogelzuges. Oekophysiologische Probleme in der Ornithologie. Sonderheft Die Vogelwarte 29, 83-91 (1977)

Bruderer, B., Steidinger, P.: Methods of quantitative and qualitative analysis of bird migration with a tracking radar. NASA SP-262, 151-167 (1972)

Gauthreaux, S.A., Jr., Able, K.P.: Wind and the direction of nocturnal songbird migration. Nature (London) 228, 476-477 (1970)

Laty, M.: Détection par le radar MELODI d'oiseaux en migration dans la région de Trappes. La Météorologie 1972, 79-91 (1972)

Sauer, E.G.F.: Die Sternenorientierung nächtlich ziehender Grasmücken (Sylvia atricapilla, borin und curruca). Z. Tierpsychol. 14, 29-70 (1957)

Williams, T.C., Williams, J.M.: Orientation of transatlantic migrants. Symposium on animal migration, navigation and homing. Tübingen, August 17-20 (1977)

Wiltschko, W.: Über den Einfluß statischer Magnetfelder auf die Zugorientierung der Rotkehlchen (*Erithacus rubecula*). Z. Tierpsychol. 25, 537-558 (1968)

Wiltschko, W.: The influence of magnetic total intensity and inclination on directions preferred by migrating European Robins (*Erithacus rubecula*). NASA SP-262, 569-578 (1972)

Wiltschko, W.: Der Magnetkompaß der Gartengrasmücke (*Sylvia borin*). J. Ornithol 115, 1-7 (1974)

Wiltschko, W.: Der Magnetkompaß der Zugvögel und seine biologische Bedeutung. Oekophysiologische Probleme in der Ornithologie. Sonderheft Die Vogelwarte 29, 76-82 (1977)

Wiltschko, W., Merkel, F.W.: Zugorientierung von Dorngrasmücken (Sylvia communis) im Erdmagnetfeld. Vogelwarte 26, 245-249 (1971)

Zink, G.: Der Zug europäischer Singvögel. Ein Atlas der Wiederfunde beringter Vögel. 1. und 2. Lieferung. Vogelwarte Radolfzell (1973, 1975)

Zink, G.: Richtungsänderungen auf dem Zuge bei europäischen Singvögeln. Oekophysiologische Probleme in der Ornithologie. Sonderheft Die Vogelwarte 29, 44-54 (1977)

Session V

Chairman:

FLORIANO PAPI, Istituto di Biologia Generale, Università di Pisa, Via A. Volta 6, 56100 Pisa, Italy

Angle Sense in Painted Quails: A Parameter of Geodetic Orientation*

F. W. MERKEL, Fachbereich Biologie, Universität Frankfurt, 6000 Frankfurt, FRG

Abstract

When running away, quails (*Excalfactoria chinensis*) try to maintain their initial direction by compensating angle deviations forced upon them by obstacles in their course. This behavior was observed in birds that passed actively through gangways as well as in birds that were transported in an open or closed railway vehicle, regardless of whether the angle deviation was caused by a turntable or by a curved rail. Handedness, especially left-handedness, superimposes angle compensation in numerous individual birds. To investigate angle compensation after walks of longer distances, we used a new apparatus whereby the experimental bird is forced to walk on a motor-driven, continuously moving belt whose direction can be changed.

Chinese painted quails (*Excalfactoria chinensis*) exhibit an orientation behavior that enables them to reestablish an initial course they were following before an obstacle forced them to leave it. Ants do this with the help of sun compass orientation. Quails also show this ability in closed rooms, as we showed with the help of a corridor system leading into the center of an arena. The birds were forced (a well-trained quail enters voluntarily) to pass through a gangway that consists of at least two sections, G_1 and G_2, before entering the terminal section, G_E, which is 50 cm long and a part of the arena. The first part of the gangway, G_1, with a compass direction chosen freely by the experimentor, is connected with the second part, G_2, by an abrupt deviation that can be a curve or an acute angle of up to 90°. G_2 brings the bird directly or via a further angle into the terminal G_E and into the center of the arena where the animal can choose its direction freely. The result is marked by its footprints on a registration paper on the floor of the arena. The quail can then leave the arena again by one of eleven gates and enter a ring corridor leading to one of four cages that are doubles of the normal living cage (Fig.1 A).

In a first paper (Merkel et al., 1973) Mrs. Fischer-Klein and I published the following main results worked out in tests with only 6 quails, which were well-acquainted with the apparatus. In corridors with lengths of 1 m to a maximum of 3 m (G_1, G_2, G_E) with deviations of 15°, 30°, and 60°, the birds compensated smaller angles better than larger ones, which means, e.g., that an angle of 15° between G_1 = 1 m and G_E = 0.50 m is fully compensated, while a deviation of -60° is responded to only by reversed turns averaging $+43^\circ$. In addition, the birds are able to balance countercurrent deviations installed in the corridor. They reply to a deviation of 60° to the left (-60°) followed by a turn of 30° to the right ($+30^\circ$) with a free course of 23° to the right ($+23^\circ$).

*Supported by "Deutsche Forschungsgemeinschaft".

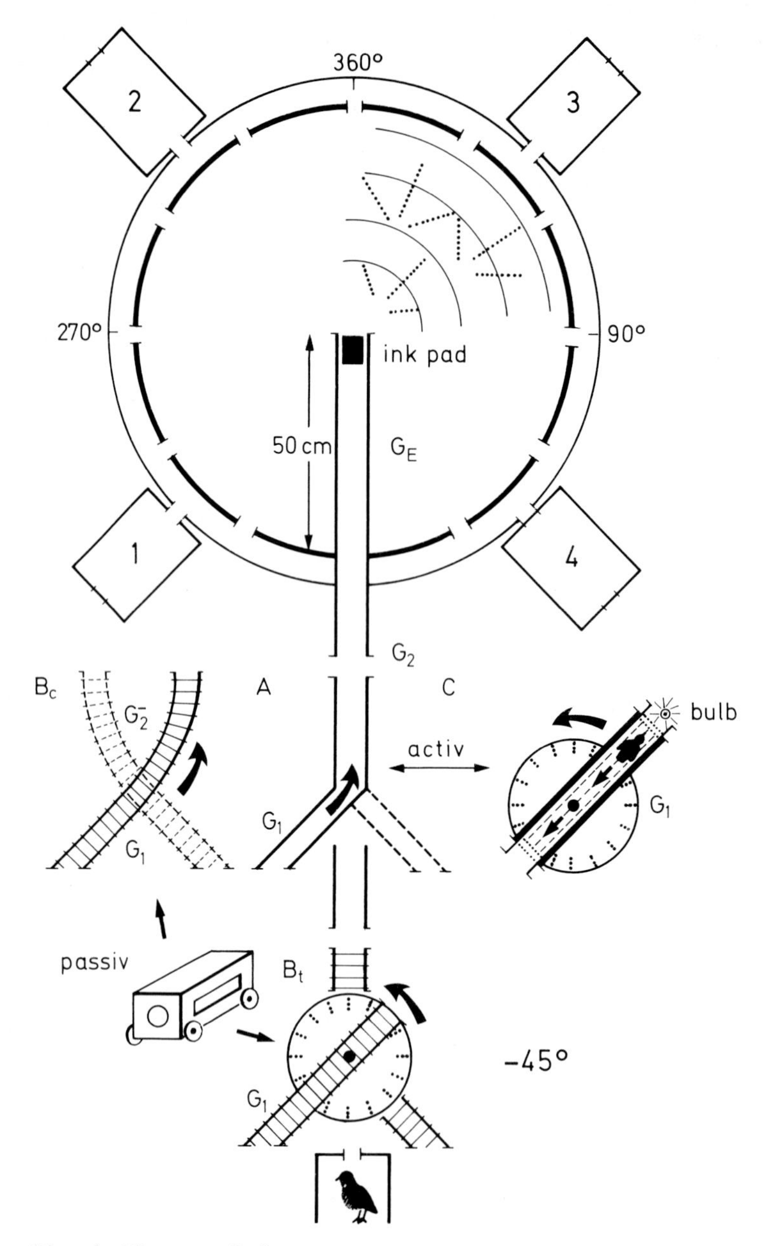

Fig. 1. Diagram of the arena test system with the different possibilities of angle deviation - here: -45°:

B_c, Railtrack with curves } for passive transport
B_t, Railtrack with a turntable

A, Gangways with acute angle } for active walks
C, Continuously moving belt

G_1, *G_2*, *G_E*, Parts of the track system. *G_1* and *G_2* used in different lengths (or time)
1-4, Trap cages

Later, we enlarged the number of experimental series by introducing deviations of ±45° and ±90° and longer parts of gangways, with corresponding results. During the initial investigation, Mrs. Fischer, after having enlarged G_1, obtained results that seemed to justifiy the statement that the length of this part of the gangway may influence the amount of reverse turning. A comparison of two further series with the same deviations of ±45° and G_2 = 200 cm in both of them, but with different lengths of G_1 - 200 cm and 600 cm - yielded contradictory results. Should it prove true that the lengths of the different parts of the corridor system affect the amount of angle compensation, it would be indicative of the birds' ability to measure distances. In my opinion, this is an important question in view of the recent developments in our knowledge of bird orientation.

We discussed whether the more incomplete compensation of larger deviation angles could be the result of an incapacity of the sensory system of the quail or a pertinent biologic behavior. The latter would be the case if the quail, so to say "already begins to think at the beginning of her run about how to return to the starting point." Such a compensation behavior, in connection with the ability to measure the ratio of covered distances $G_2 : G_1$, would secure a faster return to the starting point, e.g., the nest from which the quail was driven. From this point of view, the angle of compensation not only depends on the angle of deviation, but also on the distances covered before and after the deviation.

Furthermore, we tried to determine which sources of reference a quail uses to reestablish the initial direction given by G_1. One possibility is that, at the beginning of the experimental walk through the gangway system, the animal notices and maintains the angle between the direction of G_1 and an axis of the Earth's magnetic field. In our experiments two Helmholtz coils (Ø = 180 cm) were installed such that magnetic north in the arena could be changed by +60° to ENE without changing the normal field intensity (0.47 gauss). The compass direction of G_1 was +40°. After a deviation of -40°, one would expect - as found in the earlier experiments - that, under the Earth's normal magnetic field, the quail would compensate this angle by about +20°, while under the changed magnetic field - supposing that quails use the magnetic field for angle compensation - the bird should compensate by an angle of +20° (-40° by the forced deviation and +60° by the changed magnetic field) with a direction of the resulting walk of -20°. Since the walking directions of 3 individual quails were in the mean almost the same under both sets of conditions (quail No. 1: +23° or +22°; No. 2: +36° or +29°; No. 3: +21° or +23°), we may conclude, at least under the conditions of this special experiment, that quails do not refer to the Earth's magnetic field when performing angle compensation.

To obtain more information about the mode of perception of angle deviation, the following experiments were performed by several students. In series 3 and 4 (Table 1), the active walk through a system of gangways with deviations of ±60° in the form of a curve was compared with passive transport in a closed railway vehicle, under otherwise the same conditions. There were no significant differences between the two series, but it must be said that on the whole, the angle compensation was poor. In series 6, 7, and 8 we investigated the effect of passive transport in closed vehicles under different conditions. Series 7 with a curved deviation was compard with series 8, where the deviation was performed with the help of a turntable and with series 6 under standard conditions, i.e., an active walk through a gangway system with an acute deviation angle. Series 7 with passive transport shows the best compensation results. The probable higher acceleration on the turntable seems to have no influence on compensation.

Table 1. Experiments and their results: Different angle deviations and active and passive transport in the short distance range (see also text)

	Lokomotion Act.	Lokomotion Pass.	Transport Ope.	Transport Clos.	Distance (m) G1	Distance (m) G2	Distance (m) GE	Angles of Deviation	Angles of Reaction	N. Quail (s)	N. Test (s)	T (sec)	Wats. Will. Test	LcR-	L-Rc	L-R-	LcRc	CO-Worker
1	+				1		0.5	0°	0°30'	15	300		----					Fischer
2		+			1.9		0.5	0°	8° a=0.96	10	100		----					Inselmann
3	+				1.9		0.5	−60	a=0.88 2°	10	100	5		2	5	1	2	
	+							+60	a=0.91 356°	10	100							
4		+			1.9		0.5	−60	a=0.92 9°	13	130		▼	4	4	0	2	
		+						+60	a=0.93 359°	10	100		▲					
5	+				1		0.5	0	a=0.97 350°	13	260		---					Hänelt Römer
6	+				1		0.5	−60	a=0.91 358°	13	260			2	9	0	2	
	+							+60	a=0.93 345°	13	260							
7		+			1	1.2	0.5	−60	a=0.90 2°	13	260	20	▼	2	5	1	5	
		+						+60	a=0.93 344°	13	260	6	▲					
8		+			1	0.5	0.5	−60	a=0.90 355°	13	260	20		4	7	1	1	
		+						+60	a=0.89 346°	13	260	6						
9	+				1		0.5	0	a=0.87 349°	13	200		►10+ 12+ ►11+ 13+					Blees Giehl
− 10		+			1.5	1	0.5	−90	a=0.71 355°	10	200	20	▼ ↓	0	5	0	5	Blees
+		+						+90	a=0.87 322°	10	200		▲ ▽ ↓					
− 11		+			1.5	1	0.5	−90	a=0.67 355°	10	200		▼ △ ↑	1	5	0	4	
+		+						+90	a=0.80 324°	10	200		▲ ↑					
− 12		+			1.5	1	0.5	−90	a=0.82 354°	10	200	5	▼ ↓	2	5	0	3	Giehl
+		+						+90	a=0.78 334°	10	200		▲ ▽ ↓					
− 13		+			1.5	1	0.5	−90	a=0.82 352°	10	200		▼ △ ↑	1	5	1	3	
+		+						+90	a=0.74 322°	10	200		▲ ↑					
														18	50	4	27	

In the last four series, two questions stand in the foreground: (1) Does the possibility of an optical orientation influence angle compensation, and (2) Do the reactions of quails transported with different speeds over the same distance of 250 cm differ? The quail in the open vehicle has the possibility of looking through a bull's-eye in front and through slits on both sides of the car, and is able to see, e.g., the bulb above the center of the arena. The almost identical data of the four series prove that neither the optical orientation nor the speed takes part in angle compensation.

With increasing numbers of quails, the results became poorer, and the number of compensated walks decreased. As shown on the right side of the Table, we classified the quails of each series according to their compensation ability and got four groups: The first group includes all quails that compensate a deviation to the left (Lc) but not to the right (R-). To the second group belong all animals that compensate a deviation to the right (Rc) but not to the left (L-). The members of the third group compensate neither left nor right deviation (L-, R-) and follow the direction of the deviation. The last group includes all quails that compensate both the left and the right deviations. The data show that the animals of the second group ($\frac{L-}{Rc}$) are the most numerous. This could be interpreted as predomination of left handedness. In other experiments with other individual quails, right-handedness may also dominate. The number of birds in the third group is always very small.

In the different control series (0° runs), the quails also show in most cases a certain handedness, and there is evidence that handedness superposes and masks angle compensation. Therefore we speak of individual 0 points in contrast to the objective 0 point given by the construction of the arena. Since we have examples that individual quails exhibit the same 0 point several times during one year, we thought about calculating the other test runs with regard to the individual 0 points. However, a comparison showed that the standard deviation of the directions of 9 comparable runs of 20 out of 22 individual quails decreases in the course of one year, 1976 to 1977.

Because the first experiments with a small sample and only a few well-trained birds brought significant results, I hoped to obtain precise information in a relatively short time. Unfortunately the use of an increasing number of birds, seasons with courting and egg-laying, and the fact that different individuals worked with the same quails, resulted in nervous and more timid birds, and in poorer results. In addition, elongation of the distances or of the different parts of the corridor system resulted in difficulties with those birds that had to pass through the gangways by active runs. Therefore, we tried to expand in particular the experiments with passive transport, since longer distances can be simulated by prolongation of time. During passive transport, birds of course need smaller amounts of energy than during active runs. Thus, as a compromise, my wife and I decided to place the experimental birds on a continuously moving belt whose direction could be changed; thereby the direction of the running bird could also be altered.

Thus, we recently began to do experiments based on three methods appropriate for longer distances:

1. A railway system in which the experimental bird is transported in an electric-driven vehicle over distances of 10 m or more with curves or turntables (Fig. 1, B_c and B_t).
2. The quails must walk or run on a motor-driven, continuously moving belt at speeds of 0.35 km/h up to 3 km/h, which is within the normal range of speed of a running quail, as determined previously (Fig. 1, C).
3. A Y-like gangway system to prove handedness in individual quails.

At present I can only establish that the quails respond to angle deviations as well when being transported by the railway as when walking on the moving belt. The results optained thus far concerning the four groups of compensation behavior show almost the same distribution pattern as in the experiments with shorter distances (Table 1).

Finally, I would like to make some remarks about the implications of the various experiments reported during this symposium.

In the literature on orientation in animals, regardless of the language, the term navigation is used increasingly in a diffuse manner, sometimes in fact to mean orientation in general, including directional and compass orientation. The same is true of the term goal, which may refer to an exactly defined place, e.g., the nestbox of an individuum, as well as to the extensive wintering area of the European Starling on both sides of the English Channel. Therefore, the term navigation should be limited to the determination of the geographic location of a place, i.e., it should be used to determine true bicoordinate orientation, using astral cues, chronometers, and angle-measuring instruments In addition to bicoordinate navigation, Schmidt-Koenig (1965) enumerated two other modes of navigation - vector navigation and reverse displacement navigation. Both of these modes should be included under the term geodetic orientation. This term encompasses all the geodetic methods used in the past centuries to survey the earth and to draw the first maps. Using a compass, a pencil, and sheets of paper, together with methods for measuring the distances covered, as aids to memory ancient explorers discovered unknown parts of the earth and were able to return to their starting point. The olfactory orientation of pigeons described by Papi et al. (1974) can also be included under the concept of geodetic orientation, as can, if proved true, angle compensation for location, and homing, in terms of *zentraler Umschaltung* (central commutation).

From the work of v. Frisch it is known that bees inform other foragers not only about the direction but also about the distance to a food source. v. Frisch and coworkers gave evidence that bees measure distances by their energy consumption. Wether birds are also able to measure distances is not known for sure. Eyperiments with warblers led Gwinner (1968) to the hypothesis that a migratory bird could estimate the distance to its winterquarters with an endogenous time program. But it seems to me that birds using only such endogenous time cycles are not adaptable enough to compensate delays forced by bad weather conditions during the long migration routes. However one could suppose that the highly evolved bird during the endogenous time cycle of Zugdisposition should be able to estimate the distance - like the bee - in measuring the amount of fat produced and consumed during its migration flight.

References

Frisch, Karl v.: Tanzsprache und Orientierung der Bienen. Berlin-Heidelberg-New York: Springer, 1965

Gwinner, Eberhard: Artspezifische Muster der Zugunruhe bei Laubsängern und ihre mögliche Bedeutung für die Beendigung des Zuges im Winterquartier. Z.f.T. 25, 843-853 (1968)

Merkel, F.W., Fischer-Klein, K.: Winkelkompensation bei Zwergwachteln (*Excalfactoria chinensis*). Die Vogelwarte 27, 39-50 (1973)

Papi, F., Ioalè, P., Fiaschi, V., Benvenuti, S., Baldaccini, N.E.: Olfactory navigation of pigeons: the effect of treatment with odorous air currents. J. Comp. Physiol. 94, 187-193 (1974)

Schmidt-Koenig, Klaus: Current problems in bird orientation. Advances in The Study of Behavior 1, 217-278 (1965)

Concept of Endogenous Control of Migration in Warblers

PETER BERTHOLD, Max-Planck-Institut für Verhaltensphysiologie, Vogelwarte Radolfzell, Schloss, 7760 Radolfzell-16, FRG

Abstract

On the basis of comprehensive studies with warblers, the current concept of the endogenous control of migration in this group is presented. The control of migratory events by circannual rhythms, the detailed endogenous control of migratory patterns, the endogenous mechanisms involved in finding the winter quarters, and aspects of the interaction of endogenous and permissive environmental factors are treated.

Early concepts of the control of bird migration assumed that only environmental factors, such as, e.g., food shortage and low temperatures, were the causative factors. Beginning in the eighteenth century ideas changed and, first on an entirely hypothetical basis, there was a steady increase in the number of ideas partly involving endogenous factors (e.g., Berthold, 1975). Since 1967, when Gwinner described the first circannual rhythms of birds of the genus *Phylloscopus*, there has been a rapidly growing body of experimental evidence for the participation of endogenous factors in the control of bird migration (for review, see Berthold 1977a).

A comprehensive concept of the endogenous control of bird migration could be obtained for *Sylvia* species in our laboratories in Radolfzell and Erling-Andechs. The most important results of the more recent studies in this group are presented here.

If a migratory Sylviid specimen is hand-reared and kept under constant conditions, migratory events will nevertheless occur at about the appropriate time and for about the expected duration and to the expected extent. Figure 1 shows such data for a Garden Warbler (*Sylvia borin*) kept in a daily light-dark ratio of 10:14. Like its free-living conspecifics, at the end of juvenile molt, the bird simultaneously started to deposit fat and to display nocturnal, i.e., migratory activity. During the winter, spontaneous fat depletion occurred, followed by a timely winter molt, and in the following year there were further rhythmically occurring periods of fat deposition, migratory activity, and molt.

We register: Under constant experimental conditions without cyclic environmental information such as photoperiodism, temperature cycles, etc., Garden Warblers exhibit migratory disposition and migratory activity in a fairly correct manner. The period lengths of corresponding annual migratory events, however, deviate somewhat from 12 months, indicating that these events are controlled by truly endogenous, so-called circannual rhythms.

The control of migratory disposition and activity on the basis of circannual rhythms is now established in five *Sylvia* species: besides the typical migrants, Garden Warbler and *Sylvia cantillans*, in the less

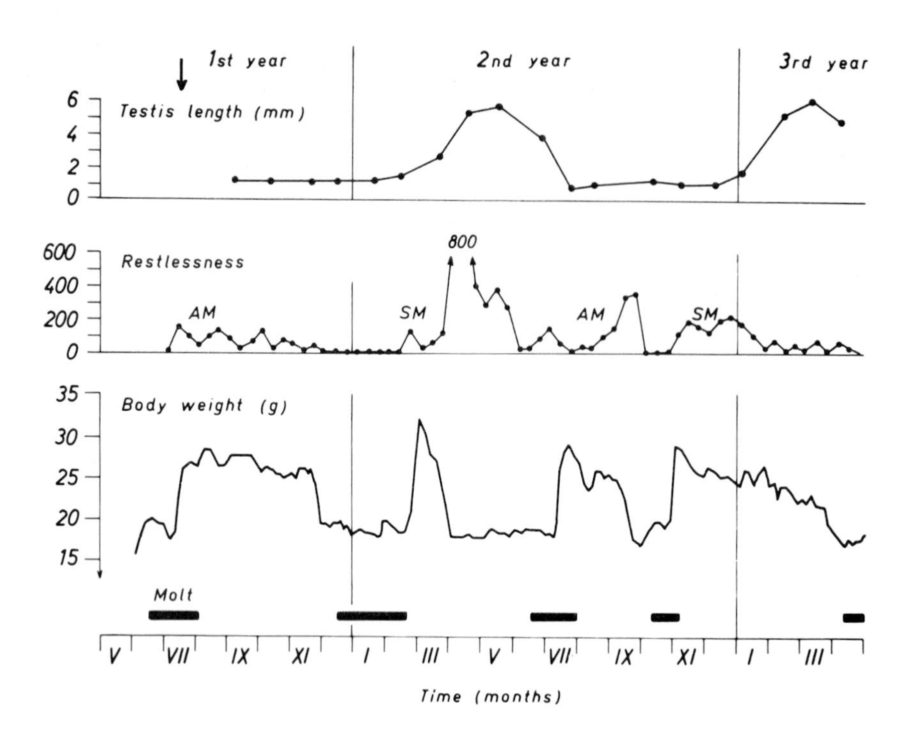

Fig. 1. Circannual rhythms of testis length, migratory restlessness, body-weight changes and molt in a *Sylvia borin* in LD 10:14 during a long-term experiment. *Arrow* indicates date of transfer of the hand-raised bird to constant photoperiodic conditions. *AM*, *SM*, autumn, spring migratory period. From Berthold et al. (1971)

typical migratory species *S. atricapilla* and in the partial migrants *S. melanocephala* and *S. undata*. In addition to these, there is thus far only one further proof for *Phylloscopus trochilus* (for review, see Berthold, 1977a).

In a long-term study under constant conditions we showed that in the Garden Warbler, more or less cyclic changes of body weight and periods with fat deposition may well occur at least into the (present) ninth experimental year (Berthold, 1977a). This longevity of rhythmic occurrence of periods with migratory disposition suggests that (1) migratory disposition in Garden Warblers is endogenously controlled for life and that (2) the underlying physiologic basis has the character of a self-sustained rhythm.

Comparisons of the nocturnal restlessness as an expression of migratory activity in caged birds and of actual migration in free-living conspecifics yielded surprisingly good conformity with respect to the time of onset and end as well as of the pattern (distribution over time) (e.g., Berthold, 1977a). Naumann (1797) believed that these nocturnal activity patterns displayed in captivity could be related to the distance to be traveled.

In a series of experiments we examined first the obvious question of whether only the onsets of these activity patterns are driven by internal rhythms or whether they are endogenously controlled in detail. The same was done with respect to the pattern of fat deposition.

Figure 2 shows an experimental group of Garden Warblers in which, by two interposed periods of starvation, fat deposition was temporarily reduced. As a consequence, the amount of migratory activity displayed during that time dropped significantly compared with a control group. The deficit of both fat deposition and migratory activity due to the temporary starvation, as compared with the control group, was neither compensated for at the end of both events by some prolongation nor after the starvation period by some exaggeration. When the birds were refed, the affected values approached those expected from the control group. In another experiment, in an experimental group of Garden Warblers, body weight was kept low from the beginning of juvenile development for several weeks. That loss of fat deposition was not made up, but migratory activity developed nevertheless at the right time, as comparisons with a control group showed (Berthold, 1977b).

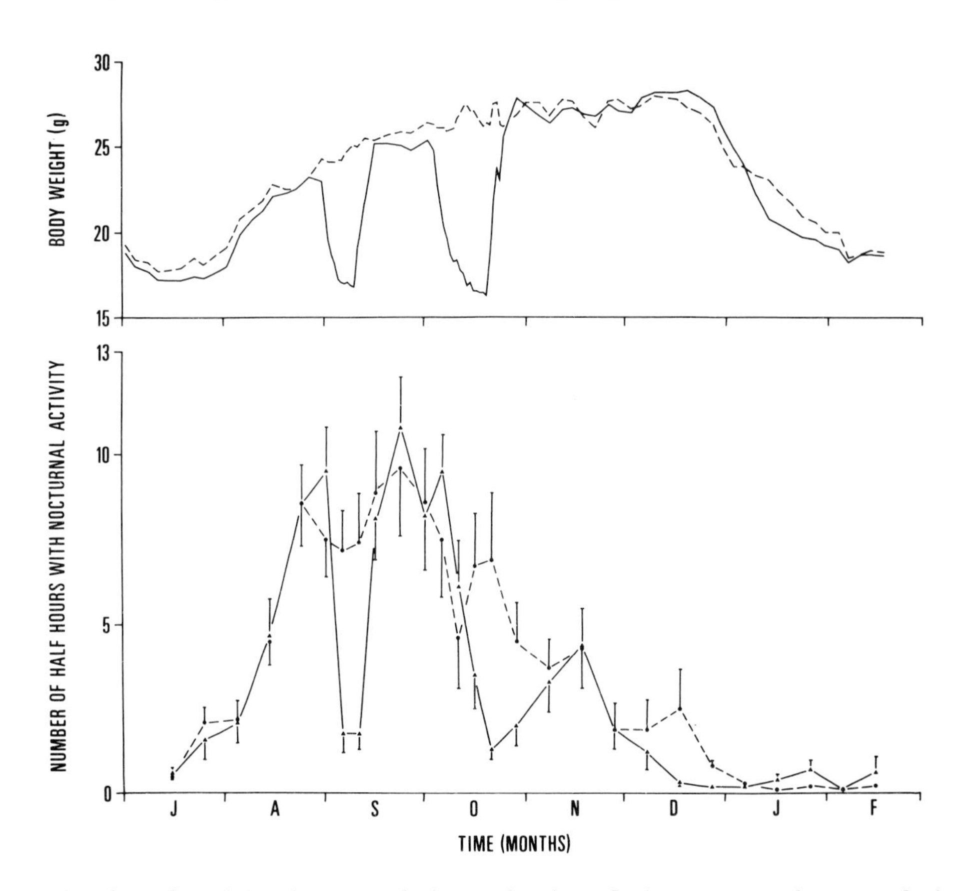

Fig. 2. Body-weight changes and the production of migratory restlessness during the period of the first autumn migration of *Sylvia borin* under natural light conditions. *Solid lines*, ▲, experimental group, with two interposed periods of body-weight reduction; *broken lines*, ●, control group, without reduction of body weight. *Vertical lines*, standard error of the mean. The values of migratory restlessness at the end of September are highly significantly and significantly different, respectively ($p < 0.001$, $p < 0.01$); the values in mid-October and at the end of October are probably significantly different ($p < 0.05$). From Berthold (1977d)

From these and some similar experiments (s. Berthold, 1977a) it must be concluded that the patterns both of migratory disposition and of migratory activity are to a large extent endogenously controlled in detail. We know, on the other hand, that they also can undergo some modification by various photoperiodic conditions (e.g., Berthold et al., 1972).

Since migratory activity of caged warblers is a fairly good reflection of actual migration, and is largely endogenously controlled, close relationships between preprogrammed migratory activity and actual migration including the distance to be traveled, could be assumed. For two *Phylloscopus* and six *Sylvia* species, it could be shown that the amount of nocturnal activity of caged groups is closely related to the migratory distances (e.g., Gwinner, 1968, Berhold, 1973): (1) In both groups the relative ratio of the average amount of restlessness exhibited corresponds closely to that of the distance to be traveled (e.g., Fig. 3).

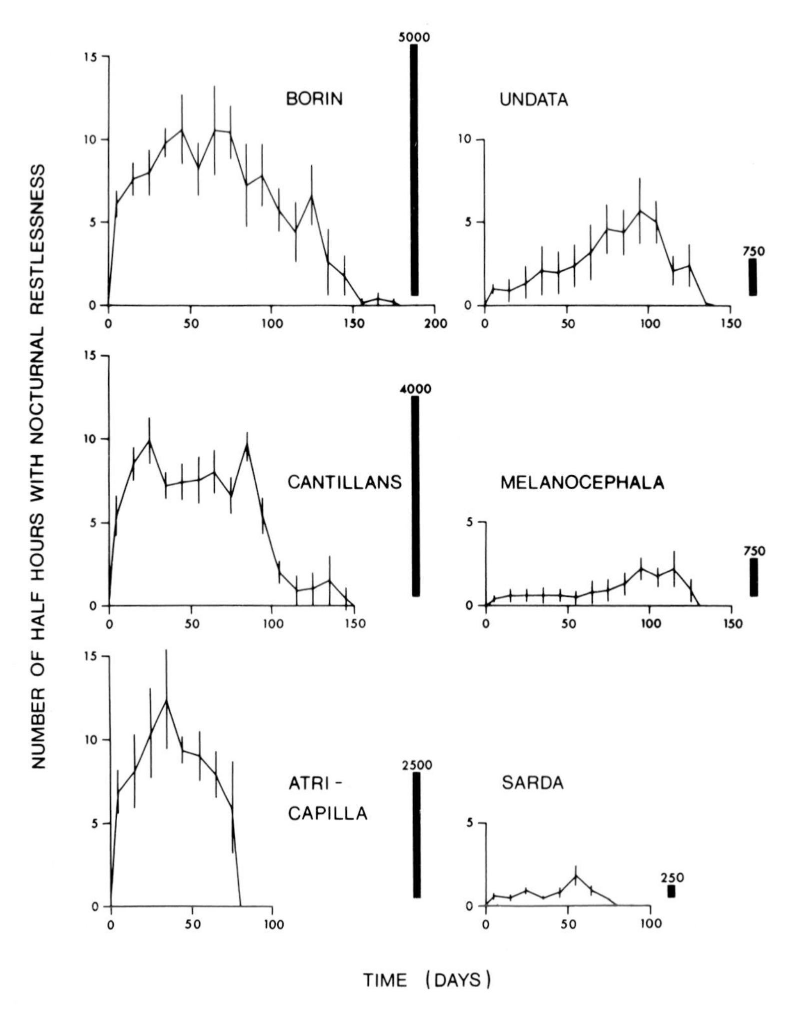

Fig. 3. Patterns of migratory restlessness (mean values and standard errors of means for 10-day periods) and estimated migratory distances of six *Sylvia* species. From Berthold (1973)

For instance, the Garden Warbler as a long distance migrant with a distance to cover of about 5000 km, displayed about 20 times the restlessness of the almost resident Marmora's Warbler with a migratory range of only 250 km, and the other less typical migrants intervene between these species according to their migratory distances. (2) With the aid of some calculations, Gwinner (1968) showed that the two *Phylloscopus* species Willow Warbler and Chiffchaff, if they had been able to use the migratory activity displayed as caged birds for actual migration, would just have reached their species-specific winter quarters.

From these results and other supporting findings (Gwinner, 1972) the so-called *Vectoren-Navigations-Hypothese*, coined by Schmidt-Koenig (1973), has been derived. It states that, at least in young inexperienced birds in their first autumn migratory period, migration is based (1) on endogenous time programs for initiation, carry-through, and termination of migratory activity, and (2) on instinctively determined routes. It is hypothesized that the birds start migration when the internal impulse ensues, that they migrate as long as migratory activity is produced according to the internal program, and that they stop migration when the internal time program fades. Migrating along species-specific routes as vectors, after a distinct endogenously programmed time, they "automatically" reach the species-specific winter quarters.

This hypothesis now also fits data for different populations of the same species (Berthold, 1977c). As Figure 4 shows, Blackcaps (*Sylvia atricapilla*) from Finland and southern France, which travel very different distances, also develop very different amounts of migratory restlessness in constant conditions: Birds of the Finnish population, as typical, long-distance migrants, produced 2.4 times as much migratory restlessness as did the short-distance, partial migrants from France.

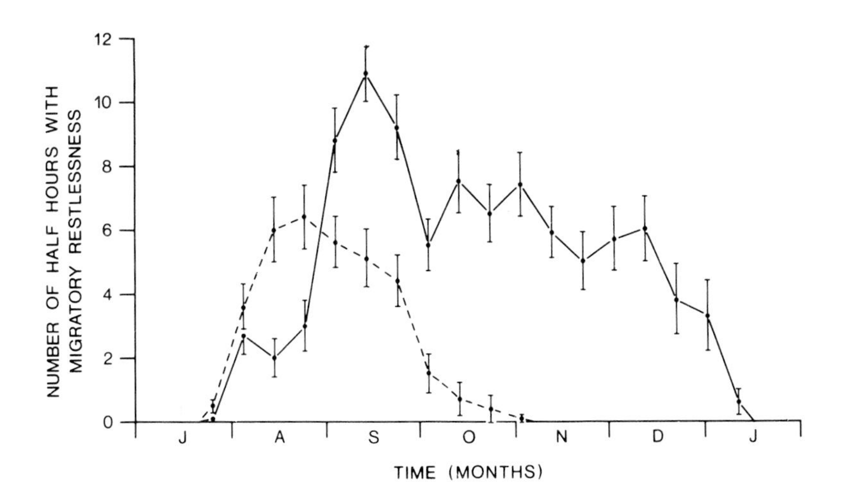

Fig. 4. Migratory restlessness of Finnish (*solid line*) and French (*broken line*) experimental *Sylvia atricapilla*. Mean values and standard errors of means for 10-day periods. From Berthold (1977c)

The question arises as to how the endogenous programs for migration are fitted into the environmental conditions favorable to and permissive for migration. As is known, mainly from radar and trapping studies, considerable short-term variations in the intensity of migration are common and obviously related to specific weather conditions. These relationships touch not only so-called weather migrants but probably also typical migrants including the warblers in question (e.g., Bruderer, 1977).

When patterns of migratory restlessness are considered in detail, it can be seen that birds, even when kept under constant conditions, do not exhibit restlessness in a completely uniform manner. On the contrary, there are regular short-term changes in the intensity of restlessness, even between maximum values and nothing from one night to another. Or there is a sequence of greater intervals with and without restlessness, respectively, which occur more erratically than rhythmically (Fig. 5). The following is hypothesized: The endogenously controlled bursts of restlessness observed in caged birds could, in free-living conspecifics, lead immediately to migratory movements if the environmental conditions are permissive, i.e., favorable during their appearance. If conditions are worse, the endogenous bursts could be temporarily suppressed and shifted to some later days when conditions have improved. This adjustement could function without any alteration of the total amount of the endogenously programmed migratory restlessness. This hypothesis will shortly be tested in a series of experiments

In summary, it can be said: Endogenous factors play an important role in the control of migration in various warbler species. In the Garden Warbler, the most intensively studied species in this regard, onset and end of the period of migratory disposition as well as of migratory activity are controlled by circannual rhythms, and the temporal patterns of both events are also endogenously controlled. Furthermore, as recently shown in detailed comparisons between south Finnish and southwest German birds (but not discussed further here), different migratory habits of different populations are obviously genetically preprogrammed. In the migratory disposition, hyperphagia and the ingestion of nutrient-poor vegetable and nutritive animal diets have an endogenous basis (for review, see Berthold, 1977a). With respect to finding the species- or populationspecific winter quarters, a simple straight migratory direction as well as necessary shifts in the direction during migration, seem to be programmed in the migrating bird, as shown recently by Gwinner and Wiltschko.

At present it cannot be excluded that the first migration of young inexperienced Garden Warblers is (almost) exclusively based on endogenous factors. The participation of modifying factors is, however, probable.

Since endogenous factors in the control of migration have been detected in other groups, including less typical migrants such as the Chaffinch and the Crossbill (for review, see Berthold, 1977a), it can be assumed that the picture seen today will certainly be much extended by current and future studies.

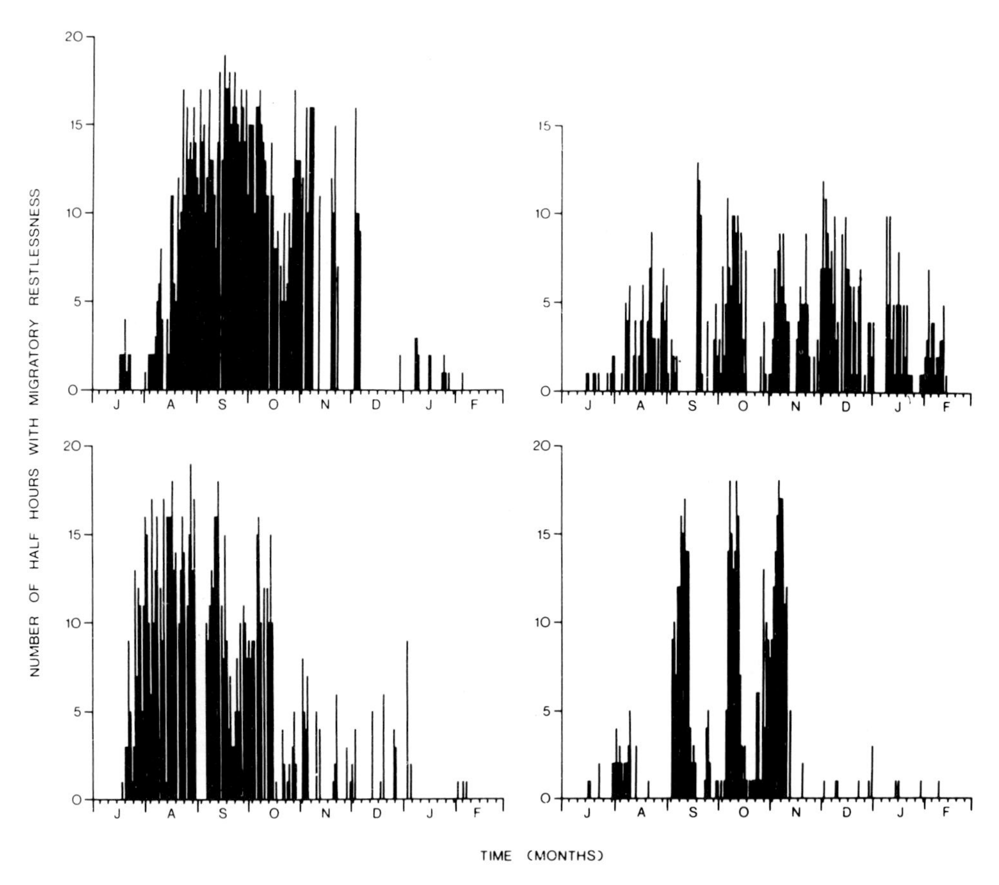

Fig. 5. Patterns of migratory restlessness of four individual *Sylvia borin* in LD 12:12 with numbers of half-hours with restlessness for each individual night. From Berthold (1978)

References

Berthold, P.: Relationships between migratory restlessness and migration distance in six Sylvia species. Ibis 115, 594-599 (1973)

Berthold, P.: Migration: Control and metabolic physiology. In: Avian Biology Farner, D.S., King, J.R. (eds.), Vol. V, 77-128, (1975)

Berthold, P.: Endogene Steuerung des Vogelzuges. Vogelwarte 29, Sonderheft, 4-15 (1977a)

Berthold, P.: Über die Entwicklung von Zugunruhe bei der Gartengrasmücke Sylvia borin bei verhinderter Fettdeposition. Vogelwarte 29, 113-116 (1977b)

Berthold, P.: The "Vektoren-Navigations-Hypothese": Its validity for different populations of the same species. Naturwissenschaften 64, 389 (1977c)

Berthold, P.: Über den Einfluß der Fettdeposition auf die Zugunruhe bei der Gartengrasmücke Sylvia borin. Vogelwarte 28, 263-266 (1977d)

Berthold, P.: Das Zusammenwirken von endogenen Zugzeit-Programmen und Umweltfaktoren beim Zugablauf bei Grasmücken: eine Hypothese. Vogelwarte 23, 153-159 (1978)

Berthold, P., Gwinner, E., Klein, H.: Circannuale Periodik bei Grasmücken (Sylvia). Experientia 27, 399 (1971)

Berthold, P., Gwinner, E., Klein, H., Westrich, P.: Beziehungen zwischen Zugunruhe und Zugablauf bei Garten- und Mönchsgrasmücke (Sylvia borin und S. atricapilla). Z. Tierpsychol. 30, 26-35 (1972)

Bruderer, B.: Beitrag der Radar-Ornithologie zu Fragen der Orientierung, der Zugphysiologie und der Umweltabhängigkeit des Vogelzuges. Vogelwarte 29, Sonderheft, 83-91 (1977)

Gwinner, E.: Circannuale Periodik der Mauser und der Zugunruhe bei einem Vogel. Naturwissenschaften 54, 447 (1967)

Gwinner, E.: Artspezifische Muster der Zugunruhe bei Laubsängern und ihre mögliche Bedeutung für die Beendigung des Zuges im Winterquartier. Z. Tierpsychol. 25, 843-853

Gwinner, E.: Endogenous timing factors in bird migration. In: Animal Orientation and Navigation Galler, S.R., Schmidt-Koenig, K., Jacobs, G.J., Belleville, R.E. (eds.) 321-338. NASA, Washington (1972)

Naumann, J.A.: Naturgeschichte der Land- und Wasservögel des nördlichen Deutschlands und angränzender Länder, nach eigenen Erfahrungen entworfen, und nach dem Leben gezeichnet 1, 192-193 (1797)

Schmidt-Koenig, K.: Über die Navigation der Vögel. Naturwissenschaften 60, 88-94 (1973)

Orientation Strategies Used by Free-Flying Bird Migrants: A Radar Tracking Study

STEPHEN T. EMLEN and NATALIE J. DEMONG, Division of Biological Sciences, Section of Neurobiology and Behavior, Cornell University, Ithaca, NY 14853, USA

Abstract

A new technique for studying the orientation of free-flying migrants following experimental manipulation is described. Birds are taken aloft, rereleased at migratory altitudes, and their flight departures accurately plotted by tracking radar.

Experiments on the relative importance of different cue systems in the spring migratory orientation of White-throated Sparrows suggested: (1) Sparrows rapidly selected appropriate, straight, migratory tracks when released under *clear night skies* or during periods of *sunset glow*. (2) Under *total overcast* sparrows still were able to select meaningful headings although with much less accuracy. They frequently adopted a sinusoidal, "zigzag" flight strategy under these conditions. (3) Landmark cues of the Atlantic coastline were used minimally if at all by the migrating sparrows.

Introduction

Most investigators agree that migratory birds do not rely upon a single navigational cue when selecting a flight direction, but rather on a complex array of multiple, and often redundant, directional information (see Keeton, 1974; Emlen, 1975). This realization leads one to ask new questions: How are these different directional inputs integrated with one another? Is there a hierarchy of importance of different cues, and does this change with the age or experience of the bird, the meteorologic or magnetic conditions during flight, or the geographic position of the traveling bird?

Conventional methods of studying bird migration and migratory orientation are not well-suited for answering these types of questions. Laboratory experiments, although rigorously controlling the cue information available to the bird, are performed in confined and unnatural environments; they do not allow access to the full range of potential directional information that is available to a free-flying migrant. Field studies, on the other hand, are fundamentally descriptive. Because of the inability of manipulating cue information, they frequently yield only correlations between the directions of migration (generally of unknown species) and general flight conditions aloft.

What is needed is a method that allows for experimental alteration of the orientational information available to an individual migrant while, at the same time, accurately tracking or following the bird as it selects its course and departs on migration. In 1971 we developed a technique that permitted such experimentation. The experimental subjects were White-throated Sparrows (*Zonotrichia albicollis*) capture on their wintering grounds in South Carolina in late April and held in captivity until the spring migration season. They were then experimentally rereleased at migratory altitudes and allowed to initiate a migratory flight while being followed by tracking radar. In an actual experiment a sparrow was placed in an especially designed cardboard box equipped with a fuse-operated opening device. This box was attached to a helium-filled weather balloon which carried the bird to the desired altitude. The fuse then allowed the floor of the box to swing open, thereby releasing the bird. (A movie showing details of the release technique was shown at the conference; additional details are available in Emlen, 1974, and Demong and Emlen, in preparation.) Through the cooperation of the National Aeronautics and Space Administration, each bird was tracked by a FPS-16 tracking radar that gave accurate position information in three coordinates (geographic position over the ground and altitude) at a frequency of one location per second. In this way the behavior of the birds in the airspace was recorded as they started out for the first few (3 - 28) kilometers of their journey. (The FPS-16 is a 1-MW, 5-cm radar with 0.25, 0.5, and 1.0-μs pulse potential and an angular beam width of 1.2°; for further details on radar specifications, see Vaughn, 1974.)

The last 1.6 km of each track (presumably after the bird had made its directional decision) was selected for analysis here. Only those birds that maintained level flight (descent rates less than 1 m/s) were included. The balloon carrying the bird aloft during its ascent was also tracked by radar, providing precise information on wind direction and speed at all altitude levels. These values were vector subtracted from the track (= direction of progress over the ground) and ground speed values to calculate the birds' actual heading (= direction of the bird's body axis) and airspeed.

Using these techniques we released known individual sparrows under different meteorologic conditions and looked for differences in their flight behavior. In this paper we report some of our results obtained from 64 White-throated Sparrows released and tracked during the springs of 1971 and 1972.

A. Behavior Under Clear Night Skies

When released under clear skies, a bird frequently would hover for a period of 5 - 15 s before striking out on a course that was generally maintained for the entirety of the flight (or until we terminated tracking in order to initiate a new experiment). Thus the bird's "decision" of a direction for flight was made rapidly and was maintained or altered only slightly as the flight continued. The tracks were quite straight although some individuals exhibited a sinusoidal flight pattern about the track direction that became damped over a period of several minutes. When present, these sinusoidal variations had peak-to-peak amplitudes of 10 - 45 m and a periodicity of 2 - 4 "cycles"/min.

The orientation of sparrows released when the winds aloft were slight to moderate in strength is shown in Figure 1. The birds adopted northerly headings ($\Phi_H = 355°$; $r = 0.75$; $N = 30$), and generally achieved northeasterly tracks ($\Phi_T = 18°$; $r = 0.81$). An independent prediction of the expected direction of migratory flight for White-throated Sparrows was obtained from recapture data fo banded birds. These values also showed a dominant direction to the north-northeast, with a mean of 31°, a value not significantly different from the departure track directions of our artificially released birds.

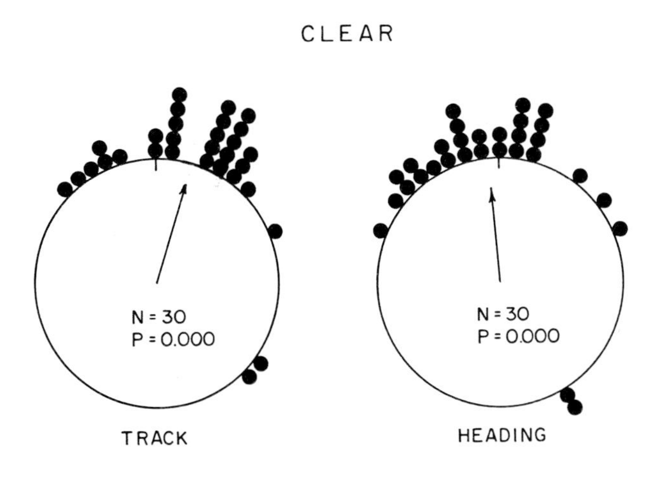

Fig. 1. Vector diagrams of the track and heading of 30 White-throated Sparrows released and tracked under clear night skies. Included are all experiments where the wind speed aloft was less than the average airspeed exhibited by migratory sparrows (9.15 m/s). The *arrow* in the center of each diagram points to the mean direction of the data distribution and has a length proportional to r. The sample size (N) and the probability that the data distribution deviates from uniform by the Rayleigh test (p) are given in the center of each diagram

Some birds were released when unfavorable winds aloft were stronger than the average airspeed of White-throated Sparrows. Under these conditions the birds are blown on tracks to the east, out over the Atlantic Ocean. Interestingly, even under these "extreme" conditions, the sparrows adopted northerly headings that were indistinguishable from those in Figure 1 ($\Phi_H = 1^o$; $r = 0.78$; $N = 15$). The birds did not show tendencies to decrease in altitude in order to land or to head to the west in order to make landfall rather than being carried out over the ocean (Fig. 2).

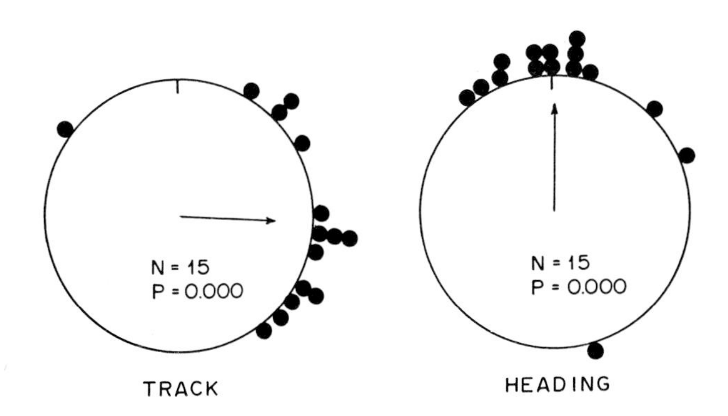

Fig. 2. Tracks and headings for White-throated Sparrows released under clear skies but when stiff winds aloft precluded the attainment of meaningful track directions. Included are all cases in which a sparrow flying at the average airspeed (9.15 m/s) would be unable to achieve forward progress along the expected track of 31°

In Summary, the birds did not exhibit a tendency to fly downwind such as has been reported in some studies of migrants conducted in the southeastern portion of the United States. Rather, the sparrows adopted northerly headings regardless of the strength or direction of the winds encountered aloft.

B. The Importance of Sunset

In a recent review article (Emlen, 1975) I speculated that the position of sunset might be an extremely important directional cue for nocturnally migrating birds. This was based partly on the fact that the sun itself has been shown an important cue for compass information in a wide variety of organisms, and partly on the fact that the position of sunset is an excellent integrating cue during the transition between day and night. Nocturnally migrating passerines typically initiate flight some 30 min to 1 h after the beginning of twilight, at a period when the last sunset cues are just fading and the first bright stars are becoming visible in the night sky. Certain directional cues become unavailable after dark; others assume importance at this time. Temporary physical landmarks in the immediate vicinity of the grounded bird, and the position of sunset, could both serve as stable references during this transition, possibly helping to integrate or redundantly check certain diurnal with nocturnal reference cues. Since sunset is a dominant visual cue occurring at precisely the time prior to most migratory departures, a series of experiments was conducted in which birds were released during the short transition period after the sun's disc had dropped below the horizon but before the stars were visible to the human eye. The glow of sunset was clearly visible at such times.

Six sparrows released under these sunset conditions all exhibited headings to the north-northeast (see Fig. 3). When the wind strength was "manageable," tracks were to the northeast. All birds rapidly selected their direction and maintained very straight tracks. These results suggest that the horizon glow from the setting sun may be used as a directional cue.

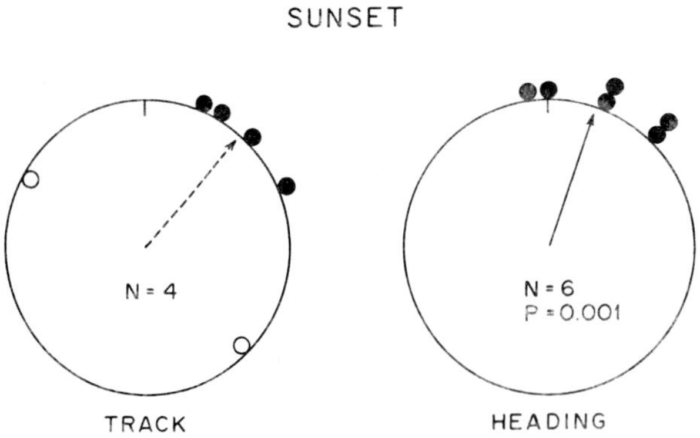

SUNSET

N = 4

TRACK

N = 6
P = 0.001

HEADING

Fig. 3. Tracks and headings of sparrows released during late twilight, when the sun's disc was below the horizon but stars were not yet visible to the human eye. *Open circles* represent tracks when the wind speed aloft exceeded 9.15 m/s

C. Behavior Under Overcast

In recent years numerous studies have provided convincing evidence of the importance of nonvisual cues in the migratory and homing behavior of birds (e.g., this Symposium). These studies are complemented by a large number of field/radar studies that have documented that nocturnal migration continues and that migrants are well oriented on overcast nights when visual cues from the sky are obscured. We were therefore extremely interested in attempting to release birds under conditions of heavy cloud.

The meteorologic station at the radar installation provided a continuous ceilometer monitoring of the height of the bottoms of cloud layers, and direct visual inspections from the ground allowed us to monitor the extent of the cloud cover. By adjusting the length of the fuses on the release box, we could preselect the altitude of release and thus position the experimental birds beneath the cloud layers.

Sparrows released under overcast showed several differences in behavior when compared to those released under clear skies. Many hovered or circled for several minutes after relase before actively initiating a flight. This could be deduced from the sequential circling of the birds' heading positions (see later) and also by the minimal airspeeds of the birds during these periods. Even after adoption of a flight direction, marked sinusoidal deviations from a straight track were common. These had both a higher frequency and continued for a longer time than those recorded from clear releases.

As a bird turns its body while circling or following a sinusoidal path, successive values of its heading will differ greatly from one another. Techniques of circular statistics can be applied to calculate a value for r_h where r is the length of the mean vector of heading values of the bird's body axis calculated at 1-s intervals. Note that the data points entering this calculation are not independent of one another and hence levels of significance cannot be applied to the resulting r_h s. However, the length of these mean vectors nonetheless provides an index of straightness that allows comparison of the flight behavior of birds released under different meteorologic conditions. During the first minute post release, when the bird presumably is actually selecting its course, the straightness index for birds under overcast was 0.55 while that for birds under clear skies was 0.76. The continuation of the irregular flight strategy by birds under overcast is indicated by the straightness indices occurring during the last mile of each track. For overcast birds this r_h was 0.82 compared to 0.96 for birds flying under clear skies.

Another characteristic of overcast birds was a tendency to fly at greatly reduced airspeeds. Using the last mile of active flight as the sample period (in order to avoid emphasizing the early decision-making phase of the experiment), the mean airspeed of overcast birds (AS = 6.64 m/s) was significantly lower than that of birds flying under clear (AS = 9.15 m/s) ($F_{1,41}$ = 11.49; $p < 0.01$). Only one of the 13 birds flown under overcast had an airspeed that exceeded the average value for birds flying under clear.

Despite these behavioral differences in flight strategy, sparrows flying under overcast still exhibited a northward tendency in their headings. (Fig. 4) The sample size is small (N = 13) and the variance large (r = 0.43), but the mean heading of 12° is significant by the V test (p = 0.02 with the expected heading = 0°). It would appear that the birds released under overcast were able to determine a northerly heading but that the process took longer and was less accurate than when selected under clear skies.

We can only speculate on the possible function, if any, of the sinusoidal or "zigzag" flight pattern. Conceivably it could be useful in sampling information that might be of use for directional purposes. Several hypotheses of magnetic detection and referencing, as well as

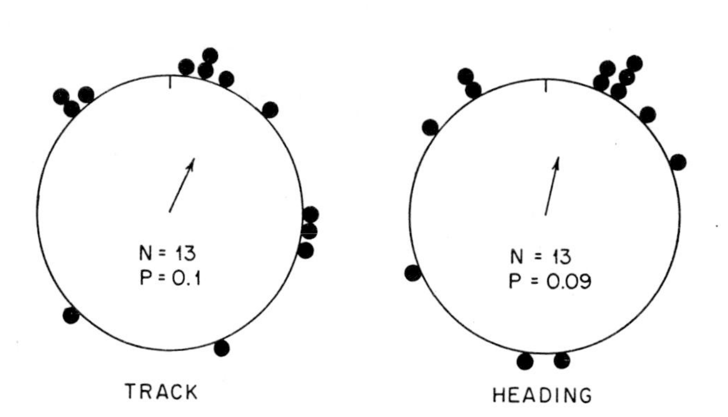

Fig. 4. Tracks and headings of White-throated Sparrows released under totally overcast skies. Details are given in the text

detection of wind direction from patterns of microturbulence, require comparisons of information obtained when the bird's body axis is in different positions. Zigzagging might also be a method of slowing progression over the ground during periods of difficulty or indecision. The small scale of the zigzags precludes their detection by most surveillance radars. However, it is interesting to note that small amplitude irregularities similar to those reported here were described by Griffin (1972, 1973) from nocturnal migrants that he tracked by radar as they were flying through cloud.

D. Use of Topographic Landmarks

The location of the NASA tracking station was at Wallops Island, Virginia, on the coast of the Atlantic Ocean. The dominant direction of the coastline runs north-northeast - south-southwest. Thus the visual features of the landscape might provide a strong orientational cue for birds migrating aloft. To investigate this, the flight paths of the sparrows were overlaid onto accurate maps of the topographic landscape. Figures 5 and 6 depict the tracks of sparrows flying under clear conditions and in manageable winds in the springs of 1971 and 1972 respectively. Although there is general agreement between the axis of the coastline and the departure tracks of the birds (the predicted migration direction based upon banding returns was 31°; the orientation of the coastline approximates 24°), a close examination of the Figures provides little evidence that the birds were actually paralleling the coast. Birds whose flight direction deviated only slightly from "up coast" (e.g., number 306 in Fig. 5, and 8 and 59 in Fig. 6) found their positions becoming increasingly distant from the shore as their flight proceeded. Such individuals showed no tendency to alter their course in any manner that would bring them closer to or more parallel with the coastline.

Recall that birds released under clear night skies but in stiff winds aloft rapidly adopted northward headings (Fig. 2). Such winds usually had a strong eastward component that caused these birds to be carried out over the ocean almost perpendicularly from the coast. Had the sparrows shifted to more westward headings, most could have avoided being carried out to sea or at least made safe landfall. Instead they

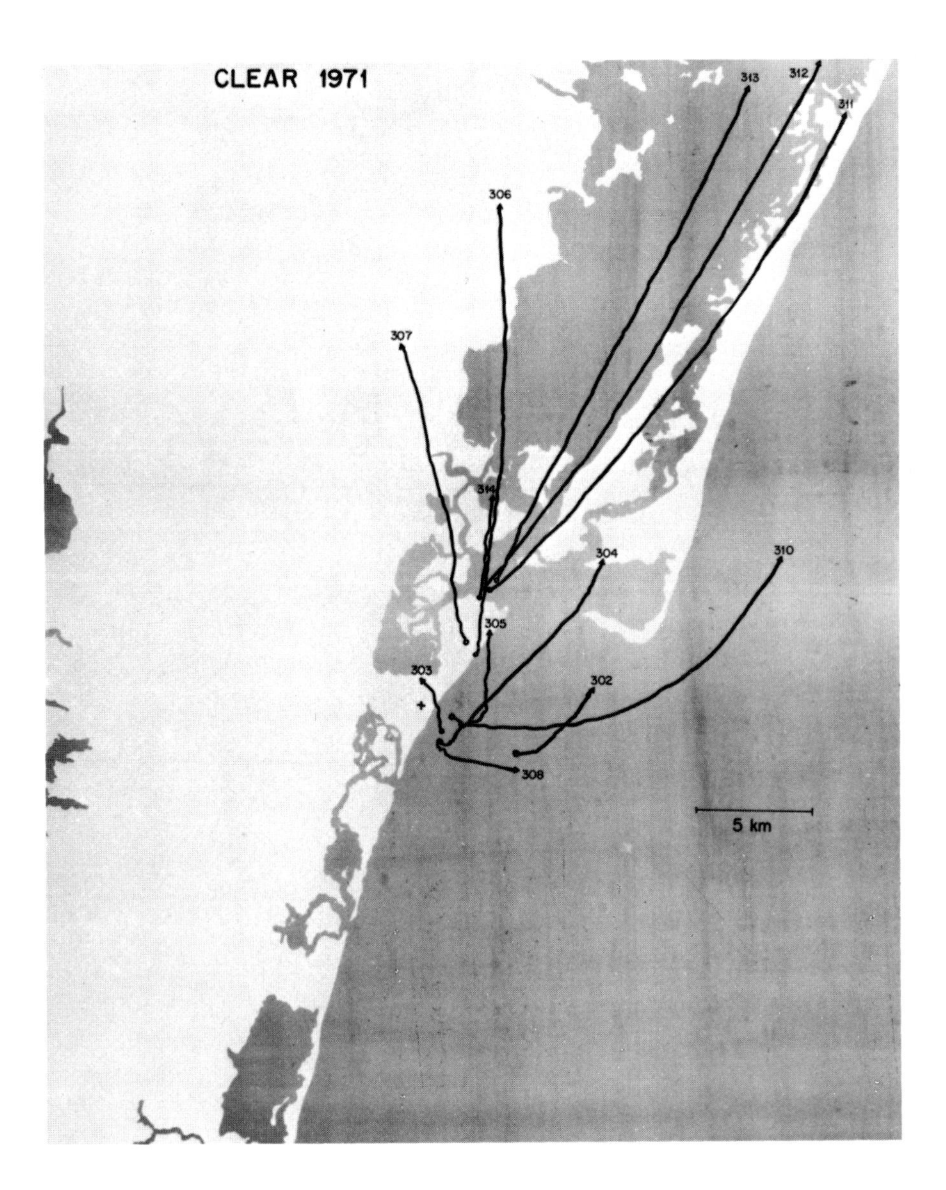

Fig. 5. Map overlay of the flight paths of White-throated Sparrows released under clear conditions in the spring of 1971. *Shaded areas* represent the waters of the Atlantic Ocean and inland bays. Geographic north is up

maintained altitude and kept on migratory headings, suggesting that the landmarks of the coastline were of minimal if of any importance.

Perhaps the most striking evidence for the lack of importance of topographic landmarks comes from examination of the tracks of birds released under overcast (Fig. 7). With visual information from the sky obscured, one might expect visual cues from the ground to assume greater importance. Yet there is no tendency for these overcast tracks to converge along the coast. Many birds are slowly heading out to sea

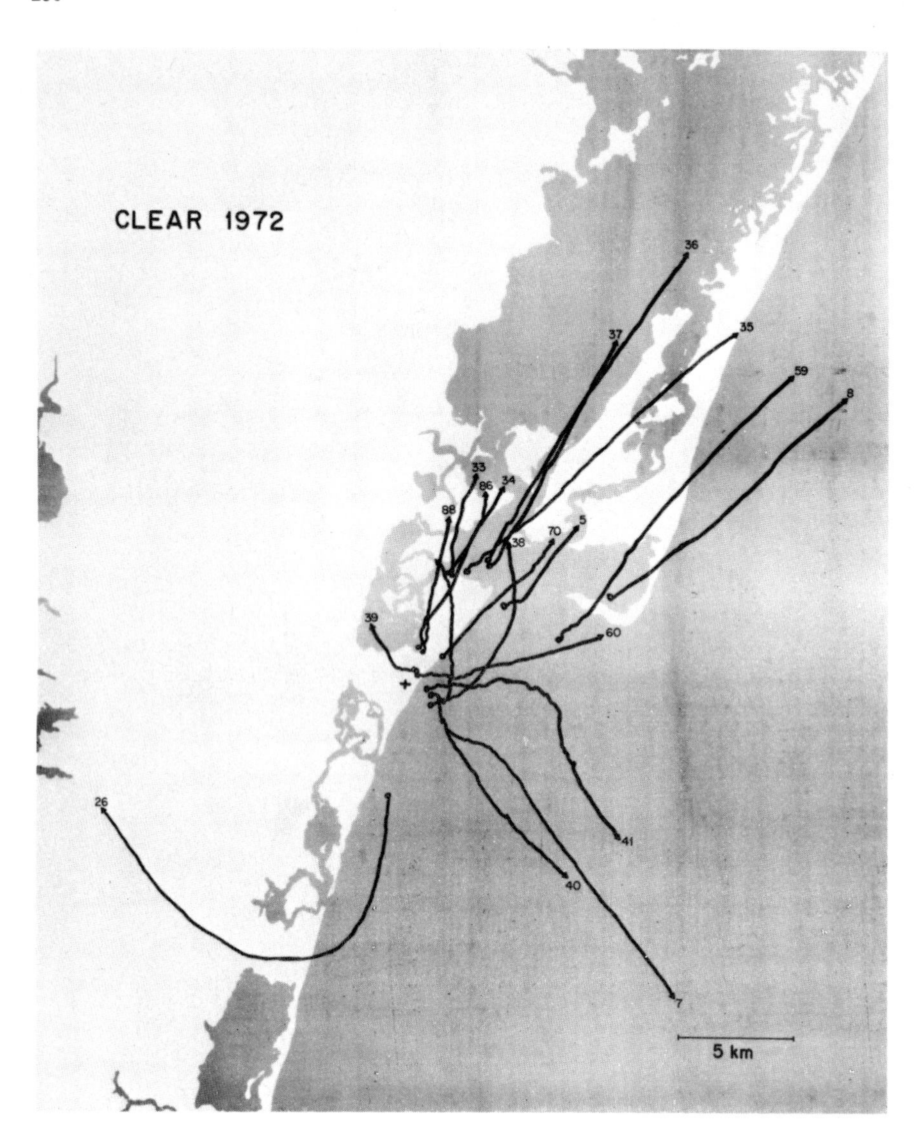

Fig. 6. Map overlay of the flight paths of White-throated Sparrows released under clear conditions in the spring of 1972

while others are traveling in various directions over the mainland. This is all the more remarkable considering the behavioral differences shown by the overcast birds (discussed above), all of which could be indicative of a greater difficulty in selecting a direction, a lower motivation to migrate, or both. If landmarks were important, either for determining a migratory orientation of for avoiding being blown over the sea, we would not have expected the tracks to "scatter" over the landscape, as is evident in Figure 7.

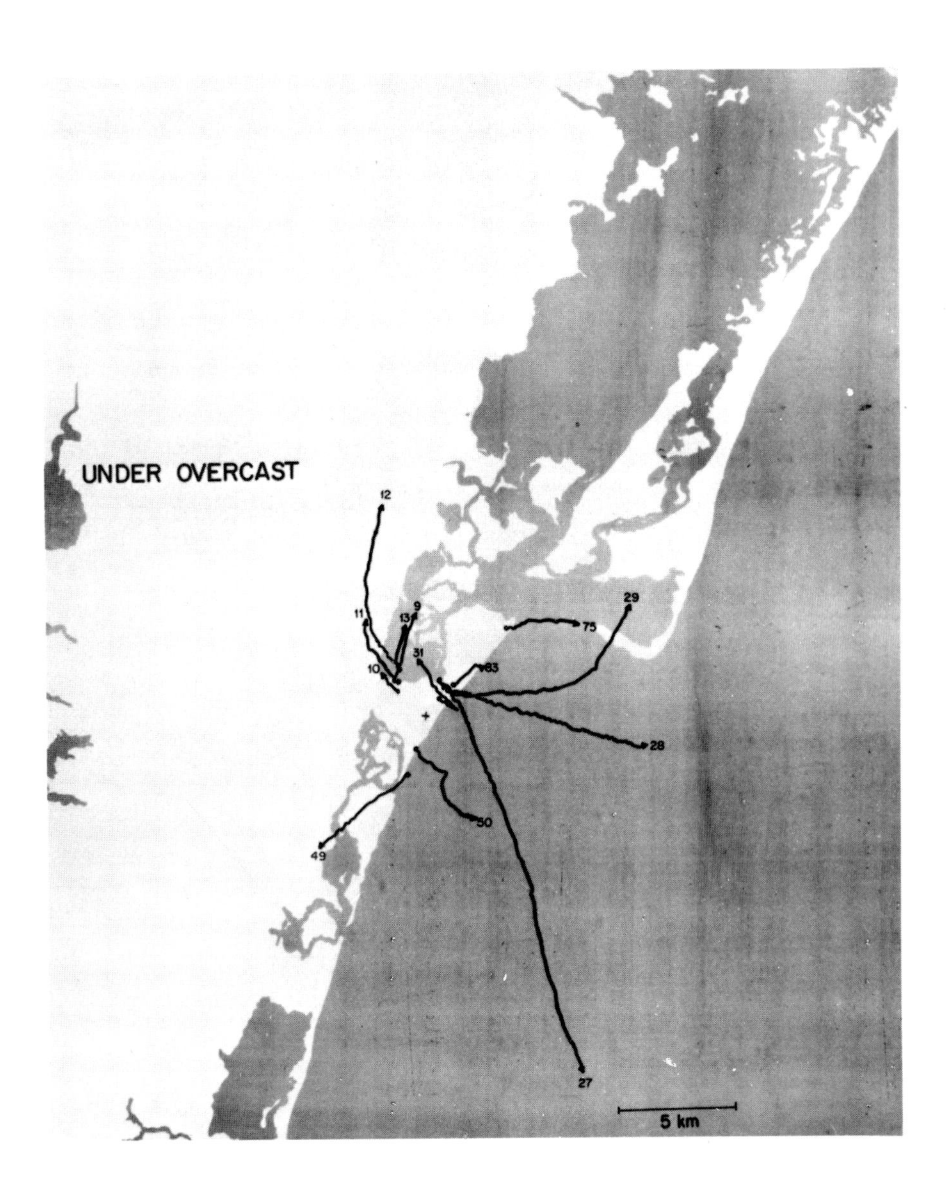

Fig. 7. Map overlay of the flight paths of White-throated Sparrows released under overcast

We hypothesize that even when flying under the potentially difficult orientational conditions imposed by total overcast, the birds were largely ignoring the dominant visual cues from the coastline below.

E. Conclusions

How then are different orientational systems integrated and weighed by migratory White-throated Sparrows? Although our results are based on a small sample of birds, it is clear that the birds had little difficulty in selecting a northward heading when released under clear skies. The same was true when birds were released during the period of twilight, providing some of the first evidence that migratory sparrows probably can use the glow of the position of sunset as a direct orientational cue. When visual information from the sky was obscured by heavy cloud, the birds were still able to adopt northerly headings, but appeared to do so with less accuracy and more difficulty. We hypothesize that the task of rapidly selecting a migratory direction under such conditions is more difficult than for a bird departing under overcast *from the ground* where greater time and additional cues are available. Finally, throughout the study the sparrows appeared to rely minimally if at all upon the visual landscape of the coast beneath them.

These results are consistent with the model that celestial cues are of strong importance to migratory White-throated Sparrows; that sufficient alternate information is available for the selection of the correct migratory direction in their absence; and that visual features of the ground below are near the bottom of any orientational hierarchy for these birds.

Acknowledgments. We wish to thank Mr. Kreiger and NASA/Wallops Island for their tremendous cooperation during these studies. S.A. Gauthreaux, Jr. and R.E. Muller helped capture the birds and I. Brown, S.A. Gauthreaux, Jr., W.T. Keeton, C.J. Ralph, and W.J. Richardson helped with the experiments. The work was supported by grants from the National Science Foundation (GB-35199X and BMS 75-18905) and the National Aeronautics and Space Administration (Project Nos. KG-5597, KL-5737, and KL-5919).

F. Summary

To bridge the gap between controlled (but unnatural) laboratory investigations of orientation, and natural (but uncontrolled) field observations of migration, we developed a technique for studying the orientation behavior of free-flying bird migrants following experimental handling and manipulation. The technique involved taking individual birds aloft at night and rereleasing them at migratory altitudes. A NASA FPS-16 tracking radar then plotted each bird's course as it selected a direction and departed on a presumed migratory flight.

Our experiments were conducted with White-throated Sparrows (*Zonotrichia albicollis*) and were designed to study the relative importance of different orientation systems and flight strategies in birds migrating under a variety of meteorological conditions.

When released under clear, starry skies, the sparrows rapidly selected appropriate migratory bearings, maintaining altitude and departing to the northeast.

When released at sunset (when the sun's disc was below the horizon and the major magnitude stars were not yet visible to the human eye), the sparrows again rapidly and correctly selected migratory bearings to the northeast.

Sparrows released under total overcast were able to select meaningful headings although with less accuracy. Under these conditions the birds adopted a different flight strategy, frequently engaging in much circling and sinusoidal "zigzagging", and also flying at significantly lower airspeeds.

Throughout the study, the prominent landmark cues of the Atlantic coastline appeared to be used minimally if at all by the migrating sparrows.

References

Emlen, S.T.: Problems in identifying bird species by radar signature analyses: Intra-specific variability. In: The Biological Aspects of the Bird/Aircraft Collision Problem. Gauthreaux, S. (ed.), Air Force Office of Scientific Research, Clemson, South Carolina, 1974, pp. 509-524

Emlen, S.T.: Migration: Orientation and Navigation. In: Avian Biology. Farner, D.S., King, J.R. (eds.) V, 1975, pp. 129-219

Griffin, D.R.: Nocturnal bird migration in opaque clouds. NASA Spec. Publication SP-262, 169-188 (1972)

Griffin, D.R.: Oriented bird migration in or between opaque cloud layers. Proc. Am. Phil. Soc. 117, 117-141 (1973)

Keeton, W.T.: The orientational and navigational basis of homing in birds. In: Recent Advances in the Study of Behavior 5, 47-132. New York: Academic Press, 1974

Vaughn, C.R.: Intra-specific wing beat rate variability and species identification using tracking radar. In: The Biological Aspects of the Bird/Aircraft Collision Problem Gauthreaux, S. (ed.) Air Force Office of Scientific Research, Clemson, South Carolina, 1974, pp. 443-476

Environmental Variables and the Nightly Emigration Ratio of the Robin *(Erithacus rubecola)* on the Island Hjelm, Denmark

JØRGEN RABØL and KJELD HANSEN, Zoological Laboratory, Universitetsparken 15, 2100 Copenhagen, Denmark

Abstract

Several weather factors, and the number and weight of the resting Robins were investigated for influence on the subsequent nightly emigration ratio, which was taken as an appropriate measure of "migratory readiness." Cloudiness influenced the emigration ratio in a negative way, and the number and weight of the birds influenced the emigration ratio in a positive way. Contrary to expectation, a following wind in the "standard direction" was found to be of no significant positive influence - the sign of the correlation was even negative.

A. Introduction

The purpose of this paper is to throw light on the causal relationships between the magnitude of the "migratory readiness" and several meteorologic and biologic variables.

Several authors have dealt with this problem - including the following who make use of multivariate statistics: Lack (1960, 1963a, b), Nisbet and Drury (1968), Richardson and Gunn (1971), Richardson (1974a, b), Able (1973), Bruderer (1974), Geil et al. (1974), and Alerstam (1976). The dependent variable in all these analyses was the migration intensity as measured by the number of echoes on the radar screen.

It is generally believed that environmental factors may have both short-term (instantaneous) and long-term (delayed) effects on the magnitude of "migratory readiness." Here we investigate the short-term or nearly short-term effects.

If the nightly number of birds emigrating from and immigrating to a small well-defined area could be estimated (Fig. 1), the emigration

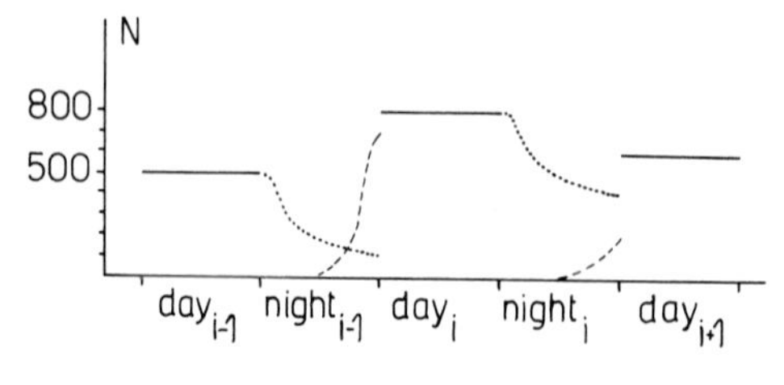

Fig. 1. Example showing the concepts of immigration and emigration ratio. On day_{i-1} 500 birds ($N_{i-1} = 500$) are resting on a small isolated area such as an island. N_{i-1} is constant through the day. Emigration proceeds during the $night_{i-1}$ and 400 birds leave the island (*dotted line*). The emigration ratio for $night_{i-1}$ is thus 400/500 = 0.80. During $night_{i-1}$ 700 birds immigrate (*hatched line*). N_i becomes 100 + 700 = 800. In the course of $night_i$ 400 birds emigrate, and the emigration ratio is 400/800 = 0.50; 200 birds immigrate and N_{i+1} is 400 + 200 = 600

ratio should certainly be a powerful measurement of the "migratory readiness." Furthermore, we come close to the environmental factors that are supposed to influence the magnitude of "migratory readiness." These variables are with certainty experienced by the birds, and we have avoided the general problem of selecting an appropriate weather station giving the weather variables supposed to be representative of the weather variables experienced by the birds in the recruitment area.

We selected the Robin (*Erithacus rubecola*) on the island of Hjelm during three spring seasons as an appropriate system for estimating the emigration ratios [for a much more elaborate report see Hansen and Rabøl (1977)]. Hjelm is a small uninhabited island (area 63 ha) about 6 km off the eastern coast of Jutland and is considered a good resting and feeding place for most passerine migrants including the Robin.

B. Materials and Methods

The Robin was the most common passerine trapped, and the total number of Robins banded and included in the analyses was 2100 (1972), 2600 (1973), and 5100 (1974). The birds were trapped in mist nets that were scattered all over the island and that were visited in a regular sequence every hour. Every day the trapping was started at sunrise and lasted 10 h. At the first capture the bird was banded with a ring and all subsequent recaptures were noted.

For each day a capture distribution was set up ($x_i = 1, 2, --- n$). The number of banded birds (R) was found to be $\sum_{i=1}^{n} f_i$, where f_i is the number of birds captured i times. The total number of captures (S) is $\sum_{i=1}^{n} f_i \cdot x_i$. On the basis of R and S, the number of resting birds (N), the number of immigrating birds, and the emigration ratio could be estimated.

N is calculated by the method of Craig (1953): $S/N = \log_e \{N/(N-R)\}$.

The emigration ratio is estimated as:

$$e.r._{i-1} = 1 - [G_{i-1} / (R_{i-1} \cdot S_i/N_i)]$$

$e.r._{i-1}$ is the emigration - ratio on $night_{i-1}$ (the night between day_{i-1} and day_i). G_{i-1} is the number of recaptures on day_i of birds banded on day_{i-1}. R_{i-1} is the number of birds banded on day_{i-1}. S_i is the total number of captures on day_i. N_i is the calculated number of resting birds on day_i.

The number of immigrating birds, I_i, is easily calculated when $e.r._{i-1}$, N_{i-1} and N_i are known.

Instead of e.r. a transformation $(e.r.)^2$ is used as the dependent variable in the analyses.

The emigration from Hjelm is thought to occur especially in the period from sunset to midnight (Geil et al., 1974). Most of the environmental variables compared to $(e.r.)^2$ are measured just preceding or at the start of this emigration period (at 1800-2100). The following variables were included (see also Table 1):

Table 1. Corr. coeffs. between the emigration ratio $(e.r.)^2$ (1972+1973+1974, n=57), and an array of variables (treated in more detail in Materials and Methods)

Variables	Corr. coeff., r
1. Date	-0.11
2. Log I	0.38[b]
3. t_{max}	0.38[b]
4. Cloudiness	-0.37[b]
5. Barometric pressure	0.13
6. Visibility	0.20
7. Precipitation	-0.31[b]
8. 0° wind (departure)	0.18
14. Wind force (departure)	-0.15
19. 120° wind (arrival)	-0.41[b]
21. Wind force (arrival)	-0.39[b]
22. $\bar{X}$ weight	0.28[a]
23. ΔY weight	0.41[b]

[a] $p < 0.05$
[b] $p < 0.01$

1. Date:

2. Log I: The logarithm of the number of immigrating birds during the preceding night.

3. t_{max}: The maximum temperature during the period 1000 - 1400 on the same day as the following emigration. As t_{max} obviously gave the best description, none of several other temperature variables investigated for was included in further analyses.

4. Cloudiness: Measured in -/8.

5. Barometric pressure: Measured in millibars.

6. Visibility: Measured in km, with an upper limit of 20 km.

7. Precipitation: The following scale was used: 0, 0.5, 1, 1.5, and 2: 1 = showers, and 2 = heavy rain.

8 - 13. Wind direction: The wind direction is a cyclic variable and transformations are needed to convert to linearity. We used the following six transformations: *8. cos $(0^\circ - v)$, 9. cos $(30^\circ - v)$, 10. cos $(60^\circ - v)$, 11. cos $(90^\circ - v)$, 12. cos $(120^\circ - v)$, 13. cos $(150^\circ - v)$.* v is the actual wind direction. Only the transformation with the numerically highest corr. coeff. is shown in Table 1.

14. Wind force: Measured in m/s. The following seven variables were measured or transformed from measurements at 0000 and 0600 - i.e., in the period of immigration to the island about 15 - 18 h before the actual emigration. The reason for the inclusion of these variables is to investigate the possible influences of the wind during the period of immigration on the subsequent emigration ratio.

15. - 20. Wind direction: The same six transformations as mentioned for Nos. 8 - 13.

21. Wind force: Measured in m/s.

22. $\bar{X}$ weight: The average weight of birds captured on rounds 6 - 10, i.e., from 5 h after sunrise and in the next 5 h. The weight during this period is normally constant, and constitutes the "afternoon level" after the normal increase in weight during the morning hours.

23. ΔY weight: The deviation of the daily $\bar{X}$ weight from the regression line on the $\bar{X}$ weight as a function of the season. This transformation corrects the general decrease of weight through the season.

Two of the analytic methods we used are described briefly in the section on Results.

C. Results

1. We calculated the parametric *corr. coeffs.* between $(e.r.)^2$ and several variables (Table 1). The differences between the three years were mostly small and insignificant (Hansen and Rabøl, 1977), and so we have presented the corr. coeffs. for all three years pooled.

2. *Multiple regression* was applied in the general form:

$$Y = k_o + \sum_{i=1}^{n} k_i \cdot X_i$$

We standardized the variables ($\bar{X}_i = 0$, $s_i^2 = 1$, $k_o = 0$). Following this procedure the constants k_i are fairly nonambiguous expressions of the relative influences of the independent variables in the equation. The k_i may be perceived cautiously as a sort of partial corr. coeffs.

We have not - as in Geil et al. (1974) - made a step up/step down multiple regression analysis with all the independent variables included in the analysis from the beginning. We have only included some of the variables, mostly those with a high simple corr. coeff. We have also included some variables (such as cloudiness) with an insignificant corr. coeff. on the suspicion of a "concealed" influence.

Table 2 shows the results, i.e., especially the k_i and their statistical significance. Compared to the simple corr. coeff. (r), log I keeps its influence, the cloudiness usually also. The reduction of "influence" especially in t_{max} and wind force (arrival) is very pronounced.

D. Discussion

Figure 2 shows the general conclusions concerning the influences of the independent variables.

1. The positive correlation between *log I* and $(e.r.)^2$ probably points toward a causal relationship between crowding and "migratory readiness." Log I is primarily influenced by the arrival weather (there is a strong negative correlation between log I and the arrival visibility and

Table 2. Multiple regression between $(e.r.)^2$ and several independent variables. k denotes the constants of the standardized variables. For comparison the corr. coeff. r is also shown R^2 denotes how much of the variation in $(e.r.)^2$ is "explained" by the combined variation in the number of independent variables included in the analyses

Variables	1974, n = 30		1972 + 1973 + 1974, n = 57		
	r	k_I	r	k_I	k_{II}
1. Date	-0.48^{b}	-0.28	-0.11	-0.10	
2. Log I	0.33^{a}	0.31^{a}	0.38^{b}	0.43^{b}	0.37^{b}
3. t_{max}	0.34^{a}	0.05	0.38^{b}	0.21^{a}	0.07
4. Cloudiness	-0.38^{a}	-0.34^{a}	-0.37^{b}	-0.39^{b}	-0.36^{b}
18. 90° wind	-0.60^{b}	-0.28			
19. 120° wind			-0.41^{b}	-0.28^{b}	-0.32^{b}
21. Wind force	-0.52^{b}	0.19	-0.39^{b}	0.08	0.01
22. $\bar{X}$ weight	0.60^{b}	0.31	0.28^{a}	0.15	
23. ΔY weight			0.41^{b}		0.24^{a}
R^2		0.63		0.52	0.53

$^{a}p < 0.05$
$^{b}p < 0.01$

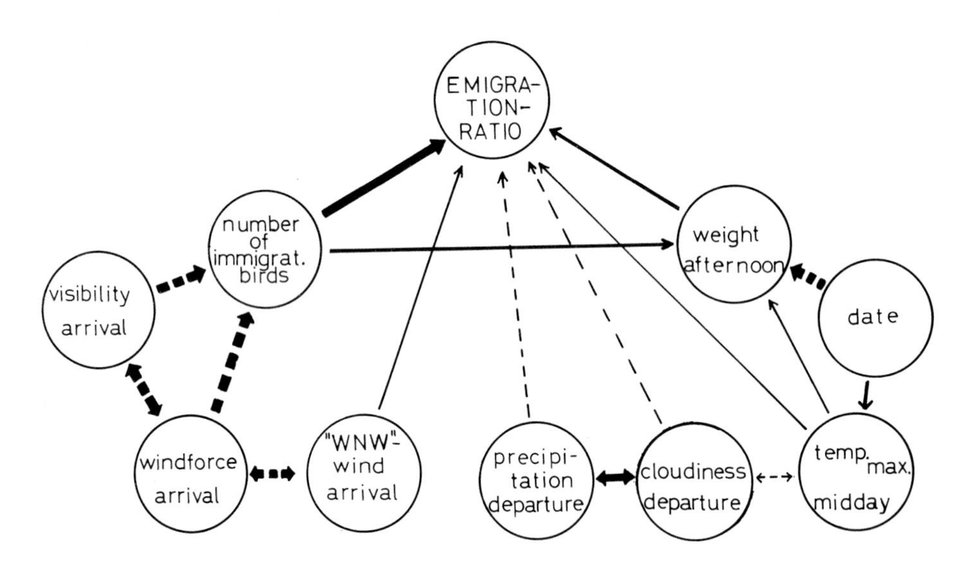

Fig. 2. The influences of different environmental variables on the emigration ratio. A unidirected *arrow* denotes a presumed causal relationship (influence), whereas a *two-sided arrow* just denotes a correlation. *Fully drawn* and *hatched arrows* denote positive and negative influences (correlations) respectively. The *thickness of the arrow* is a measure of the degree of influence (correlation)

wind force - r is about -0.50), and there seems to be no connection between the most important arrival weather factors and the subsequent variation in the "migratory readiness."

The corr. coeff. between log N and $(e.r.)^2$ is much more insignificant (r = 0.17). If N_i consists mainly of birds that immigrated to the island on $night_{i-2}$ or earlier, the migratory readiness is relatively low. Birds with a higher migratory readiness had already emigrated on $night_{i-1}$, and remaining behind on the island is a tail consisting of established and weak birds (cf. Rabøl and Petersen, 1973).

2. High *weight* in the afternoon also seems to have a positive influence on the subsequent migratory readiness.

3. The influence of *cloudiness* on migratory readiness is probably consistently negative, i.e., a clear sky promotes and/or an overcast sky inhibits the migratory readiness. This conclusion fits well with the positive correlation between the migratory restlessness of caged migrants and a clear sky (Wallraff, 1965; Rabøl, 1975). The view of the stars probably increases the migratory readiness.

4. There seems to be no influence of *temperature* in itself. The temperature, however, influences the weight (indirectly, by increasing the activity and thus availability of the insect food).

5. The influences of *visibility*, *wind force*, and *barometric pressure* on migratory readiness are probably negligible.

6. The influence of *wind direction* on migratory readiness is obscure.

The corr. coeff. between $(e.r.)^2$ and the departure wind direction (Nos. 8 - 13) is weak and insignificant (Table 1). The highest positive correlation is to a NNW wind (see also Fig. 3). The corr. coeff. between

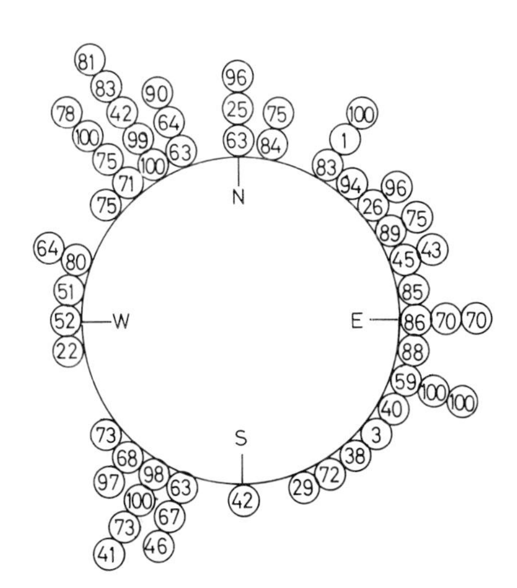

Fig. 3. Departure wind directions (Nos. 8-13, Table 1) and $(e.r.)^2$. 100, 1972 + 1973 + 1974. If we calculate the average $(e.r.)^2$ in each of the 12 sectors, 350° - 0° - 10°, 20° - 30° - 40°, - - -, 320° - 330° - 340° and make a summation of these 12 vectors, the mean vector is 6° and 0.055. This mean vector direction should be compared to the highest positive cos-transformation (Table 1), which is between 330° and 0° (closest to 0°). The accord between the two estimates of the most positive wind direction correlated with a high e.r. is reasonably good

$(e.r.)^2$ and the arrival wind direction (Nos. 15 - 20), however, is much higher (except in 1972) and peaks about WNW for the most positive wind direction (Table 1). The high coupling between the arrival and departure wind direction (in the vicinity of r = 0.50) probably "explains" much of the weak positive correlation between $(e.r.)^2$ and NNW departure winds. The picture is even more "disturbed" by the positive correlations between log I and $(e.r.)^2$, and between log I and SSE departure winds in the supposed recruitment area south of Hjelm (r = 0.30).

The conclusion should be that the wind direction as experienced by the Robins while resting on the island is probably without any influence on the magnitude of migratory readiness in the subsequent evening.

This is a surprising conclusion, since almost all authors (see the survey of Alerstam, 1976) stress the high positive correlation between migratory readiness (as measured by the intensity of echoes on the radar screen) and a following wind in the "standard direction," which should be NNE for a Robin at Hjelm in spring.

The positive correlation between the following wind and the migration intensity as measured on the radar may first become established in the air - and much of the correlation should certainly be due to wind drift (a passive downwind component) or as an artifact (Richardson, 1972) because of the MTI-filtering and the greater chance of registration of a fast-moving object. The two following disturbing influences should also be noticed. Both are based on the assumption of a circular normal distribution in the headings of the population of migrating birds. The mean vector of this distribution is in the standard direction. (a) The number of echoes passing across the radar screen (or observation area) per unit time will be maximum with a following wind in the standard direction - because of the higher flying speed in the following wind. (b) Migratory movements with a following wind are proceeding at higher altitudes than movements in opposed winds (e.g., Rabøl and Hindsbo, 1972), and normally a radar is more effective in the registration of a high-altitude object. Such conditions in combination with the scatter in headings with a mean vector in the standard direction will produce a positive correlation between the migration intensity as measured on the radar screen and a following wind in the standard direction. The positive correlations (a) and (b) are, however, both spurious. There is no positive influence of a following wind in the standard direction on the magnitude of migratory readiness.

Most authors seem to have too simple an idea of the influence of wind direction on migratory readiness, and it should be emphasized that a solely energetic point of view is biased. The proper adjustment of the migration in area and time is much more important for survival than always moving downwind in the standard direction with maximum speed at minimal energy expenditure. During most days and in most places a Robin should be able to collect enough food to build up fat reserves sufficient for a subsequent migratory movement during the night. A stereotype downwind preference - or a preference of other factors associated with downwind - should be a good adaptation for a Blackpoll Warbler (*Dendroica striata*) when crossing the Atlantic Ocean. Everybody can understand and accept such a system. However, it should not be transferred uncritically as a model for every kind of migratory movement - such as a Robin moving slowly up from the Mediterranean through Germany and Denmark toward the Fennoscandinavian breeding ground. During such circumstances the bird should be cautious about moving too quickly northward in a "favorable" downwind.

References

Able, K.: The role of weather variables and flight direction in determining the magnitude of nocturnal bird migration. Ecology 54, 1031-1041 (1973)

Alerstam, T.: Nocturnal migration of thrushes (Turdus spp.) in Southern Sweden. Oikos 27, 457-475 (1976)

Bruderer, B.: Multiple regression analysis of weather and migration data in Switzerland. Sempach (1974)

Craig, C.C.: On the utilization of marked specimens in estimating populations of flying insects. Biometrika 40, 292-306 (1953)

Geil, S., Noer, H., Rabøl, J.: Forecast models for bird migration in Denmark. Zoological Laboratory, Univ. of Copenhagen (1974)

Hansen, K., Rabøl, J.: Spring migration of the Robin (*Erithacus rubecola*) on the island of Hjelm, Denmark. Influences of weather on the emigration-ratio and the number of immigrating birds. Report. Can be obtained by request at the authors' address (1977)

Lack, D.: Migration across the North Sea studied by radar. Part 2. The spring departure 1956-59. Ibis 102, 26-57 (1960)

Lack, D.: Migration across the Southern North Sea studied by radar. Part 4. Autumn.. Ibis 105, 1-54 (1963a)

Lack, D.: Migration across the Southern North Sea studied by radar. Part 5. Movements in August, winter and spring, and conclusion. Ibis 105, 461-492 (1963b)

Nisbet, I.C.T., Drury, W.H.: Short-term effects of weather on bird migration: A field study using multivariate statistics. Anim. Behav. 16, 496-530 (1968)

Rabøl, J.: The orientation of night-migrating Passerines without the directional influence of the starry sky and/or the earth magnetic field. Z. Tierpsychol. 38, 251-266 (1975)

Rabøl, J., Hindsbo, O.: A comparison of the bird migration recorded by radar and visible field observations in the middle of Sjaelland, Denmark, spring 1971. Dansk Ornith. Foren. Tidsskr. 66, 86-96 (1972)

Rabøl, J., Petersen, F.D.: Lengths of resting time in various night-migrating Passerines at Hesselø, Southern Kattegat, Denmark. Ornithol Scand. 4, 33-46 (1973)

Richardson, W.J.: Temporal variations in the ability of individual radars in detecting birds. National Research Council Ottawa, Canada (1972)

Richardson, W.J.: Multivariate approaches to forecasting day-to-day variations in the amount of bird migration. Proc. Conf. on the Biological Aspects of the Bird/Aircraft Collision Problem. Gauthreaux, S.A. (ed.), Clemson, South Carolina, 1974a

Richardson, W.J.: Spring migration over Puerto Rico and the Western Atlantic, a radar study. Ibis 116, 172-193 (1974b)

Richardson, W.J., Gunn, W.W.H.: Radar observations of bird movements in east-central Alberta. Can. Wildl. Serv. Rep. Series No. 14, 35-68 (1971)

Wallraff, H.G.: Versuche zur Frage der gerichteten Nachtzugaktivität von gekäfigten Singvögeln. Verh. Deut. Zool. Ges. in Jena 1965, 338-356 (1965)

Further Analysis of the Magnetic Compass of Migratory Birds

WOLFGANG WILTSCHKO, Fachbereich Biologie der Universität, 6000 Frankfurt, FRG

Abstract

The orientation behavior of migratory European Robins (*Erithacus rubecula*) during *Zugunruhe* was used for a further analysis of the characteristics of the birds' magnetic compass. Tests with Robins adapted to field intensities outside the normal functional range of the magnetic compass show that the process of adaptation to fields outside this range is neither a shifting nor a simple enlargement of the functional range. Tests in several 1-Hz alternating magnetic fields with rectangular and sine-shaped impulses indicate that a certain constant portion of the impulse is necessary to enable the birds to use it for orientation.

A. Introduction

Since the sonsory basis for the perception of the magnetic field by birds is still unknown, investigating behavioral responses to different types of magnetic fields appeared to be the best approach to gain more information about how the birds' "gaussmeter" might function. One behavioral response to magnetic stimuli that is easily studied is the directional orientation of caged migratory birds - especially of European Robins - during *Zugunruhe*. Their orientational reaction (1) corresponds to the migratory direction of their free-living conspecifics as revealed by the banding recoveries and is thus biologically meaningful, (2) is directly correlated to the direction of the magnetic field applied, and (3) indicates whether a given magnetic field is adequate for the birds, since it becomes random if the field cannot be used (Wiltschko 1968, 1972). So it seemed appropriate to continue to use this response for the further analysis of the characteristics of the birds' magnetic compass.

B. Materials and Methods

All experiments reported here were conducted in the spring. The test birds - European Robins (*Erithacus rubecula*) - were either (1) captured during the preceding autumn migratory season, and their *Zugunruhe* was induced ahead of time by photoperiodic manipulations (making it possible to start the tests in January, comp. Wiltschko, 1968), or (2) freshly trapped and tested during the natural spring migratory season. As in former experiments, octagonal cages with eight radially positioned perches were used for recording the birds' activity. From the amount of activity per perch, the headings of the test nights were calculated. They were comprised in the mean vector of the series with the direction α_m and the length r_m; r_m was tested for statistical significance using the Rayleigh Test (Batschelet, 1965). A more detailed description of the experimental procedure, the registration method, and the statistical evaluation is given in Wiltschko (1968) and Wiltschko and Wiltschko (1976).

The magnetic fields to be tested were produced by pairs of Helmholtz coils (2 m in diameter, 1 m clearance). All statements pertaining to the characteristics of the magnetic fields apply to the resultant magnetic field as measured inside the orientation cage.

For the alternating fields, a pair of such coils with two independent sets of windings was positioned such that the coils' axis lay in the course of the field lines, i.e., in the direction of inclination. One set of windings was then charged with direct current to cancel out the local field; the other set was charged by a Wavetek frequency generator producing the pulses to be tested, the output of which was checked by an oscilloscope.

C. Results and Discussion

I. Intensity Range of the Magnetic Compass

Former experiments (Wiltschko, 1972) had shown that the magnetic compass normally functions only in a relatively narrow intensity range. Robins, captured and kept in Frankfurt at the local Earth's magnetic field of 0.46 gauss, could spontaneously orient at 0.43 gauss and at 0.54 gauss, but at 0.34 gauss and 0.60 gauss, random behavior was observed. Thus, the normal functional intensity range of these birds' magnetic compass appears to include ca. 0.2 gauss; its lower limit being at approximately 0.38 gauss and the upper limit at ca. 0.57 gauss (Table 1).

Table 1. Robins trapped and living at 0.46 gauss

Filed intensity (Gauss)	Inclination	Season	n	Vector length r_m	Significant mean direction
0.16	31°	Spring	79	0.11	——
		Autumn	52	0.04	——
0.34	57°	Spring	104	0.07	——
		Autumn	58	0.17	——
0.43	42°	Spring	62	0.39	mNNW
		Autumn	38	0.38	mSSW
0.46	66°	Spring	96	0.39	NNE
		Autumn	64	0.38	SSW
0.46	48°	Spring	73	0.35	NE
0.54	37°	Spring	49	0.42	mNE
		Autumn	25	0.35	mSW
0.60	45°	Spring	31	0.12	——
0.68	25°	Spring	68	0.09	——
0.81	8°	Spring	42	0.11	——
0.82	30°	Spring	28	0.06	——
1.05	23°	Spring	50	0.12	——
		Autumn	32	0.27	——

Yet Robins can orient well in field intensities outside this range, when they are permanently exposed to these intensities for 3 days or longer. This was known for fields of 0.16 gauss and 0.81 gauss (Wiltschko, 1972); in recent experiments in a 1.5-gauss magnetic field, Robins living in this intensity also showed oriented behavior (cf. Fig. 2c). Thus the potential range in which the magnetic compass of the Robins can operate in principle is considerably larger than the actual functional range that is normally encountered by Robins that we capture in our local geomagnetic field of 0.46 gauss.

To obtain some information about the nature of the processes underlying the adaptation to different intensities, robins adjusted to 0.18 gauss were tested in the local Earth's magnetic field of 0.46 gauss, and they showed a significant ($p < 0.01$) directional preference of their migratory direction (Fig. 1). Corresponding results were obtained with robins adjusted to 1.5 gauss (Fig. 2a). So far the results might

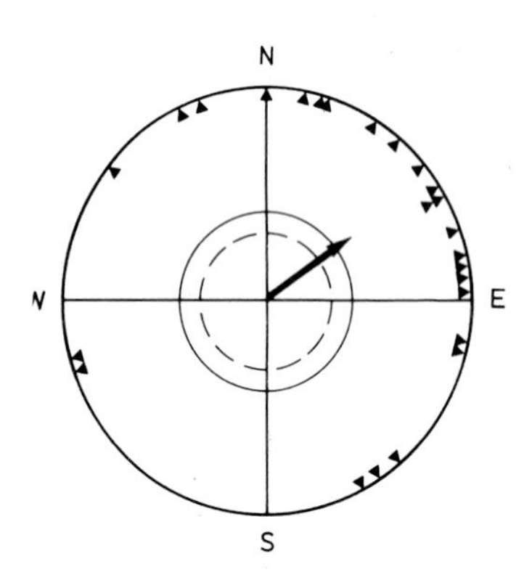

Fig. 1. Orientation behavior of robins that lived in a weak magnetic field of 0.18 gauss and were tested in the Earth's magnetic field. The headings of the individual test nights are shown as *triangles* at the periphery of the circle The mean vector is represented as an *arrow* with the length r_m in relation to the radius of the circle = 1. The two inner *circles* are the 1% and the 5% (*dotted*) significance border of the Rayleigh Test

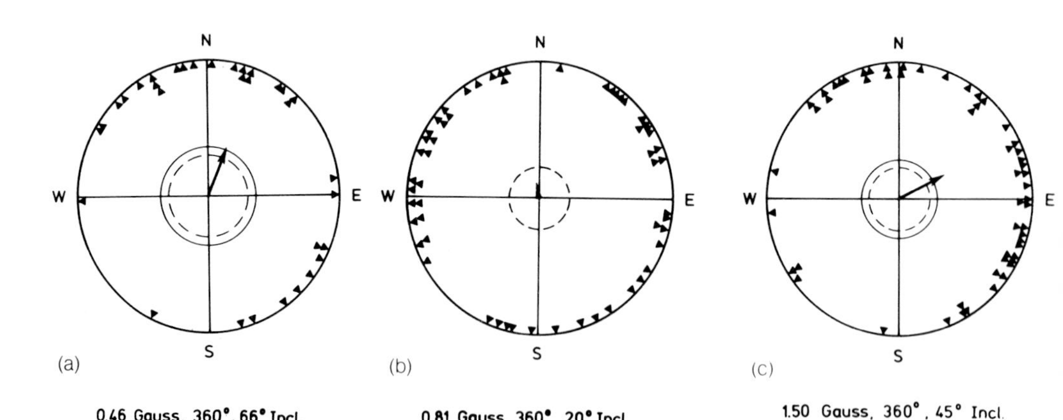

Fig. 2 a - c. Orientation behavior of robins living in a strong field of 1.5 gauss: (a) in the natural Earth's magnetic field; (b) in an intermediate field stronger than the local geomagnetic field; (c) in the strong field they are now living in. (Data presentation as in Fig. 1)

indicate an enlargement of the birds' sensitivity range, but when we additionally tested the robins adapted to 1.50 gauss in a 0.81 gauss field, a directional preference was no longer observed (Fig. 2b).

So when robins were adjusted to field intensities outside their normal functional range, they could use these fields for orientation, while maintaining the ability to read the natural gemomagnetic field they had been accustomed to earlier. However, the range between the Earth's field intensity of 0.46 gauss and the intensity of 1.50 gauss is not automatically included in the functional range when the birds are adapted to 1.50 gauss. Birds seem to be able to use only those magnetic fields with which they have become familiar (Table 2).

Table 2

Robins trapped at 0.46 gauss, living at 0.16 gauss

Field intensity (gauss)	Inclination	Season	n	Vector length r_m	Significant mean direction
0.16	31°	Spring	25	0.35	N
		Autumn	43	0.43	SW
0.46	66°	Spring	25	0.51	NE

Robins trapped at 0.46 gauss, living at 1.50 gauss

Fild intensity (gauss)	Inclination	Season	n	Vector length r_m	Significant mean direction
0.46	66°	Spring	33	0.38	NNE
0.81	20°	Spring	57	0.11	——
1.05	45°	Spring	56	0.38	ENE

The adaptation to intensities outside the normal range thus appears to be neither a shifting nor a simple enlargement of the normal functional range of the birds' magnetic compass. Instead it seems that exposing the robins to 1.50 gauss had created a new, separate, functional range at that intensity. The birds seem to have to adjust to each given intensity separately. Under natural conditions adaptation to different intensities must occur, however, in very small steps over a number of days when the intensity gradually changes during a bird's migration flight.

II. Experiments in Alternating Magnetic Fields

Former experiments have shown that the magnetic compass of the European Robin functions as an "inclination compass," not using the polarity of the magnetic field, but the inclination of the axial course of the field lines (Wiltschko, 1972; Wiltschko and Wiltschko, 1972). Indications for the same type of mechanism have also been found in the Garden Warbler (Wiltschko, 1974) and in the homing pigeon (Wallcott and Green, 1974). This means that the orientation behavior of birds is the same when the field's polarity is pointing northward and downward and when it is pointing southward and upward, provided the intensity is inside the adequate range. This characteristic of the bird's magnetic compass allows a meaningful study of the effect of alternating magnetic fields on the orientation behavior.

In the following tests in alternating fields, the axis of inclination remained constant and was identical with that of the local geomagnetic field (+66° and to the north, to -66° and to the south).

In a first experimental series a group of robins was tested alternately in a 1-Hz field with rectangular-shaped impulse and in a 1-Hz sine wave. The rectangular field had an intensity of 0.46 gauss toward north and downward for 0.5 s that changed to 0.46 gauss toward south and upward for the next 0.5 s and vice versa. The sine wave was of equal energy, which resulted in a maximum intensity of 0.72 gauss (= 0.46 x $\pi/2$).

We found normally oriented behavior in the field with the rectangular impulse (Fig. 3a, comp. Table 3), which was comparable to the orienta-

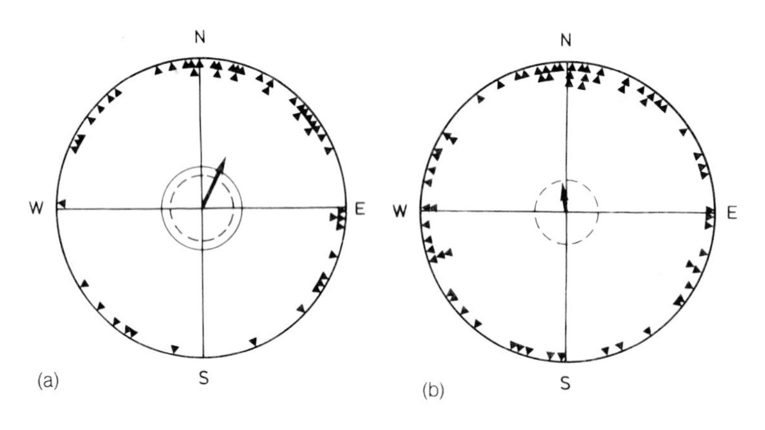

Fig. 3 a and b. Orientation behavior of robins in alternating magnetic fields. (a) 1-Hz rectangular impulse (cf. Fig. 4a); (b) 1-Hz sine wave impulse (cf. Fig. 4b)

Table 3. Results of the test series in alternating magnetic fields

Impulse	Frequency	n	α_m	r_m	Significance
Rectangle, 0.46 gauss (cf. Fig. 4a)	1 Hz	50	27°	0.38	$p < 0.01$
Sine, max. 0.72 gauss (cf. Fig. 4b)	1 Hz	65	352°	0.21	$p > 0.05$
Sine, birds kept at 0.18 gauss (cf. Fig. 4c)	1 Hz	30	326°	0.20	$p > 0.05$
"Saw", max 0.46 gauss (cf. Fig. 4d)	0.8 Hz	46	1°	0.34	$p < 0.01$
Sine, min 0.40 gauss, Max. 0.50 gauss (Fig. 4e)	1 Hz	76	7°	0.16	$p > 0.05$
Local geomagnetic field	Static	43	20°	0.35	$p < 0.01$

tion behavior observed in the Earth's magnetic field[1]. Thus these results are consistent with our previous findings about the inclination compass not using the polarity: For the robins, a field going northward-down and a field going southward-up appearently mean the same (cf. Wiltschko and Wiltschko, 1972). In the sine-shaped field, however, the robins seemed to have serious difficulties in using the magnetic field for orientation; a significant directional preference could not be found (Fig. 3b, cf. Table 3).

Looking for the reasons why the orientation was so much worse in the alternating field with the sine wave impulse than in the field with rectangular impulses, it must be considered that the two fields differ in three respects (cf. Fig. 4 a, b):

1. The strength of the field produced by the rectangular impulse stays inside the normal functional range of the magnetic compass (for more than 95% of the time), whereas the sine wave impulse is in that range only ca. 25% of the time.

2. For the birds, a 1-Hz rectangular impulse presents a frequency of 2 Hz, while the 1-Hz sine wave presents frequency of 4 Hz - it passes through the functional range of the magnetic compass 4 times per second.

3. In the rectangular field, the intensity remains constant for a considerable time, whereas in the sine wave it is constantly changing.

In analyzing which of the characteristics of the sine wave causes the bad orientation in our birds, we tried three approaches demonstrated in Figure 4 c, d, e; the results are summarized in Table 3:

1. We adapted the birds to 0.18 gauss, knowing that they would still be able to orient at 0.46 gauss as stated above, and tested them in the same 1-Hz sine field (Fig. 4c). By adaptimg the birds, we "enlarged" the functional range of the magnetic compass, thus the sine wave remained in the effective range for a considerably greater percentage of the time. Nevertheless the Robins' orientation did not improve.

2. We tested birds in a "saw"-shaped impulse of 0.8 Hz (Fig. 4d). Here the intensity remained at zero for ca. 0.20 s, increased rapidly to 0.46 gauss, where it remained for ca. 0.25 s; after that it broke down to zero for another 0.20 s to increase again to 0.46 gauss in the opposite direction. The intensity of this field is in the functional range of the birds' magnetic compass about 50% of the time. The results show that the robins could make use of this type of alternating field for direction finding.

3. We produced a magnetic field with magnetic north = 360° and 66° inclination, the intensity of which varied in a sine mode between 0.40 gauss and 0.50 gauss (Fig. 4e), which means that this field was permanently inside the functional range of the birds. Yet when the birds were tested, no significant directional preference could be found.

These results indicate that neither the relative amount of time the alternating magnetic field remains in the functional range, nor apparently the frequency alone (the field between 0.40 gauss and 0.50

[1] The same sample of birds in the Earth's magnetic field was clearly oriented in their migratory direction: $n = 43$, $\alpha_m = 20°$, $r_m = 0.35$, $p < 0.01$.

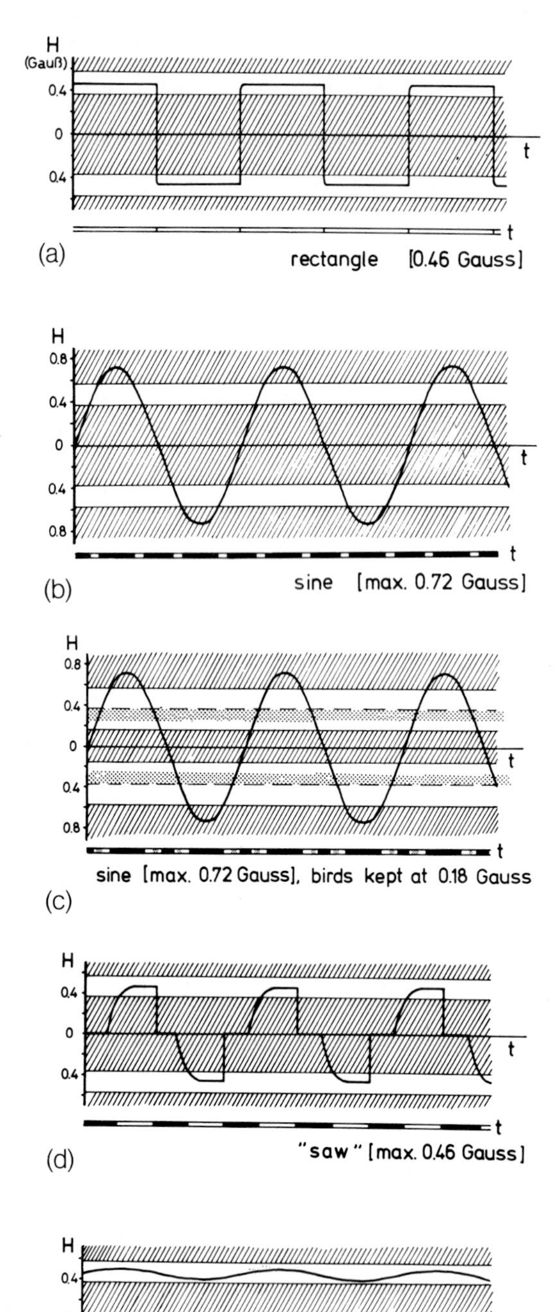

Fig. 4 a - e. Impulse shapes of the alternating magnetic fields tested. In all five fields, the axis of inclination was identical with that of the local field (to 360° north and +66° down versus 180° south and -66° up). *Abscissa:* time; *ordinate:* field intensity in gauss; *dark-hatched area:* intensities outside the normal functional range of the birds' magnetic compass. *Below*, the times when the impulse is outside the effective range are black. (a) 1-Hz rectangular impulse; (b) 1-Hz sine wave of equal energy; (c) same sine wave; situation for the birds adapted to 0.18 gauss; (d) 0.8 Hz saw-shaped impulse; (e) 1-Hz sine wave inside the functional range

gauss never left the functional range; high-frequency rectangular fields, of course, still have to be tested) plays an important role in determining whether the Robins can use an alternating field for orientation. The shape of the impulses themselves, however, was critical: Poor orientation was observed in all three sine fields, while in the rectangular field and in the saw-shaped field the robins' orientation was comparable to normal. This suggests that a magnetic field must have at least some portions of constant intensity to be used for orientation; future experiments will have to give more information about the relative and absolute length of these constant portions necessary for orientation.

D. General Discussion

When we use the behavioral response of migratory behavior to analyze characteristics of the birds' magnetic compass, we must realize that we observe a whole chain of functions resulting in this complex behavior. So when we find poorly oriented or nonoriented behavior associated with a certain field characteristic, it is not possible to discriminate whether this is caused by the receptor (still unknown) not functioning, or by the processing centers (equally unknown) in the central nervous system not being able to recognize the signals coming from the receptor. This dilemma must be remembered when our findings about the magnetic compass are compared with a model of how the physiologic detection of the magnetic field might take place in birds.

Very recently, Leask (1977) suggested that the perception of the magnetic field in animals is effected by a resonance phenomenon in the rhodopsin molecules of the retina. As he states in his paper, his hypothesis is in good agreement with the characteristics of the birds' magnetic compass known so far, i.e., that the response to the magnetic field is axial rather than polar and that it normally takes place in a rather narrow range around the Earth's field intensity. He explains the phenomenon of adaptation to higher or lower intensities by the resonance conditions now being satisfied in different regions of the retina and the birds learning to interpret this new situation in the course of a few days. Our results, i.e., that birds adapted to higher fields can still orient in the Earth's field, but that they cannot use intermediate field intensities, would also fit this model: Birds remember the former situation, but they cannot interpret an intermediate situation with which they have not become familiar. Correspondingly, according to Leask's hypothesis, an alternating field with a sine wave impulse would result in the area of resonance constantly flickering over different regions of the retina, a situation that could create problems for the nervous centers processing the information. Thus our new findings, too, can be interpreted in the sense of Leask's hypothesis and to not contradict it.

Many future investigations seem necessary before we finally will definitely know the mechanisms of magnetic field perception in animals; in the meantime I hope that gaining more information about the functional characteristics of the magnetic compass may help to stimulate theoretical considerations and experimental investigations aimed at solving this problem.

Acknowledgments. This study was supported by the Deutsche Forschungsgemeinschaft; the computer work was carried out by the Hochschulrechenzentrum of the Universität Frankfurt. I wish to thank G. Fleißner for his valuable help in setting up the experimental alternating fields, H. Golle for his assistance in trapping the test birds and conducting the experiments, and S.T. Emlen for critically reading the manuscript.

References

Batschelet, E.: Statistical methods for the analysis of problems in animal orientation and certain biological rhythms. Am. Inst. Biol. Sci., Washington, D.C. (1965)

Leask, M.J.M.: A physicochemical mechanism for magnetic field detection by migratory birds and homing pigeons. Nature (London) 267, 144-145 (1977)

Wallcott, C., Green, R.P.: Orientation of homing pigeons altered by a change in the direction of an applied magnetic field. Science 184, 180-182 (1974)

Wiltschko, W.: Über den Einfluß statischer Magnetfelder auf die Zugorientierung der Rotkehlchen (*Erithacus rubecula*). Z. Tierpsychol. 25, 537-558 (1968)

Wiltschko, W.: The influence of magnetic total intensity and inclination on directions chosen by migrating European robins. In: Animal Orientation and Navigation. NASA SP-262, U.S. Gov. Print. Off., Washington, D.C., 569-578, 1972

Wiltschko, W.: Der Magnetkompaß der Gartengrasmücke (*Sylvia borin*). J. Orn. 115, 1-7 (1974)

Wiltschko, W., Wiltschko, R.: Magnetic compass of European robins. Science 176, 62-64 (1972)

Wiltschko, W., Wiltschko, R.: Interrelation of magnetic compass and star orientation in night-migrating birds. J. Comp. Physiol 109, 91-99 (1976)

Orientation Responses of Ring-Billed Gull Chicks: A Re-Evaluation

WILLIAM E. SOUTHERN, Department of Biological Sciences, Northern Illinois University, DeKalb, IL 60115, USA and The University of Michigan, Biological Station, Pellston, MI 49769, USA

Abstract

The mean bearing for Ring-Billed Gull chicks tested in orientation cages correlates favorably with the direction in which they will head to reach the winter range. Previous studies indicated that geomagnetic cues alone were of importance in this process. Reanalysis of data from 3500 trials has shown that under some conditions (1) directional shifts rather than disorientation may occur, and that (2) solar cues may be used as a supplemental source of information. Severe magnetic disturbance usually results in disorientation, regardless of sky conditions. Changes in apparatus design may significantly influence results, and this problem must be entertained during data interpretation.

The primary winter range for Ring-Billed Gulls (*Larus delawarensis*) banded at the Calcite Colony near Rogers City, Presque Isle County, Michigan lies southeast of the colony (Southern, 1974a). Gull chicks (about 3-10 days old) and juveniles capable of flight usually express directional preferences for a series of bearings appropriate for reaching their winter range (Southern, 1969, 1971, 1972, 1974b, 1976). Results from my studies have shown that Ring-Bills apparently derive directional information from the earth's magnetic field, or that disturbances in the normal geomagnetic field disrupt the ability of young gulls to respond in a nonrandom fashion.

On the basis of this evidence, I tentatively postulated that juvenile Ring-Billed Gulls use geomagnetic cues as a primary component of their migrational orientation system. During, and subsequent to, their first migration they probably perfect skills associated with use of supplemental or redundant cues that are then integrated into their orientational system. Experiences during the orientation process assist in the fine tuning of the system that permits increased accuracy, efficient orientation under a variety of environmental conditions, and eventually movement between precise goals.

Recently I pooled my data from about 3500 orientation cage trials, and calculated the mean direction for groups of birds tested under similar conditions. Consideration was given to the possible effect of several factors that could have influenced the directional responses of gull chicks. In this paper, I have attempted to account for some of the variation that occurs in trial results, and to relate the responses of test subjects to the availability of supplemental cues that may influence orientation or improve the quality of the experimental environment to the point that subjects are more likely to respond. My findings lend a new element to my previous conclusions about Ring-Billed Gull orientation and, although the implications are not fully understood, the new evidence is discussed.

A. Methods

The procedures and apparatus used in these studies were described in detail in earlier papers (see Southern, 1974b and 1976). All trials were conducted in Southern-type orientation cages of 2.4 m diameter. Cage design was modified for some trials so as to alter cue types available to birds. Characteristics of the various cages are (1) standard cage - opaque aluminum walls 0.6 m high (2) open cage - transparent plastic wall or no wall (3) stockade cage - standard cage surrounded by a supplemental translucent or opaque wall 2.4 m high (4) covered - an opaque or translucent cover was placed over the cage to disrupt vision of the sky. Trials were also conducted in an artificially lighted prefabricated room.

The possible effect of naturally occurring disturbances in the earth's magnetic field was considered. K indices provided by the World Data Center served as a measure of the level of geomagnetic disturbance (Table 1). In addition, some birds were tested under the influence of superimposed magnetic fields. In these trials, birds had small ceramic magnets glued to their heads or backs, or magnets were arranged symmetrically about the cage floor (Southern, 1974b).

Table 1. Value of K indices (Fredricksburg Oberservatory)

K	gamma[a]				K	gamma	
0	0 - 4				6	120 - 199	
1	5 - 9	Quiet (K 0–2)			7	200 - 329	Major storms (K 6–9)
2	10 - 19		Unsettled (K 0–3)		8	330 - 499	
3	20 - 39				9	500 plus	
4	40 - 69	Minor storm (K 4–5)		Active (K 0–5)			
5	70 - 119						

[a]1 gamma = 10^{-5} oersted.

Statistical procedures were the same as used previously (Southern, 1974b). The final headings of chicks used in each trial type were pooled for analysis. When analyzing data for one or more years, I combined the raw data for final headings, and calculated a mean for the new sample size.

B. Results

I conducted 1715 of the 3500 trials in standard orientation cages. About 8% of the trials were omitted from calculations, as the chicks failed to make any attempt to move across the cage during a 2-min trial. The mean angle ($\bar{a}$) of 174° for 1533 trials conducted when geomagnetic conditions were quiet to unsettled (Table 1) was highly significant (z 7.6). This bearing corresponds favorably with the direction predicted for the population on the basis of evidence stemming from band recoveries (Southern, 1976).

Only 50 trials were run when variations in the earth's magnetic field were classed as active (K4). The 39 chicks that responded under these conditions expressed a significant (z 9.54) mean bearing of 298°. Although this sample is small, the results suggest that minor distortions in the normal geomagnetic field may cause a directional shift rather than disorientation. Vector lengths were not plotted for these data because of the overwhelming difference in sample sizes.

I. Results from Trials Sorted by Sky Condition and K Value

Separation of the composite sample (1533 trials) into two groups, those conducted under clear skies (n 825), and those run during overcast conditions (n 708), produced some noteworthy differences. There is a statistically significant (z 22.3) mean bearing of 180° for the clear trials, whereas the distribution of headings taken by chicks tested under overcast approached randomness ($\bar{a}$ 130°; z 1.06).

The differences in results from trials conducted under clear and overcast conditions remain apparent when the sample is divided into trials executed under a particular sky condition and levels of geomagnetic activity ranging between K0 and 4 (Table 2). Each of the five clear-sky groups showed a statistically significant mean heading, whereas only one group of overcast trials (K1) had a significant mean bearing. The other four overcast groups (K0, 2, 3, and 4) exhibited a tendency toward randomness. Although both groups (clear and overcast) showed a significant bearing during K1 disturbances, their mean bearings differed by 100°. Under clear skies and low levels of geomagnetic disturbance (K0-2) the mean headings ranged between 95° and 191° ($\bar{x}$ 132°). The mean angles for trials during K3 and four levels of magnetic disturbance were 215° and 292°, respectively (Table 2). In these instances, it appears that while an oriented response persisted, it was shifted to a westerly bearing, possibly as a result of the difference in geomagnetic conditions.

Table 2. Directional responses of gull chicks[a] tested under various sky and geomagnetic conditions

K value	Sky	n	$\bar{a}$	SD	R	z
0	Clear	126	109	63	10.34	4.12
0	Overcast	136	96	72	7.50	1.56
1	Clear	103	95	74	17.72	3.05
1	Overcast	108	196	70	27.10	6.80
2	Clear	156	191	68	19.80	5.94
2	Overcast	162	7	79	4.02	0.16
3	Clear	162	215	74	26.18	4.23
3	Overcast	67	46	78	5.19	0.40
4	Clear	33	292	53	18.68	10.57
4	Overcast	17	46	73	3.26	0.63

[a]Only 1070 trials are included in this table as the exact K value during some was not recorded.

None of the groups tested during clear skies and low intensity geomagnetic disturbance (KO-2) was strongly oriented toward the southeast. During KO-1, the mean bearings were only slightly south of east and during K2 they were somewhat west of south. This pattern resembles that depicted in my earlier studies (Southern, 1969, 1971, 1974b). The spread of mean angles over about 90° may represent the range of directional preferences that will be expressed by birds from this population during migration. Although the primary winter range for Ring-Billed Gulls nesting at Rogers City lies to the southeast, portions of the population winter on the lower Great Lakes, near Chesapeake Bay on the Atlantic coast, along the coast of the Carolinas, and on the Gulf coast west to Texas. If my premise that the spontaneous directional responses of Ring-Billed Gull chicks represent an innate preference for their future migrational heading is correct, then it is likely that portions of the population will select bearings other than southeast. Studies are being conducted that may elucidate the nature of this apparently innate mechanism.

Some of my earlier findings (see Southern, 1974b) also suggested that the directional responses of gull chicks being tested for the effect of differing levels of magnetic disturbance were influenced by the sky being clear or overcast. At that time, however, I discounted the possibility of chicks possessing a redundant cue system, as there was no consistent pattern in the occurrence of disorientation during overcast. Now that the pattern has occurred repeatedly in data sets for several seasons and in the pooled data, the possibility that Ring-Billed Gull chicks are responding to solar cues appears more tenable. Although it remains unclear what role the sun is playing in the orientation process of Ring-Billed Gull chicks, my results show that disorientation is more frequent under overcast conditions.

If this finding is correct, it appears that chicks develop some level of proficiency in using a sun compass prior to the time they are able to fly. It is likely that this ability develops post-hatching, and that an improved performance rate would be evident as the chicks age. However, I have not been able to observe any change in directional response that can be correlated with age, although this possibility has been considered during trials and data analysis.

II. The Effect of Magnetic Anomalies

The possibility that solar cues were being used as a back-up system was not evident when 1333 chicks were tested during simultaneous exposure to a superimposed magnetic field and clear or overcast skies. Of these, 723 were tested under clear skies, and 610 during overcast conditions. The directional tendencies for these birds were calculated for the entire sample, subgroups based on sky conditions (Table 3),

Table 3. Directional responses of gull chicks tested under the influence of sky conditions and superimposed magnetic field

Sky condition	Source of Superimposed Field	n	$\bar{a}$	SD	R	z
Clear and overcast combined	Ceramic magnets on head, back or cage floor	1333	344	80	37.13	1.03
Clear	As above	723	33	79	33.71	1.62
Overcast	As above	610	273	80	14.29	0.35

and subgroups according to sky and K value. In no instance was there a significant departure from randomness. It seems, therefore, that the more intense magnetic distortion caused by a superimposed field of 0.5 oersted will result in disorientation, regardless of sky condition.

III. Effect of an Altered or Obscured Sky

To determine if an absence of solar cues or a distorted view of the natural sky and the horizon influenced directional responses of gull chicks, I conducted trials in covered cages, stockade cages, and in a wooden room. Superimposed magnetic fields were involved in some trials.

When deprived of a view of the natural sky, yet exposed to brightly lighted conditions in a test apparatus, chicks expressed a directional preference. The mean angle, however, was shifted northward to 1° and 8° (Table 4). Chicks tested under the same conditions while wearing magnets failed to show a significant mean heading. It appears that a view of the sun or horizon is not necessary for an oriented response, but the absence of these cues causes a noteworthy directional change away from southeast. Magnetic anomalies beyond those produced by K2-3 levels of natural disturbance apparently disrupt the tendency of chicks to show a preference for any particular heading.

Table 4. Results of trials in which vision of the natural sky or horizon was prevented or distorted

Cage type	K value	Magnets or brass	N	$\bar{a}$	SD	R	z
Translucent top	0 - 3	Neither	90	1°	37°	18.71	3.89
	2 - 3	Brass	80	8	70	20.34	5.17
	2 - 3	Magnets	79	61	74	13.87	2.44
Stockade							
Clear	0 - 3	Neither	67	165	62	26.97	10.85
Overcast	0 - 3	Neither	65	123	62	26.79	11.05
Combined	0 - 3	Neither	132	144	64	50.16	19.06
Wooden room	2 - 3	Neither	69	111	76	7.70	0.86
	2	Neither	13	114	73	2.35	0.43
	3	Neither	56	110	77	5.35	0.51

When provided with a view of the clear or overcast sky directly overhead, but not a direct view of the sun, chicks demonstrated a highly significant preference for 144° (Table 4). A significant bearing was selected under either clear or overcast conditions.

Results from this battery of tests reinforce the possibility that gull chicks are more prone to express a directional preference when they can view the sky, and have little or no disturbance in the geomagnetic field. The fact that significant mean headings exist in the stockade data for clear and overcast conditions is perplexing.

The headings of chicks tested in the wooden room under artificial lighting were randomly distributed. It is possible that the noticeable difference in light intensity inside and outside the chamber influenced chick behavior. The results, therefore, may not be indicative of the actual orientational ability of gull chicks that are prevented from viewing the natural sky. It seems reasonable to expect gull chicks, or other birds, to exhibit better response rates when test conditions more closely approach those in their natural environment.

IV. Effect of a General Landscape Being Visible

During trials in an open cage, chicks were able to see the landscape surrounding the test apparatus, the sky, and the uninterrupted horizon. They could not see the colony or the observer who was hidden in a blind.

When the landscape was visible throughout a 2-min trial, gull chicks in seven out of eight data subsets expressed a statistically significant mean bearing (Table 5). In most instances, the mean angle for

Table 5. Results for trials conducted in open cages

Sky condition	K value	Magnets?	N	$\bar{a}$	SD	R	z
Clear	1 - 2	No	96	140°	70°	23.81	5.90
Overcast	1	No	34	181	61	14.76	6.40
Combined	1 - 2	No	130	156	69	36.25	10.11
Clear	1	No	27	135	63	10.47	4.06
Clear	2	No	69	144	73	13.41	2.61
Clear	0 - 4	Yes	102	160	72	22.03	4.76
Clear	0 - 3	Yes	62	127	67	19.59	6.19
Clear	4	Yes	40	222	68	11.93	3.56

the sample approximated that predicted for the population on the basis of banding data. These data are not interpreted as indicating that gull chicks are using landmarks in their selection of a course to follow within the test apparatus. Such a possibility seems unlikely, as none of the test subjects had prior experience at or near the test site. It is more likely that the more natural surroundings resulted in a higher response rate (100% in most cases) and possibly in an increased tendency for them to give an orientational response rather than some other type of reaction (e.g., fear) to experimental procedures.

Representatives of only one data set, those tested under clear sky at K2 (Table 5), failed to show a significant heading. I cannot account for this discrepancy, as it seems inlikely that this low level of geomagnetic disturbance would result in disorientation if the birds are capable of using one or more supplemental cues.

It is interesting, as well as perplexing, to note that each of the three test groups for superimposed fields (i.e., involving magnets) had statistically significant mean headings. Two of these were for directions comparable to those taken by birds without magnets. The third group, which had a mean bearing of 222°, was tested during the highest concurrent level of geomagnetic disturbance (K4). This latter factor may account for this being the only group that showed a south-west heading. It is, however, difficult to justify emphasizing such a possibility, since the other magnetic subgroups expressed a heading consistent with that predicted for their future migration. It is possible that the combined effect of K4 natural disturbance, and a superimposed field produced by magnets was sufficient to override any benefit obtained from the other available cues.

This group of trials in which birds were permitted an unhindered view of the sky and natural landscape was the only one in which magnets consistently failed to produce disorientation. It should be noted, however, that these trials were conducted under clear skies. It is possible, therefore, that chicks in these data sorts were using the sun as a supplemental cue during trials. Since a cage wall was absent, they were provided with a view of the natural horizon, which was not the case in any of the other trials. This could account for the directional preference being more consistent with that predicted for the population.

References

Southern, W.E.: Orientation behavior of Ring-Billed Gull chicks and fledglings. Condor 71, 418-425 (1969)
Southern, W.E.: Gull orientation by magnetic cues: a hypothesis revisited. Ann. N.Y. Acad. Sci. 188, 295-311 (1971)
Southern, W.E.: Influence of disturbances in the Earth's magnetic field on Ring-Billed Gull orientation. Condor 74, 102-105 (1972)
Southern, W.E.: Seasonal distribution of Great Lakes Region Ring-Billed Gulls. Jack-Pine Warbler 52, 154-179 (1974a)
Southern, W.E.: The effects of superimposed magnetic fields on gull orientation. Wilson Bull. 86, 256-271 (1974b)
Southern, W.E.: Migrational orientation in Ring-Billed Gull chicks. Auk 93, 78-85 (1976)

Primitive Models of Magnetoreception

M. J. M. LEASK, Clarendon Laboratory, Oxford, OX1 3PU, Great Britain

Abstract

It is suggested here that the physical requirements for magnetic field detection by biological systems may be satisfied by a phenomenological model based on Optical Pumping (as understood by physicists) in the visual system of birds and possibly other creatures. The model as outlined appears to provide a qualitative explanation of much of the existing data, and discussion centres on the possibilities for putting it to direct experimental test.

In a recent letter (Leask, 1977) a suggestion was made that geomagnetic field detection in biological systems might proceed via a process based on optical pumping, as an adjunct to the normal visual process. In this paper I offer a brief review of the basic physics of other more obvious possibilities for detection, firstly in order to show the extent to which they are unsatisfactory, and secondly in order to relate the comparatively unfamiliar concept of optical pumping to more familiar notions of magnetic behaviour.

The extensive literature on magnetoreception in migratory birds and homing pigeons yields the results of many very elegant experiments concerned with a number of different aspects of behaviour. Here, however, we single out those few results which are significant from the viewpoint of basic physics (1) magneto-reception takes place in the bird's head (2) the response to field is axial, rather than polar - i.e., interchanging N and S poles cannot be detected (3) the response to field is best at values about equal to the earth's field, but falls off at both lower and higher values of field (Wiltschko and Wiltschko, 1972; Walcott and Green, 1974). One further requirement of any fundamental mechanism, if the magnetoreception is to be useful orientationally, is that the receptors should be in ordered array, and that the response of individual receptors should be anisotropic with respect to field direction.

Confining the discussion to magnetic fields that are spatially uniform over volumes large compared to the biological system in question, the physics of the interaction between a moving electric charge and a flux of induction (magnetic field) B is enshrined in the Lorenz force:

$$F = q\,(v \wedge B)$$

where q is the electric charge moving with velocity v. Of relevance also is Faraday's law of induction, which follows from one of Maxwell's equations:

$$\mathrm{Curl}\ E = -\dot{B}$$

and gives rise to an emf when B is time-dependent. These two related effects lead us to *macroscopic models* of magneto-recpetion. Firstly we consider a length of conductor (nerve fibre perhaps) of suitable macroscopic dimensions - length ~ 1 cm - within the organism. Movement at 1 m s^{-1} through the geomagnetic field would produce $\sim 10^{-6}$ V, and a maximum when the directions of movement, field and conductor axis were mutually perpendicular. That is, the response would display the required anisotropy, but it would be polar rather than axial, and would increase linearly with field strength and with velocity. The strongest evidence against this model comes from the experiments (Walcott and Green, 1974) in which an additional vertical field was produced in coils on the bird's head. On this induction model the bird should have been unaffected by this field, since it was always stationary with respect to it. A second *induction model* possibility is to consider a loop of nerve fibre - ~1 cm^2 area - producing an emf (and anisotropic response) by transient rotation of the loop about an axis normal to the field direction. Such emfs would however be much smaller, $\sim 10^{-8}$ V.

We next consider a *microscopic induction model*, more usually reffered to as *diamagnetism*. Invariably considered a negligibly small effect, the diamagnetism of planar molecules can be regarded as a consequence of the Lorenz force acting on the continuously moving electrons in all atoms and molecules. For a planar molecule, the effects of this force would be highly anisotropic and also axial, which is promising, but the energy change in a diamagnetic system (and therefore presumably the response) increases as the square of the magnetic field strength.

Next we consider a *paramagnetic model* of magnetoreception. This is in fact a rather unlikely possibility at the outset, since paramagnetism depends on atoms or molecules possessing a permanent magnetic moment by virtue of unpaired electron angular momentum (usually spin angular momentum). Few molecules of biological significance are paramagnetic; the two obvious ones are oxygen and haemoglobin, and the requirement for an ordered array of receptors effectively rules out the latter. However, paramagnetic oxygen merits consideration, and we consider paramagnetism here firstly for this reason, and secondly because it is basic to an understanding of the optical pumping mechanism which we shall come to finally.

Paramagnetic atoms or molecules posses a permanent magnetic moment due to the angular momentum of the electric charges. One is therefore led to expect that the behaviour of each such magnetic moment would be similar to that exhibited by a compass needle; namely, that the application of a magnetic field would cause the magnetic moments to align along the field direction. This is a misleading analogy, for the reason that the molecule possesses angular momentum, as well as magnetic moment. The immediate effect of a magnetic field applied to such moments is not to produce alignment at all; it is rather to cause precession of the individual moments about the field direction. The analogy is more that of a spinning top in a gravitational field, and the precession frequency is simply related to the field strength, as follows:

$$\omega_p = (e/m)B$$

where e,m are the charge and mass respectively of the electron. The moments align with the field because this is a state of lower energy, and they are able to transfer energy to vibrational modes, which are themselves thermally excited. This energy transfer is reciprocal, in the sense that thermal excitation tends to disrupt the state of alignment, and the degree of alignment can be calculated from a comparison of the thermal energy (~kT, where k is Boltzmann's constant and T is the absolute temperature) and the alignment energy due to the field ($\sim \beta B$, where β is the Bohr magneton). For a typical moment in the geomagnetic field, kT is approximately 10^7 times larger than βB, so that the degree of alignment is vanishingly small. This is the first major problem facing such a model. Others are (2) the degree of alignment increases linearly with field strength, and so presumably should the response (3) O_2 molecules have a spin triplet ground state (S = 1), whose response to field would certainly be isotropic, and therefore of no use directionally. (4) the response to field is polar, rather than axial as observed experimentally. However, the point should be made here (as also for the induction models discussed earlier) that the next stage of detection in the biological system could always convert polar response into axial response by detecting the signal modulus. On the other hand it might be surprising that nature should so readily discard orientational input of such obvious value to the biological system. One might consider however that axial response, being independent of N - S polarity, would certainly have conferred survival advantage during the geomagnetic field reversals known to have occurred during the geological past. However, we conclude here that none of the models discussed so far provides an explanation of magnetoreception. Indeed, for both the diamagnetic and paramagnetic models, it appears that the problem of detecting the geomagnetic field has been replaced by the problem of detecting the field produced by the alignment of a few molecular magnetic moments, which is much more difficult.

Next we consider the basis of the proposed optical pumping mechanism, namely a *resonance model* of magnetoreception. First we recall that the primary effect of an applied magnetic field on a system of atomic or molecular magnetic moments is to cause precession of the moments at a well-defined frequency ω_p, which is a linear function of field strength. Partial alignment of the moments occurs because the orientational effect of the field competes with the randomizing effect of the thermal agitation. The essential feature of a resonance experiment on such a system (using the term in the physicist's sense) is that the randomizing process may be enormously enhanced by irradiating the system with electromagnetic radiation of the same frequency as ω_p. The partially aligned state is of lower energy than the randomized state, so energy is absorbed from the incident radiation when the frequencies match. Since ω_p depends of field strength, the energy absorption as a function of field strength displays a sharp peak at the appropriate value of field for which ω_p matches the incident frequency. The detection of such absorption peaks is the basis of a very wide variety of experiments under the headings nmr (nuclear magnetic resonance) epr (electron paramagnetic resonance) endor (electron-nuclear double resonance), to name a few.

It is not obvious however that this process is likely to be useful in magneto-reception, since, as already pointed out, nearly all molecules of biological interest have zero magnetic moment in the ground state (i.e., a singlet state S = 0). For this reason we consider (Leask, 1977) the lowest energy triplet state (S = 1) which does have a moment, but which, being an excited state of the molecule, is not normally populated. In order to achieve magnetoreception via the triplet there are a number of requirements (1) the three triplet sub-levels must be populated, presumably by excitation transfer from the excited singlet (broadband) levels normally reached by transitions from the singlet ground state (2) the populations thus achieved must be dissimilar (3) the effect of the geomagnetic (or any other) field on the sub-levels must be anisotropic, in order to achieve a directional response (4) there must exist a local well-defined frequency source in the MHz range, which will come into resonance with the triplet sub-levels and cause population equalization between them.

A wide variety of experiments on a number of different molecules have shown that conditions (1) - (3) above may be achieved under laboratory conditions (El-Sayed, 1974; Moore and Kwiram, 1974). There is no direct evidence at all for (4), but it is possible in principle. The attractive feature of the *optical pumping model*, as just outlined, is the possibility that the excited triplet state resonance may be detected, not directly by the absorption of energy as in nmr or epr, but indirectly by monitoring (in the next stage of the process) the decay rate from the individual triplet sub-levels back to the ground state. For radiative decay, the individual rates may in principle be distinguished from each other because the decay is in opposite senses of polarization. Figure 1 attemps to portray the main features outlined above.

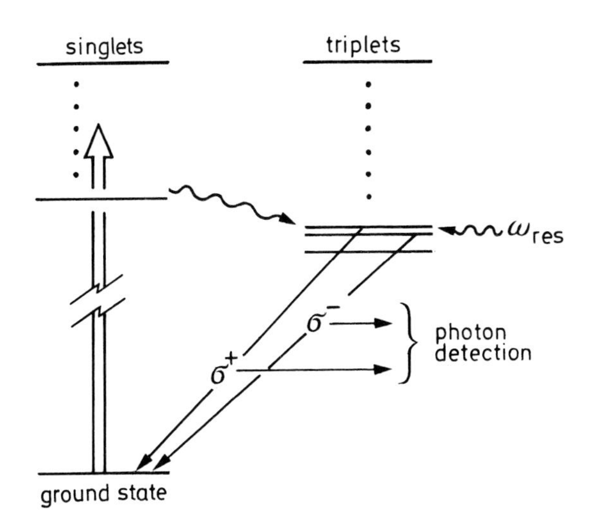

Fig. 1. Schematic diagram of optical pumping process in a molecule having a singlet (S = 0) ground state. Photon absorption is a broad-band process to the singlet excited states. Excitation transfer to triplet states (S = 1) is via (anisotropic) spin-orbit interaction, and splittings shown in lowest triplet state are due to spin-spin (dipolar) interaction

The optical pumping model is therefore of potential interest since it transforms the problem of magnetic moment or field detection into one of photon detection, which is very much simpler. Being a resonance process, the response of such a system will be a maximum at a particular value of field, and will fall off at both lower and higher fields. There are a number of other respects in which the model is at least qualitatively in accord with the experimental data; (for further details see Leask, 1977).

However, it must be admitted that in various matters of detail the model faces considerable difficulties. For instance, the parameters D and E which describe the anisotropy in the triplet state are typically two orders of magnitude larger than one would like, though one makes no prediction at present as to which molecules are likely to be involved in the process. The essence of the difficulties is simply that a resonance process is dynamic, and depends for success on a whole host of pumping rates, relaxation rates and the like being of just the right magnitude. Prediction of such factors is difficult, if not impossible. Fortunately the model is much more directly testable, in that the optical pumping process requires incident radiation (possibly visible light) and the simple issue to be decided by experiment is whether or not light (or perhaps even IR) is essential to the process of magnetoreception.

It would be premature to investigate possibilities for the more detailed physicochemical requirements of the model until this rather direct question has been settled one way or the other. Even if visible light (or IR radiation) is found to be unnecessary, however, it does not follow that the entire model need be discarded immediately. The case for magnetodetection in migratory birds and homing pigeons being a resonance process is reasonably strong, and it might be useful to draw a distinction between the resonance process itself, and the means whereby it is set up at the molecular level. Optical pumping is an obvious initial suggestion, and has been discussed here in some detail, but one could presumably begin to look for chemical excitation processes that might involve a resonance condition at some intermediate stage. In a more general sense, the resonance would then be viewed as a "points-switching" operation somewhere in the midst of a chemical process involving energy transfer.

References

El-Sayed, M.A.: Double resonance techniques and the relaxation mechanisms involving the lowest triplet state of aromatic compounds. Excited States 1, 35-77 (1974)

Leask, M.J.M.: A physico-chemical mechanism for magnetic field detection by migratory birds and homing pigeons. Nature (London) 267, 144-6 (1977)

Moore, T.A., Kwiram, A.L.: An investigation of the triplet state of flavines and flavoproteins by optical detection of magnetic resonance. Biochemistry 13, 5403-7 (1974)

Walcott, C., Green, R.P.: Orientation of homing pigeons altered by a change in the direction of an applied magnetic field. Science 184, 180-2 (1974)

Wiltschko, W., Wiltschko, R.: Magnetic compass of European robins. Science 176, 62-4 (1972)

Session VI

Chairman:

JAMES T. ENRIGHT, Scripps Institution of Oceanography,
POB 109, La Jolla, CA 92093, USA

Directional and Discriminatory Responses of Salamanders to Weak Magnetic Fields

John B. Phillips and Kraig Adler, Section of Neurobiology and Behavior, Cornell University, Langmuir Laboratory, Ithaca, NY 14853, USA

Introduction

A number of studies now exist pointing to the use of magnetic information in vertebrate orientation especially for birds, but the evidence among lower vertebrates is difficult to interpret. Several studies with fish have demonstrated a response to magnetic fields (e.g., Branover et al., 1971; Ovchinnikov et al., 1973; Andrianov et al., 1974; Tesch, 1974). However, in an aquatic environment electrical events can be induced by a body moving through a magnetic field and, therefore, it is not clear whether magnetic perception is accomplished directly or indirectly through induction. Several fish have been shown to be sensitive enough to electrical stimuli to be able to perceive such induced fields (reviewed in Fessard, 1974) even in some species which lack specialized electroreceptive organs (Rommel and McCleave, 1972, 1973). In studies with the cave salamander Phillips (1977) provided evidence for a learned directional response to the earth's magnetic field. Since this species is terrestrial, it seems more likely that the magnetic field is perceived directly. The present paper considers an additional source of variation observed in the original data from cave salamanders and also provides supporting evidence for discriminatory responses to weak magnetic fields in a second species of salamander.

A. Orientation in Cave Salamanders

The procedures used in the experiments with cave salamenders (*Eurycea lucifuga*) have been dealt with more extensively elsewhere (Phillips, 1977); only the more salient features will be mentioned here. Two groups of adult salamanders (15 in each) collected from four caves were confined in two separate training corridors (Fig. 1). Each corridor consisted of two compartments filled with pieces of limestone, connected by a darkened central passageway. Salamanders were trained by alternately supplying moisture to one of the two end compartments of each corridor, at 2-day intervals, forcing the animals to move bidirectionally through the central passageway. During training, the two corridors were aligned along the same topographical axis. The *a* corridor was enclosed in a cubically shaped coil which rotated the magnetic field 90° clockwise. This altered field resembled the natural ambient field in inclination and total intensity (see Phillips, 1977 for exact measurements). The *b* corridor was not enclosed within a coil. The movement of the A group was perpendicular to the magnetic N-S axis as a consequence of the shifted field, while the movement of the B group was parallel to the N-S axis of the earth's magnetic field.

During testing, the two groups were released simultaneously in the center of a cross-shaped testing assembly made up of the two training corridors (Fig. 2). This assembly was positioned within the cube coil

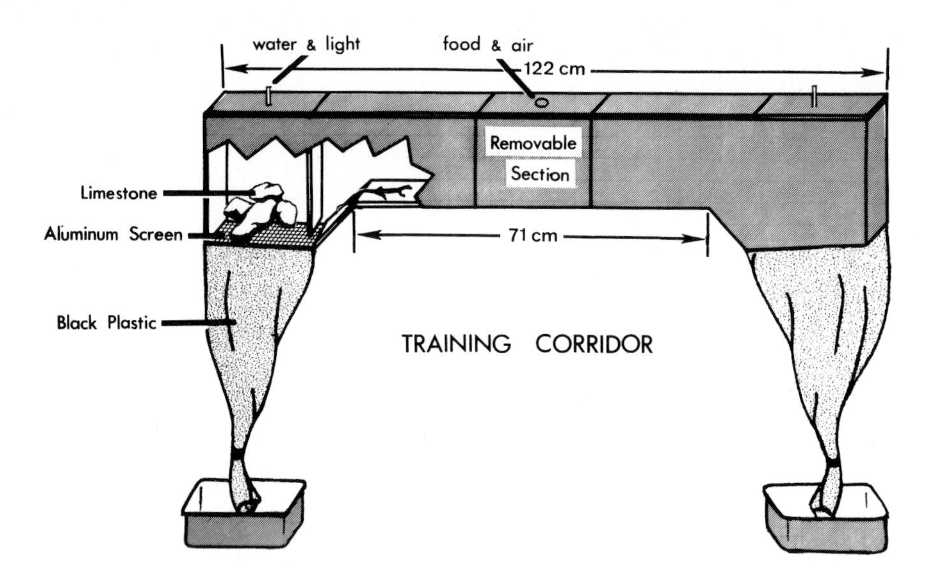

Fig. 1. Training corridor. Cut-away area shows the interior of one of the compartments located at each end. Glass tubes extending through the lid of each compartment permitted a small amount of light to enter and made it possible to moisten the rocks without opening; ventilation of the corridors with compressed air and the provision of food were carried out through the larger central opening. The removable section in the middle of the corridor was taken out to combine the corridors into the cross-shaped testing assembly illustrated in Figure 2 (from Phillips, 1977)

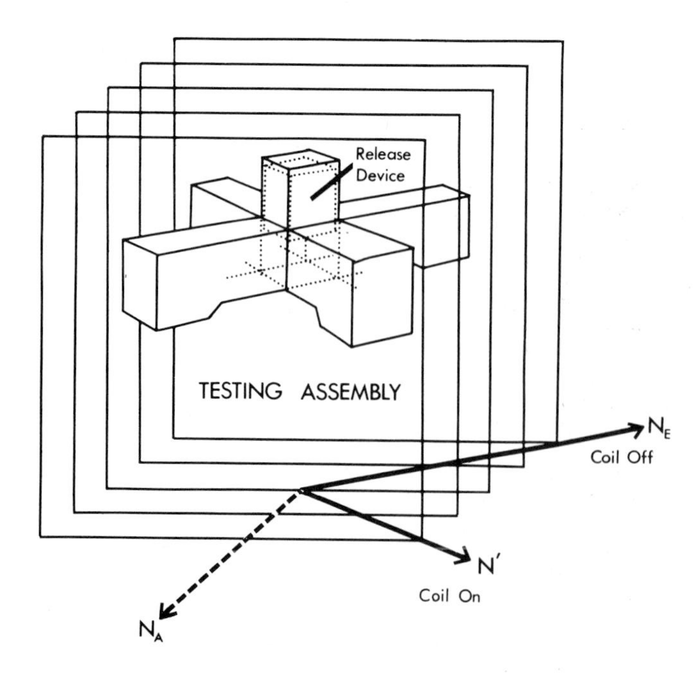

Fig. 2. Testing assembly inside the magnetic coil. A light-tight housing containing the release device fit into the central opening formed by superimposing the two training corridors. The release device was raised by means of a string (not shown) which extended out of the top of the housing. The relative orientations of the geomagnetic (N_E), artificial (N_A) and resultant or altered (N') horizontal magnetic vectors are shown (from Phillips, 1977)

for all tests. A release device placed in the central area common to both corridors was enclosed in a housing to exclude light. The apparatus was covered with three layers of opaque plastic and, in the final ten tests of the series, the glass tubes in the tops of the four compartments were also covered. It is unlikely, therefore, that light was available to the subjects inside the apparatus at least in these final tests. However, the animals were placed into the release device under dim indirect illumination, and thus were given a brief (2 - 5 min) exposure to light while in the test field. Once in the release device, the two groups of salamanders were confined together for 60 min before release, and were then allowed 40 min of free access to the interior of the cross assembly. At the end of this period, the location of each individual found in one of the compartments was recorded. Animals were returned to training after each test.

During tests the cross assembly was positioned with either the *a* or *b* corridor parallel to the "topographical" axis, along which the corridors were aligned in training. In each of the two positions of the apparatus, the salamanders were tested in both the natural and altered magnetic fields. The resultant four testing formats varied the orientation of the magnetic field with respect to the testing assembly, and to the surroundings (Fig. 3, inset diagrams).

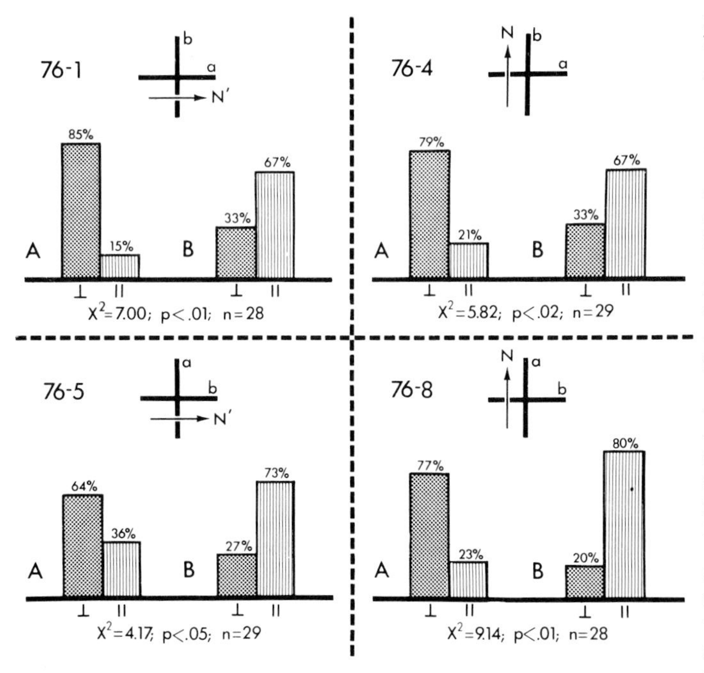

Fig. 3. Results of tests with salamander groups *A* and *B*, previously trained in corridors *a* and *b*, respectively. These tests represent the four possible combinations (*insets*) of the expected magnetic directions relative to the home corridors and the topographical axis. The histograms show the percentage of each group of animals which moved perpendicular (⊥) and parallel (||) to the natural or altered field. For example, in test 76-1 animals in group A, which were originally trained to move perpendicularly to the magnetic field, do so preferentially (85%) even though the corridor perpendicular to the magnetic axis (*b*) is not their home corridor; this preferred movement also coincides with the topographical axis (represented by the vertical axis in each inset). For statistical purposes, the combined number of A- and B-group animals which moved along the expected axis is tested (chi-square, one-tailed, one degree of freedom) against the null hypothesis of a 50:50 distribution of both groups along the two perpendicular axes (from Phillips, 1977)

Results. Sixteen tests were carried out with at least one week intervening between tests. In six tests, the two groups were significantly different in their choice between the two corridors; movement in either direction within one of the corridors was considered to be equivalent due to the bidirectional nature of the training. The difference in the distributions of the two groups was statistically significant (0.05 level in one test, 0.02 level in two, and 0.01 level in three) using a test for the difference in the proportion of each group which moved along the axis. In all six tests in which the distributions were significantly different, the majority of each group moved along the axis predicted from training with respect to the magnetic field, viz., perpendicular to the magnetic N-S axis for the A group and parallel to the magnetic N-S axis for the B group (Fig. 3). Figure 3 presents the results of four of the six tests which occurred during the final ten tests of the series and which represent each of the four testing formats. A comparison among these tests reveals that the distributions of the two groups with respect to other sources of alternative orientation cues (the testing apparatus or the surrounding room) were significantly different when the relative orientation of the magnetic field to either of these potential sources was shifted ($p < .001$, test for the difference of two proportions, one tailed). In every case the animals' distribution shifted in the manner predicted by the orientation of the magnetic field only. In one additional experiment the relative orientation of the two training groups was not significantly different ($p < 0.1$); however, the distribution of both groups was significantly different ($p < .005$) from that which occurred in other tests in which the relative orientation of the field to sources of nonmagnetic cues was altered. Again, the movement of the animals corresponded to the orientation of the magnetic fiel

Discussion. Although these results suggest that the cave salamander can learn a directional response with reference to the earth's magnetic field, the fact that this response was not evident in a majority of the tests remains to be fully explained. For reasons discussed in greater detail elswhere (Phillips, 1977), it seems likely that the A group was influenced in some tests by familiar vibrations which were present during testing, since all tests were carried out in the location in which the *a* corridor was positioned during training. In three tests, a majority of the A group oriented along the topographical axis in opposition to the expected axis of movement with respect to the magnetic field. This response was not evident in the B group in comparable tests, and did not occur in either group during the first five tests, which were conducted in a location which did not have major sources of vibrations present.

Another possible source of variation which could account for some observed variability is an influence of normal disturbance in the earth's magnetic field. A number of investigators have provided evidence that very small fluctuations in the earth's field of less than 100 gamma (1 gamma = 10^{-5} gauss) of both natural and artificial origin influence the orientation of birds (Keeton et al., 1974; Southern, 1971, 1975). In a reexamination of the cave salamander data with respect to indices of natural magnetic variation, a deterioration in the performance (expressed as the percentage of correct responses in the trained magnetic direction in each test) of the B group was found to be associated with increased disturbance in the horizontal component of the earth's field (Fig. 4). There is a significant ($p < .01$, one-tailed) inverse correlation with the magnitude of disturbance (negative DST values represent a decrease in horizontal intensity). The DST index used in this comparison (recorded at the Fredricksburg, Virginia station) provides hourly values in gamma of deviations in intensity of the horizontal component of the earth's field. The mean of the 8-hourly values prior to each test was determined empirically to produce the best fit. Although this result is clearly tentative, it may be of significance that the correlation occurs in the B group which was trained in the

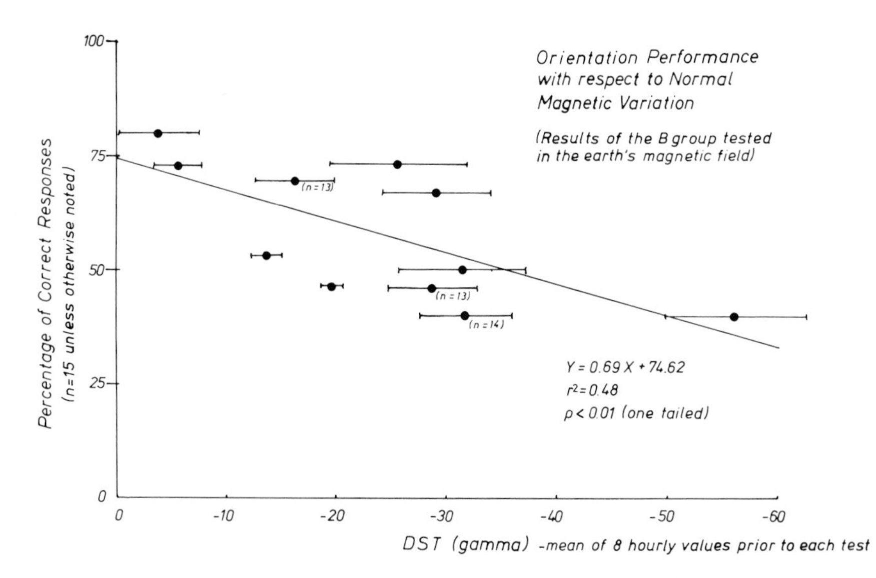

Fig. 4. Performance of the B group tested in the earth's magnetic field in relation to natural magnetic disturbance. Each point along the ordinate represents the percentage of salamanders (n = 15 unless otherwise specified) scoring in the predicted directions in a single test. Values on the abscissa are the mean hourly DST values (representing disturbance in the horizontal component of the earth's field) for the 8-h period prior to each test. Horizontal bars indicate one standard deviation on either side of the mean, to give an indication of the amount of variation in the hourly values. A probit transformation of the data did not alter the value of the regression statistic and so the untransformed data are presented

earth's magnetic field, and in tests carried out in the same field. A statistically significant correlation to DST variation was not present in the A group. In tests of the B group in the natural magnetic field, the animals were not subjected to the transition between the natural and artificial magnetic fields. This change in horizontal intensity along the earth's N-S magnetic axis is several orders of magnitude greater than that of the natural magnetic disturbance being considered. Note, however, that the total intensities are the same in the natural and altered fields.

A longer series of tests now in progress should help to substantiate whether an effect of magnetic disturbance is present. In any event, it seems likely that it will be possible to understand the principle sources of variation and to obtain a predictable group response to earth-strength magnetic fields.

B. Magnetic Discrimination in Newts

Related experiments were performed with Eastern Red-Spotted Newts (*Notophthalmus viridescens*) to determine whether there was a locomotor or positional response along a magnetic gradient. Six adult male, aquatic-phase newts were collected from a pond (waterward direction from the shore along which the animals were captured was about 120°, or ESE) in Tompkins County, New York. Each was maintained in a clear plastic dish in 2.5 cm of tap water; all dishes were kept in a room in the natural magnetic field (0.48 gauss; inclination 69° below horizontal). Tests were run in the same room in all-glass aquaria measuring 50.5 long by 9.5 wide by 15.0 cm high which were filled with tap water to a depth of 0.5 cm. Each aquarium was aligned along the magnetic N-S axis and enclosed in an opaque surround with a small observation window; the backside of each aquarium was dimly illuminated to permit data-taking. All tests were performed between 0900 and 1400 h, although a given animal was observed only for 90 min. Newts were tested singly, by placing them in the center of the aquarium at the beginning of the test; observations were made every 3 min, beginning 3 min after the animal was placed in the tank. Two kinds of data were recorded at the end of each 3-min interval (1) position of the animal - specifically, its head - along the length of the tank and (2) whether the animal was on the floor or climbing the tank walls. So-called climbing animals simply rested vertically against a wall with the hind legs in contact with the floor. Tanks were always rinsed, dried and re-filled with water for each subsequent animal tested. Every newt was tested four times in each of three conditions described below. The time interval between tests of a given newt varied from 1 to 3 days; all tests were performed during the period 14 July - 1 August (see Table 1).

Table 1. Mean positions of newts under the three magnetic conditions. Six newts (animals A - F) were tested in the sequence noted, with the bar magnet absent (Mag-O) or present at the north (Mag-N) or south ends of the tank corridor (Mag-S). Individual means are corrected to exclude positions when newts climbed on the tank wall and the mean is excluded altogether if the number of data points is less than 10. Climbing animals, i.e., those climbing on the tank walls ten or more times during a given test, are indicated by italics. Mean values are given in cm, north (N) or south (S) of the center of the tank where each animal was released. Animals A and B died of unknown causes on 1 August

Test sequence	Date	A	B	C	D	E	F	Test mean
Mag-O tests								
1	14 July	N18.47	N22.02	*N22.10*	N12.12	N15.85	N 0.84	N15.21
4	18 July	N 9.02	S 4.75	*N 1.17*	*N18.87*	S 2.54	S 1.27	N 3.43
7	21 July	N10.43	N18.77	*excluded*	*excluded*	N 9.22	*excluded*	N12.78
10	27 July	N23.06	*S24.40*	*N13.46*	S22.23	*S20.19*	N 1.35	S 4.83
Mag-N tests								
2	15 July	S 0.43	N 8.46	*N17.04*	N 9.07	N 0.97	N 5.08	N 6.68
5	19 July	N 1.45	N15.75	S13.72	*S 2.06*	*S13.54*	N 1.14	S 4.45
8	22 July	N 6.02	S11.18	N 0.18	S13.64	*N13.49*	*N 5.97*	N 0.15
11	29 July	N 2.79	*S22.10*	*S 6.02*	S23.70	S 3.18	N17.70	S 5.74
Mag-S tests								
3	16 July	N23.55	N 4.75	*excluded*	N16.08	N11.35	N15.06	N14.15
6	20 July	N 6.81	N18.47	*S 3.56*	S 9.96	N16.69	*N 5.46*	N 5.66
9	25 July	N15.32	S18.47	*excluded*	S12.19	*N23.29*	S 0.28	N 1.52
12	1 August	†	†	N13.84	*excluded*	*excluded*	S 8.89	N 2.49

Newts were exposed to three magnetic conditions: (1) magnetless tests (Mag-O) in the natural magnetic field, (2) magnetic north tests (Mag-N) in which a bar magnet was placed horizontally 5 cm beyond the north end of the tank and (3) magnetic south tests (Mag-S) in which the magnet was placed an equal distance from the south end of the tank.[1] Neither

[1] We are indebted to Thomas Quinn for suggesting this arrangement.

magnet could be seen from within the tank. A characterization of the altered fields is represented in Table 2. In both cases the north pole of the magnet was in the direction of magnetic north, and therefore increased the horizontal field intensity. The magnets were centered beyond the end of the tank, and located in the plane of the water in which the animals moved; all measurements were taken in the plane of the magnet along the central tank axis. If the magnets were to have any affect on the newts' behavior, the expectation was for them to choose preferentially the end *opposite* the placement of the magnet, where the field strength and inclination of the field from horizontal approximated the natural field.

Table 2. Characteristics of altered magnetic field used in Mag-N newt tests (bar magnet 5 cm beyond north end of tank). The natural field inside the test room had a total intensity of 0.48 gauss and inclination of the north vector of 69° below horizontal (or dip angle). The characteristics of the Mag-S field were quantitatively the same as those for Mag-N, but sequentially reversed. This produced a total intensity of approximately 10.31 gauss and inclination of a north vector below horizontal of 2.5° at position 25 South; at 25 North these values were 0.49 gauss and 66.8°, respectively

Tank position (cm)	Increase in horizontal intensity due to magnet (in gauss)	Total intensity (in gauss)	Inclination of field (north vector, in degrees)
25 North	10.13	10.31	2.5
	3.84	4.04	6.4
20	1.88	2.10	12.4
	1.06	1.32	20.1
15	0.67	0.96	28.1
	0.44	0.76	36.3
10	0.30	0.66	43.5
	0.22	0.60	49.1
5	0.17	0.56	53.1
	0.13	0.54	56.3
0 Center	0.10	0.53	58.6
	0.08	0.52	61.0
5	0.07	0.51	62.1
	0.05	0.50	63.7
10	0.05	0.50	64.2
	0.04	0.50	65.2
15	0.03	0.50	65.6
	0.03	0.49	66.0
20	0.03	0.49	66.3
	0.02	0.49	66.6
25 South	0.02	0.49	66.8

Results. In the first analysis, the mean position of each newt was averaged from the 30 positions observed during the 90-min test period. The averaged mean positions of the six animals tested four times in each of the three magnetic conditions relative to the midpoint of the tank where each animals was released, were: Mag-0 3.48 cm north (standard error *3.32*; n = 24); Mag-N 0.38 cm south (SE *2.74*; n = 24); Mag-S

8.26 cm north (SE *3.34;* n = 22). A *t*-test comparison of these means revealed a statistically significant difference only between the Mag-N and Mag-S conditions (t = 2.01; $p < .025$, one-tailed test). However, on inspection of each animal's record, it was noted that three-fourths spent varying periods of time in a climbing position against the vertical sides of the aquarium and, moreover, when in this position they were observed to remain stationary. Since such climbing positions *may* decrease the probability of further movement, and not represent a continuing selective choice on the animal's part, it was decided to recalculate the means, but treat any number of consecutive observations of an animal in the same climbing position as a single data point. In addition, after exclusion of such climbing data points, any animal was eliminated having less than ten data points per test on which to base a mean position. These corrected means for each animal are given in Table 1; the overall means for the three magnetic conditions were: Mag-O 5.76 cm north; Mag-N 0.84 cm south; Mag-S 6.48 cm north. A *t*-test comparison of these means indicates a statistically significant difference only between Mag-N and Mag-S conditions (t = 1.78; $p < .05$, one-tailed test).

Since it appeared that the climbing animals contributed proportionately more to the original observed differences, additional analyses of their climbing behavior were performed. As discussed previously, while in a climbing position newts often remained stationary and it is thus inappropriate to treat such data points as equivalent to non climbing positions. Only those newts which were observed climbing on ten or more occasions during a given test were included. Finally, the wall positions for the so-called climbing animals were tallied, and the area of most frequent occurrence was determined. Two comparisons of such animals were made, both using a 2 × 2 contingency table and a Fisher exact probability test. First, a comparison of the Mag-N and Mag-S tests was made to determine whether the two groups differ in the proportions with which they occur in the two halves of the tank. Under the null hypothesis of no difference, the Fisher exact probability was 0.048; thus, the alternative conclusion is drawn that the distributions under the two magnetic conditions (in Mag-N, 2 north end, 5 south; Mag-S, 6 north, 1 south) is different and, by inspection, the frequency of occurrence is greatest in the end of the tank *opposite* the position of the magnet. In the second analysis, a similar comparison was made between the frequency of animals climbing on side walls in the central portion of the tank (up to 22.8 cm from the midpoint) and those climbing in the last 2.5 cm at each end of the tank. For this analysis, a comparison of frequencies of occurrence was made between the Mag-O condition (sides 6, ends 3) and the combined Mag-N and Mag-S conditions (sides 2, ends 12). The Fisher exact probability was 0.016; thus, considering the observed distribution of frequencies, we conclude that when bar magnets were present, the newts significantly preferred the ends of the tank.

A comparison was also made of the overall locomotor activity of the newts in the three magnetic conditions by tallying the cumulative total distances moved between observation periods. Although the overall level of activity decreased significantly during the testing period under all three magnetic conditions, there were no statistical differences in the levels of activity among the three.

Discussion. We conclude from the analysis of these data that newts are capable of perceiving magnetic cues, and prefer positions where the field strength and inclination approximate ambient values. In the comparisons of mean positions, the Mag-O mean was never found to be statistically different from the means under Mag-N or Mag-S, although the actual value was always intermediate between them. This lack of significant difference may be due simply to the short length of the testing tank, which prevented greater discrimination by the salamanders. Our failure to find an increased activity level when the field strength was increased is in marked contrast to similar studies with other freshwater vertebrates (for example, with the eel; Vasil'yev and Gleyzer, 1973); however, since we have used a gradient and not simply a uniformly increased magnetic field, and since the animals could move to a portion of the tank where the field closely approximated the ambient field, our results are not directly comparable.

The apparent existence of magnetic sensitivity both in newts and in cave salamanders suggests that this capability may be present in salamanders generally, since these two species belong to different although phylogenetically closely related families and, perhaps more importantly, these two species exist in radically different habitats. It is not yet known whether magnetic sensitivity is universal among amphibians and vertebrates generally, or to what uses perception of the earth's magnetic field may be put, but these are clearly questions of considerable theoretical and practical interest.

Acknowledgments. We thank Thomas Quinn and Saul Lindauer for performing the newt tests, William T. Keeton for supplying the magnetic disturbance values, Melvin L. Kreithen for aid in characterizing our magnetic gradient and Howard C. Howland for advice with statistics. The research was supported in part by grant NSF BMS-75-18693.

References

Andrianov, G.N., Brown, G.R., Ilyinsky, O.B.: Responses of central neurons to electrical and magnetic stimuli of the ampullae of Lorenzini in the black sea skates. J. Comp. Physiol. 93, 287-299 (1974)

Branover, G.G., Vasil'yev, A.S., Gleyzer, S.I., Tsinober, A.B.: A study of the behavior of cells in natural and artificial magnetic fields and an analysis of its reception mechanism. J. Ichthyol. 11, 608-614 (1971)

Fessard, A. (ed.): Electroreceptors and other specialized receptors in lower vertebrates. Berlin-Heidelberg-New York: Springer, 1974, 295 p.

Keeton, W.T., Larkin, T.S., Windsor, D.M.: Normal fluctuations in the earth's magnetic field influence pigeon orientation. J. Comp. Physiol. 95, 95-103 (1974)

Ovchinnikov, V.V., Gleyzer, S.I., Galaktionov, G.Z.: Features of orientation of the European eel [*Anguilla anguilla* (L.)] at some stages of migration. J. Ichthyol. 13, 455-463 (1973)

Phillips, J.B.: Use of the earth's magnetic field by orienting cave salamanders (*Eurycea lucifuga*). J. Comp. Physiol. A, 121, 273-288 (1977)

Rommel, S.A., Jr., McCleave, J.D.: Oceanic electric fields: perception by American eels? Science 176, 1233-1235 (1972)

Rommel, S.A., Jr., McCleave, J.D.: Sensitivity of American eels (*Anguilla rostrata*) and Atlantic salmon (*Salmo salar*) to weak electric and magnetic fields. J. Fish. Res. Board Can. 30, 657-663 (1973)

Southern, W.E.: Gull orientation by magnetic cues: A hypothesis revisited. Ann. N.Y. Acad. Sci. 188, 295-311 (1971)

Southern, W.E.: Orientation of gull chicks exposed to Project Sanguine's electromagnetic field. Science 189, 143-145 (1975)

Tesch, F.-W.: Influence of geomagnetism and salinity on the directional choice of eels. Helgoländer Wiss. Meeresunters. 26, 382-395 (1974)

Vasil'yev, A.S., Gleyzer, S.I.: Changes in the activity of the freshwater eel [*Anguilla anguilla* (L.)] in magnetic fields. J. Ichthyol. 13, 322-324 (1973)

Orientation of Amphibians by Linearly Polarized Light

DOUGLAS H. TAYLOR and JILL S. AUBURN, Department of Zoology, Miami University, Oxford, OH 45056, USA

Abstract

Indoor tests of Bullfrog tadpoles and the eft and adult stages of Eastern Red Spotted Newts under linearly polarized light indicate that these two species can perceive the e-vector for spatial orientation. Other tests out of doors, under clear blue skies after sunset, indicate that both species can use the polarization patterns of the natural sky for orientation. Further, outdoor tests of Bullfrog tadpoles after sunset indicate that the reception of linearly polarized light does not reside in the eyes, but probably is associated with the pineal complex.

A. Introduction

Recently, our awareness of the perceptual capabilities of amphibians has been greatly expanded; both in terms of the receptors involved and in terms of the cues utilized for orientation. We now know that amphibians possess extraocular photoreceptors (EPOs) that can be utilized for sun-compass orientation (Landreth and Ferguson, 1967; Taylor and Ferguson, 1970; Ferguson, 1971) and that the pineal complex is strongly implicated as the EOP for this response (Taylor, 1972; Justis and Taylor, 1976; Demian and Taylor, 1977; Auburn and Taylor, 1977a; Taylor and Adler, 1977). Indoor tests under polarization filters have indicated that salamanders can perceive the e-vector of linearly polarized light for spatial orientation (Taylor and Adler, 1973).

The ability to perceive polarized light, and to use it in spatial orientation has been demonstrated in invertebrates (reviewed by von Frisch, 1967; Waterman, 1973) and in a growing number of vertebrates, including several species of fish (Dill, 1971; Waterman and Forward, 1972; Forward and Waterman, 1973; Kleerekoper et al., 1973), pigeons (Kreithen and Keeton, 1974; Delius et al., 1976), and as mentioned above, salamanders (Taylor and Adler, 1973). Prior to our present studies, there had been no investigation of amphibian orientation under clear day-time skies when the sun was not in view. Such conditions exist at sunrise and sundown, times when another potential celestial cue, linearly polarized light, is prevalent in a clear sky (Rozenberg, 1966).

Here I present evidence that indicates that both Bullfrog tadpoles (*Rana catesbeiana*) and Red-Spotted Newts (*Notophthalmus viridescens*) can perceive linearly polarized light and can use the polarization patterns in the clear blue sky for compass orientation outdoors. Our studies also indicate that the pineal complex is the probable receptor of linearly polarized light for both indoor and outdoor orientation, at least in Tiger Salamanders and Bullfrog tadpoles (Adler and Taylor, 1973; Auburn and Taylor, 1977b).

B. General Methods

The studies reviewed here can be conveniently divided into four groups according to the methodology utilized (1) indoor training and testing under polarization filters (2) animals that were either captured along their home shores or trained outdoors to a particular compass course, and then tested *outdoors* (3) animals that were either captured along their home shores or trained outdoors to a particular compass direction, and subsequently tested *indoors* under polarization filters (4) animals trained outdoors, subjected to various operative procedures designed to localize the polarized light receptor and then tested outdoors.

Data Analysis. When there was no predicted direction, scores were analyzed by the Rayleigh test (Batschelet, 1965) which tests for non-uniformity of the distribution of points around a circle by means of the test statistic z. In tests where there was a predicted direction, the V test, resulting in the test statistic u (Durand and Greenwood, 1958; Batschelet, 1972), was used to test for the clustering of points around that direction. In cases where a bidirectional centrally symmetric response was expected (most tests in Expt. 1), the angle of each score was doubled in order to convert the scores to a unimodal distribution (Durand and Greenwood, 1958; Batschelet, 1965, and pers. comm.) before applying the above-mentioned analyses. When angles were doubled before analysis, the data shown in the figures are the untransformed (not doubled) points. The "mean direction" in such figures is the mean vector calculated from the transformed data, but converted back to the corresponding bidirectional bearing. Its length in either direction, relative to the radius of the circle, is the length of the mean vector based on the transformed data.

I. Experiment 1: Indoor Training and Testing

Methods. Tadpoles were trained indoors in the apparatus shown in Figure 1A. They were housed in black plastic training tanks (45 × 24 × 20 cm) with a horizontal plastic shoreline (24 × 18 × 15 cm) at one end. Water

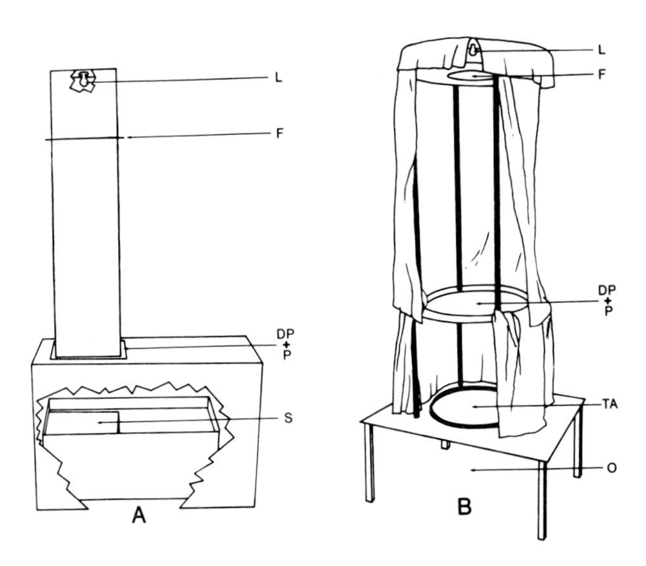

Fig. 1. Indoor training (A) and testing (B) apparatus. Symbols are as follows: *L*, light source; *F*, Fresnel lens; *DP*, depolarizer; *P*, polarizer; *S*, shorelines; *TA*, test arena; *O*, observer's position

was kept in the tank at a level such that it just covered the shoreline (1 cm deep on the shoreline, 16 cm deep away from the shoreline). A black wooden box paced around the tank contained a light source (60 W incandescent bulb), a Fresnel lens, a depolarizer (waxed paper), and a polarizer (C-1, 34% transmission, American Polarizers, Inc.), so that polarized light was projected through a circular opening (14 cm diameter) onto the shore of the training tank (11.72 lux, Gossen Panlux photometer).

The polarizer was arranged such that the e-vector of the polarized light was either parallel or perpendicular to the shoreline.

The indoor testing apparatus (Fig. 1B) consisted of a light source (200 W incandescent bulb), Fresnel lens, and large (40 cm diameter) circular depolarizer and polarizer which projected polarized light onto the surface of the test area (40.04 lux, Gossen Panlux photometer). The polarizer could be rotated within its housing to change the bearing of the e-vector in relation to the rest of the apparatus. Unpolarized light was produced by exchanging the positions of the polarizer and depolarizer. The entire apparatus was surrounded by black curtains, and the room was dark during all tests except for the light from the testing apparatus itself. For each test, tadpoles were placed individually in the center of an arena (45 cm diameter × 2 cm high) filled with water to a depth of 1 cm. A cylindrical release device (8.5 cm diameter × 4.5 cm high) held the tadpole in the center of the arena for 30 s. The experimenter raised the relase device, and scored the tadpole at the point at which it crossed the edge of a circle 30 cm in diameter.

Additional tests were made to determine whether the tadpoles, and later the newts, were perceiving the e-vector itself, or were merely responding to the brightness pattern produced by reflection, refraction and scattering of the polarized light, as suggested by Bainbridge and Waterman (1958). The procedure in this test was similar to that of Jander and Waterman (1960) and Forward and Waterman (1973): alternating dark and light quadrants were created around the test arena. Tadpoles trained to move perpendicular to the e-vector were tested in this arena under unpolarized light. Since the areas of maximum brightness produced by the action of polarized light on water are perpendicular to the e-vector, it was expected that these tadpoles would show movement toward the white quadrants in this test if their response to polarized light was due to the perception of intensity differences.

Results. Tadpoles trained to move perpendicular to the e-vector did so when the bearing of the e-vector was different from that during training (Fig. 2A; $u = 3.44 > 3.02$ $p \leq 0.001$). Rotation of the e-vector produced the predicted change in the mean angle of orientation (Fig. 2B; $u = 2.88 > 2.54$, $p \leq 0.001$). Responses under unpolarized light were not significantly different from uniform, in either a bimodal of unimodal manner (Fig. 2C; $z = 0.13 < 2.96$, $p > 0.05$), demonstrating that nothing in the testing apparatus other than polarized light was responsible for the orientation in the previous tests. In the test under unpolarized light with the alternating black and white quadrants (Fig. 2D), responses were not significantly oriented toward either the black or the white quadrants ($X^2 = 0.06 < 2.96$, df = 1, $p > 0.05$). We therefore conclude that these tadpoles can probably perceive linearly polarized light itself, and not just the concomitant intensity differences.

Tadpoles, trained to orient parallel to the e-vector, responded in the predicted direction when tested under polarized light (Fig. 2E; $u = 1.67 > 1.64$, $p \leq 0.05$) and did not orient either bidirectionally

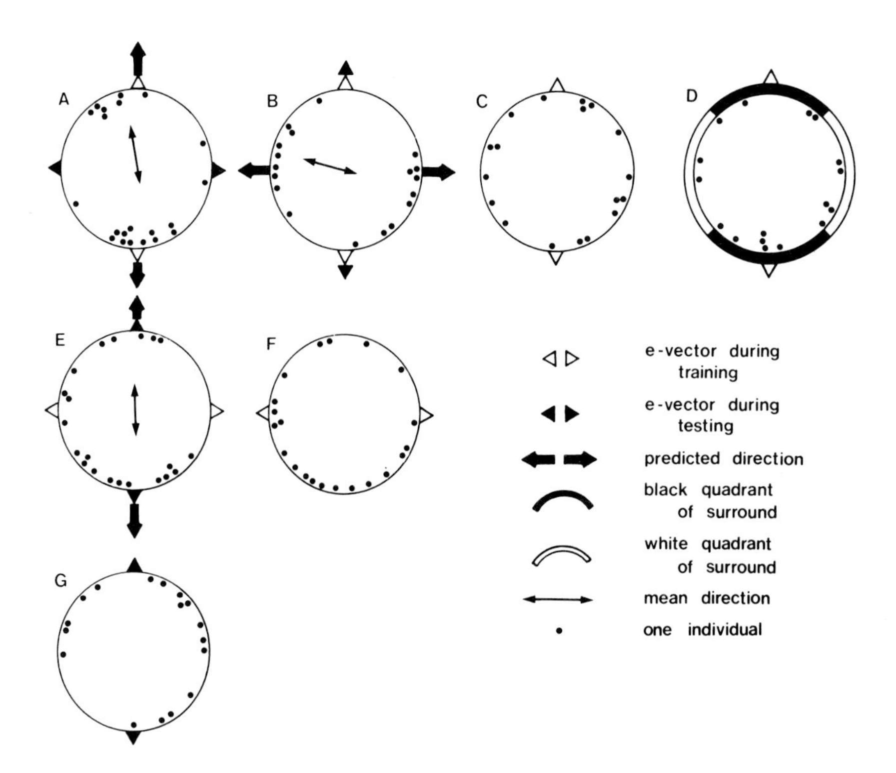

Fig. 2 A - G. Results of experiment 1. (A) through (D) show group I tested under (A), polarized light; (B), polarized light with the polarizer rotated 90^0 for test; (C), unpolarized light; (D), unpolarized light with black and white quadrants; (E) and (F) show group II tested under (E) polarized light and (F) unpolarized light. (G) shows group III tested under polarized light. Mean direction is shown only if corresponding mean vector is significant

or unidirectionally under unpolarized light (Fig. 2F; $z = 0.02 < 2.96$, $p > 0.05$). These results substantiate those obtained with group 1, and demonstrate that the response to polarized light can be modified by experience, an important finding if polarized light is to be useful in Y-axis orientation.

Tadpoles trained under unpolarized light did not orient when tested under polarized light (Fig. 2G; $z = 0.12 < 2.96$, $p > 0.05$), indicating that there is nothing in the training procedure other than the polarized light that could be reponsible for the orientation of the two previously mentioned groups.

II. Experiment 2: Outdoor Orientation

Methods. These studies were performed outdoors at sundown, when the sun was too low on the horizon to be visible. Outdoor training was similar for both tadpoles and newts. Tne animals were housed outdoors in training tanks ($1.2 \times 0.3 \times 0.3$ m) identical to those described in Justis and Taylor (1976). The animals were trained at least three times weekly for periods ranging from two weeks to two months, by removing them from the tank and placing them in the shallow end, from which they could swim back to deep water (tadpoles and adult newts), or in the case of the efts, they were placed in deep water and allowed to move back to shore. The expected direction of movement during outdoor tests was the compass direction that would have taken them either back to deep water or to the shore, depending upon the species and/or life stage, had they been in the training tank.

Testing was performed in a circular arena (3 m diameter) enclosed by an opaque wall which obscured all landmark cues, but allowed a view of celestial cues. During tests, the arena was filled with water to a depth of 2 cm, simulating the shallow water of the shoreline area. Animals were individually placed in a cylindrical release device (8.5 cm diameter × 4.5 cm high) in the center of the arena. After 30 s, the release device was raised from outside the arena, and the animal was allowed to swim free. Each animal was observed from outside the arena through periscopes placed at the N, S, E, and W compass points, and was scored when it crossed the edge of a circle 1 m in diameter. The water was stirred after testing each individual, to control for olfactory cues.

Efts were tested with a completely black surround, and also with white panels placed over opposite quarters of the test arena, either perpendicular or parallel to the outside e-vector, in a manner similar to that utilized during previously described indoor tests.

Results - Tadpoles. Tadpoles showed typical Y-axis orientation toward deep water during the day (Fig. 3A; $u = 3.00 > 2.55$, $p \leq 0.05$) and when tested at sunset under clear sky (Fig. 3B; $u = 1.87 > 1.64$, $p \leq 0.05$).

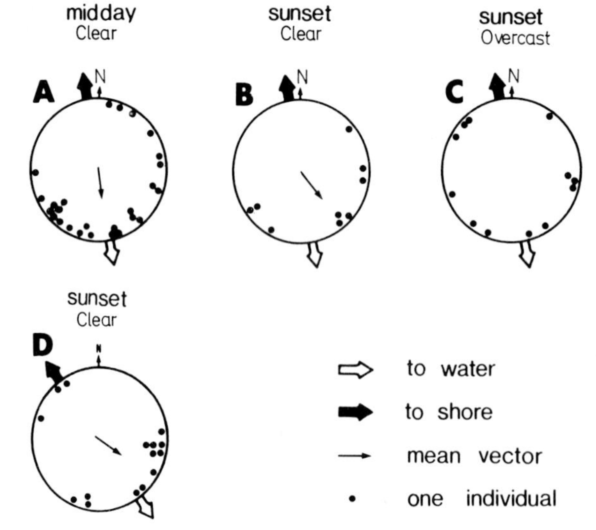

Fig. 3 A - D. Results of experiment 2 - tadpoles. All tests are from tadpoles trained outdoors except for (D) which shows results from tadpoles tested directly from the pond. Mean vector is shown only when significant

Tadpoles tested at sunset under completely overcast skies did not orient (Fig. 3C; $z = 0.17 < 2.96$; $p > 0.05$). This result is consistent with the hypothesis that orientation at sunset is guided by polarized light.

Tadpoles tested directly after collection from the pond also oriented significantly under clear sky at sundown (Fig. 3D; $u = 2.17 > 1.64$,

$p \leq 0.05$), showing that tadpoles living in a pond are capable of such behavior without any artificial training.

Results - Adult newts. A group of newts trained to North and tested outdoors responded as predicted, i.e., in a direction that would have returned them to deep water had they been in their training tank, when they were tested at midday (Fig. 4A; $u = 3.56 > 3.04$, $p \leq 0.001$). A second group of newts responded similarly when tested outdoors after sundown two days later (Fig. 4B; $u = 3.20 \geq 3.04$, $p < 0.001$). Another group of adult newts trained to West also oriented correctly when tested under clear skies after sundown (Fig. 4C; $u = 3.53 > 3.04$, $p \leq 0.001$), but did not orient when tested two days later after sundown under overcast skies (Fig. 4D; $u = 0.25 \leq 1.64$, $p > 0.05$).

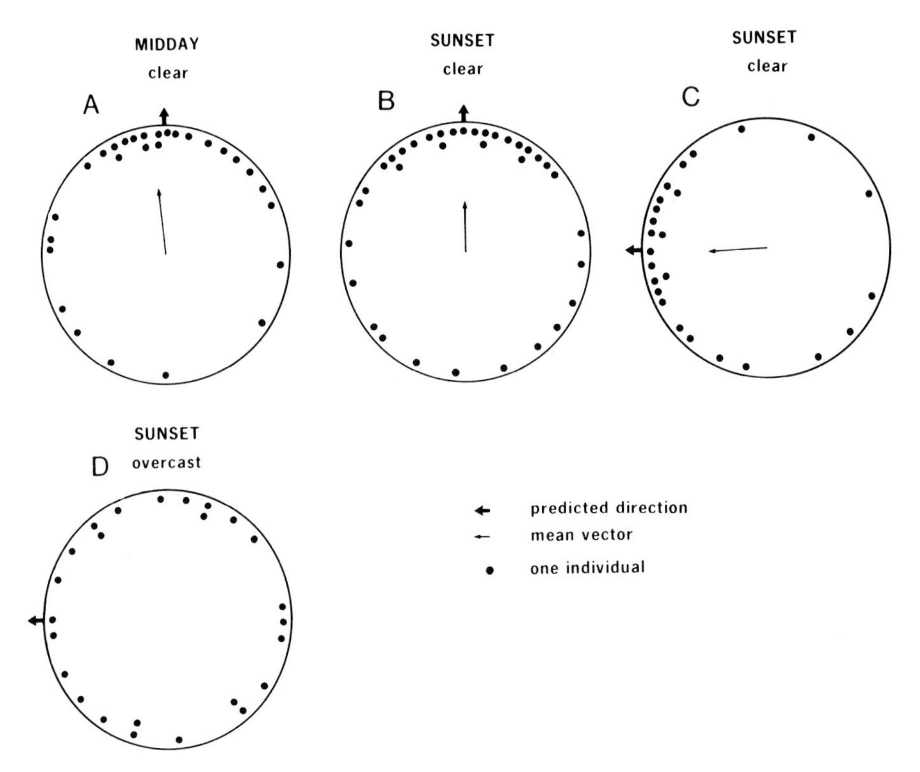

Fig. 4. Results of experiment 2 - adult newts. (A) and (B) trained to North, (C) and (D) trained to West

Results - Efts. A group of efts trained outdoors oriented in the predicted direction when tested outdoors in the circular arena after sundown (Fig. 5A; $u = 4.82 > 3.63$, $p \leq 0.0001$). This same group exhibited random orientation when tested outdoors after sundown under overcast skies one day later (Fig. 5B; $u = 0.43 < 1.64$, $p > 0.05$).

A second group of efts trained in a different direction responded in the predicted direction on two different days when tested after sundown under clear skies (Fig. 5C and D; $u = 4.46 > 3.63$, $p \leq 0.0001$; and $u = 4.01 > 3.63$, $p \leq 0.0001$). These two tests were made with the white panels in place either perpendicular to or parallel with the

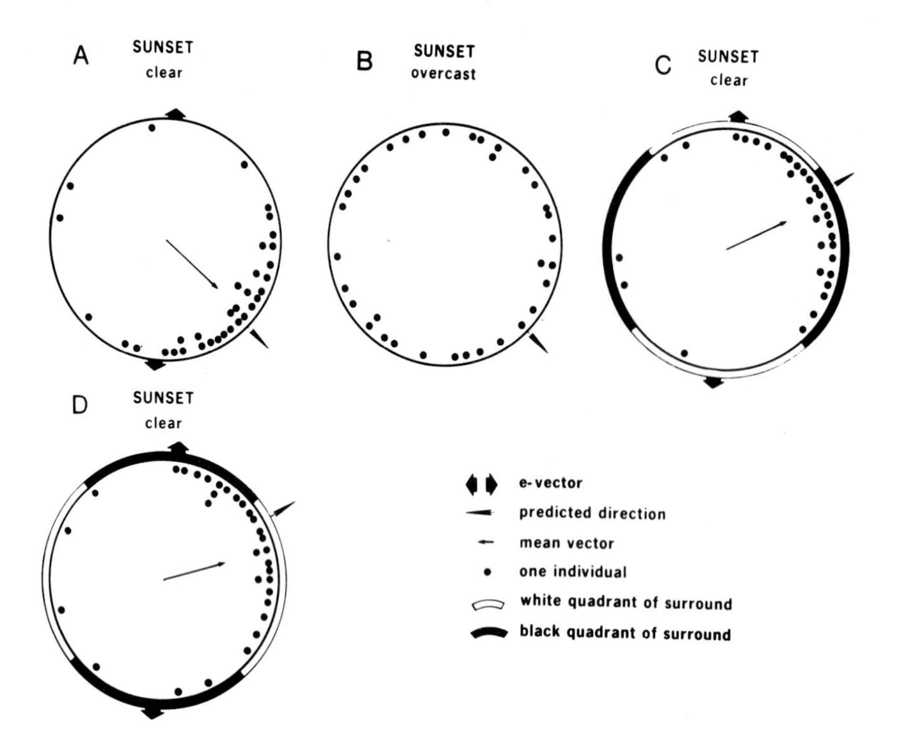

Fig. 5. Results of experiment 2 - efts. (A) and (B) trained to southeast (135°), (C) and (D) trained to northwest (60°)

e-vector. This further indicates that the efts were responding to the e-vector and not just to the brightness patterns associated with linearly polarized light.

III. Experiment 3. Trained Outdoors and Tested Indoors

As shown in Figure 6, animals trained outdoors to move to a compass direction to some angle (Θ) measured clockwise from north, and tested in the evening at a time when the bearing of the predominant e-vector in the sky was 0° - 180° (north-south) would be orienting at an angle Θ clockwise from the primary e-vector. When tested indoors at that time, they would be expected to orient at an angle Θ clockwise from the indoor e-vector, regardless of the compass bearing of the indoor e-vector. The predicted direction indoors is bidirectional due to the symmetry of the indoor polarization pattern.

Methods - Tadpoles. Two groups of tadpoles from two different ponds were tested indoors at sundown, as described in Experiment 1, with the e-vector indoors perpendicular to the e-vector outdoors at the time of the test. The same test was performed with a group of tadpoles trained outdoors, as described in Experiment 2, except with the direction to deep water equal to 80°. An evening test of the same group under unpolarized light served as a control.

Although the results of Experiment 1 showed that tadpoles trained indoors were not merely using intensity differences in their response to polarized light, additional tests were performed on a group trained

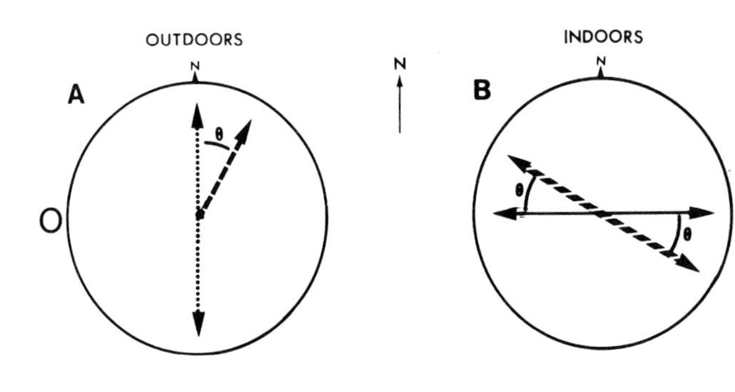

Fig. 6 A and B. Method of determination of predicted direction for Experiment 3. See test for explanation

outdoors. The group trained outdoors and tested indoors in the previous two tests was used for the following two tests. A test with unpolarized light and the black and white quadrants was performed as described in Experiment 1. These tadpoles, which should orient 60° clockwise from the bearing of the e-vector if it were present and they could perceive it, would be expected to orient 60° clockwise from the center of the black quadrants in this test if light intensity differences were the basis of their response to polarized light. This is due to the fact that the areas of minimum brightness produced by the polarized light are centered around the ends of the e-vector. An additional test was performed in which the black and white quadrants were used with polarized light, with the intensity pattern created by the quadrants conflicting with that created by the polarized light (i.e., the black quadrants were perpendicular to the e-vector). In this test, a response based on polarized light could be clearly separated from a response based on intensity patterns.

Results - Tadpoles. Both groups from the ponds oriented in the predicted direction based on the outdoor polarization pattern (Fig. 7A and B; $u = 3.46 > 3.04$, $p \leq 0.001$; and $u = 2.14 > 1.64$, $p \leq 0.05$). The group from the training tank oriented in the predicted direction when tested under polarized light (Fig. 7C; $u = 1.45 > 1.28$, $p \leq 0.10$) but did not orient, either unidirectionally or bidirectionally, under unpolarized light (Fig. 7D; $z = 0.61 < 2.96$, $p > 0.05$). When tested under unpolarized light with the black and white quadrants (Fig. 7E), the tadpoles did not orient in the direction expected if they were perceiving merely a pattern of light intensity produced by the polarized light, nor did they show a preference for either the white or the black quadrants ($X^2 = 1.19$, df = 1, $p > 0.10$). When the brightness pattern produced by the black and white quadrants conflicted with the pattern produced by the polarized light, the tadpoles oriented in the predicted direction based on the polarized light (Fig. 7F; $u = 1.55 > 1.28$, $p \leq 0.10$).

Methods - Adult Newts. Two groups of adult newts were trained to two different compass directions in the outdoor training tanks and then tested in the indoor apparatus. One group was tested with the arena surrounded just with black curtains. A second group was tested first with white quadrants placed perpendicular to the inside e-vector and then with white quadrants placed at the ends of the e-vector. This

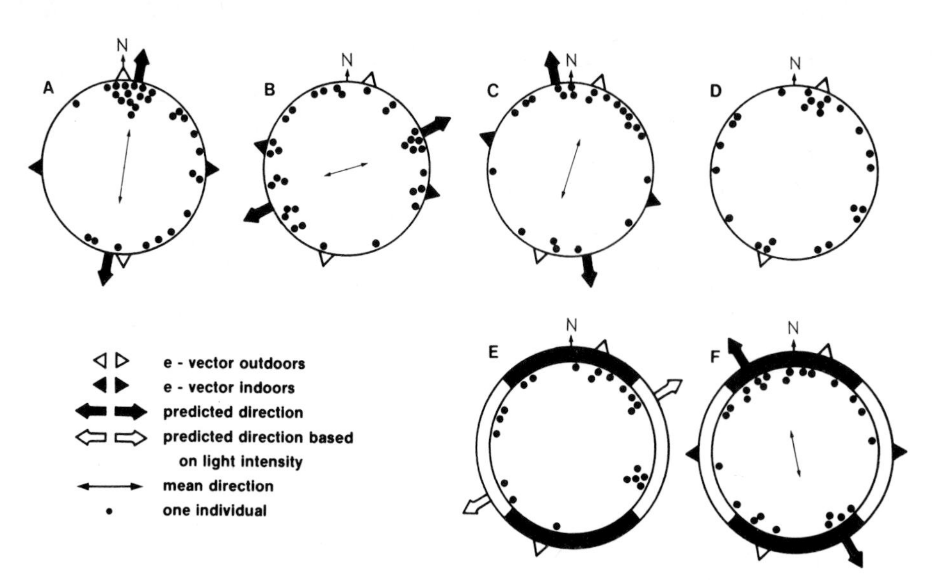

Fig. 7. Results of Experiment 3 - tadpoles. (A) and (B), tadpoles tested indoors under polarized light directly after collection from pond; (C) - (F), tadpoles trained outdoors and tested; (C) under polarized light; (D) under unpolarized light; (E) under unpolarized light with black and white quadrants; (F) under polarized light with black and white quadrants. Mean direction is shown only if corresponding mean vector is significant

second group was also tested with white quadrants in place but under unpolarized light.

Results - Adult Newts. In all three tests under linearly polarized light (Fig. 8A, B, and C; $u = 3.20 > 3.04$, $p \leq 0.001$; $u = 2.41 > 2.31$, $p \leq 0.01$; and $u = 3.16 > 3.04$, $p \leq 0.001$) the newts oriented as predicted relative to the imposed e-vector i.e., they responded to the inside e-vector as if it were the primary e-vector outdoors at that time of day. These results strongly indicate that these animals were responding to the e-vector, and not to the brightness patterns produced as a consequence of the polarization process. This point is further substantiated by the response of the newts tested with white quadrants in place, but with no e-vector present (Fig. 8D; $u = 0.24 < 1.64$, $p > 0.05$), where random orientation was observed.

IV. Experiment 4: Polarized Light Perception

Methods. Thirty-nine tadpoles (*Rana catesbeiana*) were collected on 13 April 1977 and were divided into three equal groups. They were then placed into training outdoors.

Tests were performed after sundown from 7 May to 31 May 1977. Each group was tested twice in each condition, i.e., normal and operated, but no group was tested twice on the same day. Operations were performed in a sequence that allowed all combinations of the three potential receptors (eyes, pineal, frontal organ) to be tested. The tadpoles were placed in ice water for 1 to 3 min before each operation. Eyes were removed by pushing them from their sockets and cutting the optic nerve with scissors. Frontal organs were elctrocauterized. Pineals were covered by inserting a piece of opaque black polyethylene plastic

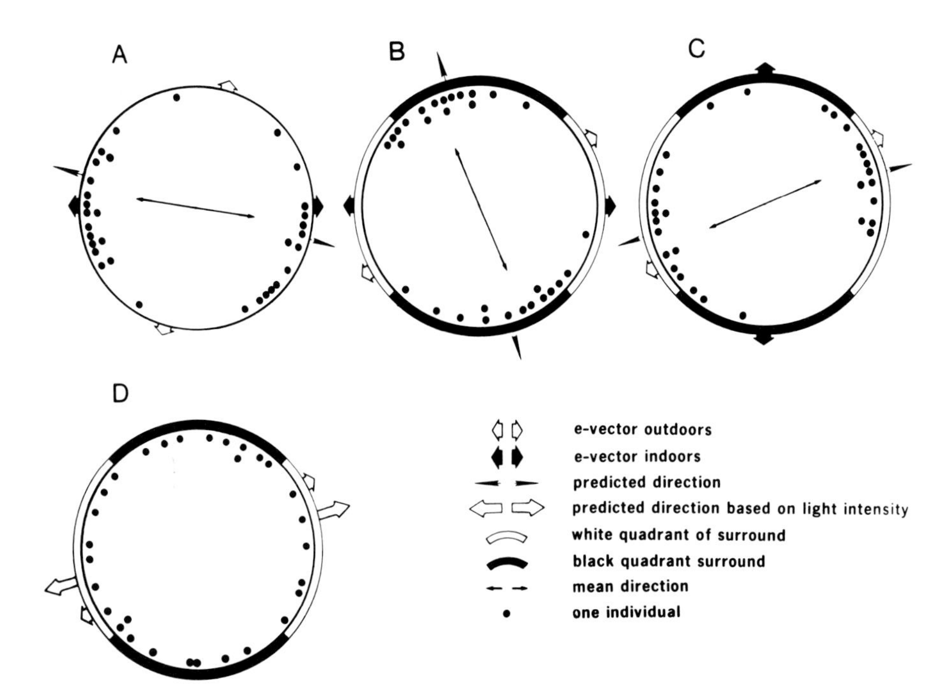

Fig. 8 A - D. Results of Experiment 3 - adult newts. (A) trained outdoors to 15°; (B), (C) and (D) trained outside to 75°

(0.3 × 0.5 mm) into a transverse incision in the skin over the area of the pineal. Frontalectomies and eye removals were done at least 24 h before testing; pineals were covered within 30 min of testing and were uncovered after each test.

The first group was made eyeless, frontalectomized and tested. Following this, their pineals were covered and they were retested. We covered the pineals of a second group and tested them. We then removed their eyes and covered their pineals before retesting them. A third group was frontalectomized and tested prior to covering their pineals and retesting them.

Results. The first group (Fig. 9A-D) oriented in the predicted direction when normal (Fig. 1A; $u = 3.44 > 3.04$, $p \leq 0.001$), when eyeless (Fig. 9B; $u = 2.53 > 2.31$, $p \leq 0.01$) and when eyeless and frontalectomized Fig. 9C; $u = 3.12 > 3.04$, $p \leq 0.001$). Eyeless, frontalectomized tadpoles with covered pineals did not orient (Fig. 9D; $u = 0.98 < 1.64$, $p > 0.05$). The second group (Fig. 9E-G) oriented in the predicted direction when normal (Fig. 9E; $u = 3.38 > 3.03$, $p \leq 0.001$) and with covered pineals (Fig. 9F; $u = 2.70 > 2.54$, $p \leq 0.005$). When these tadpoles were eyeless and their pineals were covered, however, they continued to orient, but in the opposite direction from that which was predicted (Fig. 9G; $u = 2.01 > 1.64$. using predicted direction of 280°, $p \leq 0.05$). This movement toward shore instead of toward deep water has been noted in eyeless, frontalectomized tadpoles tested at midday (Justis and Taylor, 1976), and in some groups of normal tadpoles tested in the evening as in this study (Auburn and Taylor, 1977a), but its significance is not yet known. The third group (Fig. 9H-J) oriented in the predicted direction when normal (Fig. 9H; $u = 3.07 > 3.04$, $p \leq 0.001$) and when frontalectomized (Fig.

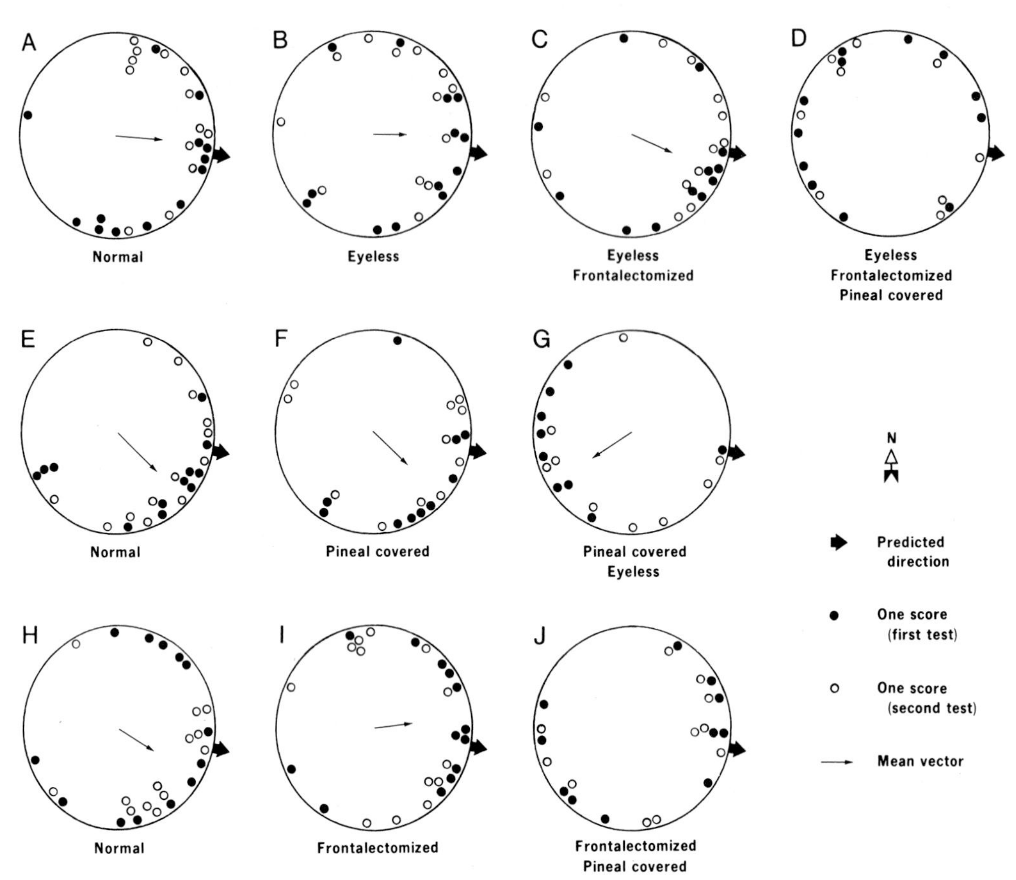

Fig. 9 A - J. Results of Experiment 4 tests. (A-D) first group; (E-G) second group; (H-J) third group. Mean vector is shown only if significant ($p \leq 0.05$)

9I; $u = 2.59 > 2.55$, $p \leq 0.005$), but not when frontalectomized with covered pineals (Fig. 9J; $u = 0.45 < 1.64$, $p > 0.05$).

While covering the pineal region, rather than removing the organ, permits the conclusion that some nearby structure other than the pineal is the relevant receptor, much evidence implicates the pineal alone (Taylor and Ferguson, 1970; Taylor, 1972, Adler and Taylor, 1973; Justis and Taylor, 1976; Adler, 1976; Demian and Taylor, 1977. Covering the pineal rather than removing it has the important advantage of leaving neural connections to the frontal organ intact.

C. Conclusions

The experiments presented in this paper strongly support the following conclusions (1) that both Bullfrog tadpoles and Red-Spotted Newts can perceive the e-vector of linearly polarized light (2) that both species can use skylight polarization patterns for compass orientation and (3) the pineal complex is the probable receptor of linearly polarized light for compass orientation in these amphibians.

These last two points represent a significant advance in our understanding of the sensory mechanism utilized by amphibians for orientation.

Acknowledgments. A major portion of the works reviewed in this paper represent a collaborative effort between one of my students (Jill S. Auburn) and myself. Jill's efforts are gratefully acknowledged. The work was supported by a NSF Grant (GB - 01259) to D.H.T. Travel support to the symposium was graciously provided by Dean C.K. Williamson, Miami University.

References

Adler, K.: Extraocular photoreception of amphibians. Photochem. Photobiol. 23. 275-298 (1976)

Adler, K., Taylor, D.H.: Extraocular perception of polarized light by orienting salamanders. J. Comp. Physiol. 87, 203-212 (1973)

Auburn, J.S., Taylor, D.H.: Polarized light perception and orientation in larval bullfrogs, *Rana catesbeiana*. Anim. Behav. (in press) (1977a)

Auburn, J.S., Taylor, D.H.: Extraocular orientation by skylight polarization in larval bullfrogs. submitted for publication (1977b)

Bainbridge, R., Waterman, T.H.: Turbidity and the polarized light orientation of the crustacean *Mysidium*. J. Exp. Biol. 35, 487-493 (1958)

Batschelet, E.: Statistical Methods for the Analysis of Problems in Animal Orientation and Certain Biological Rhythms. Washington, D.C.: Am. Inst. Biol. Sci. 1965, 57 pp.

Batschelet, E.: Recent statistical methods for orientation data. In: Animal Orientation and Navigation. Galler, S., Schmidt-Koenig, K., Jacobs, G., Belleville, R. (eds.). Washington, D.C.: U.S. Govt Bubl. NASA SP-262, 1972, pp. 61-91

Delius, J.D., Perchard, R.J., Emmerton, J.: Polarized light discrimination by pigeons and an electroretinographic correlate. J. Comp. Physiol. 90, 560-571 (1976)

Demian, J., Taylor, D.H.: The pineal gland: the effective extraocular photoreceptor for locomotor rhythm entrainment in the red-spotted newt, *Notopthalmus viridescens*. J. Herp. 11(2), 131-139 (1977)

Dill, P.A.: Perception of polarized light by yearling sockeye salmon. *Oncorhynchus nerka*. J. Fish. Res. Board. Can. 28, 1319-1322 (1971)

Durand, D., Greenwood, J.A.: Modifications of the Rayleigh test for uniformity in analysis of two-dimensional orientation data. J. Geol. 66, 229-238 (1958)

Ferguson, D.E.: The sensory basis of orientation in amphibians. Ann. N.Y. Acad. Sci. 188, 30-36 (1971)

Ferguson, D.E., Landreth, H.F.: Celestial orientation of Fowler's toad, *Bufo fowleri*. Behaviour 26, 105-123 (1966)

Forward, R.B., Jr., Waterman, T.H.: Evidence for e-vector and light intensity pattern discrimination by the teleost *Dermogenys*. J. Comp. Physiol. 87, 190-202 (1973)

Frisch, K. von.: The Dance Language and Orientation of Bees. Cambridge, Mass.: Belknap Press, 1967, 566 pp.

Jander, R., Waterman, T.H.: Sensory discrimination between polarized light and light intensity patterns by arthropods. J. Cell. Comp. Physiol. 56, 137-160 (1960)

Justis, C.S., Taylor, D.H.: Extraocular photoreception and compass orientation in larval bullfrogs, *Rana catesbeiana*. Copeia 1976, 98-105 (1976)

Kleerekoper, H., Matis, J.H., Timms, A.M., Gensler, P.: Locomotor response of the goldfish (*Carassium auratus*) to polarized light and its e-vector. J. Comp. Physiol. 86. 27-36 (1973)

Kreithen, M.L., Keeton, W.: Detection of polarized light by the homing pigeon, *Columba livia*. J. Comp. Physiol. 89. 83-92 (1974)

Landreth, H.F., Ferguson, D.E.: Newts: sun-compass orientation. Science 158, 1459-1461 (1967)

Rozenberg, G.V.: Twilight. A Study in Atmospheric Optics. New York: Plenum Press, 1966, 358 pp.

Taylor, D.H.: Extraoptic photoreception and compass orientation in larval and adult salamanders (*Ambystoma tigrinum*). Anim. Behav. 20. 237-240 (1972)

Taylor, D.H., Adler, K.: Spatial orientation by salamanders using planepolarized light. Science 181, 285-287 (1973)

Taylor, D.H., Adler, K.: The pineal body: site of extraocular perception of celestial cues for orientation in the tiger salamander, *Ambystoma tigrinum*, submitted for publication (1977)

Taylor, D.H., Ferguson, D.E.: Extraoptic orientation in the southern cricket frog, *Acris gryllus*. Science 168, 390-392 (1970)

Waterman, T.H.: Responses to polarized light: animals. In: Biology Data Book. 2nd ed. Vol. II. (Ed. by Altman, P.L., Dittmer, D.S. (eds.) Bethesda, Md.: Fed. Am. Soc. Exp. Biol., 1973, pp. 1272-1289

Waterman, T.H., Forward, Jr., R.B.: Field demonstration of polarotaxis in the fish *Zenarchopterus*. J. Exp. Zool. 180. 33-54 (1972)

Experimental Evidence of Geomagnetic Orientation in Elasmobranch Fishes

AD. J. KALMIJN, Woods Hole Oceanographic Institution, Woods Hole, MA 02543, USA

Abstract

Marine sharks, skates, and rays are endowed with an electric sense that enables them to detect voltage gradients as low as 0.01 μV/cm within the frequency range of direct current (DC) up to about 8 Hz. Their electroreceptor system comprises the ampullae of Lorenzini, which are delicate sensory structures in the snouts of these elasmobranch fishes. Sharks, skates, and rays use their electric sense in predation, sharply cueing in on the DC and low-frequency bioelectric fields of their prey. Swimming through the earth's magnetic field, they also induce electric fields that may provide them with the physical basis of an electromagnetic compass sense. Their ability to orient magnetically has in fact been demonstrated in recent training experiments.

A. Introduction

Sharks, skates, and rays are endowed with a remarkably keen electric sense that enables them to zero in on the DC and low-frequency bioelectric fields of their prey (Kalmijn, 1966, 1971). Responding to voltage gradients as low as 0.01 μV/cm, these ancient marine fishes have added an electrical dimension to their sensory world. The pertinent receptor system comprises the ampullae of Lorenzini, which are mainly located in the snout, and connect to the outside through small pores in the skin (Murray, 1962; Dijkgraaf and Kalmijn, 1963).

When swimming through the earth's magnetic field, sharks, skates, and rays induce electric fields that depend on the direction in which they are heading (Fig. 1). Since these fields are well within the

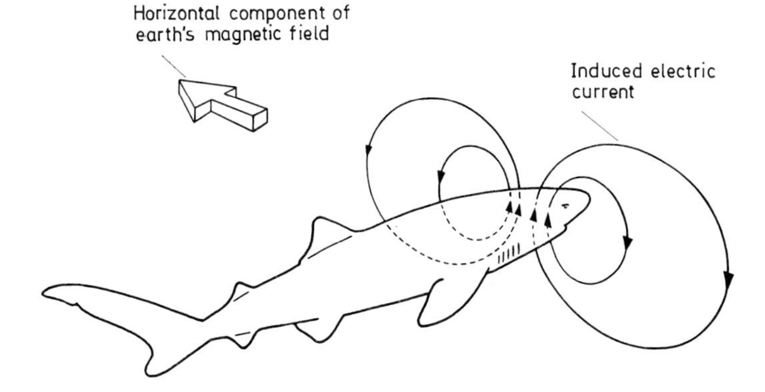

Fig. 1. A shark swimming through the earth's *magnetic* field induces an *electric* field giving the animal its compass heading. (From Kalmijn, 1974)

fishes' sensitivity range, they may form the physical basis of an electromagnetic compass sense (Kalmijn, 1973, 1974). To demonstrate the biologic feasibility of the inferred orientation mechanism, I trained the stingray *Urolophus halleri* to orient with respect to the earth's magnetic field. The present paper reports on the results of these difficult, but eventually successful experiments.

B. Physical Principles

When a shark, skate, or ray swims to the east and thereby crosses the horizontal component of the earth's magnetic field, it generates - according to Faraday's law - an internal, dorso-ventral emf of induction $\int_V^D (\bar{v} \times \bar{B}_h) \cdot d\bar{s}$, with $\bar{v}$ the swimming velocity, $\bar{B}_h$ the horizontal component of the earth's magnetic induction, and $\bar{s}$ any internal path from the ventral (V) to the dorsal (D) surface of the animal (Fig. 2).

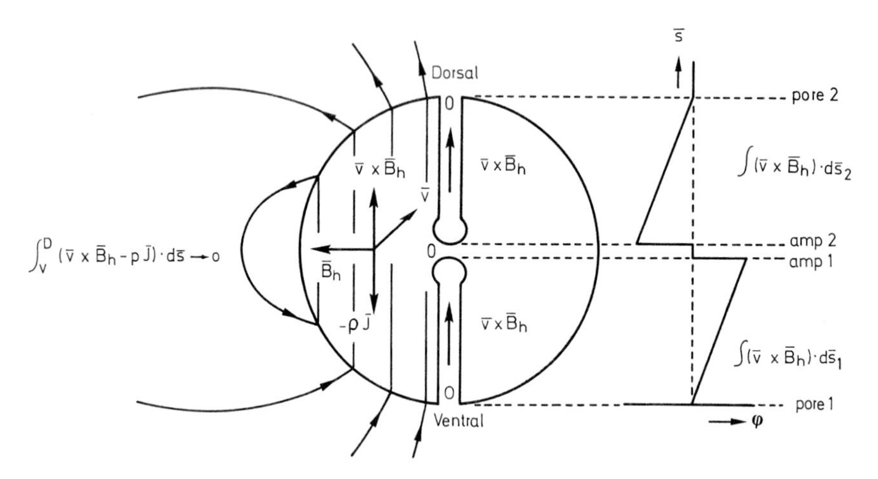

Fig. 2. Cross-section of a shark heading east with a velocity $\bar{v}$ in a magnetic field of horizontal induction $\bar{B}_h$. In the shark, a ventro-dorsal voltage gradient $\bar{v} \times \bar{B}_h$ is induced. This in turn causes an electrical current of density $\bar{J}$ to flow: ventro-dorsally within the *moving* fish and back through the *stationary* environment. With ρ the resistivity along its path, the current develops an ohmic voltage gradient $-\rho \bar{J}$ opposing the induced voltage gradient $\bar{v} \times \bar{B}_h$. Since the fish is virtually short-circuited by the highly conductive seawater environment, the average $-\rho \bar{J}$ of the body tissues effectively counteracts the $\bar{v} \times \bar{B}_h$, and the dorso-ventral potential difference tends to zero. In the blind ampullae of Lorenzini, however, $-\rho \bar{J}$ is negligibly weak. That is, $\bar{v} \times \bar{B}_h$ remains uncompensated and, integrated along the ampullary canals, gives rise to the pertinent electrical stimuli across the sensory epithelia that form the bases of the ampullae proper. (On the *left*, internal and external current field $\bar{J}$; on the *right*, potential distribution ϕ along the ampullary axes $\bar{s}_1$ and $\bar{s}_2$. (After Kalmijn, 1974)

The induced emf gives rise to electrical currents that flow ventro-dorsally through the *moving* fish and loop back through the *stationary* environment (with respect to which the motion takes place). Within the fish, these currents lead - according to Ohm's law - to a dorso-ventral ohmic potential difference $\int_V^D -\rho\, \bar{J} \cdot d\bar{s}$, with ρ the resistivity of the tissues, $\bar{J}$ the current density, and the minus sign indicating that it opposes the induced emf. Thus, the *total* dorso-ventral potential difference (the induced emf *plus* the ohmic voltage drop) equals $\int_V^D (\bar{v} \times \bar{B}_h - \rho\, \bar{J}) \cdot d\bar{s}$. However, because of the relatively large volume and low resistivity of the seawater, the fish is electrically almost completely short-circuited. That is, the ohmic potential difference effectively counteracts the induced emf; $\int_V^D - \rho\, \bar{J} \cdot d\bar{s}$ approaches $-\int_V^D (\bar{v} \times \bar{B}_h) \cdot d\bar{s}$ and the *total* potential difference between the dorsal and ventral receptor pores $\int_V^D (\bar{v} \times \bar{B}_h - \rho\, \bar{J}) \cdot d\bar{s}$ becomes negligibly small. On the other hand, the ampullae of Lorenzini move through the earth's magnetic field with the same velocity as the fish, inducing along their jelly-filled canals internal emf's of $\int_{\text{pore 1}}^{\text{amp 1}} (\bar{v} \times \bar{B}_h) \cdot d\bar{s}_1$ and $\int_{\text{amp 2}}^{\text{pore 2}} (\bar{v} \times \bar{B}_h) \cdot d\bar{s}_2$ for the ventral and dorsal sense organs respectively. Since the ampullae of Lorenzini act as high-ohmic voltmeters (Waltman, 1966; Kalmijn, 1974), there is practically no flow of current through their highly conductive canals, and the emf's induced between the skin pores and the blind ampullary endings develop without appreciable ohmic loss. Hence, across the sensory epithelia forming the bases of the ampullae in Fig. 2, potential differences of nearly $-\int_{\text{pore 1}}^{\text{amp 1}} (\bar{v} \times \bar{B}_h) \cdot d\bar{s}_1$ and $-\int_{\text{amp 2}}^{\text{pore 2}} (\bar{v} \times \bar{B}_h) \cdot d\bar{s}_2$ appear. When the fish turns north or south, the potentials vanish; when the fish turns west (instead of east), potentials of opposite polarity are induced. As at a cruising speed of 1 to 2 m/s the induced voltage gradients may be as high as 0.5 to 1 µV/cm, they certainly would suffice to explain the elasmobranchs' magnetic orientation reported on in the present paper. Note that through interaction with the vertical component of the earth's magnetic field the fish also induce motional electric fields parallel to their transverse body axes. By sensing these fields with the laterally oriented ampullae of Lorenzini, the elasmobranchs may in addition determine the magnetic latitude of their position on the globe. For a more comprehensive discussion of the physical principles involved, see Kalmijn, 1973, 1974.

C. Preliminary Evidence

The first, most simple magnetic tests were performed on the leopard shark, *Triakis semifasciata,* in outdoor, all-fiberglass pools at the Scripps Institution of Oceanography, La Jolla, California. With the fish steadily swimming along the circumference of their circular habitat, my students and I introduced a local magnetic field into the water by passing an electrical current through a small induction coil (20 cm in diameter) held externally to the tank. The field was turned on when the animals were at the far side of the pool. At that distance, the magnetic strength was too low for the sharks to respond, and they quietly continued their lap. Seconds later, however, upon swimming into the region of the coil, the sharks suddenly turned away from the imposed field and veered off to the center of the tank, even though the coil current did not distort the earth's ambient magnetic field by more than 25%.

Next, I noticed that each morning the leopard sharks rested at a particular location along the periphery of the tank in a sector a little off magnetic north. To eliminate the possibility of visual orientation, we covered their 7-m pool with a large sheet of black plastic. We also took all experimental structures out of the water, rotated the whole set up, and even moved it (for other reasons) to another site. Surprisingly, this did not change the sharks' early morning spatial preference. However, when we roughly neutralized the earth's ambient magnetic field with two large coils mounted to the outside of the tank, the animals apparently lost their sense of position and dispersed randomly.

Though consistent with the hypothesis of geomagnetic orientation, neither of these tests was fully conclusive. The avoidance reactions proved the sharks' sensitivity to fields of geomagnetic strength, but were not of obvious biologic significance. The animals' homing tendency was biologically more interesting, but at the time we were not technically prepared to reverse the field in order to randomize any remaining alternative cues, and thus to verify the magnetic nature of the response. These early observations did, however, greatly incite me to pursue this new, promising line of research.

D. Final Experiments

After moving to Massachusetts, I constructed specially designed magnetic facilities on the Quissett campus of the Woods Hole Oceanographic Institution. To scale down the technical problems of controlling the ambient magnetic field, I looked for a good experimental animal of smaller size, which one of my students found in the round stingray, *Urolophus halleri*. The round stingray is a hardy, alert, and very lively elasmobranch of subtropical and tropical seas, reaching an average fin-span of 25 cm. We selected specimens of only 15 to 20 cm to fit them to the size of our tanks. To establish their magnetic abilities, we trained the animals to seek reward and avoid punishment at locations predetermined by the direction of the earth's magnetic field.

The stingrays were tested in a circular fiberglass pool surrounded by a light-tight, twelve-sided hut devoid of any ferromagnetic materials (Fig. 3). The pool measured 1.8 m in diameter, and rested on Teflon blocks to insulate it from ground. It was filled with filtered, natural seawater to a depth of 15 cm. Coarse sand covered the bottom, and two air-stones maintained a slow internal circulation. The seawater was kept at 20°C by regulating the air temperature in the hut. To control the horizontal component of the magnetic field, two north-south-oriented Helmholtz coils, each 5 m in diameter, were erected outside the hut. We chose for our experiments the horizontal induction of the Southern California region from where the animals were taken (0.26 gauss).

During the 1- to 2-h training sessions, twelve concealed incandescent lights illuminated the ceiling over the experimental tank to produce an even, low-level light distribution. For the rest of the day, the lights were programmed to simulate the sunshift and daily variation in the brightness of the sky in accordance with the direction of the magnetic field (either normal or reversed). Lights, heaters, and coils were DC-powered from distantly located voltage and current sources.

Fig. 3. Magnetic test facility on the Quissett campus of the Woods Hole Oceanographic Institution. The two large, north-south-oriented Helmholtz coils control the horizontal component of the magnetic field in the hut. The actual training experiments are conducted in the top tank. The lower one holds extra stingrays in reserve

The wiring was tightly twisted and judiciously installed so as to prevent unwanted electric and magnetic fields from straying into the animals' habitat. Extraneous electromagnetic and vibrational noise levels were low at our test site in the undisturbed woods of the Quissett campus.

Each morning, two observers conducted a series of 10 to 20 trials after first turning on the lights, shutting off the flow of air, and checking the direction of the magnetic field. Then, they simultaneously introduced two circular enclosures into the pool, one in the magnetic East, the other in the magnetic West (Fig. 4). The enclosures consisted of

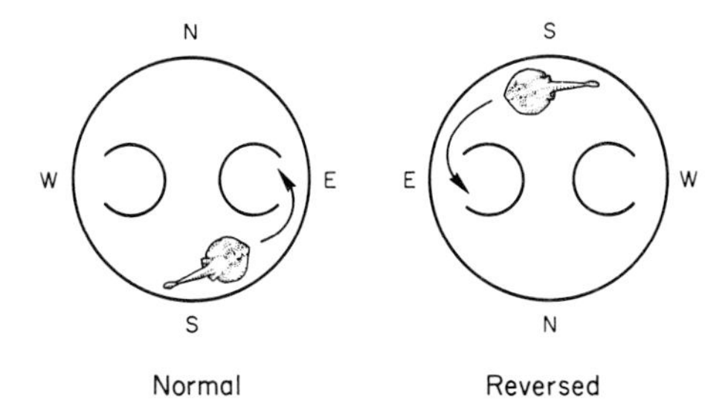

Fig. 4. To receive reward and to avoid punishment, the stingray *Urolophus halleri* ignores the enclosure in the magnetic West to enter the one in the magnetic East. The magnetic nature of the animal's response becomes evident upon reversal of the field: the stingray again enters the enclosure in the magnetic East, though it is now located at the opposite side of the pool

plastic tubs, 30 cm in diameter, with a 20-cm-wide opening near the bottom for the stingrays to enter. After positioning the enclosures at a distance of 25 cm from and with the openings toward the wall of the pool, the observers stepped back to watch the behavior of the animals from behind a black felt screen. When one of the stingrays entered either the east or west enclosure, both observers blocked off the opening of their tub by lowering a gate. If the animal chose the east, by definition the correct enclosure, it was rewarded with a small piece of herring; if the animal took the west or incorrect enclosure, it was gently prodded with a blunt Plexiglas rod as a form of punishment. Eventually, the enclosures were moved to the magnetic North and South of the pool, and the animal was set free for the next trial to start 1 to 2 min later.

It took the animals some time before they learned - or us some time to teach them - to rely on the magnetic field in deciding which side of the pool to avoid and where to go for food. Once conditioned, however, two of the stingrays made the highly significant scores of 56 and 164 correct versus 22 and 84 incorrect choices respectively ($p < 0.001$, calculated with the chi-square, corrected for continuity). A third ray had to be removed, because it could not compete fast enough and was actually losing weight. After the initial training, the stingrays' orientational performances did not significantly change, either during the individual sessions, or from day to day. This, we felt, allowed us to treat the trials as independent choices, despite the fact that the field was reversed only once a day and in a *systematic* order, that is, each night before the next morning's series of 10 to 20 trials. In a follow-up series, we changed the direction of the field again on a daily basis, but this time in *random* order. Under this regime, our most active stingray without further training made 120 correct versus 64 incorrect choices within a period of 15 days ($p < 0.001$).

In early tests, frequent field reversals appeared to confuse the animals. After refining both the set up and experimental procedure, and progressively gaining more experience, we recently began a series in which the direction of the magnetic field was altered *randomly from trial to trial* so as to provide the strongest evidence possible. In these most crucial experiments, the field was set by a third person, while the two observers did not know whether to feed or punish until after the stingrays had made their choice. The change of the field took place during the commotion of feeding or punishment by first slowly turning the coil current down to zero and then up again in either the normal or reversed direction. After being released at the North or the South of the pool, the animals usually swam about briefly, often stopping at the entrance of not only the east but also the west enclosure before making their final decision. Under these entirely double blind conditions, our fastest performing ray at times attained scores as high as 9 out of 10, and upon completion of the series totaled 101 correct versus 53 incorrect choices, which again is significant at the extremely conservative level of $p < 0.001$.

E. Discussion

As the electric fields that elasmobranchs induce when swimming through the earth's magnetic field are well within the sensitivity range of the animals' electric sense, and depend uniquely on the compass direction in which they are heading, it seems reasonable to assume that the stingrays' geomagnetic orientation shown in the present paper is indeed based upon the principles of electromagnetic induction. The discovery of a magnetic compass sense in elasmobranch fishes not only contributes to our understanding and appreciation of these marine predators, but also is of great survival value to the present author who predicted the animals' magnetic abilities on theoretical grounds (Kalmijn, 1973, 1974). It should be emphasized that the electromagnetic orientation mechanism hinges upon the conductivity of the medium with respect to which the motion takes place. Therefore, the electromagnetic principles - so relevant to marine elasmobranchs - are less promising for freshwater fish, and only applicable to land animals, insects, and birds if one thinks in terms of capacitive loading, relative motion of body parts (e.g. in wing-flapping), or *turning* movements.

Since the electric fields are induced by the elasmobranchs' own swimming movements, the inferred electromagnetic compass sense should be considered as a form of *active* electro-orientation. Elasmobranchs may also orient with regard to the electric fields of ocean currents flowing through the earth's magnetic field, thereby operating their electric sense in the *passive* mode. In recent training experiments, the stingray *Urolophus halleri* in fact readily learned to orient in strictly uniform electric fields of oceanic strengths* (0.05 to 0.5 μV/cm; Von Arx, 1962). These active and passive orientation mechanisms are not mutually exclusive, but may work side by side, providing marine sharks, skates, and rays with two invaluable sets of electromagnetic directional cues.

**Note added in proof:* Down to voltage gradients of 0.01 μV/cm and possibly lower.

Acknowledgments. I thank William L. Ackerman and Vera Kalmijn for their assistance during the training experiments and Richard T. Nowak for reading and discussing the manuscript. These studies were conducted under contract with the Office of Naval Research (N00014-74-C-0262).

References

Arx, W.S. Von: An Introduction to Physical Oceanography. Reading-London: Addison-Wesley, 1962

Dijkgraaf, S., Kalmijn, A.J.: Untersuchungen über die Funktion der Lorenzinischen Ampullen an Haifischen. Z. Vergl. Physiol. 47, 438-456 (1963)

Kalmijn, A.J.: Electro-perception in sharks and rays. Nature (Lond.) 212, 1232-1233 (1966)

Kalmijn, A.J.: The electric sense of sharks and rays. J. Exp. Biol. 55, 371-383 (1971)

Kalmijn, A.J.: Electro-orientation in sharks and rays: theory and experimental evidence. Scripps Institution of Oceanography Reference Series, Contr. No. 73-39, pp. 1-22 (1973)

Kalmijn, A.J.: The detection of electric fields from inanimate and animate sources other than electric organs. In: Handbook of Sensory Physiology. Fessard, A. (ed.), Berlin-Heidelberg-New York: Springer, 1974, vol. III/3, pp. 147-200

Murray, R.W.: The response of the ampullae of Lorenzini of elasmobranchs to electrical stimulation. J. Exp. Biol. 39, 119-128 (1962)

Waltman, B.: Electrical properties and fine structure of the ampullary canals of Lorenzini. Acta Physiol. Scand. 66, Suppl. 264, 1-60 (1966)

The Magnetic Behavior of Mud Bacteria

AD. J. KALMIJN[1] and RICHARD P. BLAKEMORE[2], [1] Woods Hole Oceanographic Institution, Woods Hole, MA 02543, USA, [2] University of New Hampshire, Durham, NH 03824, USA

Abstract

When separated from the substrate, Blakemore's mud bacteria swim back to the bottom of the sea following the earth's magnetic field lines. Their magnetotactic response appears to be due to the presence of internal ferromagnetic dipole moments of single-domain properties.

While microscopically examining marine and freshwater muds collected from the Woods Hole area, Blakemore observed that various kinds of anaerobic or micro-aerophilic bacteria (as yet unnamed) consistently swim to the north when separated from the sediments. Their direction of swimming was readily changed by approaching the microscope slide with a small bar magnet. In Kalmijn's magnetic facilities at the Woods Hole Oceanographic Institution, we jointly established that the bacteria also orient to the north in strictly uniform fields of geomagnetic strength, and actually follow the steeply inclined earth's magnetic field lines. Their north-seeking tendency became particularly evident upon reversing the ambient magnetic field with large Helmholtz coils. The bacteria immediately diverged from their course to make 180° U-turns, several cell diameters wide. They again aligned themselves with the field in a matter of seconds, this time heading in the opposite direction. On the basis of their magnetic behavior, and the presence of the iron-rich particles that Blakemore discovered in these organisms, we tentatively assumed them to be equipped with permanent magnetic dipole moments. Recently, we have substantiated this

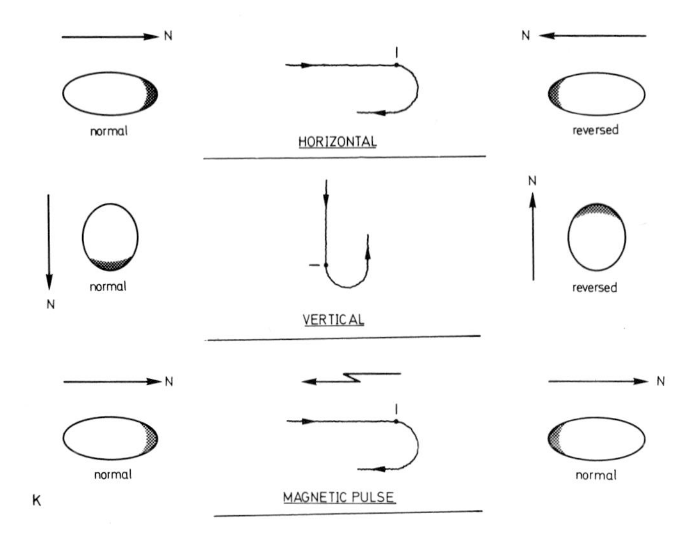

Fig. 1. *Top line*, Bacteria collect at the north side of a water droplet. Reversal of the horizontal component of the earth's magnetic field (at *dot* along track) makes them migrate to the opposite side of their small puddle. *Middle line*, Similar results are obtained in the vertical plane, despite gravity. *Bottom line*, A strong magnetic pulse of short duration, applied antiparallel to a steady background field, causes the bacteria to turn around and swim *south*

novel hypothesis experimentally through the application of relatively strong, monophasic magnetic pulses of short duration. These pulses were superimposed upon the steady earth's magnetic field, of which the vertical component was nulled so as to have the bacteria, for convenience, move in the horizontal plane. The magnetic pulses were generated by discharging a 4 μF high-voltage capacitor through a series resistor and two single-turn Helmholtz coils, 4 cm in diameter, forming part of a critically damped RCL circuit, designed and implemented by Kalmijn in collaboration with Richard T. Nowak. The discharges were triggered by means of a fast-switching silicon-controlled rectifier (SCR), which after the capacitor was charged maximally to 1200 V allowed a peak current of up to 1350 A to develop within 1.75 μs, yielding a virtually uniform field as strong as 600 gauss in the center of the coil system. When antiparallel to the horizontal component of the steady earth's magnetic field, and sufficiently strong, the magnetic pulses instantly caused the bacteria to turn around, and from then on swim in the opposite direction, i.e., straight to the south, as one would expect from the reversal of their intrinsic magnetic dipole moments. To affect 50% of the freshwater specimens from Cedar Swamp, a pulse strength of approximately 375 - 400 gauss was required. For the prevailing marine species from Eel Pond, this value was significantly higher, viz about 525 - 550 gauss. Depending on the strength of the imposed magnetic pulses, the bacteria either remained northbound or became southbound - none were completely depolarized. In that respect, they exhibit single domain properties. Although the role of their response remains to be determined, the magnetic behavior of these bacteria provides direct evidence of ferromagnetic orientation in biological specimens.

Acknowledgments. These studies have been conducted under contract with the Office of Naval Research (Kalmijn, N00014-74-C-0262).

References

Blakemore, R.P.: Magnetotactic bacteria. Science 190, 377-379 (1975)

Kalmijn, A.J., Blakemore, R.P.: Geomagnetic orientation in marine mud bacteria. Proc. Int. Union Physiol, Sci. 13, 364 (1977)

Olfactory Imprinting in Coho Salmon *(Oncorhynchus kisutch)**

ARTHUR D. HASLER and ALLAN T. SCHOLZ, Laboratory of Limnology, University of Wisconsin, Madison, WI 53706, USA

Abstract

Taken together, we believe the results of our recent, unified studies provide conclusive evidence for olfactory imprinting. In addition, since our studies were conducted in the field, they provide direct evidence that Coho Salmon retain and use artificial chemical information to achieve successful homing. It seems likely that salmon possessing this ability would also use it in the natural environment.

Our findings have direct practical applications for salvaging endangered stocks of salmon as well as being interesting from a purely scientific viewpoint.

A. Introduction

The olfactory hypothesis for salmon homing (Hasler and Wisby, 1951) states: (1) before juvenile salmon migrate to the sea they become imprinted to the distinctive odor of their natal tributary, and (2) adult salmon use this information as a cue for homing when they migrate through the home-stream network to the home tributary.

Restating this in more precise terms, the use of olfaction for homing by salmon requires that:

1. Each stream must have a characteristic and persistent odor perceptible by the fish.
2. Fish must be able to discriminate between the odors of different streams.
3. Fish must be able to retain an "odor memory" of its home stream during the period that intervenes between the downstream migration and homing migration.

Here we review the behavioral and physiologic experiments conducted at our laboratory that support these three postulates.

B. Laboratory Conditioning and Sensory Impairment Experiments

The first step in testing the odor hypothesis was to determine if fish could discriminate two different streams by smell (Hasler and Wisby,

*A more detailed version of this report was published in *American Scientist* by Arthur D. Hasler, Allan T. Scholz, and Ross M. Horrall.

1951). Using reward (food) and punishment (electric shock) for conditioning, we trained groups of coho salmon to discriminate between waters collected from two streams. However, when their nasal sacs were cauterized, trained fish were not able to discriminate these waters. This study implied that each stream has a characteristic odor that is discernible by fish.

This study also provided information on two additional points. First, trained fish were not able to identify a water sample if the organic fraction was removed. Second, fish that were trained to discriminate water obtained from a stream during one season were able to discriminate water from the same stream collected in a different season. This result suggests that the factor that the fish could detect was long-lasting and present in the stream throughout the year. This, we believed, was an important point because salmon may reside in the ocean for several years before returning to their natal stream - thus the imprinting factor must be long-lasting. This experiment was repeated by other investigators (McBride et al., 1964; Idler et al., 1961; Tarrant, 1966; Walker, 1967) and their results were similar to ours.

Our second step in testing the odor hypothesis was to conduct sensory impairment experiments in the field (Wisby and Hasler, 1954). The purpose of these experiments was to determine if fish deprived of their sense of smell could locate their home stream. The research site was a small Y-shaped tributary located 25 km from Seattle, Washington: Issaquah Creek and its East Fork (Fig. 1). Each branch had its own native stock of coho salmon. We captured 322 fish in traps as they entered their home branches and plugged the nasal sacs of half of them to block their olfactory sense. The remaining fish were left unplugged to control for olfactory impairment. We also tagged each fish to tell

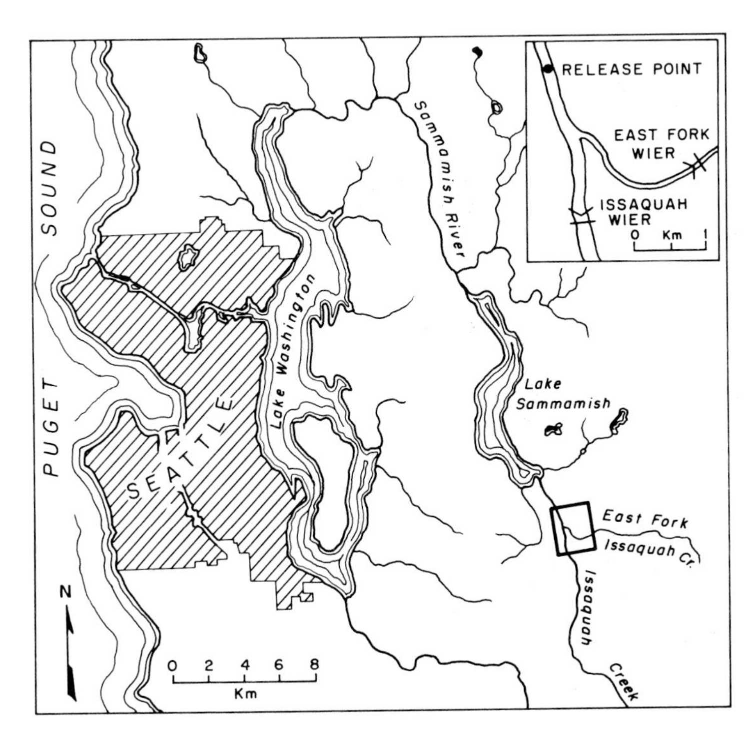

Fig. 1. Study area for olfactory impairment experiments in Lake Washington watershed, adapted from Wisby and Hasler (1954). *Inset* shows detail of study site at Issaquah Creek and its East Fork

us which treatment it had received and in which branch it had been caught. All of the fish were then released 1.6 km below the fork and allowed to repeat their upstream migration. The control fish returned to their home tributary almost without error, but many of the fish that had been deprived of their sense of smell entered the wrong branch of the stream. The results are summarized in Table 1.

Table 1. Results of nose-plugging experiments on coho salmon conducted at Issaquah Creek in 1954 (from Wisby and Hasler, 1954)

Stream of origin	Treatment of fish	Number released	Number recaptured in	
			Issaquah	East Fork
Issaquah	Controls	121	46	0
	Nose plugged	125	39	12
East Fork	Controls	38	8	19
	Nose plugged	38	16	3

This type of sensory impairment experiment has been repeated twenty times by other investigators (reviewed by Hasler, 1966 and Stasko, 1971). Species tested include coho, chum (*O. keta*), chinook (*O. tschawytscha*) and Atlantic salmon (*Salmo salar*), and cutthroat (*S. clarki*), rainbow (*S. gairdneri*), and brown trout (*S. trutta*). The results of these studies are remarkably consistent and agree closely with the results of our own experiment. For 16 experiments the olfactory sense appeared to be necessary for correct homing. In addition, two studies demonstrated that blinded fish homed nearly as well as control fish, and thus that vision was not so important for relocating the original stream, at least during the upstream migration (Hiyama et al., 1966; Groves et al., 1968).

Up to this point our experiments showed that each stream has a characteristic and persistent odor, and that salmon can discriminate between the odors of different streams, thus supporting the first two tenets of the olfactory hypothesis. However, these studies did not provide conclusive evidence for olfactory imprinting and long-term olfactory memory, hence we modified our plan of research to test this particular point of the hypothesis.

C. Evidence for Emprinting - Transplantation Experiments

Central to our basic disign were the results obtained from experiments with fish transplanted from their natal tributary to a different tributary during the smolt stage, i.e., at the time they begin to make their seaward journey.

When Rounsefell and Kelez (1938) transferred young, pre-smolt coho salmon from their native river to a different river where they marked and released them, the fish migrated to sea and returned as adults to the second river. Donaldson and Allen (1957) also found that coho salmon raised in a hatchery and transplanted before undergoing smolt transformation returned to the river of release. There is evidence that this process is rapid. Jensen and Duncan (1971) and Carlin (1968) transplanted coho salmon and Atlantic salmon, respectively, just as they began to smolt; the fish left the river within two days and returned to it during the spawning migration.

In contrast, when Peck (1970) did not transplant hatchery-raised coho salmon in a Lake Superior tributary until several weeks after smolt transformation, the return to the stream of release was poor and many fish were recovered in other streams. Peck concluded that the fish may have spent the sensitive period in the hatchery (water supply not connected with Lake Superior) before stocking and suggested that imprinting terminates soon after smolt transformation begins, thereby preventing fish from being imprinted to other tributaries during their downstream migration. Additional support for this view comes from a study by Stuart (1959) working with brown trout at Dunalastair Reservoir in Scotland. He held fish in their native tributary through the smolt stage before transferring them to a different tributary. During the spawning migration the adults returned to their natal stream. However, fish transplanted before undergoing smolt transformation returned to the river of release.

Thus, results from transplantation experiments suggest that homing is connected with a period of rapid and irreversible learning (or "imprinting") of the cues that identify the home stream at the time the young salmon begin their downstream migration, i.e., smolt transformation.

D. Artificial Imprinting Experiments

We reasoned that a decisive test for odor imprinting was to substitute a man-made chemical for the natural scents of the home stream (Hasler and Wisby, 1951). We could then expose salmon to the chemical during the smolt stage (the "critical" or "sensitive" period for imprinting - see transplant experiments above) and try to decoy them to a stream scented with it during the spawning migration.

We decided that the chemical had to be an organic compound since our previous work had demonstrated that the identifiable component of stream water was contained in the organic fraction. In addition, it would have to be chemically stable, soluble in water, not normally found in natural waters, and one that would neither naturally repel nor attract the fish. Finally, we felt that it was important for the chemical to be detectable in small quantities in order to minimize any possible damage it might have on a natural stream system. Wisby (1952) screened likely substances and his tests turned up an appropriate compound - morpholine (C_4H_9NO), a heterocyclic amine, which could be detected by unconditioned coho salmon at a concentration of 1×10^{-6} mg/l.

We then conducted a preliminary electrophysiologic experiment using this compound (Dizon et al., 1973). Two groups of coho salmon were raised at a hatchery until the smolt stage, then one group was exposed to a concentration of 1×10^{-5} mg/l morpholine for one month. The fish were retained at the hatchery for six additional months and then brought to our laboratory for testing. The tests were performed by presenting .01% and 1% morpholine to restrained fish and recording EEG signals from an electrode inserted through a hole cut in the skull over the forebrain into the olfactory bulb; 14 fish from each group were tested. Our results showed that fish that had been exposed to morpholine displayed significantly higher EEG activity than unexposed controls.

We started artifically imprinting salmon on a large scale in the spring of 1971 (Cooper et al., 1976; Madison et al., 1973; Scholz et al., 1973). We transported 16,000 coho fingerlings, of the same genetic stock and hatched and raised under uniform conditions, from a Wisconsin State Fish Hatchery in central Wisconsin to holding tanks at a water-filtration plant in South Milwaukee (Fig. 2). We held the fish there for about 30 days during their smolting period. All the fish were held in tanks with water piped in from Lake Michigan; thus, during their entire early life history they were not exposed to water from any Lake Michigan tributary.

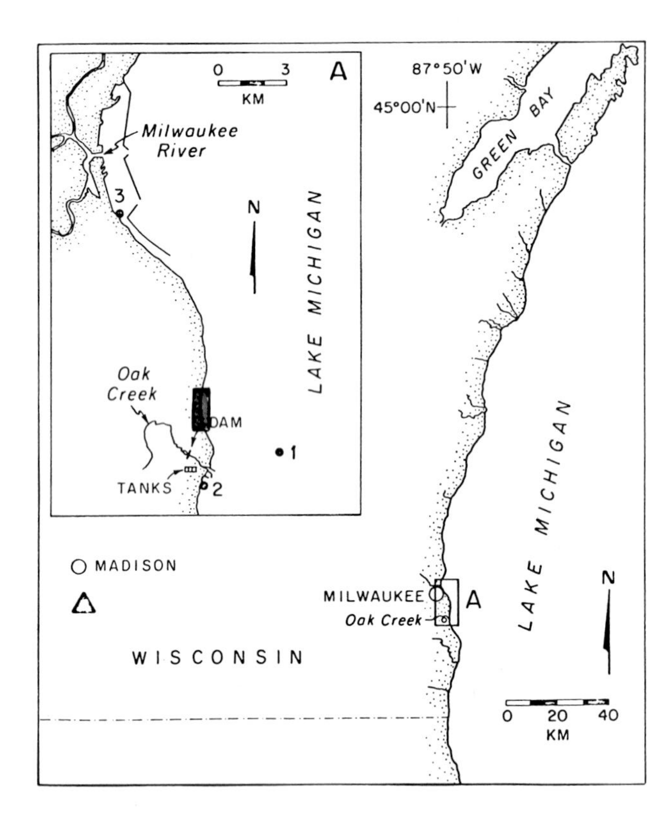

Fig. 2. Research area, South Milwaukee, Wisconsin. Inset shows detail of: (*1*) the water intake for the tanks at the South Milwaukee water filtration plant, (*2*) the Oak Creek stocking site, and (*3*) the Milwaukee Harbor stocking site

Morpholine was metered into a tank containing half of the fish. A concentration of 5×10^{-5} mg/l was maintained in the tank throughout the exposure period. The rest of the fish, held in a second tank, were not treated, to serve as a control group. The fish from each group were then marked with identifying fin clips and released into Lake Michigan 0.5 km south of the mouth of Oak Creek, the stream which was to be scented during the spawning migration. We released them directly into the lake in order to reduce the possibility that they might learn additional cues about the test stream.

During the spawning season in the fall of 1972, 18 months after the fish were released, we created an artificial home stream for the homing adults by metering morpholine into Oak Creek at approximately the same concentration to which they had been exposed. We hypothesized that if salmon use a chemical imprinting mechanism for homing, morpholine-exposed fish would return to the scented stream in larger numbers than unexposed fish. Unexposed fish served as controls to determine if fish would return to the stream independently of the chemical cue. The stream was monitored by creel census surveys, gill netting, and electro-fishing. We captured a total of 216 of the fish that had been exposed to morpholine and only 27 from the control group (Table 2). This result supported our hypothesis but we decided to repeat the experiment.

Table 2. Census record of Coho Salmon caught at Oak Creek in fall 1972 (from Scholz et al., 1973, 1975, and Cooper et al., 1976)

Treatment	Stocking location	Number released	Date released	Number recovered	Percent of fish stocked
Morpholine	0.5 km south of Oak Creek	8000	May 1971	218	2.58
Control	0.5 km south of Oak Creek	8000	May 1971	28	.35

During the spring of 1972, we started experiments to replicate the 1971-1972 series (Scholz et al., 1975; Cooper et al., 1976). We worked with larger numbers of smolts: 18,200 were exposed to morpholine at a hatchery and 20,000 were left unexposed. This time we released the fish at two different points along the Lake Michigan shoreline, 0.5 km south and 13 km north of Oak Creek (see Fig. 5). In the fall of 1973, we caught 1515 morpholine-exposed fish and 169 controls - a ratio of nearly 10 to 1 (Table 3). Morpholine-exposed fish released as smolts 13 km north of Oak Creek homed to Oak Creek in a manner similar to those released near it (Table 3).

Table 3. Census record of Coho Salmon caught at Oak Creek in fall 1973 (from Cooper et al., 1976)

Experiment	Treatment	Stocking location	Number released	Date released	Number recovered	Percent of fish stocked
1	Morpholine	0.5 km south of Oak Creek	5000	May 1972	437	8.74
	Control	0.5 km south of Oak Creek	5000	May 1972	49	.95
2	Morpholine	0.5 km south of Oak Creek	5000	May 1972	439	8.78
	Control	0.5 km south of Oak Creek	5000	May 1972	55	1.10
3	Morpholine	13 km north of Oak Creek	8000	May 1972	647	7.89
	Control	13 km north of Oak Creek	10,000	May 1972	65	.65
TOTAL	Morpholine		18,000		1515	8.42
	Control		20,000		169	.85

During a third, control experiment conducted in 1973, we exposed 5000 smolts to morpholine and left an equal number unexposed (Cooper et al., 1976; Scholz et al., 1975) but in the fall of 1974, when the fish were expectal to return, morpholine was not added to Oak Creek. The results of this experiment were different from the others; exposed and non-exposed fish were captured in equally low numbers (51 vs. 55) at about the same rate as control fish from previous experiments (Table 4). This experiment illustrates the importance of morpholine to the return of imprinted salmon.

Table 4. Census record of Coho Salmon caught at Oak Creek in fall 1974 (from Scholz et al., 1975, and Cooper et al., 1976)

Treatment	Stocking location	Number released	Date released	Number recovered	Percent of fish stocked
Morpholine	0.5 km south of Oak Creek	5000	May 1973	51	1.02
Control	0.5 km south of Oak Creek	5000	May 1973	55	1.10

Morpholine was not present in Oak Creek.

In a more refined test, one group of coho smolts held at a hatchery was exposed to morpholine, a second group to phenethyl alcohol (PEA, $C_8H_{10}O$), and a third left unexposed (Scholz et al., 1976). All three groups were released in Lake Michigan midway between two test streams that were located 9.4 km apart (Fig. 3). The artificial imprinting for these experiments was done in the spring of 1973 with 5000 fish in each group and repeated in the spring of 1974 with 10,000 fish in each group. During the spawning migration 18 months later, in the fall of 1974 and again in the fall of 1975, morpholine and PEA were metered into separate test streams - the Little Manitowoc and East Twin Rivers, respectively (Fig. 3). The streams were surveyed for marked fish by

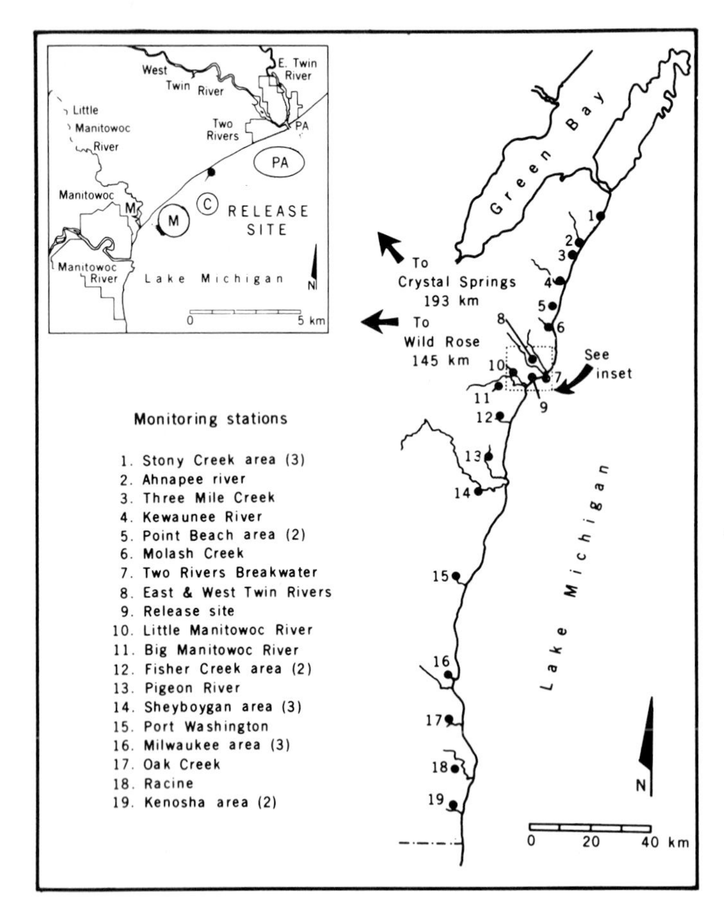

Fig. 3. Research area, Wisconsin shore, Lake Michigan. *Numbers in parentheses* represent the number of streams in the general area of the monitoring station that were surveyed. *Inset* shows detail of the release site, the morpholine-scented Little Manitowoc River (*M*) and the phenethyl alcohol-scented breakwater area at Two Rivers (*PA*)

Table 5. Total number morpholine-exposed (M), phenethyl alcohol-exposed (PEA), and control (C) salmon captured at individual locations. Data from the Little Manitowoc River (morpholine scented) and Two Rivers area (PEA scented) are printed in boldface type. Fishing effort is summarized by type and number of collecting trips at each location (from Scholz et al., 1976)

Location	1974						1975					
	Effort-No. Trips			No. Recovered			Effort-No. Trips			No. Recovered		
	Creel census	Gill net	Electro-fishing	M	PEA	C	Creel census	Gill net	Electro-fishing	M	PEA	C
1 Stony Creek (3)	90	13	13	1		4	40					12
2 Annapee River	138	3	5		2	7	224	4	14	6	1	37
3 Three Mile Creek	27	5	5	2	1	1	26					2
4 Kewaunee River	71		5				9					
5 Nuclear Power Plants (2)	123			1		4	3					2
6 Molash Creek	8					2	1					
7 TWO RIVERS BREAKWATER	*184*	*3*	*1*	*3*	*118*	*15*	*126*	*14*	*1*	*3*	*192*	*12*
8 EAST AND WEST TWIN RIVERS	*123*		*9*		*15*	*7*	*17*		*14*	*3*	*8*	*21*
9 Stocking Site	90	1		1		7	30					1
10 LITTLE MANITOWOC RIVER	*189*		*8*	*207*	*6*	*24*	*135*			*452*	*14*	*52*
11 Big Manitowoc River	44		5	2	3	31	7				1	26
12 Fisher Creek (2)	44					3	2					1
13 Pigeon River	23											
14 Sheboygan River (3)	75		1	1		3	1					3
15 Port Washington	38											
16 Milwaukee area (3)	65											
17 Oak Creek	306		5		1	7						
18 Racine	11					1						
19 Kenosha (2)	14											

creel census, gill-net fishing, and electrofishing. Seventeen other locations were also monitored to determine if significant numbers imprinted fish were straying into nonscented streams. The results from both experiments (Table 5) show that the majority of the fish exposed to morpholine were captured in the stream scented with morpholine and most fish exposed to PEA were captured in the stream scented with PEA. By contrast, large numbers of control fish were captured at other locations.

The results of our artificial imprinting experiments provide direct evidence for olfactory imprinting in coho salmon. We have also conducted three experiments with rainbow trout (Cooper and Scholz, 1976; Scholz et al., 1975, 1978b) and one with migratory brown trout (Scholz et al., 1978a) and in all cases significantly higher numbers of morpholine-exposed, as opposed to unexposed trout, returned to a stream scented with morpholine. In a preliminary experiment (Cooper et al., 1976), coho salmon that had been exposed to morpholine for two days at the onset of smolting returned to a simulated home stream in about equal numbers as fish exposed for 30 days. Thus, very short periods of morpholine exposure seem sufficient to imprint fish successfully.

E. Ultrasonic Tracking Experiments

Ultrasonic tracking experiments performed at the same time as the marking studies allowed us to obtain direct information about the behavioral response of morpholine-imprinted and nonimprinted salmon to morpholine. This study was conducted at Oak Creek from 1971 to 1973. Salmon captured in Oak Creek as part of the census studies were equipped with ultrasonic transmitters and displaced back into Lake Michigan to a release point located 3.2 km north of Oak Creek. They were released next to shore. A directional hydrophone connected to receiving equipment on a tracking boat was used to follow the signal from the tagged fish. The fish typically traveled within 50 m of shore in a straight line toward Oak Creek. A test area located between the release point and Oak Creek was scented with morpholine and the behavior of fish encountering this area was observed. Control experiments were conducted by tracking fish through the test area when morpholine was absent or when a different chemical was present.

During 20 experimental tracks (Fig. 4a), morpholine was introduced into the test area in a narrow band perpendicular to shore, creating an "odor barrier" through which the fish had to swim; the morpholine concentration in this area was approximately 5×10^{-5} mg/l. Water currents were measured with drogues to determine how long the chemical remained there. Twenty morpholine-exposed fish were tracked into the area when morpholine was present and in all cases they stopped migrat-

Fig. 4 A - D. Tracks of all salmon used in tracking experiments 1971 - 1973. The first two digits of each track number identify the year during which the track was recorded. Time when the track was started is recorded immediately below the track number; *dots* along the track path represent 15-min intervals. Morpholine or an alternate chemical was released in the test area (stippled in A) when the fish had moved to the position of the diamond. Tracks show responses of: (A) imprinted salmon when morpholine was present, (B) imprinted salmon when morpholine was absent, (C) nonimprinted salmon when morpholine was present, (D) imprinted salmon when n-β-hydroxyethyl-morpholine or *PEA* was present. *Arrows* represent 1-h current vectors for surface and 1-m currents. Imprinted fish remained in scented area for longer periods when currents were slow

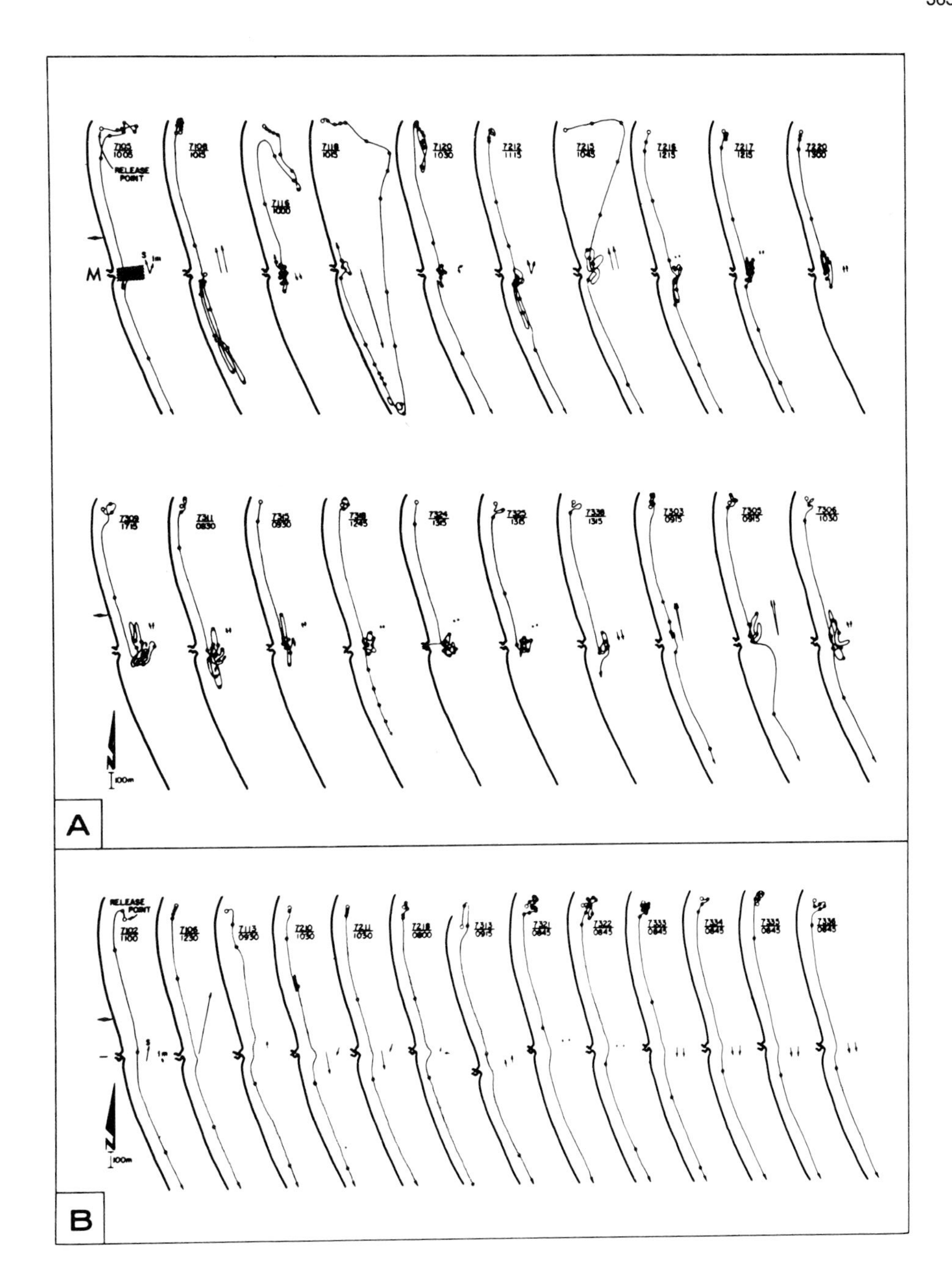

ing and milled around for 1/2 to 4 h before leaving. The time spent in the scented area was correlated with water current velocity; fish remained there until the morpholine scent had been dissipated by water currents (Scholz et al., 1975).

A series of control experiments (Fig. 4b) was conducted when morpholine was not present in the test area. Fourteen fish were tracked and in

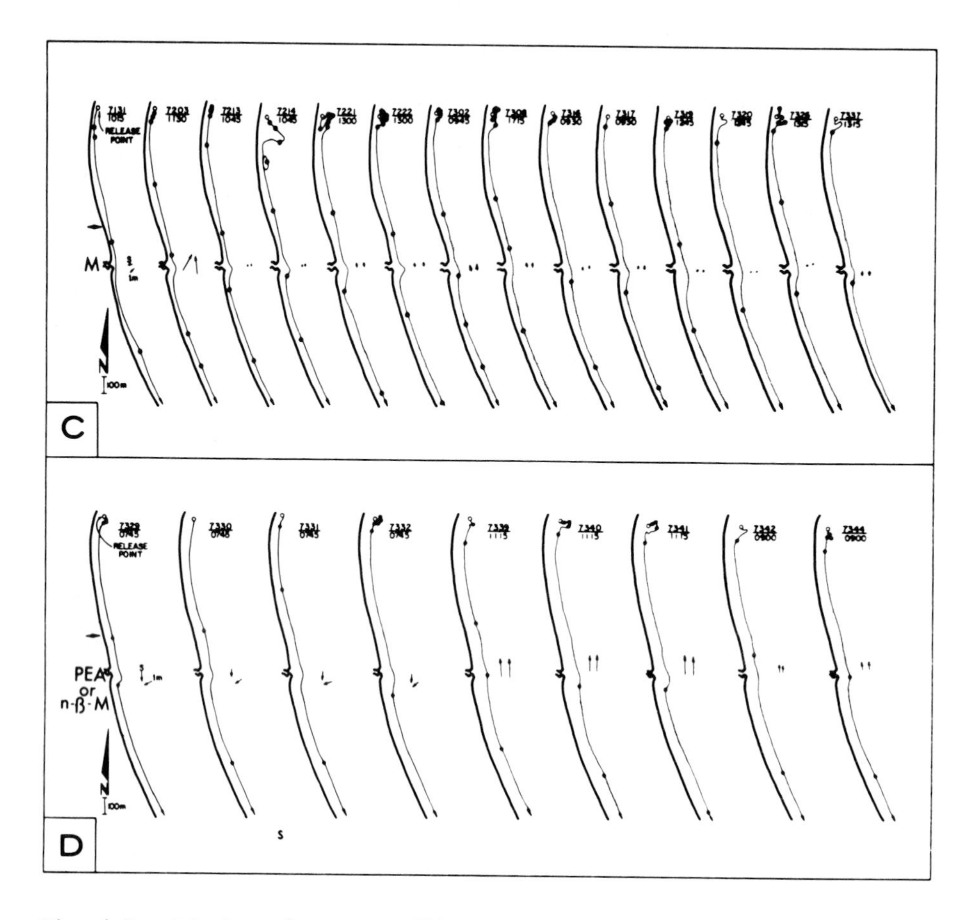

Fig. 4 C and D. Legend see page 364

all cases they moved through without stopping. Thus, morpholine-exposed fish stopped migrating when morpholine was present, but did not stop when morpholine was absent.

At this point we were concerned that perhaps the behavior of fish milling in the scented area could be explained as a result of morpholine acting as a unique odor not normally encountered along the shoreline instead of acting as a cue for homing. To test for this possibility we tracked 13 nonimprinted fish through the test area when morpholine was present and they all moved through it without stopping (Fig. 4c). In addition, nine morpholine-imprinted fish were tracked through the test area when it was scented with a different chemical (PEA or N-β-hydroxyethyl morpholine) instead of morpholine, and in all cases they did not stop (Fig. 4d). Thus, it does not seem likely that morpholine-exposed fish were reacting to morpholine because it was a unique shoreline odor. The tracking results indicate that the response of morpholine-exposed fish was related to chemical imprinting and long-term memory of morpholine. We conclude that they were using it as a cue for homing.

F. EEG Experiments

In addition to the tracking we conducted electrophysiologic experiments to determine if electric activity could be evoked from the olfactory bulb of imprinted salmon (Cooper and Hasler, 1974, 1976). EEG responses to morpholine and other water samples were recorded from imprinted and nonimprinted fish captured in Oak Creek. A total of 50 imprinted and 40 nonimprinted fish were tested. There was a significant difference in the amplitude of the EEG signals to morpholine in distilled water for imprinted compared to control fish (Fig. 5). Significant differences

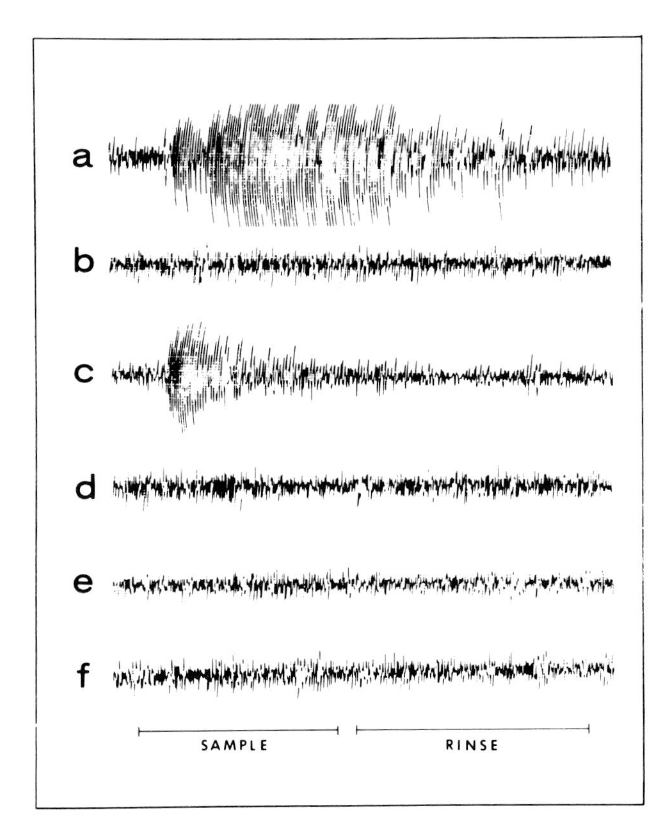

Fig. 5 a - f. EEG responses of (a) imprinted salmon to morpholine, (b) nonimprinted salmon to morpholine, (c) imprinted salmon to Oak Creek water containing morpholine, (d) imprinted salmon to Oak Creek water alone, (e) imprinted salmon to phenethyl alcohol (PEA), (f) imprinted salmon to N-β-hydroxyethyl-morpholine

were not seen when other substances such as N-β-hydroxethyl morpholine, phenethyl alcohol, and Lake Michigan water were tested. In addition, imprinted fish responded to Oak Creek water containing morpholine, but did not respond to Oak Creek water by itself. Thus, electrophysiologic experiments suggest that early exposure of juvenile salmon to morpholine significantly influences the subsequent responses of the olfactory bulb of sexually mature adults. This result correlated well with the behavioral studies.

For a summary statement see abstract on p. 356.

References

Carlin, B.: Salmon conservation, tagging experiments, and migrations of salmon in Sweden. Lecture Series, the Atlantic Salmon Assoc., Montreal (1968)

Cooper, J.C., Hasler, A.D.: Electrophysiological evidence for retention of olfactory cues in homing coho salmon. Science 183, 336-338 (1974)

Cooper, J.C., Hasler, A.D.: Electrophysiological studies of morpholine imprinted coho salmon (*Oncorhynchus kisutch*) and rainbow trout (*Salmo gairdneri*). J. Fish. Res. Board Can. 33, 688-694 (1976)

Cooper, J.C., Scholz, A.T.: Homing of artificially imprinted steelhead trout. J. Fish. Res. Board Can. 33, 826-829 (1976)

Cooper, J.C., Scholz, A.T., Horrall, R.M., Hasler, A.D., Madison, D.M.: Experimental confirmation of the olfactory hypothesis with artificially imprinted homing coho salmon (*Oncorhynchus kisutch*). J. Fish. Res. Board Can. 33, 703-710 (1976)

Dizon, A.E., Horrall, R.M., Hasler, A.D.: Long-term olfactory "memory" in coho salmon, *Oncorhynchus kisutch*. Fish. Bull. 71, 315-317 (1973)

Donaldson, R., Allen, G.H.: Return of silver salmon *Oncorhynchus kisutch* (Walbaum) to point of release. Trans. Am. Fish. Soc. 87, 13-22 (1957)

Groves, A.B., Collins, G.B., Trefetren, G.B.: Roles of olfaction and vision in choice of spawning site by homing adult chinook salmon (*Oncorhynchus tshwytscha*). J. Fish. Res. Board Can. 25(5), 867-876 (1968)

Hasler, A.D.: Underwater guideposts. Univ. of Wisconsin Press, Madison. 155 p. (1966)

Hasler, A.D., Wisby, W.J.: Discrimination of stream odors by fishes and relation to parent stream behavior. Am. Nat. 85, 223-238 (1951)

Hiyama, Y., Taniuchi, T., Suyama, K., Ishoka, K., Sato, R., Kajihara, T., Maiwa, R.: A preliminary experiment on the return of tagged chum salmon to the Otsucki River, Japan. Jpn. Soc. Sci. Fish. 33, 18-19 (1966)

Idler, D.R., McBride, J.R., Jones, R.E.E., Tomlinson, N.: Olfactory perception in migrating salmon, II: studies on a laboratory bio-assay for a homestream water and mammalian repellant. Can. J. Biochem. Physiol. 39, 1575-1584 (1961)

Jensen, A., Duncan, R.: Homing in transplanted coho salmon. Prog. Fish-cult. 33, 216-218 (1971)

Madison, D.M., Scholz, A.T., Cooper, J.C., Horrall, R.M., Hasler, A.D., Dizon, A.E.: Olfactory hypothesis and salmon migrations: a synopsis of recent findings. Fish. Res. Board Can. Tech. Rept. No. 414. 35 p. (1973)

McBride, J.R., Fagerlund, U.H., Smith, M., Tomlinson, N.: Olfactory perception in juvenile salmon, II: conditioned response of juvenile sockeye salmon (*Oncorhynchus nerka*) to lake water. Can. J. Zool. 42(2), 245-248 (1964)

Peck, J.W.: Straying and reproduction of coho salmon, *Oncorhynchus kisutch*, planted in a Lake Superior tributary. Trans. Am. Fish. Soc. 99, 591-595 (1970)

Rounsefell, G.A., Kelez, G.B.: The salmon and salmon fisheries of Swiftsure Bank, Puget Sound, and the Fraser River. Bull. U.S. Bur. Fish. 49, 693-823 (1938)

Scholz, A.T., Cooper, J.C., Madison, D.M., Horrall, R.M., Hasler, A.D., Dizon, A.E., Poff, R.J.: Olfactory imprinting in coho salmon: behavioral and electrophysiological evidence. Proc. Conf. Great Lakes Res. 16, 143-153 (1973)

Scholz, A.T., Gosse, C.K., Cooper, J.C., Horrall, R.M., Hasler, A.D., Daly, R.I., Poff, R.J.: Homing of rainbow trout transplanted in Lake Michigan: a comparison of three procedures used for imprinting and stocking. Trans. Am. Fish. Soc.: in press (1978b)

Scholz, A.T., Horrall, R.M., Cooper, J.C., Hasler, A.D., Madison, D.M., Poff, R.J., Daly, R.: Artificial imprinting of salmon and trout in Lake Michigan. Wis. Dept. Nat. Resources Fish. Mgmt. Rept. 80. 46 p. (1975)

Scholz, A.T., Horrall, R.M., Cooper, J.C., Hasler, A.D.: Imprinting to chemical cues; the basis for homestream selection in salmon. Science 196, 1247-1249 (1975)

Scholz, A.T., Horrall, R.M., Cooper, J.C., Hasler, A.D.: Homing of morpholine imprinted brown trout (*Salmo trutta*). Fish. Bull. 76(1), 293-295 (1978a)

Stasko, A.B.: Review of field studies on fish orientation. Ann. N.Y. Acad. Sci. 188, 12-29 (1971)

Stuart, T.A.: Tenth annual report of the supervisory committee for brown trout research, 1957-1958. Freshwat. and Salmon Fish. Res. 23, 6-7 (1959)

Tarrant, R.M.: Thresholds of perception of eugenol in juvenile salmon. Trans. Am. Fish. Soc. 95, 112-115 (1966)

Walker, J.C.: Odor discrimination in relation to homing in Atlantic salmon. M.S. Thesis, Univ. of New Brunswick, Fredericton (1967)

Wisby, W.J.: Olfactory responses of fishes as related to parent stream behavior. Ph. D. Thesis, Univ. of Wisconsin, Madison. 42 p. (1952)

Wisby, W.J., Hasler, A.D.: The effect of olfactory occlusion on migrating silver salmon (*O. kisutch*). J. Fish. Res. Board Can. 11, 472-478 (1954)

Energetics of Migration in American Shad*

WILLIAM C. LEGGETT[1] and CLIFFORD L. TRUMP[2], [1] Department of Biology, McGill University, Montreal, Québec, Canada H3A 1B1, [2] Department of Biology, Laval University, Québec, Canada G1K 7P4

Abstract

The energetic efficiency of the observed migratory behaviors of the anadromous American shad in fresh and salt water was examined. Swimming behavior in fresh water is inefficient, relative to the optimum strategy, and appears to be regulated by optimotor rather than energy constraints. Near the salt-fresh water interface, immediately following entry into fresh water, swimming behavior approaches the optimum energetic strategy. This shift in behavior appears to result from energy constraints related to osmoregulation.

The mean swimming speeds of shad in both salt and fresh water are near the optimum. The population consequences of these behavior patterns are discussed.

A. Introduction

Ultrasonic telemetry studies of the behavior of adult American shad (*Alosa sapidissima*), an anadromous clupeid, during the spawning migration in the Connecticut River (Leggett, 1976) have shown that, in fresh water, shad vary their swimming speed in response to changes in river current velocity created by tidal action. This behavior results in an approximately constant rate of progress with respect to the bottom (Fig. 1A). Similar behavior occurs during migration in salt water prior to entry into the home river (Dodson and Leggett, 1973, 1974). This behavior is modified during transition from salt to fresh water (Dodson, et al., 1972; Legget, 1976) when swimming speeds are slower and are largely independent of current velocity. This behavior, coupled with changes in current velocity resulting from tidal influences, results in variable rates of progress over the bottom and a slower overall rate of upriver progress (Fig. 1B).

Recent investigations of the energetics of the fresh water migrations of shad (Glebe and Leggett, 1978) have shown that the utilization of stored body reserves during the migration (adult shad to not feed in fresh water) is a major determinant of postspawning survival and of the population characteristics of individual stocks. As a consequence these stocks have evolved different patterns of energy utilization in response to the particular demands of their home environment.

In this study we examined the energetic costs of the two types of swimming behavior observed during our earlier studies, relative to

*Contribution to the program of GIROQ (Groupe Interuniversitaire de Recherches Océanographiques du Québec).

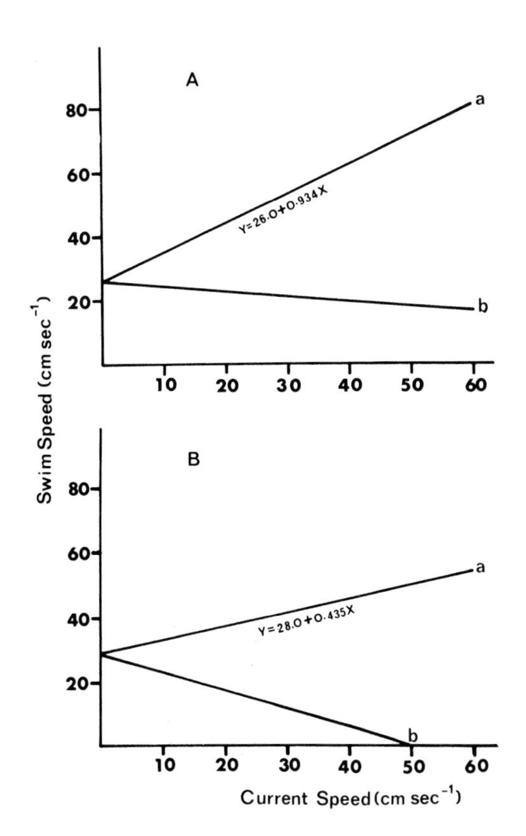

Fig. 1 A and B. Relationships between swim speed and current velocity in fresh water (A) and brackish water (B). (*a*, speed through water; *b*, speed over bottom)

an alternative hypothetical behavior of constant speed with respect to the water, and evaluated the relationship between swimming speed and absolute energy requirements in an attempt to assess their significance in the overall energetics of the fresh water migration.

B. Methods

The relative energetic efficiencies of the various swimming behaviors were evaluated using expressions 1 and 2 (Trump and Leggett, in prep.):

$$\bar{E}x = \left[abe^{bU_O+1}\right]\left[\frac{e^{bW_D}}{bW_D+1}\right]\left[1/T\int_0^T e^{bB\ \sin\ wt}\,dt\right] \qquad (1)$$

where: $\bar{E}x$ = energy in Cal/kg/cm

abe^{bU_O+1} = the minimum energy expenditure possible. This assumes a constant swimming speed of $W_{om} = U_o + 1/b$ cm/s with respect to the water [U_o = mean current speed; a and b = constants in the equation $E(t) = ae^{bw(t)}$ describing the relationship between specific energy per unit time, E(t) and swimming speed w(t) (Brett, 1965)].

$\frac{e^{bWD}}{bW_D+1}$ = factor expressing the change in energy cost of swimming at some mean speed (W_o) other than W_{om} ($W_D = W_{om} - W_o$)

$1/T \int_o^T e^{bB \sin wt} dt$ = factor expressing the change in energy cost of varying swimming speed in response to oscillations in current (resulting from tidal action); B = (dw/du) A where dw/du is linear slope or relationship between swimming speed (w) and current velocity (u) and A = amplitude of oscillations in current velocity

$$\bar{E}t = ae^{bw_o} \cdot 1/T \int_o^T e^{bB \sin wt} dt \quad (2)$$

where: $\bar{E}t$ = energy in Cal/kg/s

Empirical values of U_o, A, and B relating to behavior at the salt-fresh-water interface and in fresh water were derived from Leggett (1976) and unpublished data. Estimates of a and b were derived from Brett's (1965) study with salmon since no direct data are available for the energy cost of swimming in shad. This approximation may lead to errors in absolute energy determinations (see below), but comparisons of relative energy costs of the different swimming strategies are valid. The significance of the amplitude of variations in current velocity on the energetics of swimming was evaluated by varying A, and hence B, in simulations. The importance of body size was also examined since this is known to influence the values of a and b (Brett, 1965) and because the mean size of spawning shad varies significantly between populations. In all analyses the time period considered was one full tidal cycle.

C. Results

I. Swimming Strategies

Examination of the integral $\int_o^T e^{bB \sin wt} dt$ reveals that for all values of B other than B = 0 the integral value is > 1. Therefore, the optimum strategy, regardless of the characteristics of the current, is to swim at a constant speed W_{om} with respect to the water. Figure 2A illustrates the added energy cost of various values of B ≠ 0 relative to B = 0 for a 2000-g shad migrating to the Connecticut River (b = .0193 s/cm; A = 37 cm/s). During the fresh-water phase of the migration the mean value of B for the Connecticut River population was 0.934 (Leggett, 1976). The added energy cost of varying swimming speed in response to changes in current velocity rather than employing the optimum strategy was thus 11.8 %.

During the period of transition from salt to fresh water the mean value of B was 0.435. During this phase of the migration the differential energy cost between the optimum and observed strategies was 2.5 %.

The equivalent energy relationships for a 1000-g shad migrating under identical environmental conditions (b = 0.023 s/cm; A = 37 cm/s) are given in Figure 2B. The energy differential in fresh water is 15.8 % and at the salt-fresh-water interface 3.4 %.

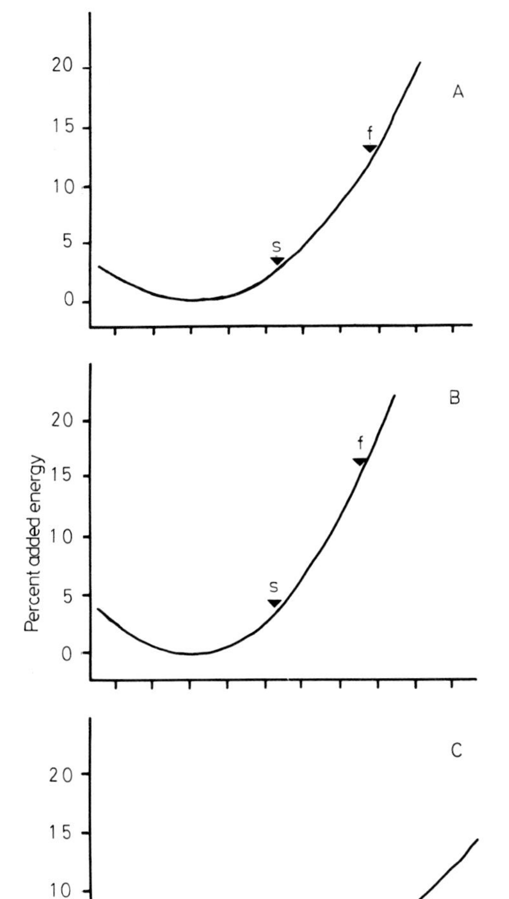

Fig. 2 A - C. Added energy cost of varying swimming speed in response to changes in current velocity relative to optimum strategy of constant speed through water. (B) (dw/du) is slope of relationship between swim speed and current velocity. (A) 2000-g shad in current oscillation of 37 cm/s. (B) 1000-g shad in current oscillation of 37 cm/s. (C) 2000-g shad in current oscillation of 25 cm/s. *s*, observed slope in brackish water, *f*, observed slope in fresh water

II. Amplitude of Current Variations

A 32 % reduction in the amplitude of current variations about the mean (A = 25 cm/s) resulted in a 6 % reduction in the energy differential between the optimum and observed swimming strategies for a 2000-g fish migrating in fresh water in the Connecticut River. The equivalent saving at the salt-fresh-water interface was 1 % (Fig. 2C).

III. Absolute Energy Requirements

In this analysis we assumed constant upriver orientation. This condition occurred during 83 % (331/401) of Leggett's (1976) observations in the fresh-water portion of the Connecticut River and during 73 % (116/159) of his observations in the brackish water portion of the estuary. We estimate U_o = 37 cm/s and A = 37 cm/s (Leggett, 1976, unpublished data). The relationships between shad swimming speed and current velocity in the Connecticut River are given in Figure 1. From these we calculate:

$W_o = W(U_o)$ = 61 cm/s (fresh water)
54 cm/s (brackish water) and

B = (dw/du) A = 34.6 cm/s (fresh water)
16.1 cm/s (brackish water)

From Brett (1965) we calculated a = 4.25×10^{-5} and b = 0.0193 for a 2000-g fish. Hence from Eq. 2 we estimate:

$\bar{E}t$ (fresh water) = 1.533×10^{-4} cal/kg/s and

$\bar{E}t$ (brackish water) = 1.236×10^{-4} cal/kg/s.

From Eq. 1 we estimated, for 1000-g (a = 4.26×10^{-5}, b = 0.023) and 2000-g (a = 4.25×10^{-5}, b = 0.0193) fish, the energy cost per unit of upstream displacement ($\bar{E}x$) of swimming at various mean speeds (W_o) in fresh water. The resulting relationships are given in Figure 3A. Figure 3B gives similar data for a 2000-g shad during the northward migration along the Atlantic coast of North America (U_O = 10 cm/s, B = 0, Leggett, 1977).

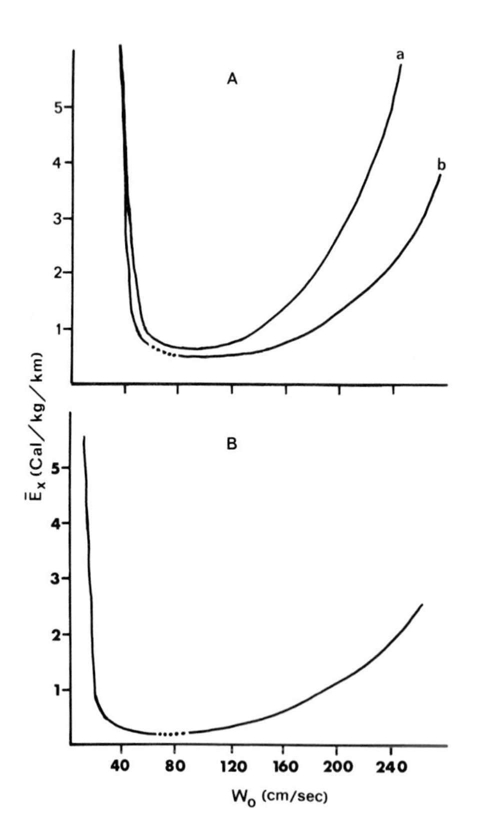

Fig. 3 A and B. Relationships between swimming speed (W_O) and energy per unit distance ($\bar{E}x$). (A) 1000-(*a*) and 2000-(*b*) g shad in fresh water (Connecticut River). (B) 2000-g shad during marine phase of migration. *Dashed lines* indicate 95% confidence limits of observed mean swimming speeds

D. Discussion

In the Connecticut, and presumably in other rivers, the spawning migration of American shad has two behaviorally distinct phases. In brackish water, and at the salt-fresh-water interface, extensive milling occurs, there is little adjustment of swimming speed to changes in current velocity, and average swimming speeds are low. In fresh water, the upstream migration is more direct, occurs at a higher average swimming speed, and speed is varied in response to tidal-induced variations in current velocity yielding an approximately uniform rate of progress over the bottom (Leggett, 1976). The adjustment of swimming speed to changes in current velocity also occurs during the final salt-water phase of the homing migration (Dodson and Leggett, 1973) and results from optomotor responses (Dodson and Leggett, 1974). This behavior is energetically inefficient, relative to the alternative strategy of maintaining a constant swimming speed through the water and allowing displacement over the bottom to vary with changes in current speed, and results in an added energy expenditure, on the part of the average shad in the population, of approximately 12 %. This aspect of the migration of shad thus appears to be regulated more by the typical optomotor response of fish to moving backgrounds (Jones, 1963; Arnold, 1974) than by energy constraints.

During the transition from salt to fresh water the observed behavior is much closer to the optimum, the difference between the two being approximately 2.5 %. The altered behavior observed during this phase of the migration serves two purposes: First it slows the rate of transition to fresh water. This is important since adult shad are severely stressed and experience high mortality if transferred rapidly from salt to fresh water (Leggett and O'Boyle, 1976). Secondly, it significantly reduces the absolute energy demands of swimming. In the specific case considered in this paper, the reduction in energy consumption per unit time, relative to the fresh-water level (19 %) is in close agreement with estimates of the added energy costs of osmoregulation during periods of osmotic stress (Rao, 1968; Farmer and Beamish, 1969). We thus conclude that the alteration of behavior observed during the transition to fresh water results from physiologic and energy constraints related to osmoregulation during the period of adjustment to fresh water, which override the normal behavioral response to current-induced changes in speed relative to the bottom. Similar behavioral responses are known to occur in other anadromous species (Groot et al., 1975). These, too, may be similarly related to the period of acclimation to fresh water.

Evaluation of the absolute energy cost/km of displacement at various swimming speeds reveals that adult American shad swim at mean speeds near the optimum both in fresh water (Fig. 3A) and during the marine migration (Fig. 3B). During the marine phase of the life history this behavior serves to maximize the energy available for growth and the accumulation of energy reserves required for the fresh-water migration (adult shad do not feed in fresh water). The extent of energy utilization in fresh water is high, and in some rivers is the direct determinant of levels of postspawning mortality (Glebe and Leggett, 1978). Migration at speeds significantly different from the optimum would lead to decreases in the levels of repeat reproduction and decreased population stability in northern populations; and could result in high mortality prior to reaching the spawing areas in some southern populations.

Absolute energy requirements of adult Connecticut River shad, based on proximate analyses of fish sampled at various stages throughout the migration (Glebe and Leggett, 1978), when adjusted to include gonad weight (17% of body wt; Leggett, unpublished data), indicate an average energy expenditure in fresh water of 4.1 Cal/kg/km. This is considerably higher than estimates derived in this analysis (0.65 Cal/kg/km). One possible reason for this difference is the assumption of constant upriver orientation. Based on an average displacement of 34 cm/s, and assuming the minimum distance path is followed, the mean migration rate expected is 29 km/day. The observed 9 year mean in the Connecticut River was 6 km/day (S:D 1.4) (Leggett, 1976). This indicates that the realized path may exceed the minimum distance route by up to 5 times. Leggett (1977) reached a similar conclusion concerning the path followed by shad at sea. Allowance for this factor increases our estimate of the absolute energy required to 3.2 Cal/kg/km. A second reason for the difference in estimates of absolute energy costs may be a basic difference in the metabolic efficiency of shad and of salmon (from which we derived our estimates of a and b). Our analysis suggests that salmon are significantly more efficient.

In the St. Johns River, Florida, migrating adult shad use approximately 25 % less energy/kg/km than do shad migrating in Connecticut, despite their smaller body size and the higher water temperatures they experience (Glebe and Leggett, 1978). The finding that a reduction in the amplitude of oscillations in current velocity significantly reduced the energy cost of migration in shad suggests that the smaller variations in current velocity, coupled with the lower mean current speed may account, in large part, for the reduced cost of migration there.

Finally, we are led to comment on the possibility of an adaptive relationship between mean body size at maturity and the energy requirements of the spawning migration. Shad native to southern rivers are smaller at maturity than their northern counterparts (Glebe and Leggett, 1978). In general, northern shad rivers have higher mean discharges, and greater variability in discharge rates, than those in the south (E. Shoubridge, McGill University, pers. comm.). They also have steeper gradients and higher average current speeds (Glebe and Leggett, 1978). From Figure 3A it is obvious that large body size, characteristic of northern stocks, provides a greater range of swimming speeds over which absolute energy requirements are minimal. This would tend to reduce the influence of environmental variability on the energy costs of migration and hence to stabilize mortality and population stability. It is important to emphasize, however, that other selection factors, related directly to life history strategies, are also important in determining size at maturity (Shoubridge, 1977; Leggett and Carscadden, in prep.).

Acknowledgments. Financial support for this work was provided by Groupe Interuniversitaire de Recherches Océanographiques du Québec (GIROQ); the Ministry of Education, Province of Quebec; and the National Research Council of Canada (Grant No. A6513 to WCL). Travel assistance for WCL was provided by the Government of Quebec, Ministry of Intergovernmental Affairs, Office of Education.

References

Arnold, G.P.: Rheotropism in Fishes. Biol. Rev. 49, 515-576 (1974)

Brett, J.R.: The relation of size to rate of oxygen consumption and sustained swimming speed of sockeye salmon (*Oncorhynchus nerka*). J. Fish. Res. Board Can. 22, 1491-1501 (1965)

Dodson, J.J., Leggett, W.C.: Behavior of adult American shad (*Alosa sapidissima*) homing to the Connecticut River from Long Island Sound. J. Fish. Res. Board Can. 30, 1847-1860 (1973)

Dodson, J.J., Leggett, W.C.: Role of olfaction and vision in the behavior of American shad (*Alosa sapidissima*) homing to the Connecticut River from Long Island Sound. J. Fish. Res. Board Can. 31, 1607-1619 (1974)

Dodson, J.J., Leggett, W.C., Jones, R.A.: The behavior of adult American shad (*Alosa sapidissima*) during migration from salt to fresh water as observed by ultrasonic tracking techniques. J. Fish. Res. Board Can. 29, 1445-1449 (1972)

Farmer, G.J., Beamish, F.W.H.: Oxygen consumption of *Tilapia nilotica* in relation to swimming speed and salinity. J. Fish. Res. Board Can. 26, 2807-2821 (1969)

Glebe, B.D., Leggett, W.C.: Bioenergetics of the freshwater spawning migration of American shad (*Alosa sapidissima*). In press, J. Fish. Res. Board Can. 35 (1978)

Groot, C., Simpson, M., Todd, I., Murray, P.D., Buxton, G.A.: Movements of sockeye salmon (*Oncorhynchus nerka*) in the Skeena River estuary as revealed by Ultrasonic Tracking. J. Fish. Res. Board Can. 32, 233-242 (1975)

Jones, F.R.H.: The reaction of fish to moving backgrounds. J. Exp. Biol. 40, 437-446 (1963)

Leggett, W.C.: The American shad (*Alosa sapidissima*), with special reference to its migration and population dynamics in the Connecticut River. Am. Fish. Soc. Monogr. No. 1, 169-225 (1976)

Leggett, W.C.: Ocean migration rates of American shad. J. Fish. Res. Board Can. 34, 1422-1426 (1977)

Leggett, W.C., O'Boyle, R.N.: Osmotic stress and mortality in adult American shad during transfer from saltwater to freshwater. J. Fish. Biol. 8, 459-469 (1976)

Rao, G.M.M.: Oxygen consumption of rainbow trout (*Salmo gairdneri*) in relation to activity and salinity. Can. J. Zool. 46, 781-785 (1968)

Shoubridge, E.: Reproductive strategies in local populations of American shad (*Alosa sapidissima*). MSc Thesis, McGill Univ. (1977)

Trump, C.L., Leggett, W.C.: Optimum swimming strategies in fish: The problem of currents (in prep.)

Horizontal and Vertical Swimming of Eels During the Spawning Migration at the Edge of the Continental Shelf

FRIEDRICH-WILHELM TESCH, Biologische Anstalt Helgoland, 2000 Hamburg 50, FRG

Abstract

Tracking experiments on migratory eels (*Anguilla anguilla* L.) on the shelf near the East Atlantic continental slope showed a NNW preferred compass direction (PCD) and confirmed earlier results in the North Sea. Eels tracked off-shelf (water depth more than 200 m) exhibited a westerly PCD with probably a slight tendency to the southwest (direction of the Sargasso Sea). Off shelf preferred hourly mean depths were 50 - 215 m at night. Short-term depth changes of 200 m occurred at a maximum. At dawn the eels showed a strong vertical downward migration. The preferred depths were probably greater than 400 m during daylight.

A. Introduction

The spawning migration of the eel (*Anguilla spec.*) in the sea has been investigated recently by ultrasonic tracking methods (Tesch, 1972, 1974a, 1977, 1978; Stasko and Rommel, 1974; Westerberg, 1975; Westlin and Nyman, 1977). After leaving the fresh-water and coastal regions the migration takes place in two completely different hydrographic areas. The first part is the route on the flat and mostly turbid shelf area; the second part begins after the eel has left the continental shelf, when it encounters great depths, water of a high transparency, and generally more homogeneity of temperature and salinity. Tracking results (Tesch, 1974a) and sporadic recapture of conventionally tagged eels (Lühmann and Mann, 1958) in a large, comparatively uniform shelf area of the North Sea reveal that migrating silver eels, upon leaving the coastal areas of the German Bight, swim in a north westerly direction. It is implied that the eels leave the North Sea, passing around Scotland to the North and not via the comparatively recent opening of the North Sea in the southwest, the Channel. The results presented here are derived from tracking experiments conducted further offshore than those reported earlier (Tesch, 1974a). Some new or recently published data (Tesch, 1977, 1978) from trackings taken on the shelf immediately in front of the continental slope (less than 200 m deep) and off the shelf at depths greater than 200 m are compared. It is especially interesting to know where or at which place the direction of the migratory route changes. Somewhere the eels have to turn southwest in order to reach the Sargasso Sea. The development of depth preferences is also of importance.

B. Materials and Methods

Origin, size, tracking area, and conditions of the experimental eels are summarized in Table 1. They were in the migratory - or the so-called silver-eel stage. The transmitters employed, including six pressure-telemetring pingers, and the receiver system were described earlier (e.g., Tesch 1974a, 1977, 1978). The tracking vessel from which hydrographic water samples down to a depth of 500 m were taken is illustrated in Figure 1. Except for eel No. 28 (lightship measurements)

Fig. 1. The tracking vessel R.K. *Friedrich Heincke* (length: 38 m)

no consistent data on currents are available. The water movements caused by tidal influences are probably small in most of the areas under consideration because of great depths and for this reason have been neglected for evaluation of tracking data. An occasional measurement (eel No. 26) of the current in a depth of 25 m over a water depth of 120 m with a drifter exhibited a current of 0.1 m s^{-1}; this is low compared with the normal migratory speed of eels of 0.5 m s^{-1} (Tesch 1974a, 1977). Calculations of the directional tendency of single specimens are presented as the direction from release to the end of the tracking on the shelf or on the slope. This method, as compared with calculating the mean direction from the different sections of the track, results in no essential difference (Tesch, 1977b). In addition Decca navigational positions obtained on the continental slope at quite a distance from the radio transmitters tended to display considerable inaccuracy. This had a strong bearing on the determination of positions for the single sections of the tracking course, which frequently resulted in a zig-zag (Fig. 4) course that was not observed during earlier trackings.

C. Results

I. Horizontal Migration

Figures 2 and 3 show the migratory direction of eels from the place of their release to the end of the tracking on the shelf and off the shelf west of Ireland, France, or Spain, including all tracking experiments previously performed in these areas. To increase the comparatively low sample sizes, eel No. 15 is presented in both Figures 2

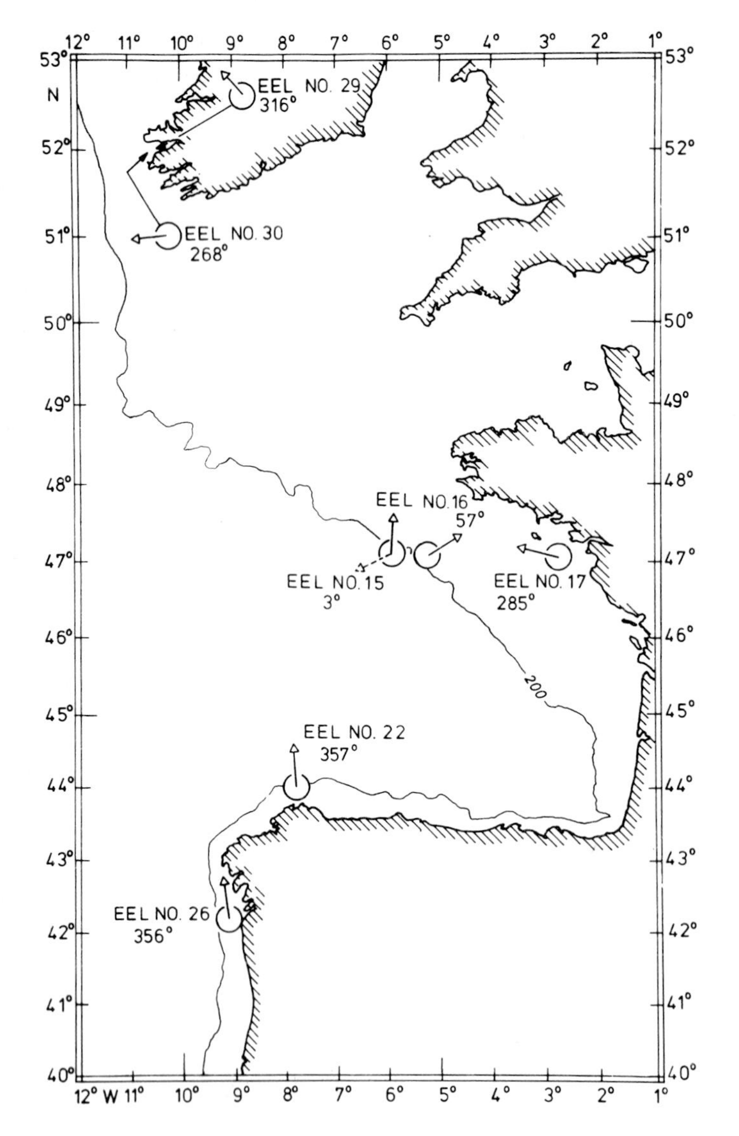

Fig. 2. Directions and positions of eel tracks on the east Atlantic shelf not far from the continental slope. *Dashed arrow No. 15* indicates the tracking course off the shelf

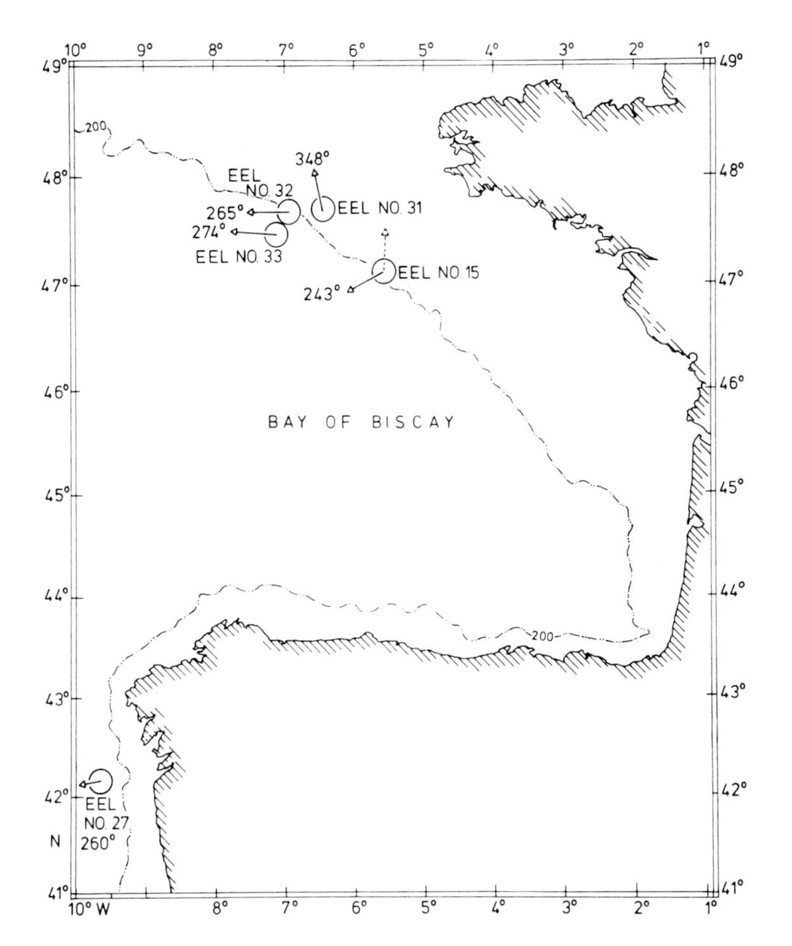

Fig. 3. Directions and positions of eel tracks on the East Atlantic continental slope. *Dashed arrow No. 15* indicates the part of the tracking course on the shelf

and 3 because it exhibited migration both on the shelf and off the shelf. Released at a depth of 200 m it swam north and turned southwest at a water depth between 200 and 400 m (Fig. 4). The map that shows the migration over the shelf (Fig. 2) includes two specimens near the Irish coast, an area that does not have free access from all directions. One of these specimens (No. 29) changed its direction when it contacted shallower water during its NW migration. For this reason only the first part of its track was used for evaluation. Out of seven specimens tracked on the shelf only one (No. 16) exhibited no significance when the Hodges and Ajne test (Batschelet, 1972) was applied on the directionality of their movements. Most courses tracked on the shelf exhibited a northwest direction. Hence the mean direction (337^{0}) is nearly equal to the mean direction of all specimens tracked earlier in the North Sea (341^{0}).

When examining the five eels tracked on the continental slope for the randomness of their tracks according to a Hodges and Ajne test, they displayed directed movements except for eel No. 27, whose track suffered from considerable inaccuracy of the Decca positions and therefore the randomness of the track might be due to technical failures. The five specimens performed a more westerly course. An exception is eel No. 31, which preferred a NNW direction. As demonstrated later, its vertical preferences were also exceptional. The Rayleigh test on the

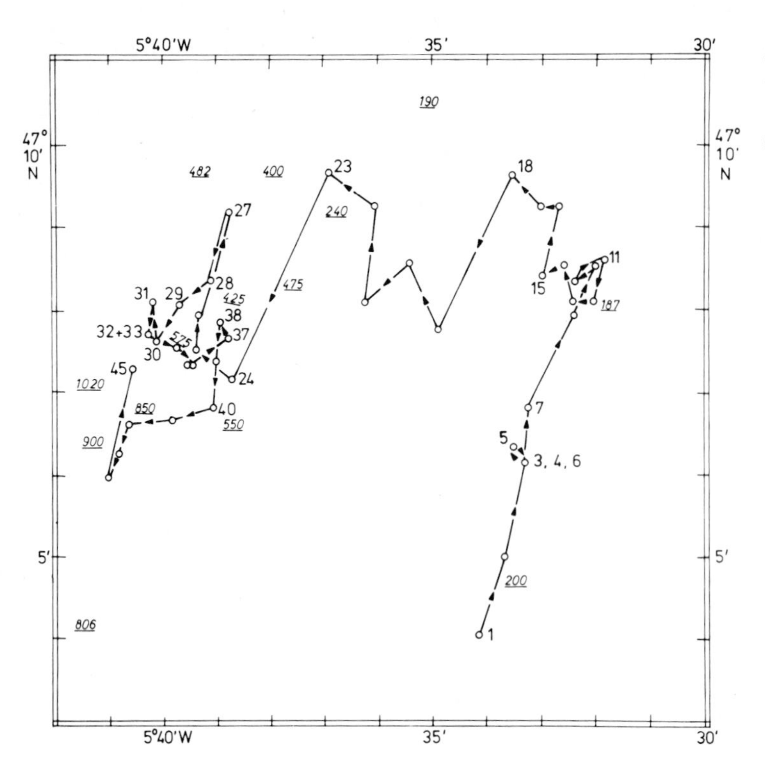

Fig. 4. Tracking positions of eel No. 15 with directions exhibited during migration on the shelf and off the shelf (deeper than 200 m)

five slope tracks exhibited significance of the mean direction at the 5 % level. If No. 31 is excluded and only the off-shelf part of track No. 15 is evaluated, the mean course is 260°, which is not much different from the direction to the Sargasso Sea (Fig. 5).

Wheter the difference between the two directions on the shelf and on the slope is significant can be calculated by using the parametric two sample test of Watson and Williams (Batschelet, 1965). Similar concentrations (k) of the two samples are required, and hence application to the five-slope specimens and the six North Sea tracks is justified (k = 3.1 and 3.4, respectively). The difference is significant at the 2.5 % level. The concentration of the seven East Atlantic shelf tracking courses is too low (k = 1.9) to be comparable to the slope trackings.

II. Vertical Preferences

An example of the depth transmitted by the pressure-sensing tag of eel No. 32 is presented in Figure 6. This specimen was released on the continental slope during daylight, and the tracking was continued until noon of the next day. It dived strongly after release (see also eel No. 17, Tesch 1977a), but later swam until dusk, usually immediately below the surface. At night it preferred much greater depths (215 m), swimming between 100 and 300 m. At dawn the preferred depth increased to 400 m and at 0930 it increased even further so that it was no longer measurable by the pressure transducer. Eels No. 27 and 33 exhibited a similar pattern of depth preferences except for the first part of their tracking, which took place during the night, and their preferences were not as deep during the night (average: 75 and 166 m respectively).

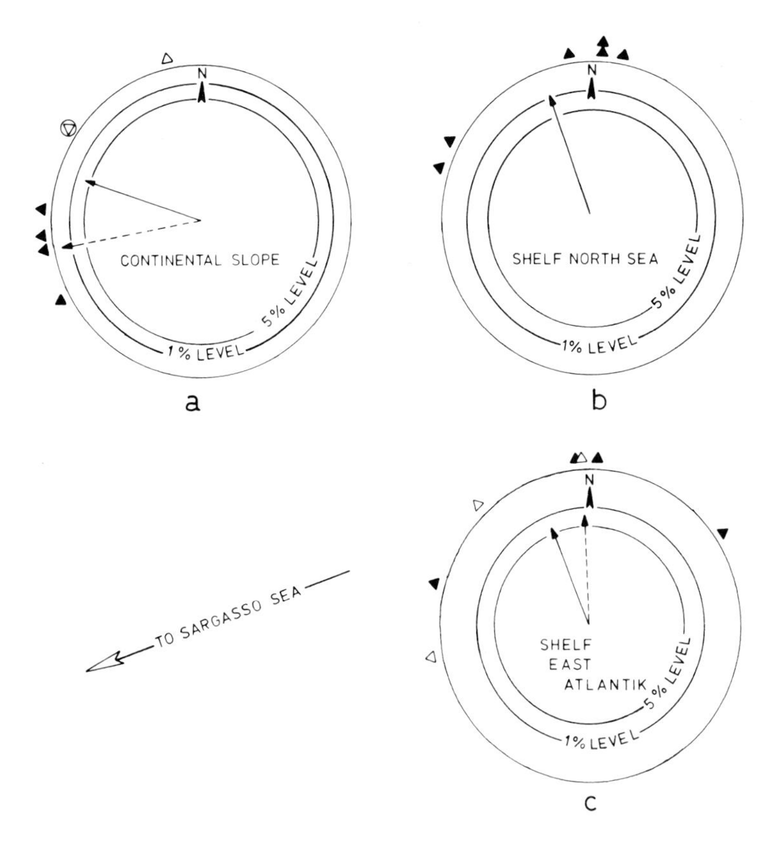

Fig. 5 a – c. Circular distribution of directions (*triangles*) of eels and mean vectors with indications of significance levels (not valid for *dashed arrows*). (a) Continental slope; *unbroken arrow:* mean of all *triangles* except the southern-most (off-shelf component of No. 15); *dashed arrow*, mean of all *black triangles;* the *white triangles* indicate individuals with part of their movements on the shelf, the *encircled triangle* direction of No. 15 from start to end of tracking; (b) North Sea (after Tesch, 1974a); (c) East Atlantic shelf; *unbroken arrow*, mean of all *triangles; broken arrow*, mean of *black triangles* (specimens are free to more in all directions)

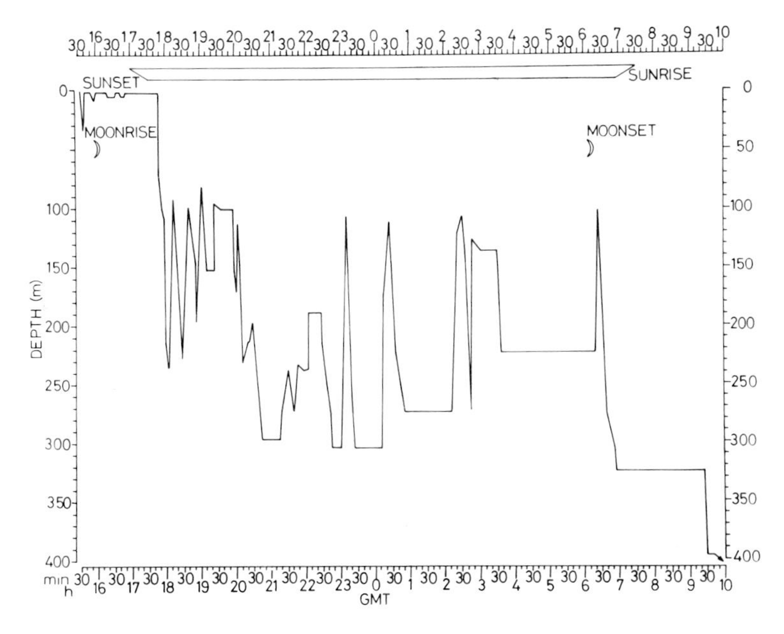

Fig. 6. Example of vertical movement pattern from release to beginning of deep diving in the morning hours (0932 GMT, Nov. 6, 1976) with rise and setting of sun and moon. From 9:32, depth exceeded the range of the pressure transducer until end of tracking (1145)

Their hourly mean depth is shown in Figure 7 together with that of eel No. 32. Eel No. 27 swam at 100 m depth during moonlight and at 50 m after the setting of the moon (Tesch, 1978). During trackings of eels No. 31, 32, and 33, the moon was above the horizon nearly all night. Eel No. 31, which was released during daylight and only one day earlier than No. 32, showed in addition to its abnormal horizontal movement a quite different pattern of diurnal depth curve (Fig. 8). Its mean hourly swimming depth is compared with the depth curve of an eel (No. 28) that was tracked six weeks earlier in the North Sea and that ex-

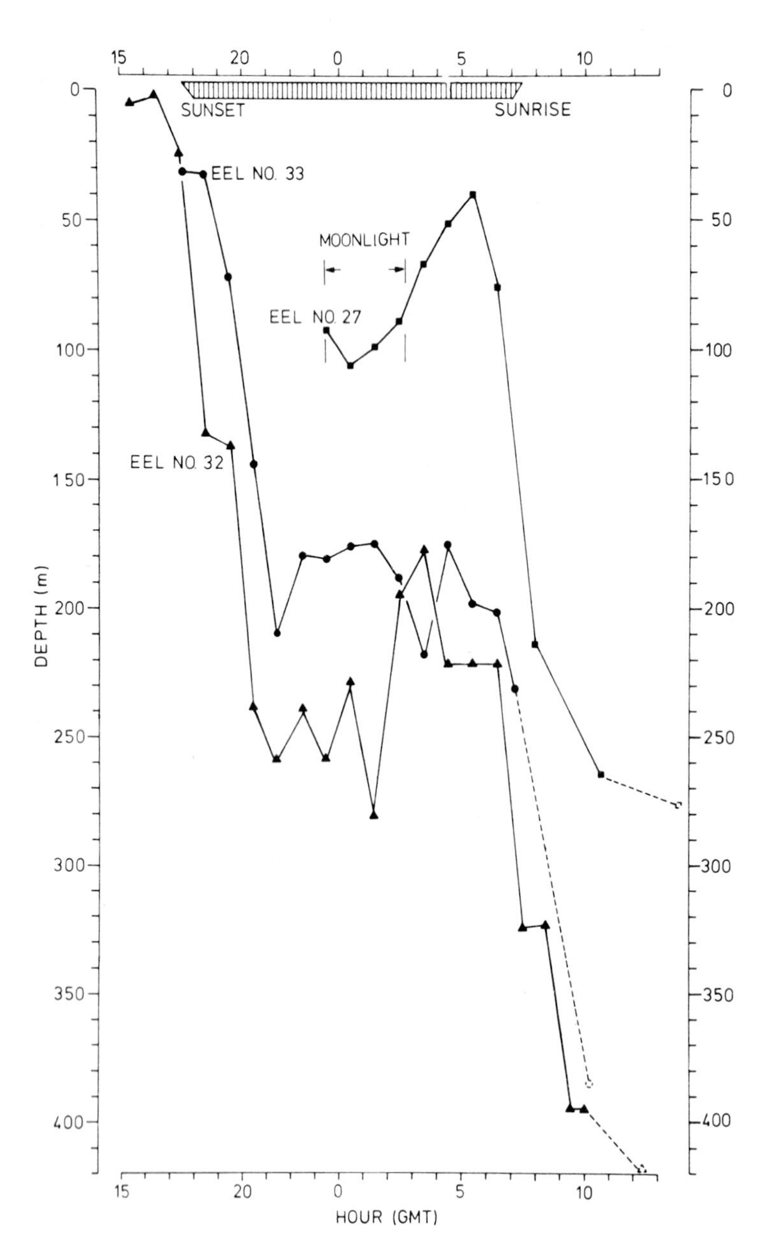

Fig. 7. Hourly mean swimming depth of eels (Nos. 27, 32, 33) at the continental slope in relation to sunrise and sunset (after Tesch, 1977b). Time of moon above the horizon is given for No. 27. *Dashed lines*, no continuous and accurate recording

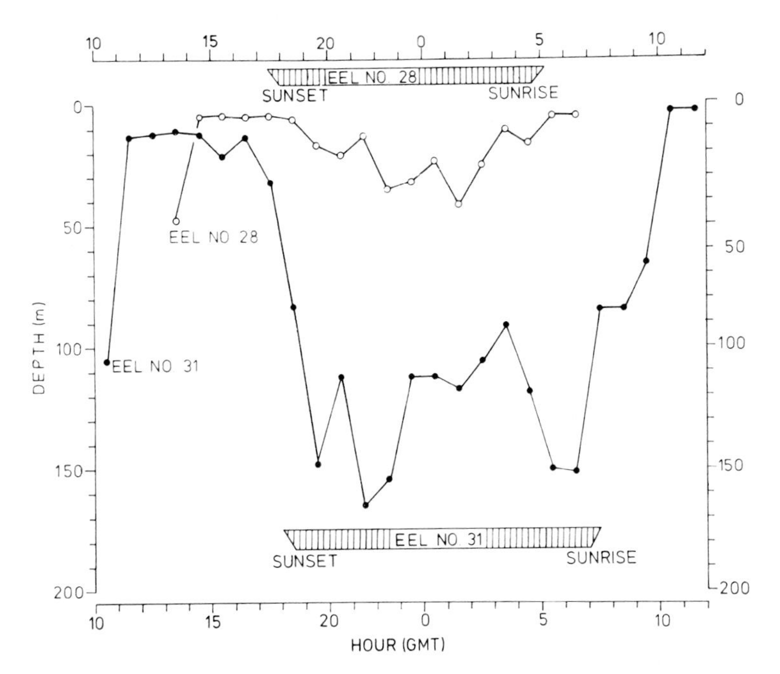

Fig. 8. Hourly mean swimming depth of eel. No. 31 at the edge of the northeast Atlantic shelf and of eel No. 28 in the North Sea in relation to sunrise and sunset

hibited random horizontal movements. The pattern of the depth curve of this eel is similar to that of No. 31 (below the surface during daylight, deeper at night) except for a generally shallower depth level (water depth 38 m). Numbers 28 and 31 might have been suffering under an especially strong experimental treatment before their release.

Hydrographic conditions during trackings on the slope are only noteworthy if temperature is considered; salinity showed only very small changes with depth (0.1 - 0.2 ‰ between 0 and 500 m). The change of temperature and depth together with the mean swimming depth of eels is demonstrated in Figure 9. At the area and time of tracking Eels No. 32 and 33, a slight thermocline is present at a depth between 75 and 100 m. Both fish swam exclusively below the termocline, and the temperature at the preferred water layers was slightly above 11° C. During tracking of No. 27 at a similar depth, a stronger thermocline occured and the eel swam generally below it except after the setting of the moon. In most instances of diving or emerging, crossing of the thermocline occurred. In conclusion, none of the eels tracked at the slope chose a special water layer characterized by a special temperature.

D. Discussion

Due to the fact that experimental sampling is extremely expensive results of horizontal movements at the continental slope are based on a minimum amount of data. The dispersion of directions in field experiments of this kind is comparatively low (Tesch, 1974a). As few as two or three specimens can give an idea of the migratory course. This was shown by the first North Sea tracking experiments (Tesch, 1972). The results of the northeast Atlantic shelf trackings have to be con-

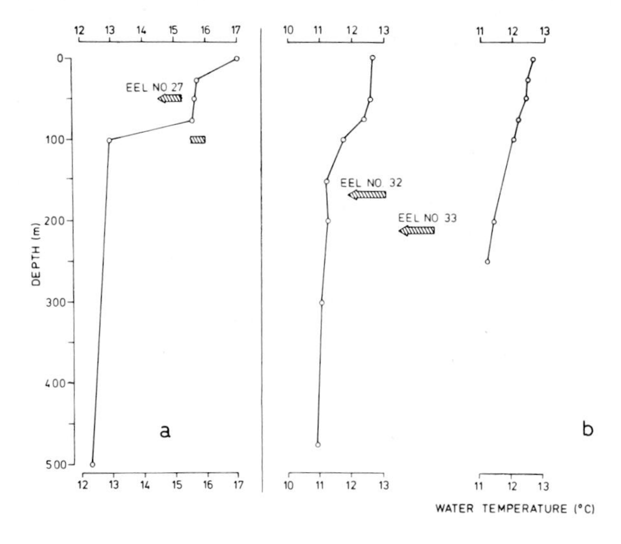

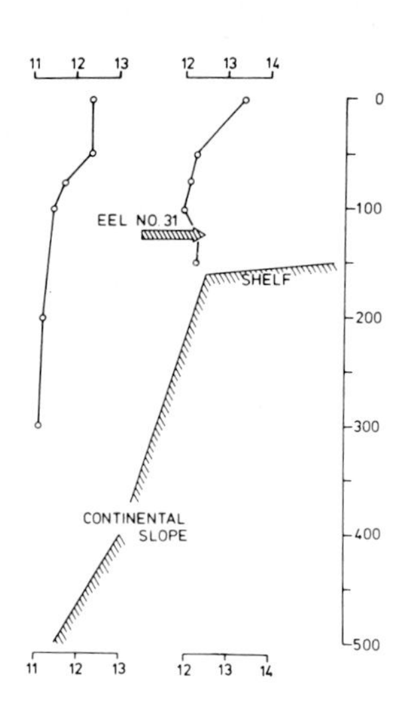

Fig. 9 a and b. Temperature-depth relationship at the water depths and time when trackings at the continental slope took place, mean swimming depth of eels (*arrow*) at night; the direction of the *arrow* indicates when the eel swam to the shelf (*to the right*) or off the shelf (*to the left*). (a) Eel No. 27 at the west Spanish continental slope with mean depth during moonlight and after setting of the moon; (b) Eels No. 32, 31, 33 at the continental slope of the Northern Bay of Biscay

sidered with more reservation (mean direction 337°). Out of seven eels (Fig. 2), only four (Tesch 1977a) were tracked in places where they had free access to all directions, provided that the first part of track No. 15 is included; their mean direction was 357° and was pointed even more northerly than the whole sample (337°) and the North Sea trackings. Therefore the probability that the migratory courses of the eels on the East Atlantic shelf not far from the slope, are equal (NNW) or similar to courses far from the slope (e.g., North Sea and Baltic, after Tesch, 1974; Westin and Nyman, 1977) is great. It is therefore probable that at increasing depths over about 200 m, a release mechanism begins to act, turning the migratory course farther south into the final direction of the Sargasso Sea. Eel No. 15 probably demonstrated this change of direction during migration from the shelf to the slope. The release for the change could be the decreasing light conditions or, more probably, the increasing pressure. Similar conditions of low irradiance can perhaps also be encountered during migration in turbid shelf areas.

The sense of pressure also seems to be important during migration on the shelf; it probably releases directional changes if shallower water is encountered and enables the eel to swim along the shoreline without leaving a certain minimum depth (Ovchinnikov, 1971; Tesch, 1977; Westin and Nyman, 1977). Under this assumption the sense of pressure must be as finely developed as it is in some bony fishes (Tsvetkow, 1969). On the other hand the body has to withstand a great range of pressure. Silver eels dive at a maximum speed of 0.6 m s^{-1} (Tesch, 1978). Our results have confirmed those laboratory experiments which showed that diving at a moderate speed and deeper than 400 m can be withstood without any permanent damage (Belaud, 1975). That *Anguilla anguilla* occurs at even greater depths is indicated by the fact that it has been found as a prey of benthopelagic predator fishes at a depth more than 700 m (Reinsch, 1968).

Table 1. Conditions and locations of capture, release, and tracking of eels tagged by pressure-sensing ultrasonic transmitters

Silver eel tracking experiment No.	Length of eel (cm)	Location of capture	Method of capture	Date of tracking	Time of tracking (GMT)	Area of release	Water depth (m)	Weather conditions during tracking, wind in Beaufort	Remarks
				Continental slope					
15	82	Coast of Fehmarn/ Baltic	Pound net	17 - 18 Nov.1973	18:10 - 17:20	Biscay 90 nm SW of Brest	200-575	Overcast, later clear, SSE 4	No depth telemetry, after Tesch, 1978
27	105	Lake Erne Ireland	?	14 - 15 Nov.1975	23:25 - 12:15	43 nm West of Vigo (Spain)	2000-2200	Clear, from varying directions	after Tesch, 1978
31	94	Coast of Fehmarn/ Baltic	Pound net	4 - 5 Nov.1976	09:58 - 11:28	Biscay 100 nm West of Brest	410-150	Overcast, westerly 2 - 5	after Tesch, 1978
32	97	River Shannon Ireland	Stow net	5 - 6 Nov.1976	15:35 - 11:45	Biscay 100 nm WSW of Brest	960->2500	Overcast, westerly 2 - 7	after Tesch, 1978
33	89	Coast of Fehmarn/ Baltic	Pound net	6 - 7 Nov.1976	17:39 - 10:20	Biscay 100 nm WSW of Brest	1450-2000	Overcast, 5 - 8	after Tesch, 1978

Table 1. (continued)

Continental shelf									
16	85	Mediter-ranean Coast of France	?	18 - 20 Nov.1973	22:15 - 11:20	Biscay 90 nm SW of Brest	200-160	Overcast, partly bright: wind from varying directions	No depth telemetry, after Tesch, 1977a
17	102	Mediter-ranean Coast of France	?	23 Nov. 1973	19:00 - 22:58	Biscay 40 nm SW of St. Nazaire	53	Fog and overcast wind easterly 1 - 2	after Tesch, 1977a
22	104	Lake near Nante, France	?	10 - 11 Nov.1974	12:05 - 10:45	North Spanish coast near La Coruna	17-94	Mostly bright, NW-SW 0 - 1	No depth telemetry, after Tesch, 1977a
26	89	Lake Erne, Ireland	?	13 - 14 Nov.1975	12:50 - 18:15	West Spanish Coast 35 nm SW of Vigo	80-120	Bright and cloudy, weak and varying directions	No depth telemetry, after Tesch, 1977a
28	89	Coast of Gotland/ Baltic	Pound net	21 - 22 Sept.1976	13:31 - 11:00	North Sea German Bight	38	Easterly 0 - 1	No depth telemetry,
29	83	River Shannon Ireland	Stow net	27 - 28 Oct.1976	20:15 - 01:00	Coast of SW Ireland	8-20	Clear, easterly 1 - 2	No depth telemetry
30	84	Coast of Fehmarn/ Baltic	Pount net	28 - 29 Oct.1976	16:35 - 06:25	Coast of SW Ireland	11-21	Clear, wind, varying directions	No depth telemetry

Light is an important factor affecting silver eel migration, because it prevents activity. This has been shown in several laboratory studies (e.g., Edel, 1975; Veen et al., 1976; Westin and Nyman, 1977). Catch statitics in fresh and salt water are in accordance with these results, and even moonlight influences activity in the laboratory as well as during migration under natural conditions (e.g., Lowe, 1952; Bräutigam, 1961; Hain, 1975; Winn et al., 1975; see also Lindroth, 1977). Field experiments have partly proved the inactivity of silver eels during daylight in the sea (Westerberg, 1975). However, in other cases turbid water conditions probably screened the light to such an extent that eels could continue to migrate during the day also (Tesch, 1977a). In the present study, eels escaped from high irradiance conditions by diving, and there was no indication that their horizontal movement stopped (Tesch, 1978). Influence of moonlight on depth preferences of other organisms (Euphausiacea) has been demonstrated by Roger (1975), and therefore it is not unlikely that the eels also have been affected by their depth preference.

Short diving excursions to considerable depths have also been observed at night. Similar observations are known from the eel shelf migration (Stasko and Rommel, 1974; Westerberg, 1975; Tesch, 1977a). It is possible that the experimental treatment of the eels, especially during the beginning of the tracking, might have caused this. Light or noise of the tracking vessel may have evoked diving as a kind of fleeing reaction. These vertical excursions, sometimes completely down to the sea bottom, might be attributed to orientational activities. Stasko and Rommel (1974) presented such an idea: "... geoelectric fields are among the more likely orientation cues (Rommel and McCleave, 1973 ...). The resulting electric currents are strongest near the surface and bottom; weakest in between ..." and the eel changes between the layers of strongest stimuli. Considering the migratory behaviour of eels observed on the European continental slope, this kind of searching for orientational cues becomes unlikely. I would rather speculate that the eels simply "measure" the depth to obtain the release for the directional change at the edge of the slope. As discussed earlier, directional cues for finding the compass course may be received from the geomagnetic field (Tesch, 1974a, b, 1977). Orientation on water-current-generated electric fields on the shelf would lead the fish in different directions because tidal currents swing between varying directions. On the slope and above the deep sea the currents are low and change with depth. In the Gulf Stream, the direction of current can change to a countercurrent (Richardson and Knauss, 1971), at a depth which during daylight is probably shallower than that preferred by eels; countercurrents also occur at the boundaries of the Gulf Stream.

Light as a cue for the compass course is even more improbable because irradiance at the preferred depth is very low, and directionality of celestial cues is practically nonexistent. Temperature has been postulated as acting as a releaser and as a directional stimulater (Westin and Nyman, 1977). My earlier investigations indicated that temperature has a releasing function for the beginning or the stopping of the migration (Tesch, 1972, 1974a, 1977). However, nothing pointed to the fact that temperature (and smell) could have been important for finding the direction. In addition the depth preferences found during the present investigation have shown that the eel obviously is not attracted by different temperature layers; thus an orientational function of temperature is unlikely.

References

Batschelet, E.: Statistical methods for the analysis of problems in animal orientation and certain biological rhythms. Monogr. Am. Inst. Biol. Sci. 1-57 (1965)

Batschelet, E.: Recent statistical methods for orientation data. In: Animal orientation and navigation. Galler, S.R., Schmidt-Koenig, K., Jacobs, G.J., Belleville, R.E. (eds.). National Aeronautics and Space Admin. Wash. D.C., 61-91, 1972

Belaud, A.: Contribution à l'étude de quelques réactions physiologique de l'Anguille (Anguilla anguilla L.) soumise à diverses conditions hyperbares. Diss. Univ. Bretagne occid. Sci. Nat. 271 pp. (1975)

Bräutigam, R.: Über Versuche zur Intensivierung des Blankaalfanges durch die Kombination von Lichtsperren und Großreusen und ihre grundsätzlichen Bedingungen. Fisch.-Forsch. Rostock, 4,9-25 (1961)

Edel, R.L.: The effects of shelter availability on the activity of male silver eels. Helgoländer wiss. Meeresunters. 27,167-174 (1975)

Hain, J.H.W.: The behaviour of migratory eels, Anguilla rostrata, in response to current, salinity and lunar period. Helgoländer wiss. Merresunters. 27, 211-233 (1975)

Lindroth, A.: Eel catch and lunar cycle on the Swedish east coast. Rapp. P.V. Réun. Cons. Int. Explor. Mer. (in Press) (1977)

Lowe, R.H.: The influence of light and other factors on the seaward migration of the silver eel (*Anguilla anguilla* L.). J. Anim. Ecol. 21, 275-309 (1952)

Lühmann, M., Mann, H.: Wiederfänge markierter Elbaale vor der Küste Dänemarks. Arch. Fisch Wiss. 9, 200-202 (1958)

Ovchinnikov, V.V.: The influence of hydrological factors upon orientation of European eel (Anguilla anguilla L.). ICES Anadromous and Catadromous Fish. Comm. M: 12, 1-2 (1971)

Reinsch, H.H.: Fund von Flußaalen, Anguilla anguilla, L., im Nordatlantik. Archiv. Fisch Wiss. 19, 62-63 (1968)

Richardson, P.L., Knauss, J.A.: Gulf stream and western boundry undercurrent observations at Cape Hatteras. Deep Sea Res. 18, 1089-1109 (1971)

Roger, C.: Effects of moon phase and moonlight upon the vertical distribution of macroplantonic Crustacea (Euphansiacea). Cah. O.R.S.T.O.M. Paris, Ser. Oceanogr. 12, 159-172 (1975)

Rommel, S.A., McCleave, J.D.: Sensitivity of American eels (Anguilla rostrata) and salmon (Salmo salar) to weak electric and magnetic fields. J. Fish. Res. Bd. Can. 30, 657-663 (1973)

Stasko, A.B., Rommel, S.A.: Swimming depth of adult American eels (Anguilla rostrata) in a saltwater bay as determined by ultrasonic tracking. J. Fish. Res. Bd. Can. 31, 1148-1150 (1974)

Tesch, F.-W.: Versuche zur telemetrischen Verfolgung der Laichwanderung von Aalen (Anguilla anguilla) in der Nordsee. Helgoländer wiss. Meeresunters. 23, 165-183. (Experiments on telemetric tracking of spawning migration of eels (Anguilla anguilla) in the North Sea. Fish. Res. Bd. Can. Transl. Ser. 2724, 1-29) (1972)

Tesch, F.-W.: Speed and direction of silver and yellow eels, Anguilla anguilla, released and tracked in the open North Sea. Ber. dt. wiss. Komm. Meeresforsch., 23, 181-197 (1974a)

Tesch, F.-W.: Influence of geomagnetism and salinity on the directional choice of eels. Helgoländer wiss. Meeresunters. 26, 382-395 (1974b)

Tesch, F.-W.: Tracking of silver eels (Anguilla anguilla) in different shelf areas of the Northeast Atlantic. Rapp. Réun. Cons. Int. Explor. Mer. (in Press) (1977)

Tesch, F.-W.: Observations on the spawning migration of the eel (Anguilla anguilla) west of the European continental shelf. Env. Biol. Fish. (in press, 1978)

Tsvetkov, V.I.: On the threshold sensibility of some freshwater fishes to the rapid change of pressure. (Russ.). Vop. Ikhiol. 9, 715-721 (1969)

Veen, Th. van, Hartwig, H.G., Müller, K.: Light dependent motor activity and photonegative behaviour in the eel (Anguilla anguilla L.). Evidence for extraretinal and extrapineal photoreception. J. Comp. Physiol 111, 209-219 (1976)

Westerberg, H.: Counter-current orientation in the migration of the European eel (Anguilla anguilla L.). Göteborg Univ. Oceanogr. Inst., Rep. 9, 1-18 (1975)

Westin, L., Nyman, L.: Activity, orientation and migration of Baltic eel (Anguilla anguilla (L.)). Rapp. P.V. Réun. Cons. Int. Explor. Mer. (in press) (1977)
Winn, H.E., Richkus, W.A., Winn, L.K.: Sexual dimorphism and natural movements of the American eel (Anguilla rostrata) in Rhode Island streams and estuaries. Helgoländer wiss. Meeresunters. 27, 156-166 (1975)

Session VII

Chairman:

ARCHIE CARR, Department of Zoology, University of Florida,
Gainesville, Fl 32601, USA

Some Experiments on Navigation in the Harbour Seal, *Phoca vitulina*

HENRY JAMES[1], and R. W. DYKES[2], [1] Department of Psychology, Dalhousie University, Halifax, N.S., Canada, [2] Departments of Surgery, Neurology and Neurosurgery, McGill University

Abstract

Seals, captured on Sable Island (43° 55' N; 60° 00' W) and released 24 h later in the interior of the island from places with which they are not familiar and from which the sea is not visible, head SSW on the shortest route to the coast. They do so irrespective of the direction from which the sound of the surf is most clearly audible. Seals prevented from seeing the normal pattern of polarized light, but with their vision otherwise unimpaired, head SSW if the surf is clearly audible anywhere in the southern hemisphere, but go due north into the dunes when the sound of the surf comes from the north. Blindfold animals head in the direction of the sound of the surf. Deaf seals go due north or due south: they are unable to discriminate the seaward from the non-seaward hemisphere even under sunny skies, and they cannot determine the shortest path to the sea even when they do go south. Anosmic seals behave normally.

An hypothesis is suggested to account for these results.

At the end of December each year, several thousand grey seals (*Halichoerus grypus*) converge on a particular beach on Sable Island, a vegetated sand bar about 160 km off the Nova Scotian coast. There they give birth to their young, nurse them, mate, and then go back to sea. We identified some of these animals individually as they arrived for one breeding season and then watched for them to arrive the next year (Boness and James, in prep.). They returned at the same time (± 24 h) to the same place (± 18 m). Feats of the same order are performed by *Callorhinus ursinus* (Kenyon and Wilke, 1953; Peterson, 1965) and, perhaps, by all those pinnipeds that combine pelagic feeding with terrestrial polygyny (Repenning, 1976).

At 400 kg apiece, grey seals are not the most suitable candidates for experiments on navigation. The harbour seal (*Phoca vitulina*), which also breeds on Sable Island, is more easily handled and, because of some unusual circumstances, has given us an opportunity to experiment on the sensory control of orientation in a phocid. During the breeding season (May - mid-June) and the moult (August), about 15% of the harbour seal population on Sable Island colonizes some small lakes (the Wallace Lakes, Fig. 1) in the interior of the island. The seals reach these lakes by crossing the dry bed of a lagoon (the Sandy Plain) from the south coast of the island, an overland journey of 0.3 km to 1 km or more, depending on the departure point. A ridge of high dunes, which extends for some 11 km from east to west along the north edge of the Sandy Plain and which borders the north edge of the lakes, makes it impossible for seals to gain access to the lakes from the north coast.

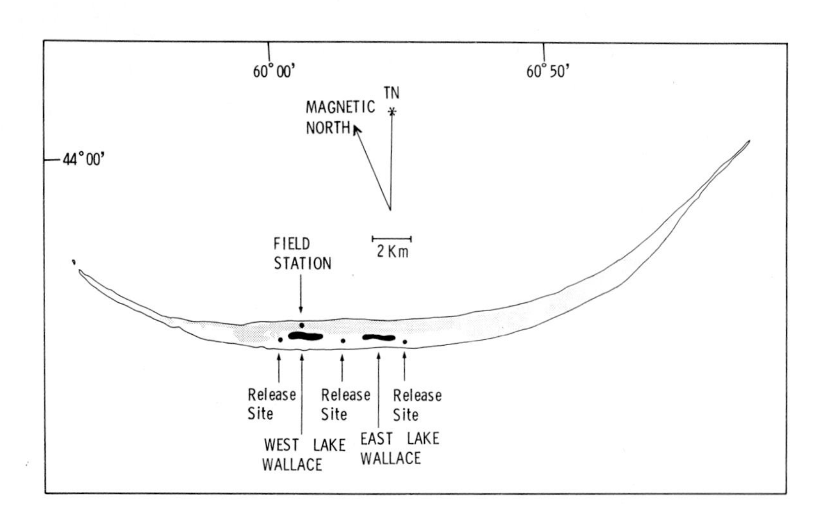

Fig. 1.
Sable Island

Apart from these dunes, which present a rolling landscape with no clues to latitude that we have been able to discern, the only landmarks are a series of less than 10 small, isolated dunes (the number has been decreasing annually) along the south edge of the Sandy Plain between E and W Lake Wallace. Neither the sea nor the lakes were visible from the places (release sites, Fig. 1) where we tested our seals, but except in very dense fog, the horizon appeared closer to the SSW (i.e., in the direction of the shortest path to the sea, approximately 205° magnetic) than it did in other directions. The sound of the surf was usually audible on the Sandy Plain, but its apparent direction is correlated with the direction of the wind ($r = 0.81$), and it is not a reliable indicator of the direction in which the south coast lies or of the shortest path to the sea.

Seals which have been captured while travelling between the lakes and the sea and then displaced immediately to one of the release sites shown in Figure 1 either go back to sea on a course that does not deviate significantly from the shortest path to the south coast or make an appropriate correction for their involuntary displacement in latitude and head straight for the nearest corner of the nearest lake (James and Renouf, in prep.). Seals which have been kept in captivity for 24 h before they are tested almost invariably head back to the south coast from the release site and never attempt to reach one of the lakes. Because some of the experiments described below made it necessary to hold seals at the field station (Fig. 1) for at least a day, we have had to confine ourselves to the question: how do harbour seals find the sea?

A. Methods

Adult harbour seals were captured either while crossing the Sandy Plain or while they were hauled out on the N or S coasts of the island. Each was taken immediately in a covered cage to the field station, where it was held in an open-air pen. The seal was then either subjected to surgery, given a sham operation, anaesthetized, and then allowed to recover without further interference, or simply left in its pen. About 24 h after capture, it was taken in a covered cage to one of several release sites (Fig. 1). On arrival, a record was made of the direction and speed of the wind, the direction and loudness of the sound of the surf, the amount of cloud cover, and the extent of horizontal visibility. The seal was then released through the N door of the cage, recaptured 10 min later, and returned to its pen at the field station. The track which the animal left in the sand made a convenient record of its movements, and while it was being recaptured, measurements were taken so that an exact map of its entire track could be reconstructed. Approximately 24 h later, most seals were given a second trial at one of the other release sites. A few escaped, dislodged their blindfolds etc., developed pneumonia, or were used as pilot subjects for other experiments after they had received only one trial. As noted below, the seals in some groups were given more than two trials.

Here we shall summarize only the experiments we have made with seals that were either anosmic, blindfold, deaf, or fitted with wave-retarding goggles, together with the results from normal and sham-operated controls. The operative procedures were as follows.

I. General Anaesthesia

The seal was immobilized with a single dose of 2 - 3 ml succinylcholine chloride (10 mg/ml) administered via the intravertebral epidural vein. One ml atropine (1 mg/ml) was injected at the same time to reduce mucous secretion. An endotracheal tube was immediately inserted and respiration begun with a respirator. Anaesthesia was induced with 4% halothane and maintained with a 1% - 1.5% mixture for a minimum of 45 min, although during some surgical procedures the animal was kept under anaesthesia for up to 4 h. The animals remained areflexic throughout surgery.

II. Destruction of the Organ of Corti

The inner ear mechanism was destroyed bilaterally in 19 seals. The ventral aspect of the bulla was exposed surgically and opened with a dental drill. The vascular membrane in the middle ear was pushed aside and haemostasis established. The petromastoid portion of the temporal bone was thus exposed and opened with a drill. The interior of the cochlear canal was disrupted with forceps. On recovery from anaesthesia, 9 of the animals were found to have symptoms of vestibular dysfunction (slight lateral tremor of the head accompanied, in some cases, by nystagmus) as well as being completely deaf. Because the deaf seals often refused to leave the cage at the release site, we gave these animals as many trials (1 - 7) as our facilities and the seal's condition would allow.

III. Attachment of Translucent Blindfolds

Nineteen seals were fitted with translucent blindfold, cut from white plastic (Fig. 2). The blindfolds were attached under general anaesthesia to the periorbital region of the skin with surgical thread. While they passed sufficient light to enable one to determine the approximate position of the sun (when this was visible to the naked eye), they prevented pattern vision and made it impossible to detect polarized light.

Fifteen of the blindfold seals were given 2 trials each, while the remaining 4 were given only one.

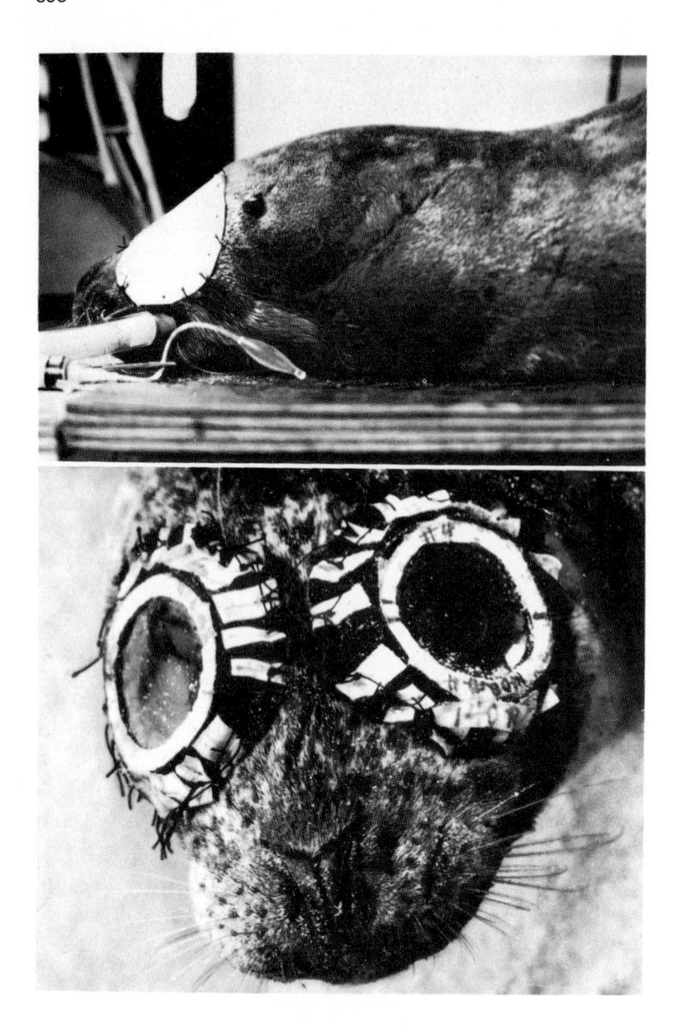

Fig. 2. Photographs of seals wearing a translucent blindfold (*top*) and goggles fitted with wave-retarding lenses (*bottom*)

IV. Attachment of Optical Retarders

Ten seals were fitted with goggles, the lenses of which were made from optically retarding plastic (Fig. 2). In 5 cases, the lenses had a retardance of 140 ± 20 μ at 560 μ, and in the other 5, the retardance was 280 ± 20 μ at 560 μ. The goggles were attached in the same manner as the blindfolds and as much care as possible was taken to place the axes of the retarders in the same orientation over both eyes, with the fast axis at approximately 90° to the long axis of each eye. The lens holders were made of absorbent material, so that they did not become clouded with condensation, and they were kept as free of sand as possible.

Nine of these animals were given 3 trials each, while the tenth was given only two.

V. Interruption of the Olfactory Tract

In 8 seals, the olfactory tract was interrupted bilaterally under general anaesthesia by crushing it against the cribiform plate with a small, curved, blunt instrument inserted through the posterior ethmoidal foramen.

VI. Unoperated and Sham-Operated Controls

It was impractical, for several reasons, to try to prepare the same number of sham-operated controls as experimental animals. Twenty seals, selected at random and tested concurrently with the appropriate experimental groups, were subjected to a sham operation under general anaesthesia: 5 were subjected to the full surgical procedure involved in deafening animals except that the cochlear canal was not opened, 5 were fitted with goggles made of clear plastic that was not dichroic, and 10 were simply allowed to recover from anaesthesia without further intervention. In addition, 30 unoperated, unanaesthetized controls, also chosen at random, were tested during these experiments.

Eleven of the controls received 1 trial, 31 were given 2 trials, and 8 were given 3 each.

B. Measurement of the Seals' Behaviour

In most cases (except when deaf), a seal would leave the cage as soon as the door was opened, turn in a shallow curve, and then set out on a straight course, which it maintained for the rest of the trial. From the map that we made of each track, we determined the compass bearing of the first 100 yards (91.4 m) of straight (± 9.1 m) track. We shall refer to this bearing as the final vector (FV) of the track. We shall describe as "failures" those trials on which a seal either travelled less than 91 m or failed to take up a straight course within this distance of the cage. No FVs were calculated for these trials.

C. Results

The FVs obtained from the various experimental groups, as well as from the controls, are shown in Figure 3, together with the number of failures occuring under each condition.

I. Controls

There were no differences between the sham-operated and the unanaesthetized unoperated controls and so their results have been combined. The mean final vector ($\overline{FV}$) for control trials under clear or partly clouded skies ((Fig. 3a); $\overline{FV}$ = 211°; coeff. dev. = 33°; r = 0.69) does not differ from that obtained under completely overcast skies ((Fig. 3b); $\overline{FV}$ = 207°; coeff. dev. = 29°; r = 0.77). Neither $\overline{FV}$ differs from the shortest path to the south coast, which had a bearing of approximately 205° from the release site.

The relation between the FVs of the controls, under clear and cloudy weather, and the provenance of the sound of the surf is shown in Figures 4 a and b respectively. It will be seen that, under both sets of conditions, the correlation is low and that, even when the sound of the surf was either inaudible to the human ear or very difficult to hear and localize (right-hand columns of Figs. 4a and b), the controls tended to head in the seaward direction.

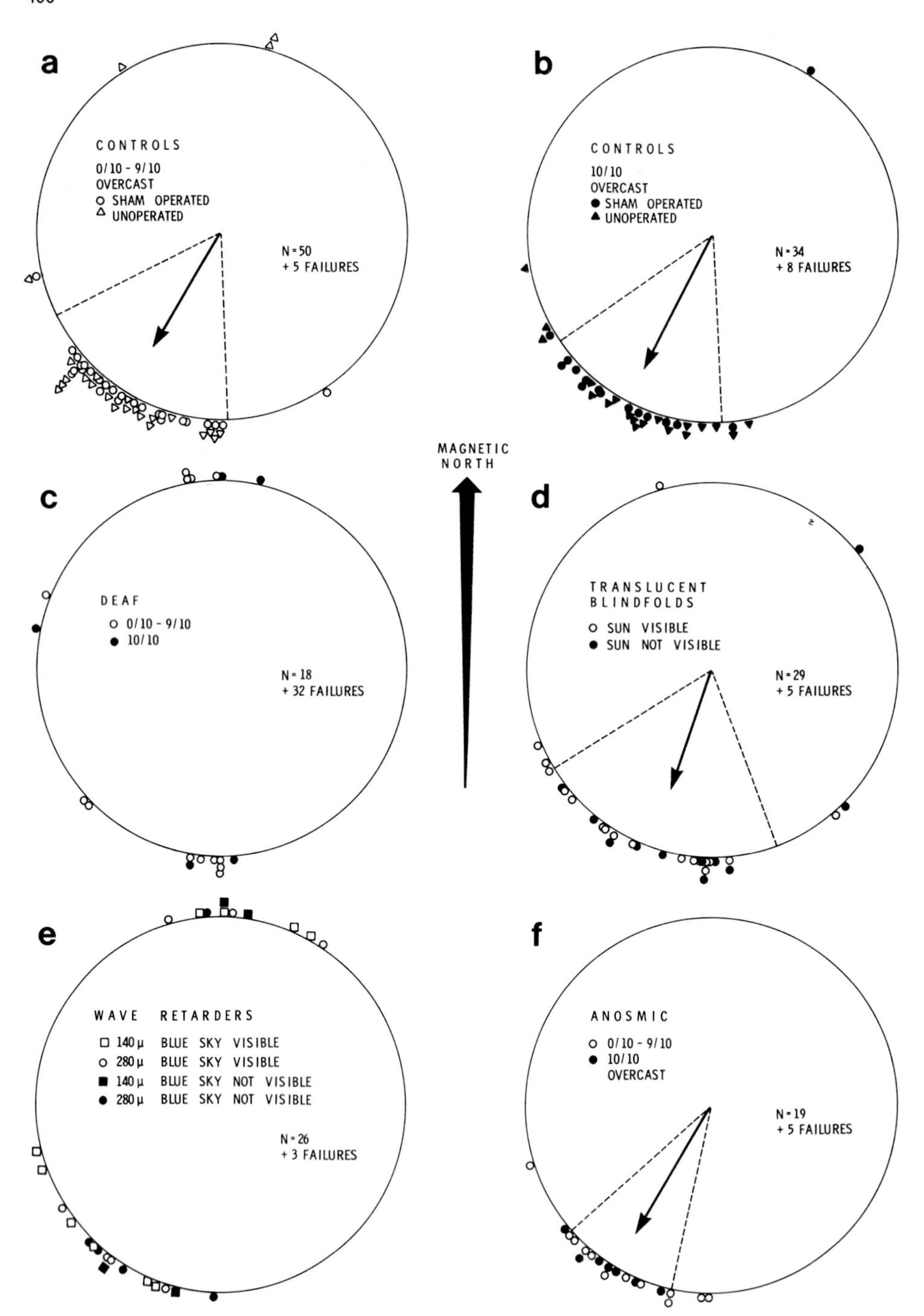

Fig. 3 a - f. Track directions (*final vectors*) of (a) controls under clear or partly clouded skies; (b) controls under completely overcast skies; (c) deaf seals; (d) seals wearing translucent blindfolds; (e) seals wearing wave-retarding goggles; (f) anosmic seals. *N*, no. of trials. *Open circles*, clear or partly clouded skies; *filled circles*, completely overcast skies. The number of trials on which a seal failed to produce a straight track (*failures*) is shown for each condition

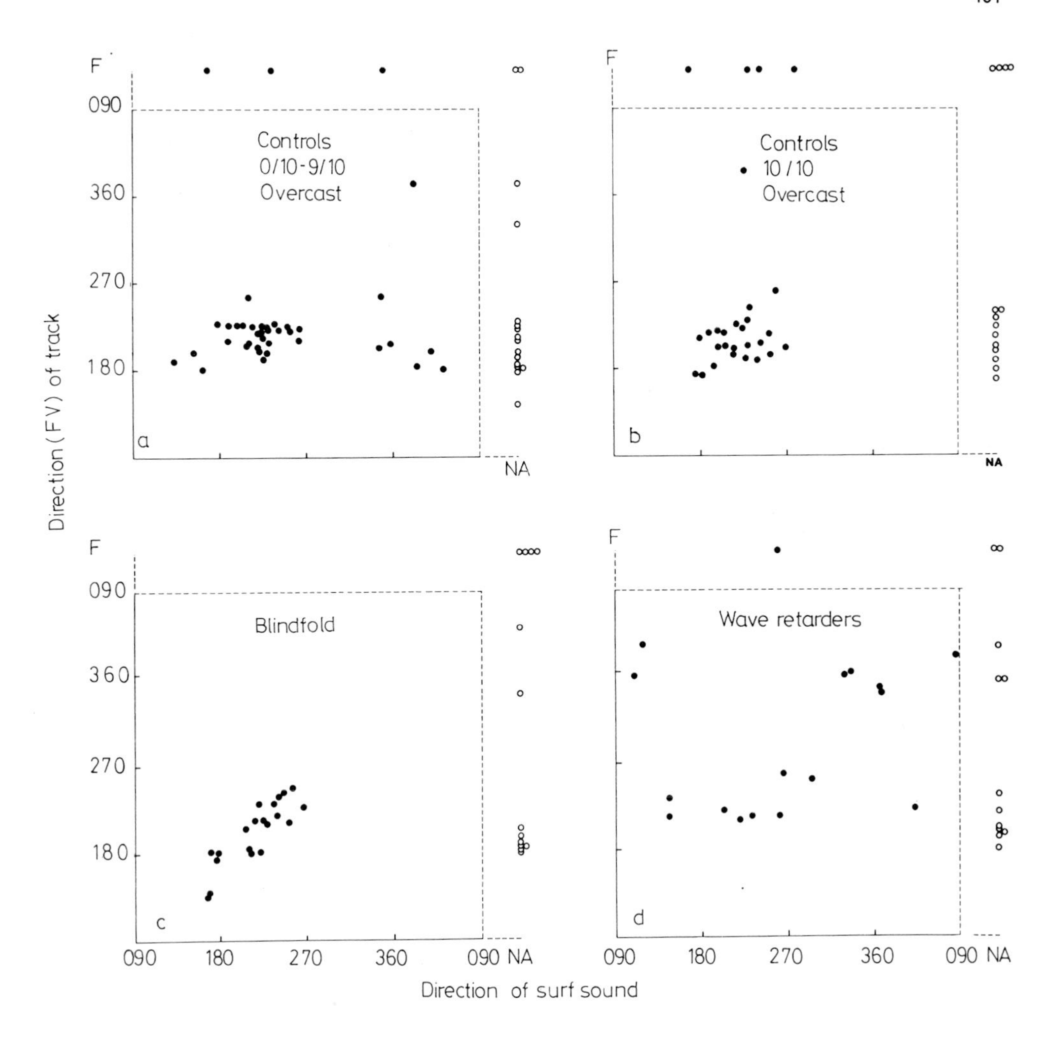

Fig. 4 a - d. Relation between track directions and the direction of the sound of the surf for (a) controls tested under clear or partly clouded skies, (b) controls tested under completely overcast skies, (c) blindfold seals, and (d) seals wearing wave-retarding goggles. *NA*, sound of surf very faint and difficult to localize. *F*, failures

Figure 4 also demonstrates one of the peculiarities of the weather on Sable Island, which may have an important bearing on the seals' ability to find the sea under overcast conditions: whereas it sometimes happens that the surf is clearly audible to the N, but not the S, under sunny skies, it is usually audible only to the S when the weather is bad. We shall return to this point later.

II. Deaf Seals

The FVs of the deaf seals (Fig. 3c) appear to be bimodally distributed about the poles of the magnetic meridian. We have been unable to find out why a given seal sometimes went north and at others went south. The choice of direction did not vary with the presence or absence of solar cues (Fig. 3c), was not correlated with wind direction ($r = 0.29$), and did not depend on the extent of horizontal visibility. When a deaf seal travelled north, it climbed into the dunes, whence a large expanse of ocean was visible to the south and nothing but even higher dunes to the north. Nevertheless, when allowed to do so, it persisted on its northerly course and fiercely resisted all attempts to turn it about.

Most of the failures amongst the deaf seals happened because the animal went straight back through the open door of the cage and stayed inside for the rest of the trial. These failures occurred predominantly, but not exclusively, in animals showing signs of vestibular dysfunction.

III. Blindfold Seals

The final vectors of the blindfold seals are shown in Fig. 3(d). Their tracks are significantly clustered in the seaward direction ($\overline{FV} = 190^{\circ}$; coeff. dev. = 39°; $r = 0.59$). However, this finding must be interpreted with caution for 2 conjoined reasons: (1) we have found a high correlation between the FVs of these seals' tracks and the direction of the sound of the surf (Fig. 4c; $r = 0.86$) and (2) we have never managed to have blindfold seals ready for testing when the sound of the surf was clearly audible from the north and not the south.

When the sound of the surf was very faint and difficult to localize (NA) blindfold seals tend to haul close to the magnetic meridian (Fig. 4c).

IV. Seals Wearing Optical Retarders

The final vectors of these animals are shown in Figure 3e. Their FVs appear to be bimodally distributed. They differ from the deaf seals in that they veer to the SSW (i.e., the seaward direction) when they are southbound, and they resemble the deaf seals by hauling close to the meridian when they go north.

The correlation between the direction of the sound of the surf and the FVs of the seals wearing wave retarders is high ($r = 0.79$) and it is apparent from an inspection of Figure 4d that this is largely because the decision whether to go N or S is determined by the provenance of the sound of the surf and not because the latter determines the precise angle of the track with respect to the meridian.

V. Anosmic Seals

The FVs of the anosmic seals are shown in Figure 3f. These animals behaved like normal controls.

D. Discussion

These results can be understood if one is prepared to make the following assumptions: (1) Harbour seals are able to detect a meridian that coincides with the magnetic meridian; (2) they can differentiate between the N and S poles of this meridian if the normal pattern of polarized light is visible; (3) their orientation with respect to this meridian is first determined by the provenance of the sound of the surf; and (4), they then search for visual clues indicating the proximity of the coast, this search being restricted to the sector from which the sound of the surf appears to come.

Under overcast skies, the sound of the surf almost invariably comes from the SW quarter (Fig. 4), and so normal seals do not have to face the problems that arise when the breakers are audible only to the north of the Sandy Plain; they simply have to orient in the direction of the sound of the surf, locate that sector to the SSW where the horizon appears to be closest, and then head for the point within that sector which is closest to the apparent source of the sound of the surf (Fig. 4b). Deaf seals cannot hear the sound of the surf, and so have no way of knowing where to look for horizon cues; thus, if our model is correct, they have no choice but to stay where they are or to head along the meridian. Since they cannot hear the surf, they cannot tell whether it is coming from the north or south even when polarized light cues are available, and so the direction of travel along the meridian is presumably chosen at random (Fig. 3c). Blindfold seals cannot use horizon cues to modify the course indicated by the sound of the surf, and our model leads one to expect the high correlation between track and surf sound directions which we actually observed (Fig. 4c). Although it is extremely rare for the sound of the surf to be completely inaudible on the Sandy Plain, it is sometimes very difficult to localize it precisely. Under these conditions, blindfold seals tend to head along the meridian. It should be added that, in experiments with completely blind seals which we have not described here because of lack of space, we have found that these animals also head in the direction of the sound of the surf (although with considerably more scatter than is the case with blindfold seals), but that they do not head along the meridian when the sound of the surf is difficult to localize. A detailed analysis of the behaviour of blind and blindfold seals may provide some interesting insights into Leask's (cf. this Symposium) optical pumping model of magnetoreception.

Normal (and anosmic) animals are able to find the seaward direction even when the sound of the surf is audible only from the N, provided that the sky is clear or only partly clouded. Under natural conditions, seals on Sable Island have ample opportunity to go inland from the south coast and thus to learn that the sea is in the opposite direction to the sound of the surf when the latter is audible only to the north. (Because of the sheer drop from the top of the dunes to the beach on the north coast, seals cannot go inland on the north side of the island and so they do not have an opportunity to learn the otherwise confusing fact that the coast may be closer to the north when the sound of the surf is from that direction.) It is possible that the seals apply this partial knowledge under our test conditions. To do so, however, they must be able both to hear the sound of the surf and to perceive the pattern of polarized light. Deaf seals cannot hear the sound of the surf, and those wearing optical retarders cannot see the normal pattern of polarization; thus neither group has the information necessary to reverse their bearings when the sound of the surf is from the north.

Acknowledgments. We particularly thank Daryl Boness, Derek Sarty, Derek Tanji, and Zoe Lucas for their assistance in the field. The research was supported by grants to H.J. from the National Research Council of Canada and the Izaak Walton Killam Fund of Dalhousie University.

References

Kenyon, K.W., Wilke, F.: Migration of the northern fur seal, *Callorhinus ursinus*. J. Mamm. 34, 86-98 (1953)

Peterson, R.S.: Behavior of the northern fur seal. Johns Hopkins Univ., Baltimore, Md., D.Sc. Thesis. (1963)

Repenning, C.A.: Adaptive evolution of sea lions and walruses. Syst. Zool. 25, 375-390 (1976)

Homing in Wild Myomorph Rodents: Current Problems

JACQUES BOVET, Department of Biology, Laval University, Québec, Canada, G1K 7P4

Abstract

Authors currently disagree as to whether random scatter, familiarity with a large area, or navigation is the basic mechanism of homing in rodents. Starting from the behavior of animals that fail to home, I present a new hypothesis based on motivational factors, which could explain the observed decreases of homing success that occur with increasing distances in rodents and other animals, irrespective of the possible mechanisms involved. I also suggest that consideration of the behavior of nonhomers is essential in attempts to explain homing.

Currently, the best evidence that we have on homing of myomorph rodents is that, in field experiments with inexperienced animals (i.e., animals which participate for the first time in a homing trial), there is an obvious drop of return percentages (homing success) with increasing displacement distances. This was discovered by Murie and Murie (1931) in their study on deer mice (*Peromyscus maniculatus*). It has been subsequently confirmed by many authors working on various species. Figure 1 gives seven examples. It is further recognized that the largest distance between any two points in the regular home range of a mouse is smaller than the distances over which homing still occurs. In the field vole *Microtus pennsylvanicus* for instance, a 44% homing success is obtained after 200-m displacements (see Fig. 1, curve 1) from animals for which observed home range length does not exceed 30 m (Robinson and Falls, 1965). Based on these facts, three general kinds of alternative hypotheses about homing mechanisms are proposed in the literature. They were already mentioned by Murie and Murie in 1931. The first is that random dispersal away from release site produces the homing performances observed. The second is that homing is dependent upon the successful mice being familiar with the area into which they are released, and using previously known landmarks to find their way home. The third is that some kind of navigation mechanism enables the mice to orient toward home without previous familiarity with the release site area. These hypotheses are still under consideration and discussion (cf. the three most recent papers on the topic: DeBusk and Kennerly, 1975; Anderson et al., 1977; Cooke and Terman, 1977).

There are two reasons for that apparent lack of progress over almost 50 years. One is that the decrease in homing success seen with increasing distances appears to be a poor indicator of homing mechanisms. Wilkinson (1952) showed that it can be the result of a random search process, linked or not to familiarity with an area possibly much larger than a standard home range. On the other hand, the data produced by Wallraff (1970) on "free-flight" inexperienced pigeons show that a decrease in homing success with increasing distance (see Fig. 1, curve 9) is compatible with the existence and use of a navigation mechanism (demonstrated for instance by the initial orientation of the birds after release). The second reason is that the other types of ex-

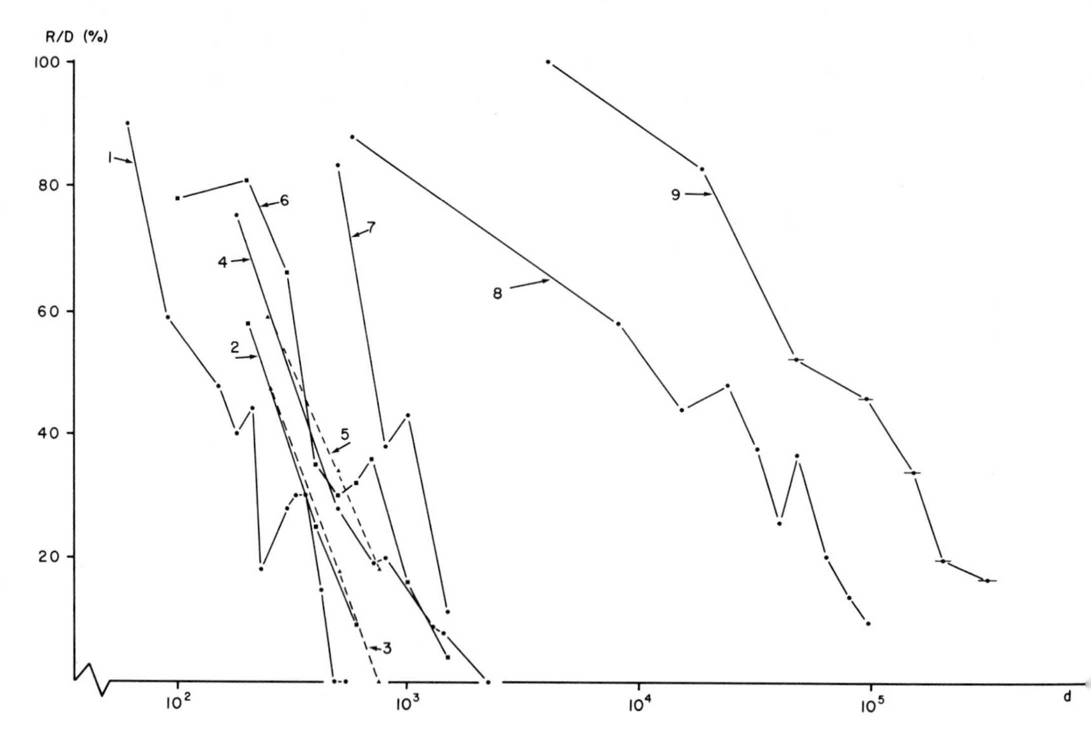

Fig. 1. Relationship between homing success (*R/D*) of inexperienced individuals and displacement distance (*d*) in several species of myomorph rodents and other animals. *1*, meadow vole, *Microtus pennsylvanicus* (from Robinson and Falls, 1965); *2*, American redback vole, *Clethrionomys gapperi* (from Bovet, in prep.); *3*, European redback vole, *Cl. glareolus* (from Bovet and Gogniat, in progress); *4*, deer mouse, *Peromyscus maniculatus* (from Murie, 1963); *5*, wood mouse, *Apodemus* spp. (from Bovet and Gogniat, in progress); *6*, cotton rat, *Sigmodon hispidus* (from DeBusk and Kennerly, 1975); *7*, Malaysian woodrat, *Rattus jalorensis* (from Harrison, 1958); *8*, little brown bat, *Myotis lucifugus* (from Mueller, 1966); *9*, homing pigeon (from Wallraff, 1970; some data obtained from different distances have been pooled, and the ranges of distances involved are indicated by the small horizontal lines)

perimental results that have been produced in addition to the relationship between homing success and distance are sometimes contradictory, and often ambiguous with respect to the three basic alternatives under debate (references and discussion in Bovet, 1972).

The major cause for our inability to solve the problem is the paucity of the information that we have on the behavior of the mice that fail to home or on the reasons why they fail. This requires an explanation for people unfamiliar with mice. Most field studies on the ecology or behavior of myomorph rodents rely heavily on live-trapping procedures (capture-marking-recapture), and most of the basic information that exists is, therefore, of the type "when and where a given mouse was found in a trap." The most common procedure in homing studies is best illustrated in the papers by Murie (1963), Furrer (1973), and DeBusk and Kennerly (1975). Trapping is conducted in a single area that comprises the home ranges of all the mice to be displaced, and the various release sites are scattered around the area in a more or less arbitrary way. Continuous trapping in the home area provides an opportunity for each displaced mouse to "prove" its homing performance in getting recaught in a trap: If it does, it is a *homer*, if it does not it is a *nonhomer*. The criterion used to define a homer is a good one: The mouse is known to be, at a certain time after release, at a well-defined place, which is the home range it occupied before displacement. However, the criterion used to define a nonhomer is vague and negative; the only thing we know about the mouse is that it did not enter a trap in the home area between the time of release and the end of trapping. This says nothing of the whereabouts of the mouse, not even that it failed to home (a mouse can home and not enter a trap, although I agree that it is unlikely). Within the framework of trapping procedures, better information on the behavior of nonhomers could be gained if they were given chances of getting recaught in traps set for extended periods of time in several directions near and/or at some distances around the release site(s). The pattern of successes and failures in recatching the nonhomers in these additional traps would then yield information on their behavior and could provide answers to such questions as: Do they move away from the release site at all? If so: What courses do they take? When, where, and possibly why do they stop traveling? Does displacement distance affect the answers to these questions? It is obvious that this kind of information is highly relevant in critical considerations of any of the three basic hypotheses mentioned above, and at the same time is likely to provide cues as to why homing success can drop with increasing distances irrespective of the mechanism involved.

Some answers are found in the literature. When trapping was conducted close to a release site (up to about 50 m) for an extended period of time after the releases had been made, it was found that a number of nonhomers had become residents of that area, as evidenced by repeated recaptures there (Fisler, 1962, 1966; Robinson and Falls, 1965; Cooke and Terman, 1977). Applying this procedure at a common release site used for deer mice displaced over three different ranges of distances, I did not find any relationship between displacement distance and the proportion of displaced mice that settle down near the release site (Bovet, 1972). This is also true for redback voles, *Clethrionomys gapperi* (Bovet, in prep.). On the other hand, a few experiments have also been performed (over relatively short displacement distances), in which mice had opportunities not only to be recaught at the home site but also at other places equidistant from the release site (Stickel, 1949; Saint Girons and Durup, 1974; Cooke and Terman, 1977). Their results showed a low rate of recapture of nonhomers, as compared to that of homers. Finally, in an experiment in which mice displaced over relatively long distances had opportunities to get recaught in several directions and at several distances around the release site, I found that most of the nonhomers that showed up in traps did so fairly close to the release site (Bovet, 1962).

To complement these data, I report the results of an experiment that we have made with European woodmice (*Apodemus sylvaticus* and *A. flavicollis;* justification for treating these two species as a single "unit" is given in Bovet, 1962). The trapping procedure is illustrated in Figure 2. Trapping stations are set north, west, south, and east at points 250 m, 500 m, and 750 m from a central point, and are operated simultaneously. Resident mice from each of the twelve trapped areas are displaced to and released at the central point, which thus represents a common release site for all mice tested. An animal that then moves away from this site has an opportunity to be recaught not only as a homer at its original place of capture, but also as a nonhomer at any of the eleven other trapping stations. We have applied that procedure in two different areas within the same forestry range

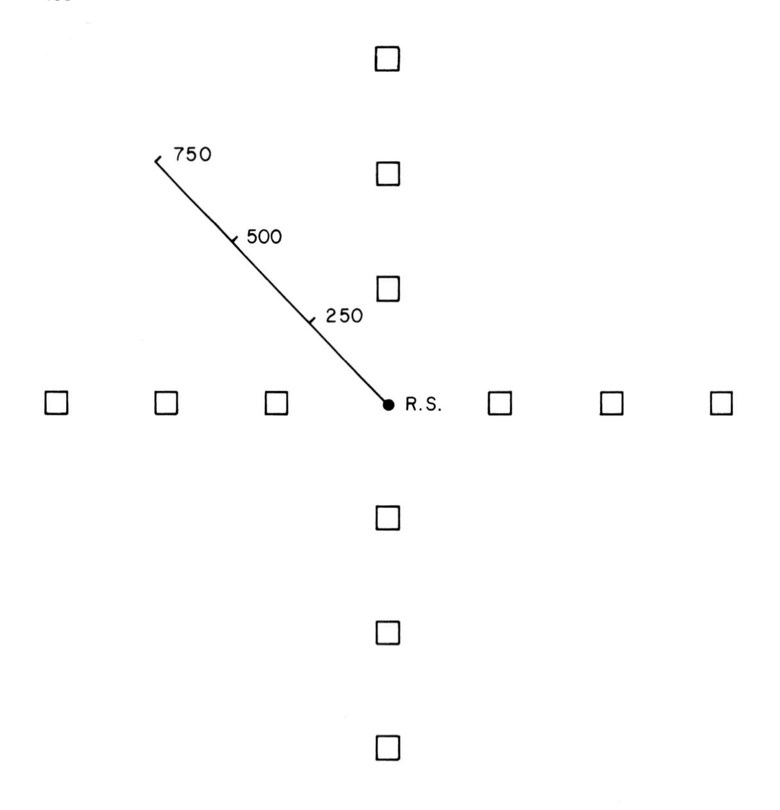

Fig. 2. Trapping procedure. The twelve trapping stations (*squares*) are distributed symmetrically around a single release site (*R.S.*). Each station has 20 traps (4 lines of 5 traps each, with 10 m between any two adjacent traps of a line and 10 m between any two adjacent lines). Scale in meters

near Lausanne, Switzerland. The results obtained from inexperienced mice in the two series are pooled in Figure 3 in such a way that the home direction is at the top of the Figure for each tested mouse. The positions of the individual symbols show the last places of recapture after displacement (only 6 of the 136 mice recaught, 3 homers and 3 nonhomers, were found at another place before they reached the position shown in Fig. 3).

We have thus obtained positive information on 51 nonhomers, i.e., 32% of the 159 mice which failed to home. The last places of recapture of these 51 animals concentrate at 250 m from the release site rather than at 500 or 750 m (43, 5, and 3 mice respectively) and are distributed fairly evenly between the four directions. An examination of the individual trapping records shows that many of these mice settled down at the places shown in Figure 3, as evidenced by their repeated captures there over periods of 48 h or more: Out of the 45 individuals that did not die in a trap at or within 48 h from first recapture, 30 (67%) are settlers according to this criterion. Table 1 gives a detailed analysis of recaptures of nonhomers at 250 m from release site, according to displacement distance. The average rates of settlement per available trapping station that we obtain for the three displacement distances (Table 1, F) are very close to each other. This suggests that the actual rates of settlement of nonhomers at 250 m from release site (including settlement in areas with no traps set) are independent of displacement distance.

In summary, we now can say that some of the nonhomers establish a new home range not far from and in all directions around the release site, and that their number is independent of displacement distance. Data

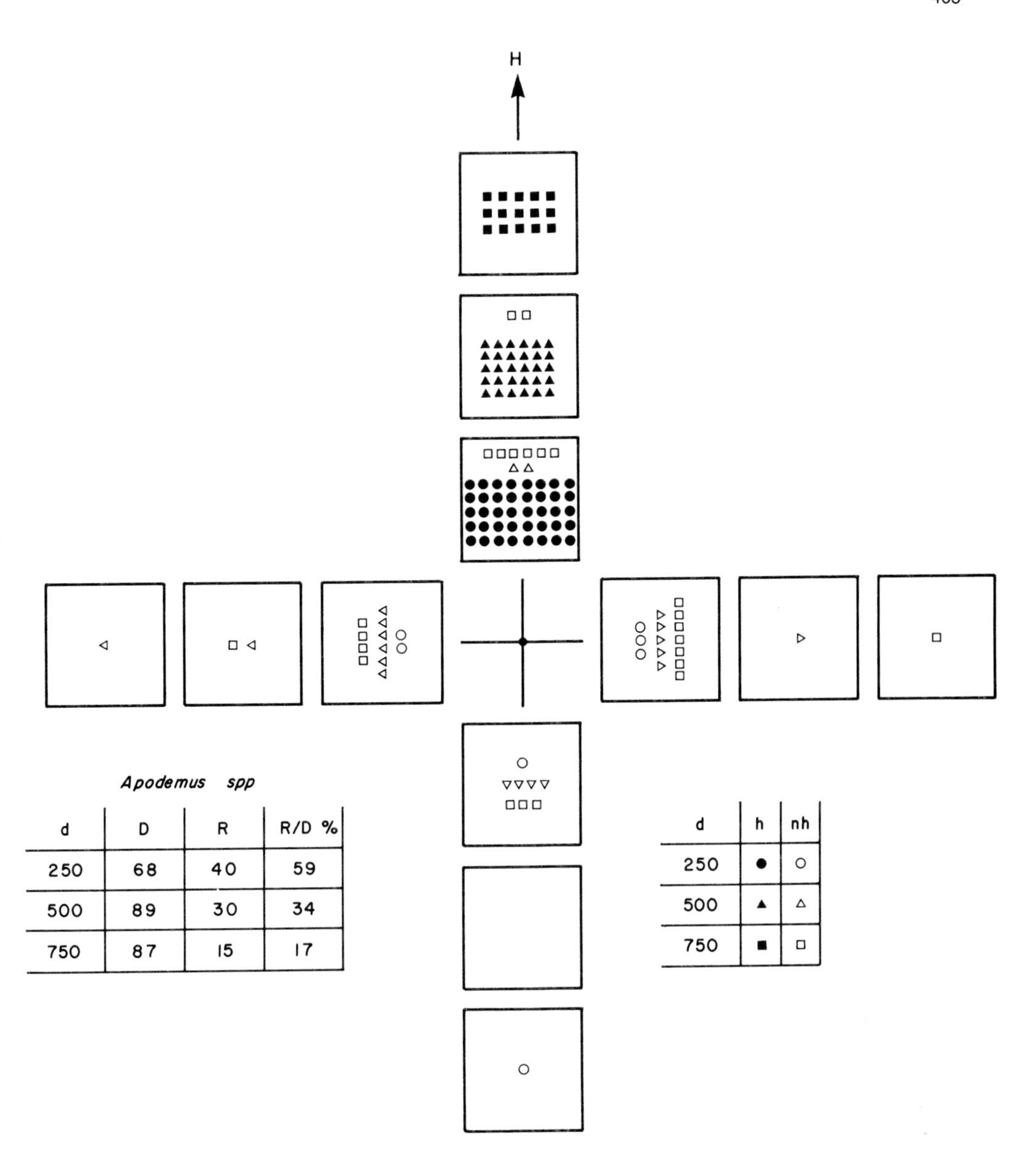

d	D	R	R/D %
250	68	40	59
500	89	30	34
750	87	15	17

d	h	nh
250	●	○
500	▲	△
750	■	□

Fig. 3. Last places of capture of displaced mice after release. For each mouse, the home direction (*H*) is at the top of the Figure. The 12 *square fields* correspond to the trapping stations (see Fig. 2), the release site is at the center. *Lower right corner inset* explains the individual symbols: *d*, displacement distance; *h*, homer; *nh*, nonhomer. *Lower left corner inset* gives homing success statistics: *d*, displacement distance; *D*, number of mice displaced; *R*, number of homers; *R/D%*, homing success

produced so far are inadequate to assess what proportion of the mice displaced do so. A rough estimate would be that, as an average, 30% of the mice settle down in this manner after a limited amount of travel. However, very little is known about the behavior and fate of the other nonhomers, other than that their number is dependent on displacement

Table 1. Recaptures of nonhomers at trapping stations 250 m from release site. A: mice found dead in traps at or within 48 h from first recapture; B: nonsettlers (found for less than 48 h at station); C: settlers (found for more than 48 h at station); D: total number of mice displaced; E: number of stations available to nonhomers; F: average rate of settlement per available station ($C \times D^{-1} \times E^{-1} \times 100$)

Displacement distance (m)	A	B	C	D	E	F
250	1	0	5	68	3	2.45
500	3	4	10	89	4	2.81
750	2	9	9	87	4	2.59

distance. Attempts to recatch them have been largely unsuccessful, despite the trapping efforts applied outside their home areas.

This leads me to the following considerations. It is implicit in virtually all explanations given in the past on possible homing mechanisms in mice that all the individuals are equally motivated to home at release time. The evidence available on settlement not far from the release site makes this now questionable. It suggests that the settling tendency shown by perhaps 30% of the mice is due to a motivational state that induces these animals to establish residence at the first convenient place encountered rather than to return to the old, familiar home range. Logically then, these animals should not be accounted for in the search for individual homing mechanisms: Their behavior is not aimed at "homing" in the usual meaning of "returning to the old home." Withdrawing them from homing success computations will have a definite, but limited effect on the kinds of curves shown in Figure 1. Let us accept the estimate of 30% settlers as a basis of discussion. Out of 100 mice displaced, only 70 will show a homing tendency. If 35 mice home, the standard homing success figure will be 35%, but the one corrected by withdrawing the 30 settlers will be 50% (35 out of 70). Generalized application of this type of correction will thus result into an overall increase of the level of the curves. However, since the proportion of mice that settle down in this manner is independent of displacement distance, the correction will not erase the decrease of homing success with increasing distances.

This last effect is necessarily due to the other nonhomers on which so little is known yet. Their number being displacement distance-dependent, it can be supposed that their behavior and motivations differ from those of the mice found to settle down near release site in numbers independent of displacement distance. Along the line of the considerations on motivation developed above, I propose the hypothesis that these animals exhibit at first a homing tendency, with the result that they behave at that stage like eventual homers, in particular that they move away from the release site. This is supported by their low rate of recapture at short distances from release site, which is similar to that of eventual homers, and which might be linked with an apparent avoidance for traps displayed by mice when they are "on the move" outside of a normally occupied home range (see Robinson and Falls, 1965). A further step in the hypothesis is that these animals do not complete their homing journey because they experience a shift from their homing to a settling tendency at some time during their search for the old home range. If and when this happens, these mice would be expected to establish a new home range near the place where the shift in tendency has occurred, in other words near the last position they reached while still under homing motivation. The pattern of scatter of the new home ranges around the line from release site to former home area would thus be a good indicator of mechanisms involved, especially with respect to the basic alternative between random scatter and goal-oriented behavior. Since the mice would have then lost their apparent avoidance for traps, this pattern could be revealed by an appropriate trapping procedure, e.g., a distribution of traps at distances around the release site more uniform than the one we had in our *Apodemus* study for instance.

The motivation hypothesis can finally be developed to provide an explanation for a feature that is apparent in Figure 1. The homing success curves of rodents shown there are distributed from the left to the right in the following order: (1) *Microtus;* (2) the two *Clethrionomys* species; (3) *Peromyscus* and *Apodemus* (two ecologically equivalent genera); (4) *Sigmodon;* (5) *Rattus*. This is the same order in which the species concerned would appear if they were ranked according to increasing sizes of home range (references in the papers quoted in Fig. 1; for *Apodemus* and *Clethrionomys*, see Bovet, 1962). This kind of consideration can be paralleled to findings by Robinson and Falls (1965) and by Anderson et al. (1977) to the effect that, within a species and for a given displacement distance, homing success is smaller in populations characterized by smaller home ranges than it is in populations with larger home ranges. This relationship between homing success and home range size is interpreted by these authors as a support of the "familiarity with the terrain" explanation, under the assumption that the larger the home range of an animal, the larger the zone of familiarity around it will be and, consequently, the greater the likelihood for the animal to be released within its zone of familiarity or to reach it by chance. A similar argument has been proposed to explain the homing performance of bats (Davis, 1966; Wilson and Findley, 1972) and, as a matter of fact, bats appear to fit very well into the pattern of relationship between home range size and homing success when compared to rodents (Fig. 1, curve 8). It should be stressed, however, that inexperienced homing pigeons fit equally well in the pattern (Fig. 1, curve 9) despite their well-known navigational abilities. This, I claim, points again to the paradoxically poor value of homing success figures as indicators of homing mechanisms (see p. 405). Within the framework of the motivation hypothesis, we can assume that, for a given species, the length or duration of travel previous to the shift in tendency varies widely between individuals, but that its average value is directly related to home range size. This would then explain the relationship between homing success and home range size in rodents, bats, and pigeons, independently of mechanisms involved.

Acknowledgments. The original research reported in this paper was carried on with technical assistance of Mr. A. Gogniat, and supported jointly by National Research Council of Canada (grants A-6639 and T0743 to the author) and by Swiss National Science Foundation (grant 3.886-1.72 to M. Dolivo and F. Schenk). Destructive (but highly profitable) criticism by Dr. G. FitzGerald on a first draft of this paper is gratefully acknowledged.

References

Anderson, P.K., Heinsohn, G.E., Whitney, P.H., Huang, J.P.: *Mus musculus* and *Peromyscus maniculatus:* homing ability in relation to habitat utilization. Can. J. Zool. 55, 169-182 (1977)

Bovet, J.: Influence d'un effet directionnel sur le retour au gîte des mulots fauve et sylvestre (*Apodemus flavicollis* Melch. et *A. sylvaticus* L.) et du campagnol roux (*Clethrionomys glareolus* Schr.) (Mammalia, Rodentia). Z. Tierpsychol. 19, 472-488 (1962)

Bovet, J.: Displacement distance and quality of orientation in a homing experiment with deer mice (*Peromyscus maniculatus*). Can. J. Zool. 50, 845-853 (1972)

Cooke, J.A., Terman, C.R.: Influence of displacement distance and vision on homing behavior of the white-footed mouse (*Peromyscus leucopus noveboracensis*). J. Mammal. 58, 58-66 (1977)

Davis, R.: Homing performance and homing ability in bats. Ecol. Monogr. 36, 201-237 (1966)

DeBusk, J., Kennerly, Jr., T.E.: Homing in the cotton rat, *Sigmodon hispidus* Say and Ord. Am. Midl. Nat. 93, 149-157 (1975)

Fisler, G.F.: Homing in the California vole, *Microtus californicus*. Am. Midl. Nat. 68, 357-368 (1962)

Fisler, G.F.: Homing in the western harvest mouse, *Reithrodontomys megalotis*. J. Mammal. 47, 53-58 (1966)

Furrer, R.K.: Homing of *Peromyscus maniculatus* in the channelled scablands of east-central Washington. J. Mammal, 54, 466-482 (1973)

Harrison, J.L.: Range of movement of some Malayan rats. J. Mammal. 39, 190-206 (1958)

Mueller, H.C.: Homing and distance orientation in bats. Z. Tierpsychol. 23, 403-421 (1966)

Murie, M.: Homing and orientation of deer mice. J. Mammal. 44, 338-349 (1963)

Murie, O.J., Murie, A.: Travels of *Peromyscus*. J. Mammal. 12, 200-209 (1931)

Robinson, W.L., Falls, J.B.: A study of homing of meadow mice. Am. Midl. Nat. 73, 188-224 (1965)

Saint Girons, M.C., Durup, M.: Retour au gîte chez le mulot, *Apodemus sylvaticus*, et le campagnol roussâtre, *Clethrionomys glareolus*. Facteurs écologiques, apprentissage et mémoire. Mammalia 38, 389-404 (1974)

Stickel, L.F.: An experiment on *Peromyscus* homing. Am. Midl. Nat. 41, 659-664 (1949)

Wallraff, H.G.: Über die Flugrichtungen verfrachteter Brieftauben in Abhängigkeit vom Heimatort und vom Ort der Freilassung. Z. Tierpsychol. 27, 303-351 (1970)

Wilkinson, D.H.: The random element in bird "navigation". J. Exp. Biol. 29, 532-560 (1952)

Wilson, D.E., Findley, J.S.: Randomness in bat homing. Am. Nat. 106, 418-424 (1972)

Orientation Mechanisms of Marine Turtles

N. MROSOVSKY, Departments of Zoology and Psychology, University of Toronto, Toronto, Canada, M5S 1A1

Abstract

The principal mechanism in the sea-finding orientation of hatchling turtles when they move from the nest to the water is based on a phototropotactic reaction to light and a complex balancing of stimulation to different retinal areas and corresponding regions of the optic tectum. Transient potentiation of parts of this system occur following abrupt changes in illumination. In contrast to guidance systems that demand fine resolution of stimuli, sea-finding depends on integration of information, integration both over space (wide field of view) and over time (averaging illumination over about 1 s). Sea-finding of turtles when on land is relevant to their movements when in the water since there may be some calibration of other guidance systems against headings initiated by the sea-finding mechanisms.

If a hatchling sea turtle is released on the beach within a shed that allows sea sounds and winds to penetrate but has an open door facing landward, it first moves away from the sea, crawling out through the open door; once outside it moves around the shed and heads toward the water. This is only a rather striking illustration of the effectiveness of a sea-finding mechanism that routinely quides turtles in more normal circumstances quickly from their nests to the water's edge.

Sea-finding has been studied for two reasons, first as a phenomenon in its own right and, second, in the hope it might be a model for the long-distance migrations of turtles (Carr and Ogren, 1960). Sea-finding is a relatively accessible robust phenomenon and considerable progress has been made in understanding it. Among many experiments the following have been especially instructive:

1. Transporting hatchlings to beaches facing in different directions from their natal one does not disrupt sea-finding, implying that turtles react to cues on the beachscape rather than to the position of celestial bodies or geomagnetic forces (Carr and Ogren, 1960).

2. Turtles respond positively toward light and fail to orient towards the sea when blindfolded, indicating that vision is the principal modality involved (eg. Carr and Ogren, 1960; Mrosovsky and Shettleworth, 1975).

3. Placing obstructions off to one side, rather than in the most seaward path, causes turtles to veer away from the obstructions, showing that they are not simply reacting to cues from the seaward direction but basing their orientation on information from a wide field of view (Mrosovsky and Shettleworth, 1968, 1975; see also Verheijen and Wildschut, 1973).

4. Unilaterally blindfolded turtles move in circles, suggesting a tropotactic reaction to light (Daniel and Smith, 1947; Ehrenfeld, 1968; Mrosovsky and Shettleworth, 1968).

Taken together these results suggest that during sea-finding, turtles balance illumination in symmetrically placed receptors. Even if the water itself is out of sight at turtle eye-level, the open brighter horizon is nearly always seaward. If placed in a room with only one open door, balancing of brightness would direct them to the middle of the doorway.

It must, however, be mentioned that there are other explanations of sea-finding, namely, that it depends on orienting toward the largest illumination vector with a wide angle of acceptance mechanism assessing where this vector is (Verheijen and Wildschut, 1973). Although the actual behavior in a number of circumstances (e.g., tests with dark obstructions) can be accounted for as well by this view as by a tropotactic reaction to light, it does not in itself explain the circling of unilaterally blindfolded turtles. However, as I have said repeatedly in my papers, there may be systems other than phototropotaxis, or even other modalities, that contribute to sea-finding or that act as backups if tropotaxis fails. Moreover even the tropotactic system is more complicated than a balancing of inputs to the left and right eyes. It will be convenient to begin by giving a model for this elaborated system; this should help make a number of diverse facts more coherent, but need not of course be taken as more than a redescription of the results.

The model proposes that both ipsilateral and contralateral turning can be associated with stimulation of a single eye. The actual turning direction shown follows from assessment of the intensities of stimulation in different parts of the retina by some sort of comparator mechanism, seaward orientation depending on stimulation of the ipsilateral turning system for the right eye together with that of the contralateral system for the left eye balancing stimulation of the ipsilateral system for the left eye together with that of the contralateral system for the right eye. These systems have spatial counterparts in the visual field, retina, and optic tectum (Fig. 1). Furthermore it is proposed that in leatherbacks (*Dermochelys coriacea*) the contralateral systems are stronger than those in green turtles (*Chelonia mydas*). Consistent with this model are the following:

1. When leatherbacks are tested soon after unilateral blindfolding, they often circle towards their covered eye, whereas green turtles seldom do this (Mrosovsky and Shettleworth, 1975).

2. When only the nasal visual field of one eye is left unoccluded, green turtles do sometimes turn toward the completely covered eye, indicating that contralateral turning systems exist in this species also. Animals with harlequin blindfolds (nasal visual field of one eye and temporal field of other eye occluded) circle as predicted by the model (Mrosovsky and Shettleworth, 1974).

3. Turning can be elicited by electric stimulation of the optic tectum of other reptiles (alligators); moreover the direction of turning depends on which part of the tectum is stimulated (Schapiro and Goodman, 1969; see also Mrosovsky, 1972a).

4. Recent experiments, in collaboration with Granda and Hay, show that in a turtle with one eye covered and a small light suspended near the other eye, the direction and tightness of turning depends on whether the light is in the nasal or temporal field of view: Circles toward

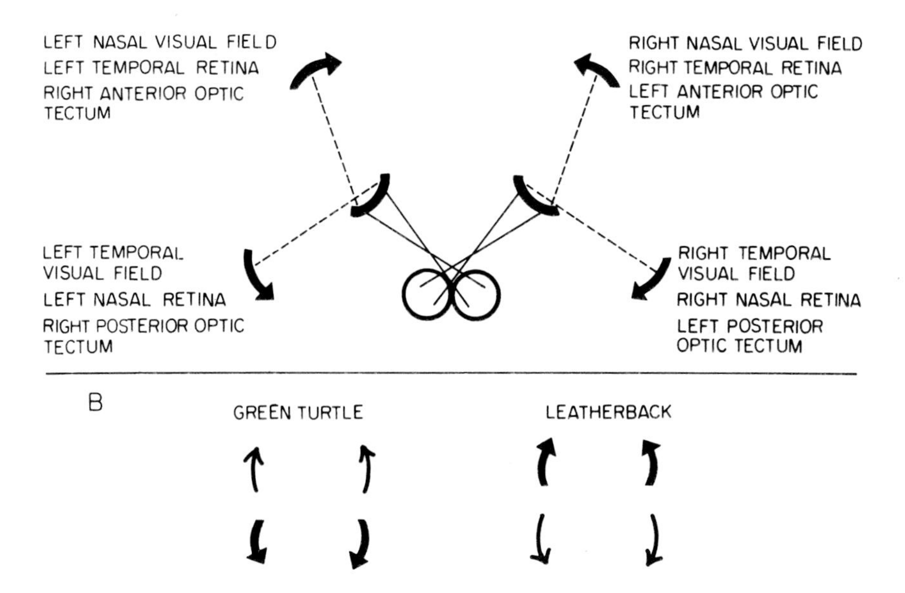

Fig. 1. (A) Model of spatial organization of balancing system for sea-finding. *Arrows* show directions of turning tendencies associated with light in different parts of the visual field and corresponding parts of the retina and optic rectum. (B) Relative strengths of turning systems in two different species of marine turtle. *Heavier arrows* indicate stronger turning tendencies relative to the other species (strengths of turning tendencies within a species are not represented)

the covered eye can be demonstrated with the light in the nasal field. Also, bilaterally asymmetric lesions of the optic tectum result in turtles circling or heading off to the side of the directions taken by control animals.

Nevertheless there are some aspects of the data that do not fit especially well with the model shown in Figure 1; in particular, the work with hemi-blindfolds on green turtles did not establish the existence of a contralateral turning system in an especially striking way. Rather few circles toward the midline, with respect to the partly uncovered eye, were actually seen. While there are various technical reasons why the methods used may not have brought out the contralateral system especially strongly (see Mrosovsky and Shettleworth, 1974), perhaps there is an additional aspect of the sea-finding mechanism that has been neglected so far.

So far only the spatial aspects of the sea-finding mechanism have been considered. However two recent experiments have shown that the temporal properties are also important:

1. When one eye of a leatherback turtle is covered, or when a cover is removed from an eye, there are transient effects that last about 2 min (Fig. 2; Mrosovsky, 1977). These effects can be summed up by saying that following the covering of one eye there is a transient potentiation of the ipsilateral turning system for the uncovered eye (of course it could be a transient inhibition of the contralateral system for the uncovered eye or a change in one of the other components

TURNING DIRECTION POTENTIATED IMMEDIATELY AFTER:

	A. COVERING RIGHT EYE	B. REMOVING COVER FROM RIGHT EYE	C. REVERSING COVER FROM RIGHT TO LEFT EYE
RESULTS	LEFT	RIGHT	LEFT

TENTATIVE EXPLANATION

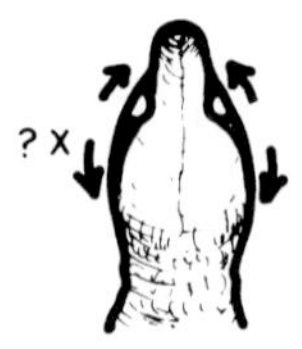

Fig. 2. *Arrows* show turning directions associated with light in nasal field of view (*upper arrows*) and temporal field (*lower arrows*). ?√ suggests turning system potentiated; ? *X* suggests turning system inhibited. Results are from Mrosovsky (1977). Inhibitory links between ipsilateral turning systems (*lower arrows*) might account for some of these effects

of the turning system that leads to an enhancement of ipsilateral turning tendencies with respect to the open eye). Following removal of a blindfold from an eye there is a transient potentiation of the ipsilateral turning systems with respect to that eye. Figure 2 shows the experimental procedures and results from which these statements are derived. Because these effects are transient, in some cases immediately after a blindfold is put on, a leatherback circles one way, while a few minutes later it may be turning in the other direction (Mrosovsky and Shettleworth, 1975; Mrosovsky, 1977).

2. With green turtles tested several hours instead of soon after unilateral blindfolding, van Rhijn (in preparation) found that circling is toward the closed rather than toward the open eye. This is also consistent with a transient potentiation of ipsilateral turning systems after covering the other eye.

Because of the transient potentiation of ipsilateral turning systems after covering an eye, testing a turtle immediately after blindfolding is the worst time to demonstrate the contralateral systems in action. This is probably the reason why Mrosovsky and Shettleworth (1974) did not find more striking effects in their experiments as they begun testing within 30 s of blindfolding. Another implication of a balancing system that combines contralateral and ipsilateral turning systems for each eye with transients is that there is no reason why unilaterally blindfolded turtles might not orient seaward, as indeed has sometimes been found (Mrosovsky and Shettleworth, 1975; van Rhijn, in prep.). The exact directions taken would depend on the time course of the transients (presumably in themselves dependent on the time course of dark adaptation and inhibitory links between components of the turning systems) and the relative strengths of the ipsilateral and contralateral turning systems (cf. Fraenkel and Gunn, 1961).

A more interesting phenomenon that transients may help explain is the "orientation circle" of leatherbacks. Sometimes leatherbacks, without any blindfolds, in natural circumstances, make complete circles on their way to the sea (for a review see Mrosovsky and Shettleworth, 1975). Green turtles virtually never do this. Given the evidence that transients do not effect both ipsi and contralateral systems equally, if contralateral systems are more developed in leatherbacks (Fig. 1), then it seems intuitively reasonable to think that along with this difference in dominance of the two systems there might go a greater susceptibility to transients. This might operate in nature to produce "orientation circles" as follows: A hatchling pausing next to a small mound of sand or a wind-blown leaf might experience a mild version of the situation produced by having one eye covered. It is also noteworthy that adult leatherbacks make "orientation cirles" more frequently when illumination is variable, on cloudy moonlit nights (Bacon, 1973). However, detailed explanations of "orientation circles" remain to be elaborated.

More generally these considerations emphasize that the temporal as well as the spatial aspects of the sea-finding mechanism must be considered. For instance, it is important that emphasis is placed on the permanent illumination cues that differentiate seaward and landward directions, namely, the dark tree line and open seaward horizon, rather than on the fleeting brightness changes that result from moving past a small object close to the eye. Experiments on green turtles confronted with a choice between flashing and continuous lights have shown that the continuous light is preferred. This preference increases as the ratio of off-to-on time on the flashing side increases but is not much influenced by the rate of flashing over ranges of 1 - 14 flashes/sec (Mrosovsky, 1978). In other words, turtles appear to base their orientation on the average amount of illumination they receive over periods of at least 1 s rather than being diverted by brief changes of illumination.

In summary then, the sea-finding mechanism integrates both over space and time. There is integration of cues from a wide field of view (probably without relying on pattern vision, which is presumed to be poor on land, Walls, 1942; Ehrenfeld and Koch, 1967), and also over wide spans of time. In the context of some of the migratory mechanisms that depend on resolution, for instance of star patterns, the sun's position, or of magnetic cues, it is almost embarrassing to talk about a system that is so apparently crude. Yet a mechanism that works so well in so many different circumstances, by storm, by starlight, or moonlight, by day or by night, with the sea visible or out of sight, uphill or downhill, on a beach with clear access to the water or one with obstructions to avoid and tidal pools to cross, has a simplicity and robustness of function that may well have, in general terms, some evolutionary appeal for other migrants.

Turning to the long-distance migrations of turtles (Carr, 1967; Pritchard, 1976), it might seem that the sea-finding system, depending, as it turns out, on simple landward-seaward visual differences, would not be a useful model as hoped, but recent work by Frick (1976) suggests that the two may not be unrelated, at least for the initial phase of the hatchling's aquatic life. Frick found that when the hatchlings swim away from shore they take a bearing that is fairly similar to that of their seaward orientation while still on land and their immediate offshore swimming direction. This bearing is maintained even when the turtles are too far away to see the shoreline. Perhaps, therefore, magnetic, olfactory, or other quidance systems are calibrated against the directions initiated when the turtles are on land, or in the immediate offshore region, where differences between seaward and landward horizons are still present. Some notable contributions to this matter will soon be forthcoming (van Rhijn, in prep.). Knowledge of sea-finding may yet turn out to be a very useful starting point and link toward understanding the guidance systems used by turtles at sea.

At the moment though, how turtles find their way about in the oceans remains completely unanswered. It is not even known if they reach the feeding grounds by swimming or by passive drift. There are no experimental data even on what modality guides the adults back to the nesting beaches, though there are speculations that olfaction might be important (Koch et al., 1969) or that sounds from breakers might assist in the final stages (Mrosovsky, 1972b). It has been argued that a major obstacle in understanding turtle migrations is not knowing what paths the turtles actually take (Carr. 1967; Koch et al., 1969). Personally I doubt that the development of techniques to track turtles over long distances at sea and then studying the cues along the route is likely to be an especially important or rewarding approach at present. Some of the most striking advances in the study of bird migration have depended on being able to bring the phenomenon into a laboratory setting. Now that the basic facts of long-distance migration of turtles have been established by the pioneering studies of Carr (1967) and others (Balazs, 1976), we need to bring the behavior into the laboratory. We should at least look, as a high priority, for photoperiodic effects or *Zugunruhe*, because if these were demonstrated then progress in analyzing the orientation systems could follow rapidly. The progress in the study of the mini-migration of hatchling turtles from the nest to the sea, now at a stage of analysis of the brain mechanisms involved, has only been possible because the phenomenon is relatively accessible; even in this case a full understanding will probably require laboratory techniques, fortunately now successfully devised (van Rhijn, in prep.). Methodologically then also, sea-finding orientation might serve as a model for the study of the more spectacular and challenging movements of turtles.

Acknowledgments. Field work was made possible through the courtesy and cooperation of the Surinam Forest Service and the Caribbean Conservation Corporation. I thank the National Research Council of Canada for support and F.A. van Rhijn and S.J. Shettleworth for stimulating discussions.

References

Bacon, P.R.: The orientation circles in the beach ascent crawl of the leatherback turtle, *Dermochelys coriacea*, in Trinidad. Herpetologica, 29, 343-348 (1973)

Balazs, G.H.: Green turtle migrations in the Hawaiian Archipelago. Biol. Conserv. 9, 125-140 (1976)

Carr, A.: Adaptive aspects of the scheduled travel of *Chelonia*. In: Animal Orientation and Navigation. Storm, R.M. (ed.), Oregon State University Press, Corvallis, 1967, pp. 35-55

Carr, A., Ogren, L.: The ecology and migrations of sea turtles. IV. The green turtle in the Caribbean Sea. Am. Mus. Nat. Hist. Bull. 121, 1-48 (1960)

Daniel, R.S., Smith, K.U.: The sea-approach of the neonate loggerhead turtle (*Caretta caretta*). J. Comp. physiol. Psychol. 40, 413-420 (1947)

Ehrenfeld, D.W.: The role of vision in the sea-finding orientation of the green turtle (*Chelonia mydas*). II. Orientation mechanism and range of spectral sensitivity. Anim. Behav. 16, 281-287 (1968)

Ehrenfeld, D.W., Koch, A.L.: Visual accommodation in the green sea turtle. Science 155, 827-828 (1967)

Fraenkel, G.S., Gunn, D.L.: The Orientation of Animals, New York: Dover Press, 1961

Frick, J.: Orientation and behaviour of hatchling green turtles (*Chelonia mydas*) in the sea. Anim. Behav. 24, 849-857 (1976)

Koch, A.L., Carr, A., Ehrenfeld, D.W.: The problem of open-sea navigation: the migration of the green turtle to Ascension Island. J. Theoret. Biol. 22, 163-179 (1969)

Mrosovsky, N.: The water-finding ability of sea turtles: behavioral studies and physiological speculations. Brain Behav. Evol. 5, 202-225 (1972a)

Mrosovsky, N.: Spectrographs of the sounds of leatherback turtles. Herpetologica, 28, 256-258 (1972b)

Mrosovsky, N.: Individual differences in the sea-finding mechanism of hatchling leatherback turtles. Brain, Behav. Evol. 14, 261-273 (1977)

Mrosovsky, N.: The effects of flashing lights on sea-finding behavior of green turtles. Behav. Biol. 22, 85-91 (1978)

Mrosovsky, N., Shettleworth, S.J.: Wavelength preferences and brightness cues in the water finding behaviour of sea turtles. Behaviour 32, 211-257 (1968)

Mrosovsky, N., Shettleworth, S.J.: Further studies of the seafinding mechanism in green turtle hatchlings. Behaviour, 51, 195-208 (1974)

Mrosovsky, N., Shettleworth, S.J.: On the orientation circle of the leatherback turtle, *Dermochelys coriacea*. Anim. Behav. 23, 568-591 (1975)

Pritchard, P.C.H.: Post-nesting movements of marine turtles (Cheloniidae and Dermochelyidae) tagged in the Guianas. Copeia, No. 4, 749-754 (1976)

Schapiro, H., Goodman, D.C.: Motor functions and their anatomical basis in the forebrain and tectum of the alligator. Exp. Neurol. 24, 187-195 (1969)

Verheijen, F.J., Wildschut, J.T.: The photic orientation of hatchling sea turtles during water finding behaviour. Neth. J. Sea Res. 7, 53-67 (1973)

Walls, G.L.: The vertebrate eye and its adaptive radiation. Cranbrook Inst. Sci., Bloomfield Hills (1942)

Nighttime Orientation of Hatchling Green Turtles *(Chelonia mydas)* in Open Ocean

LEONARD C. IRELAND[1], JANE A. FRICK[2], and DAVID B. WINGATE[3], [1] Oakland University, Rochester, MI, USA, [2] POB 364, Lincoln, MA, USA, [3] Department of Agriculture and Fisheries, Hamilton, Bermuda

Abstract

Eleven hatchlings equipped with sonic transmitters and three equipped with chemical lights were tracked following their departure from beaches facing different points of the compass. The work was carried out off Bermuda with turtles hatched from eggs brought from Tortuguero, Costa Rica. Unless confronted with opposing shorelines, the turtles' travel paths either approximated straight lines or took the form of gradual curves. These courses did not change to any great degree when land was below the horizon from turtle eye-level. The data suggest that hatchling *C. mydas* possess no inborn directional preference other than 'away from land.' The adaptive significance of this behavior may be that it serves to move the turtles rapidly beyond the reach of inshore predators.

A. Introduction

Immediately after leaving their nest, hatchling *C. mydas* travel rapidly across the beach, enter the sea, and disappear for about a year, reappearing at feeding grounds that are usually far distant from any natal beach (Carr, 1967a, b). The water-finding behavior of hatchling sea turtles has been studied in some detail over the past 60 years; the ability appears largely explicable in terms of a positive, tropotactic response to brightness cues (Mrosovsky, 1972; Hayes and Ireland, in press). The behavior of hatchlings after leaving the beach, however, has received only scant attention. Until recently, our knowledge of the behavior of newborn *C. mydas* in the ocean was restricted to the following: (1) shore observers report that hatchlings orient and swim rapidly toward open ocean, at least for the few meters that they remain in view (Carr, 1967a); (2) *C. mydas* hatchlings placed in aquaria swim vigorously for periods of 24 h or more (Carr, 1972); (3) hatchlings are only rarely seen in the vicinity of nesting beaches after the nesting season (Carr, 1967a); and (4) hatchlings have been captured at sea on more than one occasion, at distances over 15 km offshore (Caldwell, 1969). Taken together, these observations suggest that newborn *C. mydas* travel rapidly toward and into open ocean immediately after leaving the beach, and that they may occupy a pelagic habitat during their first year of life.

Frick (1976) was the first to track hatchling *C. mydas* over long distances in the ocean. She was able to both observe their behavior and determine their travel paths by either swimming behind them, at distances of 1.5 - 3 m, or by following them in a small boat as they left the beach. Swimmers were accompanied by a small boat following some distance behind. Frick's work was carried out off the island of Bermuda and off Tortuguero, Costa Rica during daylight hours. She was able to determine the travel paths of 27 hatchlings (24 followed by swimmers) over distances of 0.6 - 6.5 km and to show that (1) the turtles swam steadily on nonrandom courses even when land was below the horizon from turtle eye-level; (2) hatchlings released on beaches facing the open ocean tended to travel on courses which were roughly at right angles to the beaches; (3) turtles released on a beach facing another shoreline reoriented until they were traveling away from land and then maintained these courses with considerable accuracy; and (4) the hatchlings displayed no real preferred direction of travel other than 'away from land.'

There are two potential problems with Frick's (1976) procedures. First, the presence of a swimmer or a boat close behind a hatchling might influence its choice of direction. Second, the vast majority of green turtles emerge from their nests at night rather than during the day. Frick's study, therefore, might be considered to have been conducted under somewhat 'unnatural' conditions. The purpose of the study reported here was to examine the orientation of hatchling *C. mydas* in open ocean at night, employing tracking techniques which made it possible to maintain contact with the turtles from distances of 15 m or more.

B. Materials and Methods

The study was carried out off Bermuda during October of 1973 and 1975. The hatchlings were obtained from eggs brought from Tortuguero, Costa Rica and incubated in artificial nests, as were the turtles employed in Frick's Bermuda experiments. Eleven hatchlings were equipped with miniature acoustic transmitters, the smallest we could obtain. Transmitter frequency, 50 kHz, was above the range of auditory sensitivity reported for adult *C. mydas* (Ridgway et al., 1969). We are not aware of any studies concerning the auditory range of hatchlings. A description of the transmitter will appear in a later report. Because the transmitters proved too heavy for the hatchlings to carry for long periods of time, each transmitter was suspended from a small balsa wood float that was attached to the rear of the turtle's carapace by a fine thread. The floats were near neutrally buoyant. During daylight tests, with swimmers observing the hatchlings, the turtles appeared to ignore both the float and transmitter completely, and to have little difficulty either towing them or pulling them underwater. Contact with each hatchling was maintained by investigators in a small boat equipped with an acoustic receiver and a directional hydrophone. This equipment is described in Ireland and Kanwisher (in press). In the clear, deep water off Bermuda our telemetry system proved to have a practical range of approx. 1 km. In addition, three turtles were equipped with balsa wood floats equipped with small chemical lights. For information on this simple and inexpensive nighttime tracking technique see the paper by Buchler (1976). Hatchlings equipped with 'light floats' were also followed by investigators in a small boat.

Each turtle was tracked as soon as possible after it left its nest. Unfortunately, adverse weather conditions and rough seas often delayed the beginning of our experiments. Times between emergence from the nest and release ranged from 2 - 72 h ($\bar{x}$ = 26 h). All turtles were released between the hours of 2023 and 0400 (Eastern Standard Time). The hatchlings were released from 5 different beaches (see Fig. 1). The prevailing northwesterly winds often dictated that the experiment be carried out from the south beach of Nonsuch Island. Each of the turtles

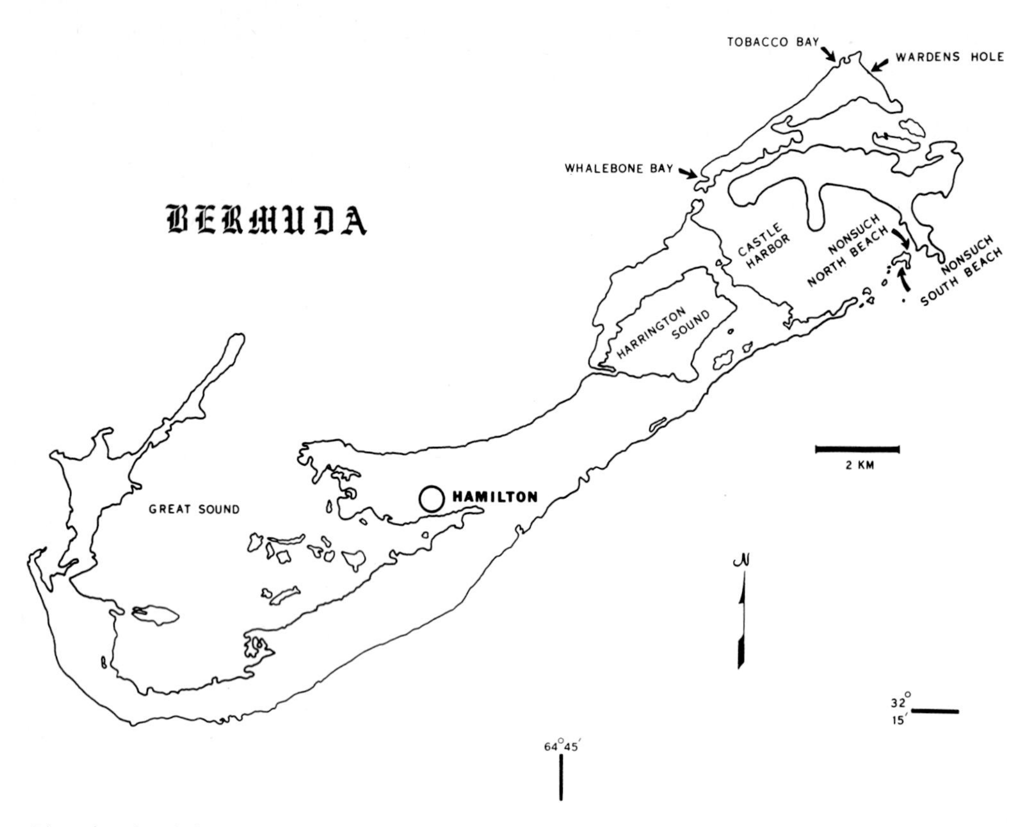

Fig. 1. The island of Bermuda. *Arrows* indicate the beaches where the hatchling *C. mydas* were released

was released on shore and allowed to crawl toward and into the water. Distances traveled across the beach were 3 m or more. Turtles with 'light floats' were followed at a distance of approx. 15 m, with the outboard motor running most of the time. In the case of transmitter-equipped turtles, we usually remained 50 m or more behind the hatchling, with the engine shut off. At regular intervals, we started the engine and, guided by the signal from the transmitter, moved to within approx. 15 m of the hatchling in order to determine its position. Once we arrived near the turtle, the engine was shut off. Distance from the turtle was estimated from maximum signal strength. Position over time for each hatchling was determined by either compass or line-of-sight bearings from the boat to points on shore. All the experiments were carried out under clear or partly overcast skies. Ocean surface temperatures ranged from 21.4 - 22.8° C.

C. Results and Discussion

We experienced little difficulty in maintaining contact with the turtles either via acoustic telemetry or by means of the 'light float' technique. The turtles were followed for periods of 50 - 292 min ($\bar{x}$ = 185 min), over distances of 1.0 - 6.2 km ($\bar{x}$ = 3.4 km). The number of bearings taken for each of the tracks ranged from 4 - 9 ($\bar{x}$ = 6.1). The hatchlings traveled at speeds of 0.8 - 3.2 km/h ($\bar{x}$ = 1.1 km/h). The tracks are shown in Figures 2-6. Turtles No. 11, 13, and 14 (Figs. 5 and 6) were equipped with 'light floats.' Tracks were terminated for a variety of reasons including transmitter failures, apparent loss of the turtles to reef predators, and poor weather. At the end of 10 of the 14 tracks, we managed to capture the turtles and remove the floats.

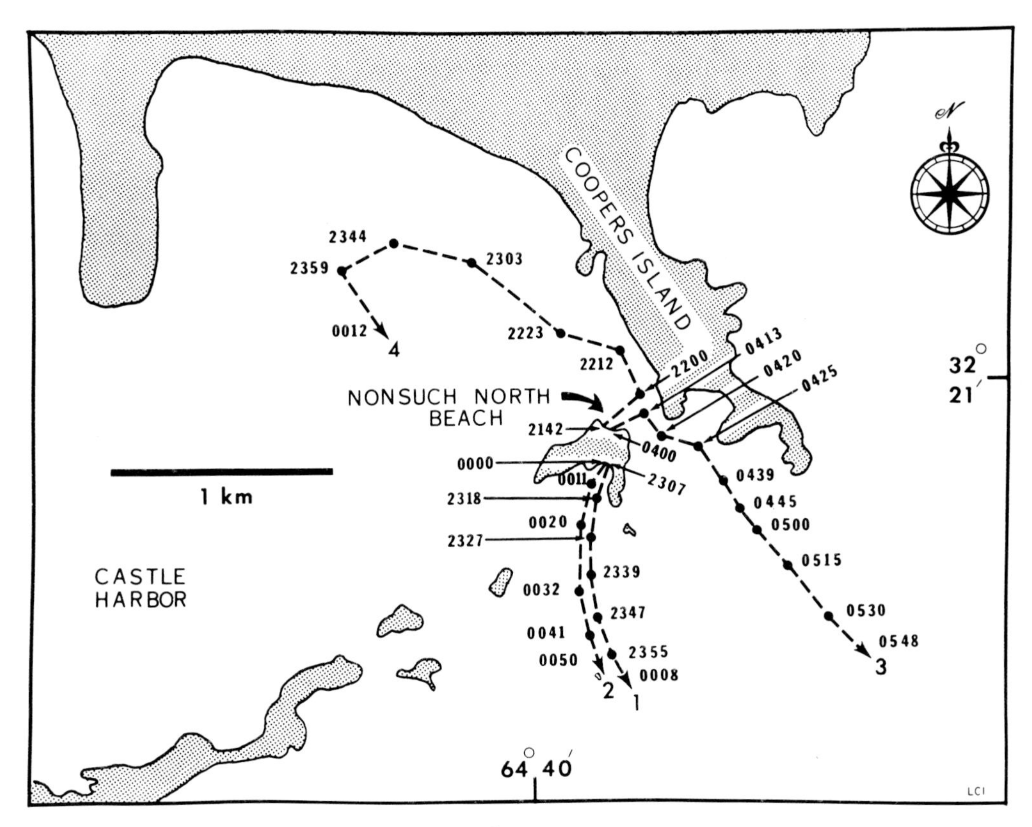

Fig. 2. Travel paths of hatchling *C. mydas* released from the south and north beaches of Nonsuch Island, Bermuda

Examination of Figures 2 - 6 reveals that the tracks of turtles released from beaches facing open ocean either approximate straight lines or take the form of gradual curves. Turtles released from a beach facing another shore change direction until they are facing open ocean (see Fig. 2). Once oriented seaward, turtle No. 3 maintained a near straight line travel path. Poor weather forced us to stop tracking turtle No. 4 before we could determine if it would behave in a similiar manner.

At present, no statistical test is able to conclusively determine whether an animal's travel path is 'oriented' or 'random'. The problem is one of independence among data; each section of an animal's travel path influences the one following. Batschelet (1972) and Schmidt-Koenig (1975) suggest the use of tests formulated by Hodges and Ajne when confronted with this problem. Unfortunately, we do not have a sufficient number of bearings per track to employ these tests. Consequently,

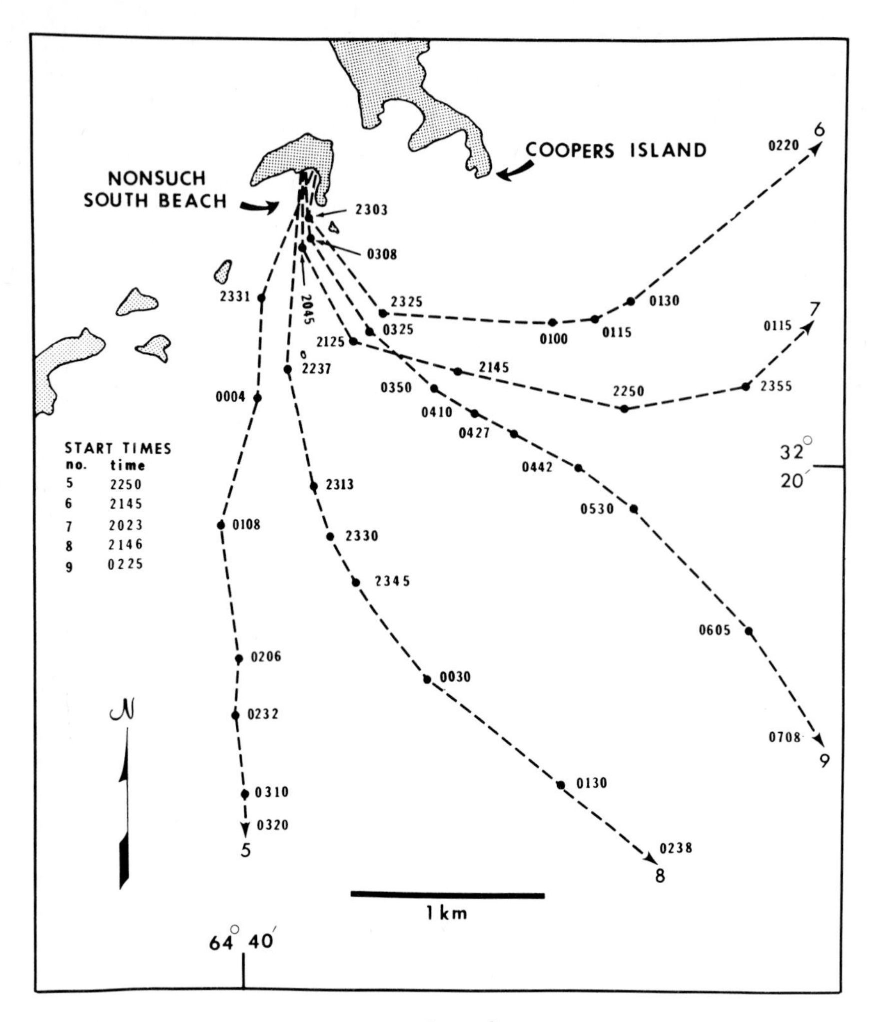

Fig. 3. Travel paths of hatchling *C. mydas* released from the south beach of Nonsuch Island, Bermuda

in order to gain some idea of the 'directedness' of the turtles' travel paths we applied Rayleigh tests (Batschelet, 1965) to each track. In using the test we make the assumption that each turtle made a new choice of direction each time that we took a bearing. The experimental hypothesis is that the turtle has a 'preferred' direction of travel. The null hypothesis is that the distribution of directions for each track is uniform. The results of the tests allow the rejection of the null hypothesis for all turtles ($p < .05$ or less) except turtle No. 4 ($p > .05$). We suggest that continual reorientation, presumably while searching for a route to open ocean, accounts for the 'randomness' of this track. According to our criterion, then, hatchling *C. mydas* appear capable of nonrandom travel at night in open ocean. Further, a human lying on a surf board 2.5 km off Bermuda is unable to see the island. It is almost certain that hatchling sea turtles lose the ability to see land before reaching this distance from shore. Since our tracks do not indicate that the hatchlings become disoriented when over 2.5

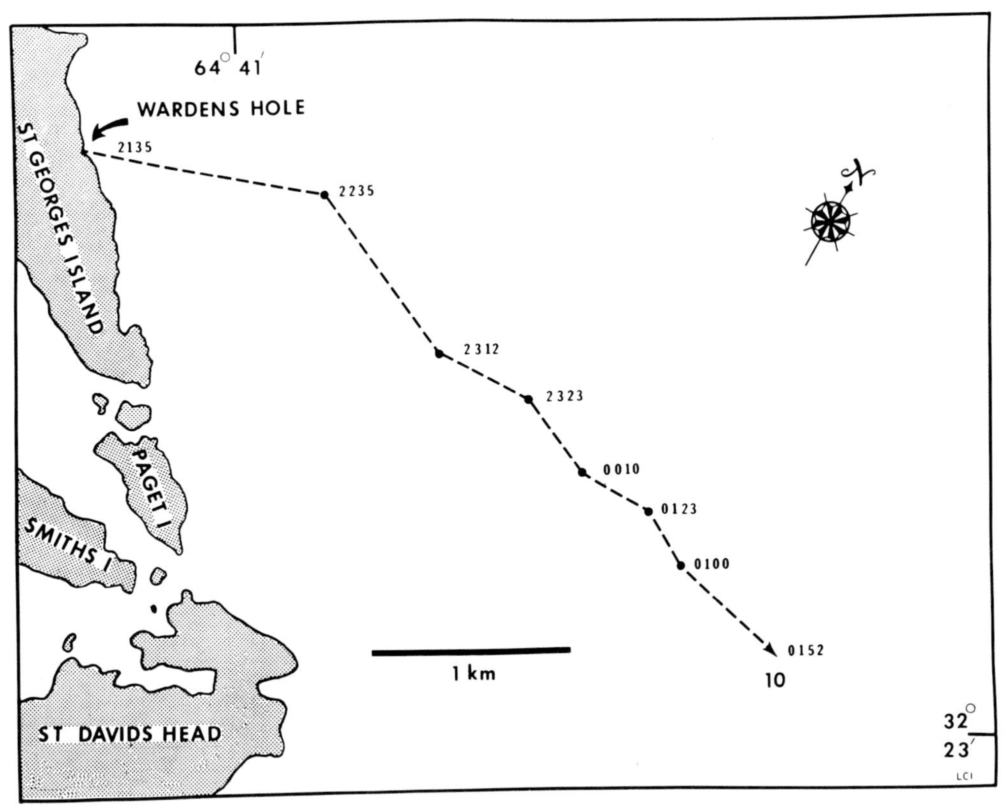

Fig. 4. Travel path of a hatchling *C. mydas* released from the beach at Wardens Hole, Bermuda

km from land, we conclude that hatchling *C. mydas* is capable of non-random travel at night when land is below the horizon from turtle eye-level.

A summary of the initial and final directions taken by the hatchlings, as determined from the first and last bearings for each track, is given in Figure 7. The initial directions taken by the turtles after leaving the beach are shown in Figure 7A. Applying Rao's test (Batschelet, 1972) to the data in Figure 7A, we are able to reject the null hypothesis of a uniform distribution of directions ($p < .05$). On the basis of the data shown in Figure 7A alone, we might conclude that hatchling *C. mydas* have an initial preference, albeit not a strong one, for travel toward the SSW. Figure 7A, however, does not take into account the orientation of the beaches from which the hatchlings were released, or the number of turtles released from each beach. These variables are incorporated in Figure 7B. To prepare Figure 7B, we first drew lines parallel to each beach from which the turtles were released. We then drew additional lines, pointing seaward, at right angles to the first set. Each of these 'seaward' lines was considered to be 'north' or '0°' for that particular beach. The initial directions taken by the hatchlings were then replotted according to these new points of reference. A Rao's test applied to the data in Figure 7B proved highly significant ($p < .01$), indicating that the initial preferred direction for the turtles was quite close to 'straight off the

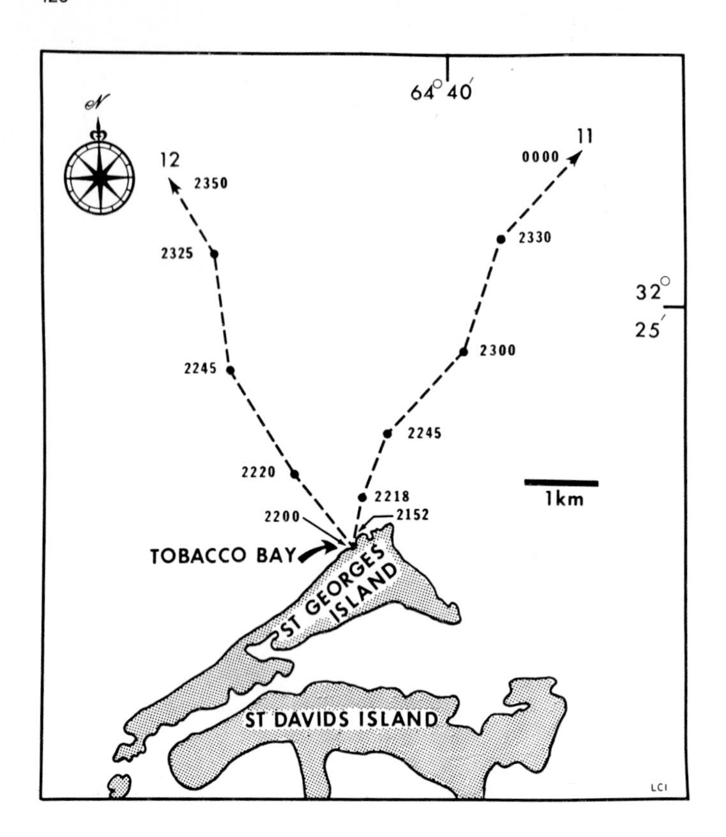

Fig. 5. Travel paths of hatchling *C. mydas* released from the beach at Tobacco Bay, Bermuda

beach.' Figure 7C shows the distribution of the final directions taken by the hatchlings. A Rao's test applied to these data proved significant ($p < .05$), indicating that the distribution of directions is not uniform. Inspection of Figure 7C reveals that it is bimodal with concentrations of directions to the NE and SE. We believe that ocean currents are a major cause of the changes in the hatchlings' initial and final directions of travel. Perhaps the primary weakness of the present study is that we were unable to measure current directions or velocities with any precision. While tracking turtles No. 6, 7, 8, 9, and 14, however, it was obvious to us, through changes in position when the engine was not running, that we were drifting in directions between E and NE. This is not to say that currents had no influence on the travel paths of the other turtles. Conversations with local fishermen confirmed that eastward tending longshore currents are a regular phenomenon off both Bermuda's north and south shores, especially during rough weather. We were unable to locate any document containing comprehensive information concerning local currents off Bermuda. Note that the travel paths of turtles 6 - 9 and 14, as well as of turtles No. 1, 2, and 13, might be explained by assuming that they are maintaining a heading close to 'straight out from the beach' but are being displaced eastward or northeastward by currents. This hypothesis receives indirect support from Frick's (1976) tracking experiments off Tortuguero, Costa Rica. Three hatchling *C. mydas* were followed by investigators in a small boat, and longshore currents were measured by means of drift bottles. When vector analysis is used to remove the effects of currents from the tracks, the turtles' headings are shown to be quite close to a right angle with the beach. Figure 7D was pre-

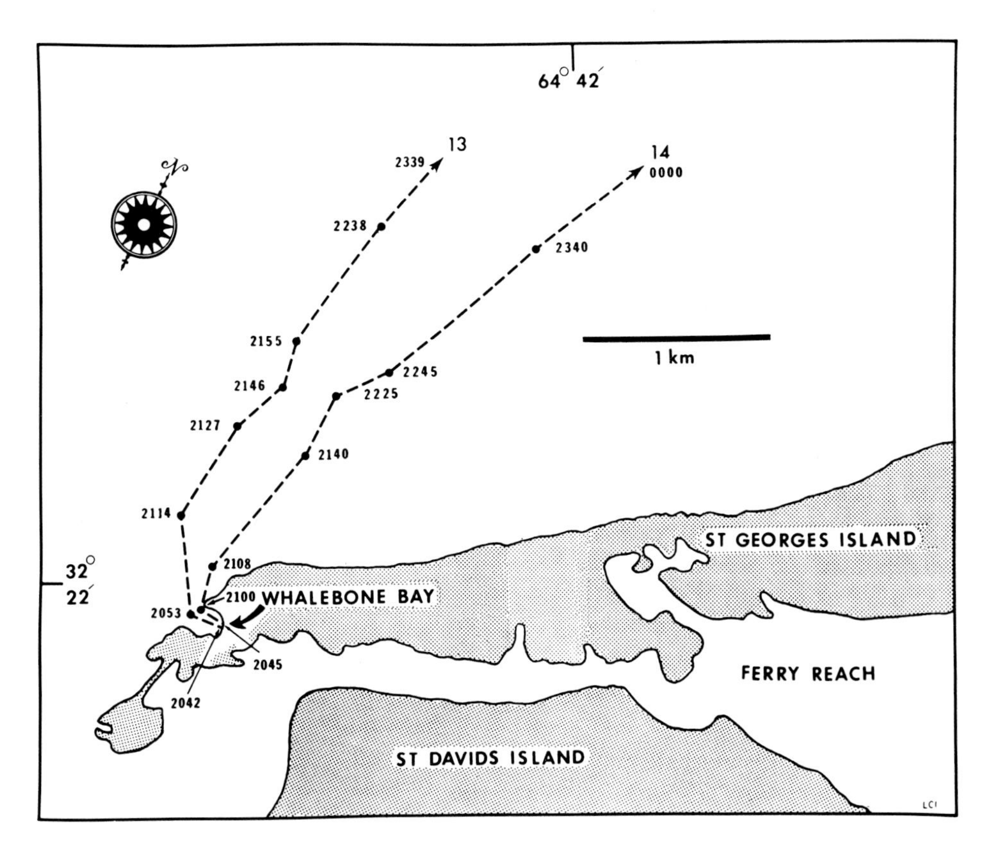

Fig. 6. Travel paths of newborn green turtles released from the beach at Whalebone Bay, Bermuda

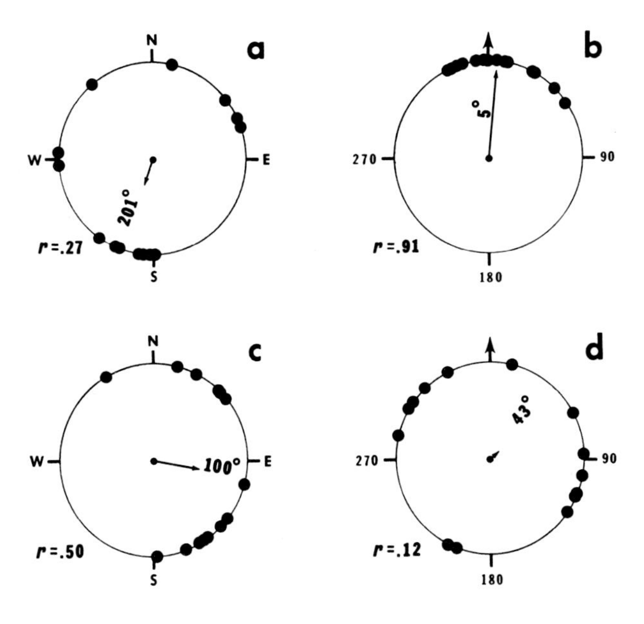

Fig. 7 a - d. Distributions of the initial and final directions of travel of the hatchling *C. mydas*. *Small arrows* indicate the mean vectors of each of the distributions; r indicates the length of each mean vector. *Large arrows* on distributions (b) and (d) indicate the direction "at a right angle to the beach" or "straight out from the beach." See text for further explanation

pared in the same manner as Figure 7B except that final rather than initial directions are shown. A Rao's test applied to these data proved insignificant (p > .05), indicating that the hatchlings no longer displayed a preference for travel at a right angle to the beach, or in any other direction. In the matter of final directions, our data differ considerably from those of Frick (1976). The final directions of the turtles Frick released from beaches facing the open ocean (n = 19) differed from right angles to the beaches by an average of approx. 26° (sd = 21°), while the turtles we released in the same situation (n = 12) differed by an average of 79° (sd = 43°). We have no explanation for this difference except to suggest that currents played a greater role in determining the travel paths of our hatchlings than those of Frick. We can think of two reasons why this might be true. First, we frequently worked in moderately rough seas while Frick could work only under the calmest conditions. Second, drag produced by the floats no doubt slowed our turtles to some degree and allowed currents to exert greater effects on our turtles' travel paths.

In summary, the results of our nighttime experiments agree with the results of Frick's (1976) research in most respects. Initial travel paths were roughly at right angles to the beaches from which the turtles were released. The majority of turtles displayed nonrandom travel paths, at least according to our criterion. Travel paths did not change to any great degree when land fell below the horizon from turtle eye-level. The sensory basis of this orientational ability is unknown. The travel paths of turtles released from beaches facing open ocean either approximated straight lines or took the form of gradual curves. Hatchlings released from beaches facing another shore reoriented until traveling toward open ocean. Our data do not agree with Frick's in the matter of final directions of travel. This is probably because ocean currents influenced the travel paths of our hatchlings to a greater degree. Further studies, concentrating on the role of ocean currents in determining turtle travel paths are indicated. The turtles displayed no preferred direction of travel other than 'away from land.' We suggest that the adaptive significance of this behavior lies in the fact that it serves to move the hatchlings rapidly beyond the reach of inshore predators.

Acknowledgments. We wish to thank the many who assisted us in tracking the hatchlings, and the Bermuda Dept. of Agriculture and Fisheries for the use of facilities on Nonsuch Island. David Han provided invaluable help with the illustrations. This research was partially supported by grants from the American Philosophical Society (Penrose Fund) and Sigma Xi.

References

Batschelet, E.: Statistical methods for the analysis of problems in animal orientation and certain biological rhythms. Am. Inst. Biol. Sci., Wash. D.C. (1965)

Batschelet, E.: Recent statistical methods for orientation data. In: Animal Orientation and Navigation. Galler, S., Schmidt-Koenig, K., Jacobs, G., Belleville, R. (eds.) 61-91. N.A.S.A., Washington, D.C., 1972

Buchler, E.R.: A chemiluminescent tag for tracking bats and other small nocturnal animals. J. Mamm. 57, 173-176 (1976)

Caldwell, D.K.: Hatchling green sea turtles, *Chelonia mydas*, at sea in the northeastern Pacific ocean. Bull. So. Calif. Acad. Sci. 68, 113-114 (1969)

Carr, A.F.: So excellent a fishe. Nat. Hist. Press, Garden City, New York (1967a)

Carr, A.F.: Adaptive aspects of the scheduled travel of *Chelonia*. In: R. Storm (ed.) Animal orientation and navigation, 35-55. Oregon St. Univ. Press. Corvallis (1967b)

Carr, A.F.: The case for long range chemoreceptive piloting in *Chelonia*. In: Animal Orientation and Navigation. Galler, S., Schmidt-Koenig, K., Jacobs, G., Belleville, R. (eds.) 469-483. N.A.S.A., Washington, D.C.

Frick, J.A.: Orientation and behavior of hatchling green turtles (*Chelonia mydas*) in the sea. Anim. Behav. 24, 849-857 (1976)

Hayes, W.N., Ireland, L.C.: Visually guided behavior in turtles. In: The Behavior of Fish and Other Aquatic Animals. Mostofsky, D. (ed.) New York; Academic Press, in press, 1977

Ireland, L.C., Kanwisher, J.W.: Underwater acoustic biotelemetry. In: The Behavior of Fish and Other Aquatic Animals. Mostofsky, D. (ed.) New York; Academic Press, in press, 1977

Mrosovsky, N.: The water-finding ability of sea turtles. Brain Behav. Evol. 5, 202-225 (1972)

Ridgway, S.H., Wever, E.G., McCormick, J.G., Palin, J., Anderson, J.H.: Hearing in the giant sea turtle, *Chelonia mydas*. Proc. Nat. Acad. Sci. USA 64, 884-890 (1969)

Schmidt-Koenig, K.: Migration and Homing in Animals. Berlin - Heidelberg - New York: Springer, 1975

Mass Migration of Spiny Lobster, *Panulirus argus* (Crustacea: Palinuridae): Synopsis and Orientation

WILLIAM HERRNKIND[1] and PAUL KANCIRUK[2], [1]Department of Biological Science, Florida State University, Tallahassee, FL 32306, USA, [2]Science Division, Oak Ridge Natural Laboratory, Oak Ridge, TN 37830, USA

Abstract

During autumnal mass migrations of spiny lobster in the northern Bahamas, large numbers move from the shallows to the areas fringing oceanic waters (e.g., the Gulf Stream). The phenomenon involves nocturnal immigration of lobsters over several weeks into the fringe. Concurrent with the first severe autumnal squall, the population moves synchronously in queue formations often both day and night. Thermal declines caused by the storms correlate statistically with onset of hyperactivity and queuing in captive lobsters. Lobsters becoming active facilitate others to queue up; queuing behavior serves migration by reducing hydrodynamic drag on each individual. Hydrodynamic stimuli, water currents and wave surge oscillations, serve as orientational guideposts under certain conditions.

A. Introduction

The mass autumnal migrations of the western Atlantic spiny lobster, *Panulirus argus*, are characterized by explosive, directional movements over several days of thousands of lobsters in single-file "queue" formations. (The synopsis is based on the following papers unless otherwise noted: Herrnkind and Cummings, 1964; Herrnkind, 1969; Herrnkind and McLean, 1971; Herrnkind et al., 1973; Kanciruk and Herrnkind, 1976 and in press). Our impression at the outset of research in 1969 was that this phenomenon was both enigmatic and unique, so we harbored few preconceptions that our findings would apply broadly to marine animal migration. The research was approached holistically; we sought to accurately characterize the migratory syndrome, determine the proximal causes and relation of environmental conditions, establish the behavioral adaptations, investigate the orientational components and, hopefully, infer the evolutionary significance of the event. Presently we have accumulated sufficient information to portray a hypothetical model and assert the directions of future research goals to more satisfactorily explain the mass migration syndrome. In addition, several features of this event may have broad applicability; namely, a mechanism of regulation of migratory state through photoperiodic and thermal events (Kanciruk and Herrnkind, in press) and an orientational guidepost based on wave surge, available to all shallow-dwelling marine animals (Herrnkind and McLean, 1971; Walton and Herrnkind, 1977).

B. Mass Migration: Synopsis

Mass autumnal migrations are reported from throughout the range of the species (Fig. 1) but typically in areas possessing extensive shallow continental shelfs; e.g., the Little and Great Bahama Banks. Observations of migrations from each area provide equivalent descriptions of the event; i.e., large numbers of lobsters move diurnally for several days in areas of shallow water near the shelf edge after a severe autumnal storm. The time of migration varies with the climactic pattern of each area so movements in the Bahamas and Florida occur in late October while movements do not occur until about mid-December in Yucatan and Belize. The direction of migration also varies with the area as does the age of the migrants. Migrations are southerly at Bimini, southerly and westerly at Grand Bahama, northerly near Boca Raton, Florida. Migrants in all areas are nonreproductive and include both males and females. The size (age) composition seems to reflect local population structure such that Bimini migrants average 82 - 87 mm carapace length (3 - 4 years) while Yucatan migrants average over 120 mm c.l. (6 - 7 years). Migrants are recorded as small as 55 mm c.l. (Bimini) and as large as 150 mm c.l. (Yucatan) indicating that migratory queuing may occur over most of their benthic life.

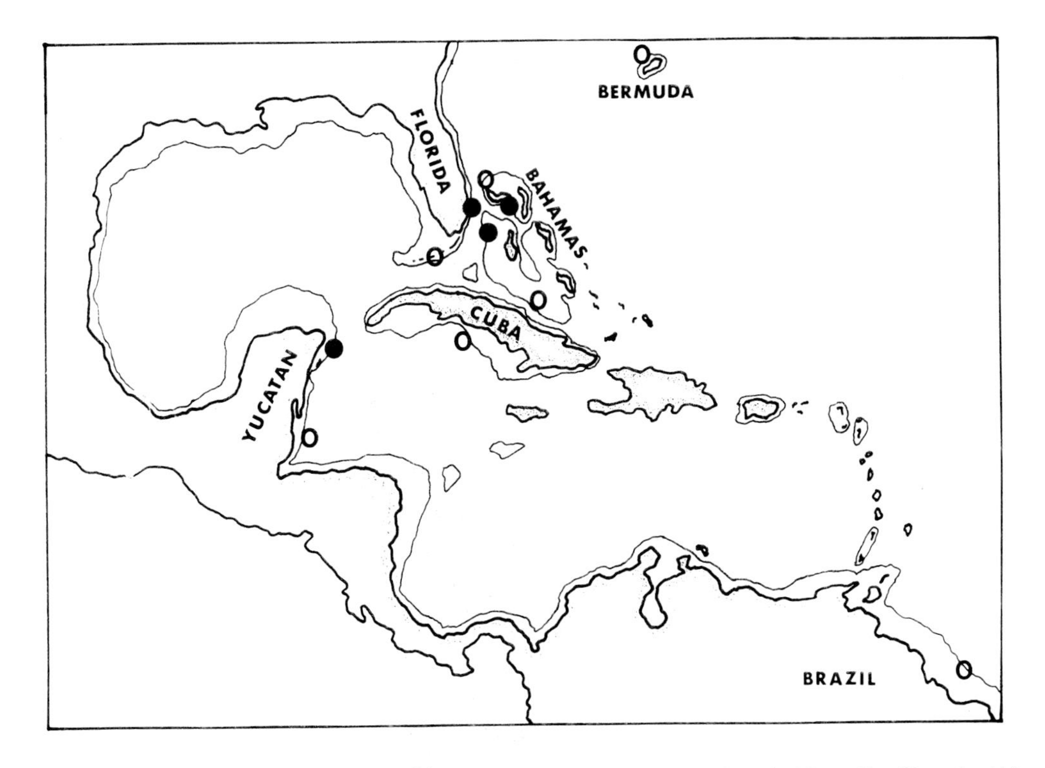

Fig. 1. The spiny lobster, *Panulirus argus*, is common to the shallow (2 - 50 m depth) tropical and subtropical marine waters of the western Atlantic. Mass autumnal migrations have been witnessed by us (*closed circles*), or reported reliably by others (*open circles*), especially adjacent to extensive shallow banks with large lobster populations

At Bimini, Bahamas, the site of our most extensive observations, the migratory period is characterized by four general phases: (1) premigration, (2) build-up, (3) mass queuing, (4) postmigratory dispersal (Fig. 2). In the early autumn, before migration, lobsters are distributed as two distinct groups. Older reproductively active lobsters reside on the deep reefs adjacent to the Gulf Stream while a much larger

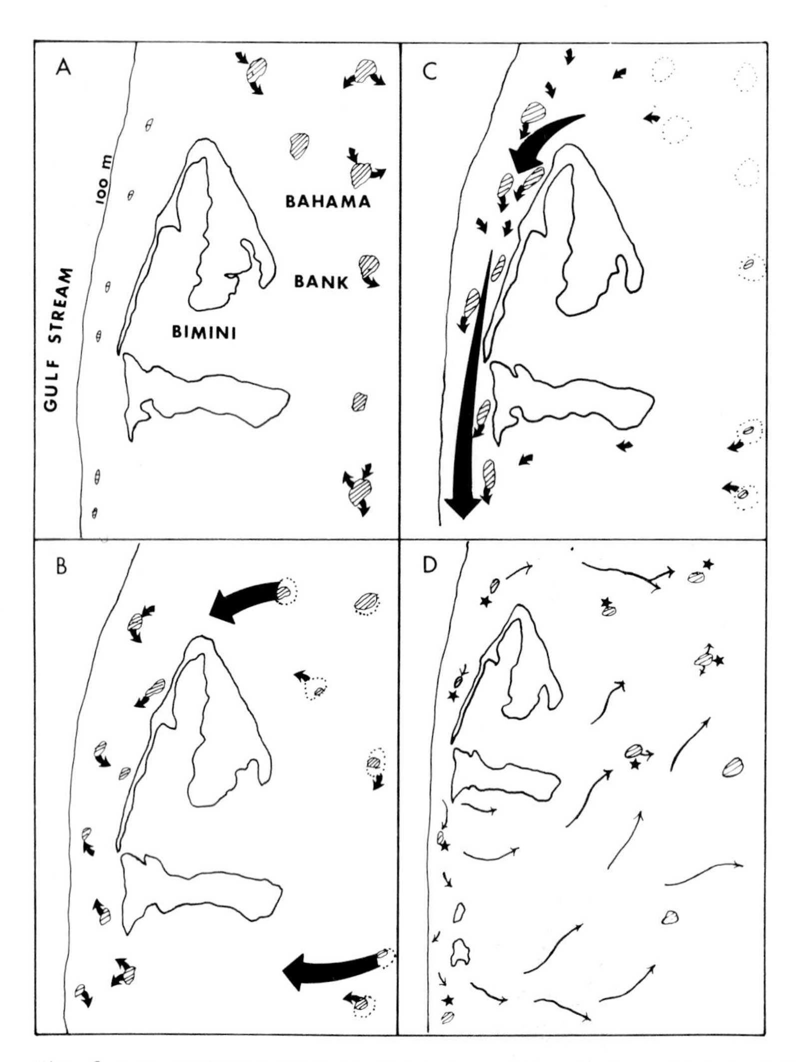

Fig. 2 A-D. Stylized summary of lobster movements near Bimini, Bahamas during autumn. (A) Premigration. Most lobsters move about nomadically on the shallow bank (3-10 m depht) while some still in reproductive condition reside near the edge of the Gulf Stream. (B) Build-up. In mid-autumn, lobsters move by nocturnal queues into and about the narrow fringe (migration pathway) adjacent to Bimini, both from the bank and deeper areas. (C) Mass migration. Following an autumnal storm, lobsters coalesce into queues moving both day and night, roughly southward along the fringe area. (D) Postmigration. Migrants disperse into available cover along the pathway; some wander back onto the bank over the winter while others remain residential. Stars indicate tag recapture locations. Scale distorted; area depicted is approximately 2500 sq. km

number of younger nonreproductive lobsters move about nomadically during noctural feeding forays over the extensive shallow waters to the east. The sparsely populated migratory pathway possesses a few postreproductive lobsters. Relatively little interchange occurs between bank and reefs until mid-autumn.

The build-up period involves an influx of formerly bank-dwelling lobsters into the pathway area with some overflowing into the deeper reefs. Simultaneously, but more gradually, postreproductive lobsters at the seasonal terminus of spawning move eastward mixing with the numerically dominant bank emigrants. Most movement at this time is by nocturnal queuing although the directions of movement vary (Fig. 2 and 4). The sparse shelter in the pathway and vicinity is now crowded with lobsters residing in large aggregations, often in "substandard" crevices rarely occupied at other times. Their behavior is much less cryptic than during nonmigratory periods such that aggregated lobsters can be easily coaxed to emerge and queue up even in midday by gentle prodding. Locomotory activity begins before darkness and any active lobster attracts a stream of followers.

The build-up condition continues for up to several weeks or until the onset of a severe polar-front storm. Storms lasting up to 4 or 5 days increase wind velocities from easterly 10 - 15 km/h to northerly with gusts up to 50 km/h and decrease the air temperature from 28° C to below 20° C. The violent atmospheric conditions quickly transfer energy to the shallow bank waters causing wave turbulence, turbidity from sediment displacement, and a drop in water temperature by as much as 5°C.

Following the storm, within 1 - 5 days, lobsters in the pathway are bolstered by large numbers of new migrants from the bank and reefs which all queue *en masse*, unidirectionally south and along the western fringe of the bank both day and night. Queues of migrants stream into shelters already overflowing while new queues form and move away. Walking rates average as high as 1 km/h over distances of up to several kilometers. Data from locomotory activity studies in large circular pools suggest nearly constant queuing for periods of 13 - 24 h per day.

Diurnal queuing may continue over several days while the shelter in the pathway and vicinity has high densities of transient lobsters. Thereafter, the queuing activity becomes sporadic and the number of lobsters in the area gradually decreases as the migrants disperse. A small proportion establish residency in the area, as shown by tag recoveries, while others continue southward. Lobsters tagged as migrants have been caught over the next several months at widely distant points back on the bank where the bulk of the population presumably originated. Our present tracking and tagging data are inadequate and we hope to examine postmigratory behavior more effectively in the future.

The occurrence of mass diurnal migration in conjunction with autumnal storms is mimicked by the behavior of captive lobsters. There, the onset of the migratory syndrome correlates statistically with the temperature reductions accompanying the storms (Kanciruk and Herrnkind, in press). Within the period of cooling, captive lobsters initiate locomotory activity in daylight and often remain active into daylight hours the next morning; we recorded up to 48 h of continuous activity. The periods of activity also include a marked increase in sociality such that queuing becomes the dominant behavior.

In addition to the strong inference that temperature triggers mass migration, other research showed an apparent influence of photoperiod on seasonal character of daily locomotory rhythms (Kancurik and Herrnkind, 1973). Hence, we propose to experimentally test the hypothesis that autumn decreases in photoperiod induce the onset of *Zugunruhe*, observed in the field as nocturnal queuing, and a subsequent temperature decrease of the order of 5° C over 2 - 3 days triggers further hyperactivity.

Other stimuli and factors probably significant to the degree of migratory activity are illustrated in Figure 3. Storm swells cause violent oscillations of water near the substrate (surge) which drastically curtail the locomotion of lobsters. Thus, despite strong motivation to move about, lobsters are probably held captive. Conceivably, removing this constraint itself results in a hyperactive state. This species of spiny lobster becomes less active as light intensity increases, being most active normally on moonless nights (Sutcliffe, 1956). Our observations show that diurnal queuing often occurs under conditions of reduced light levels caused by clouds and turbid water (unpublished data). Then, crowding and lack of shelter further augment the tendency to form queues (Herrnkind, 1969; Berrill, 1975).

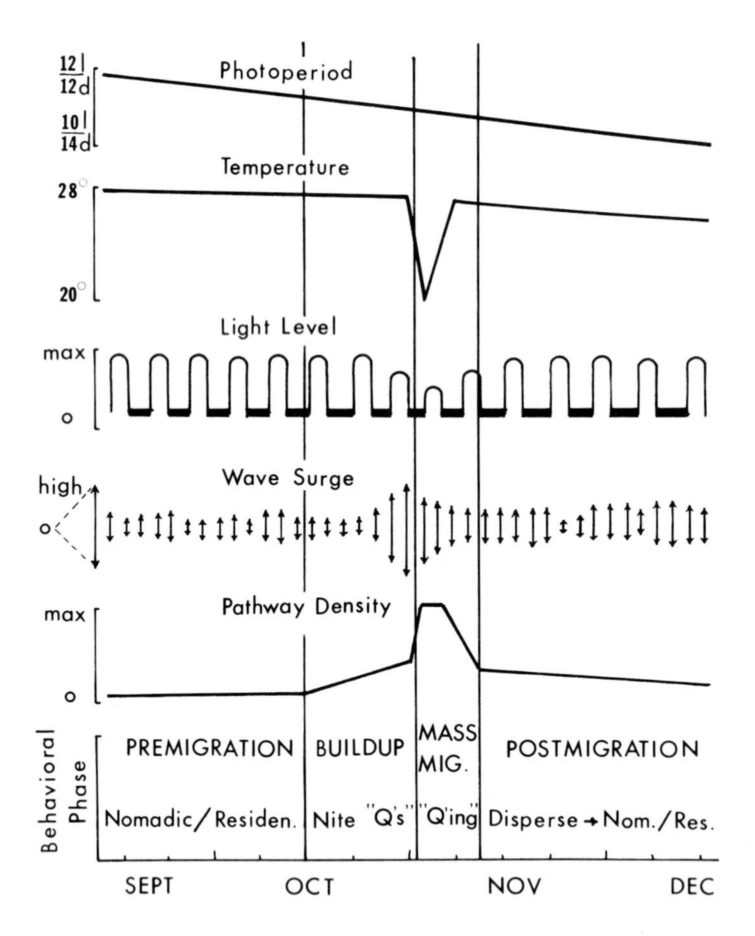

Fig. 3. Hypothetical model depicting the chronology of environmental factors influencing mass migration. The decreasing autumn photoperiod brings on *Zugunruhe* manifested as nocturnal queuing during the build-up phase. Mass diurnal migration is triggered by the sharp decline in water temperature caused by an autumnal storm. The amount of migratory activity is possibly augmented by the decrease in daily light levels resulting from water turbidity and atmospheric cloud cover, by the entrapment or agitation effect of wave surge from storm swells, and by the overcrowding of available shelter in the migration pathway. The postmigratory period involves waning of the *Zugunruhe* and reestablishment of a nomadic/residential phase of individual behavior

Queuing behavior appears to be a natural consequence of the general increase in sociality brought on with *Zugunruhe* (Herrnkind, 1969). Formation movement in tandem results in drag reduction of approximately 50% for groups of 6 or more at the mean migratory walking pace (1 km/h; Bill and Herrnkind, 1976). The lead position, which sustains highest drag, is exchanged frequently, often without change in bearing (Herrnkind and McLean, 1971). Queuing theoretically provides a mechanism of energy conservation facilitating efficient long-distance locomotion.

Discerning the biological significance of the autumnal mass migration to the species has been especially problematic. The fall movements are not explainable merely as escape responses to the severe hydrodynamic conditions created by the storms since the build-up movements occur in calm periods, and winter storms later in the year are not associated with equivalent mass movements at Bimini. Storms, including hurricanes, are anecdotally associated with mass migrations other times elsewhere, but no adequate information is available to assess the condition of the population in the period immediately preceding these movements. In fact, lobsters can be induced to queue at any time by a variety of causes although for only short duration (Herrnkind, 1969; Berrill, 1975). We here confine discussion to the autumnal migratory syndrome.

The event is obviously not a prereproductive movement nor does it appear to be primarily a postreproductive exodous. The fact that migrants in some regions are largely mature (Yucatan) argues against the migrations serving as an ontogenetic shift in habitat. Food supply and lobster density do not appear to change markedly in the primary source areas prior to the build-up movements.

The hypothesis most closely fitting present data is that the mass migration is a concentrated seasonal movement adapted to moving the population from the shallow banks, subject to severe cooling, to the oceanic fringe where conditions are suitable to overwintering. Clearly, shallow waters over much of the range of the species cooled to temperatures of 10° C or less for long periods during glacial winters (Climap, 1975). In fact, temperatures in this range were recorded in 1976 in Florida (25° N lat.) and are probable for periods of recorded "fish kills" in the 1930 s. The spiny lobster becomes moribund, cannot feed effectively, and is unable to complete molting below 12 - 15° C (Witham, 1973; G. Davis, pers. comm.; D. Wynne and W.H., unpublished data). Hence, selection would strongly favor the tendency to emigrate from the areas most subject to winter cooling. *Zugunruhe* brought on by photoperiod and migration triggered by the harbinger of winter, the first autumnal storm, would be effectively served by queuing behavior. During cold periods, a thermal barrier prevents dispersal back to the shallows but water temperature during mild winters is sufficiently high that postmigratory nomadism broadly redistributes many of the migrants back over the banks. Redistribution and dispersal *per se* may be sufficiently advantageous to augment the selective value of such migrations. In summary, the mass migration of *P. Argus* is possibly a highly specialized type of seasonal movement characteristic of many marine animals including several other spiny lobsters (Olla and Studholme, 1971; Dorgelo, 1976; Herrnkind, 1977).

C. Orientation

Queues observed in the pathway during build-up periods show considerable variation in instantaneous bearings (Figure 4). We noted that queues initially forming at dusk made circuitous paths and frequently split up or mixed with other queues. The variable headings of queues sighted during the build-up period suggested mixing of lobsters throughout the area during this period.

By comparison, diurnal queues seen after storms showed strong clumping of bearings between 180° and 195°, or slightly west of south, roughly parallel to the run of the Bahama Bank edge. These bearings at Bimini were consistent from year to year, at widely spaced locations and at

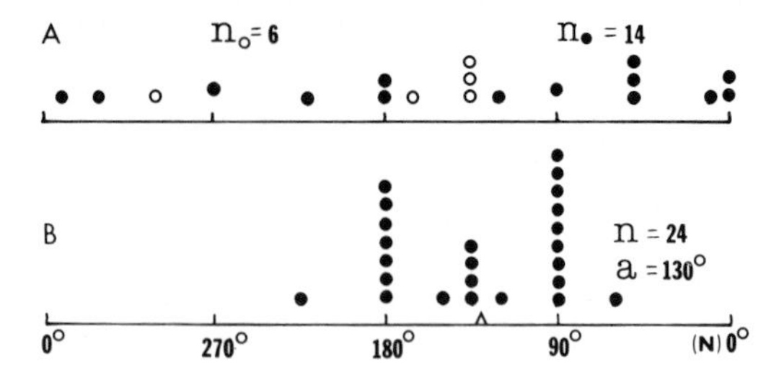

Fig. 4 A and B. Instantaneous headings of queues observed in the pathway area during build-up in (A) 1969, *open circles*, 1973, *closed circles*, and (B) 1974, show a variety of directions. Lobsters apparently move about within the area after arrival from the deep reefs to the west and, especially, the shallow bank to the east

different times of day (Fig. 5; statistical treatment of all orientation data after Batschelet, 1965, 1972).

The question of a group influence on orientation was given preliminary examination by comparing the grouped headings of individuals, queues of two, three, four, five, six-ten, eleven or more (Fig. 6). Rayleigh (<.05) and V-test values (<.05) show significant clumping and equivalent bearings, respectively, for all queues of three or greater. Individuals showed a dispersed array of headings although the nonsignificant mean

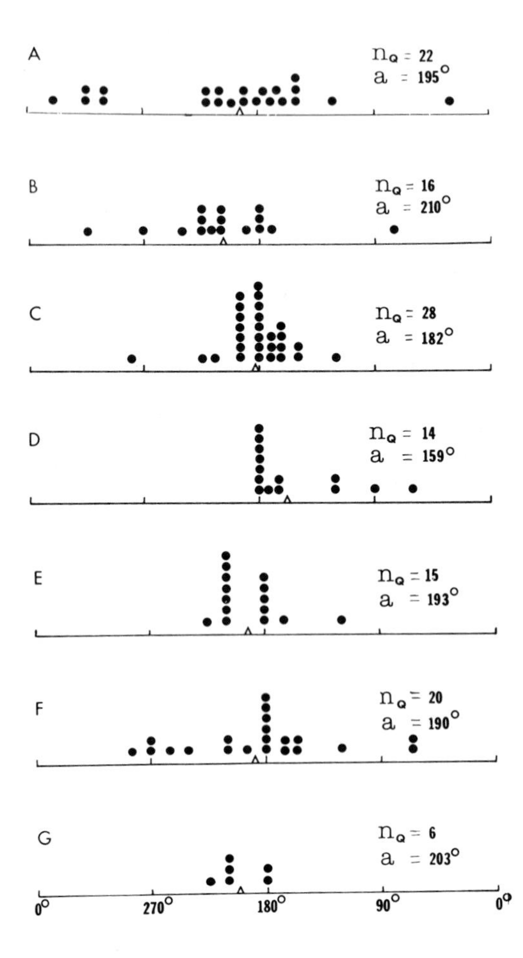

Fig. 5 A - G. Instantaneous headings of queues (*NQ*) of mass migrants observed during daylight over several days at various locations near Bimini in 1969. South Bimini (A) 3 November; (B) 4 Nov.; (C) 6 Nov., 1 km south of South Bimini (D) 3 Nov.; (E) 4 Nov., North Bimini, south end; (F) 7 Nov., North Bimini, north end; (G) 4 Nov. All headings are statistically grouped; mean directions (*a*) significant by Rayleigh test at least < .05

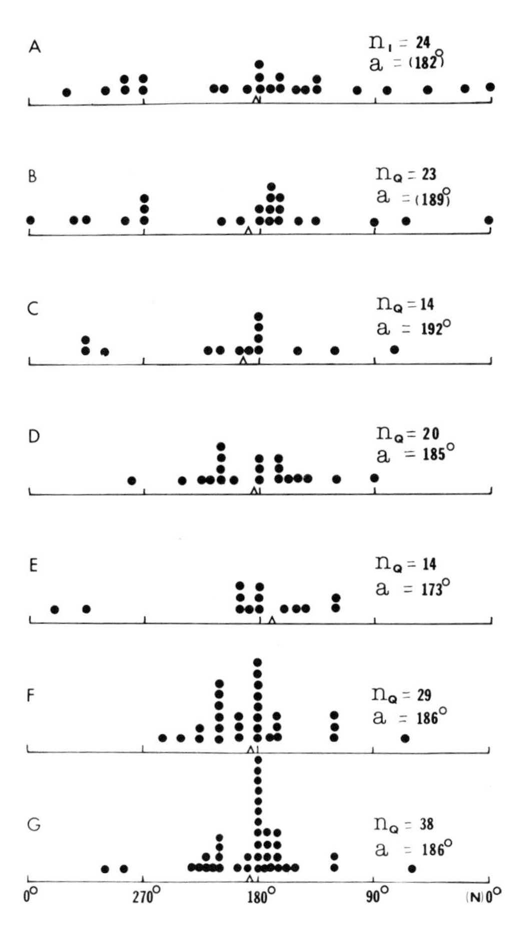

Fig. 6 A - G. Instantaneous headings of (A) solitary migrants, n_I; (B) queues of 2, (C) 3, (D) 4, (E) 5, (F) 6 - 10, and (G) 11 or more. Solitary migrants show weak grouping (Rayleigh $<$ .10). Queues of 2 show slightly stronger clumping (Rayleigh $>$ .05, $<$ .10) while queues of 3 and greater show strong directionality (Rayleigh, $<$ .05 or better)

direction was southward. Queues of two showed somewhat less scattering and approached significance at the 5% level (Rayleigh, <.10, >.05) with a southward bearing. Lone migrants were seen infrequently (38 of 1638 lobsters or 2.5%) and we observed several to join passing queues. We cannot determine whether the solitary lobsters were merely poor orienters or whether the random bearings serve as a strategy to increase the probability of an individual intercepting a queue.

The potential consequence of being queue leader was briefly examined by selectively removing the lead lobster from moving queues. The diver slowly approached from just ahead and to the side of a queue, deftly netted the leader, and then retreated. Four of eighteen removals resulted in scattering of the former queue members while another four caused no change in queue direction. The overall distribution of bearings was significantly clumped in the original queue bearing (Rayleigh, <.05; V test, <.05). Because queues frequently change leaders spontaneously, we suggest that many of the queuing lobsters are capable of appropriate orientation. Further experiments should examine the question of orientational pooling and other experiential influences on mass migrants.

Thus far we know that spiny lobsters orient directionally during homing and when displaced to unfamiliar areas as well as during mass migration (Herrnkind and McLean, 1971). The directions taken are appropriate to

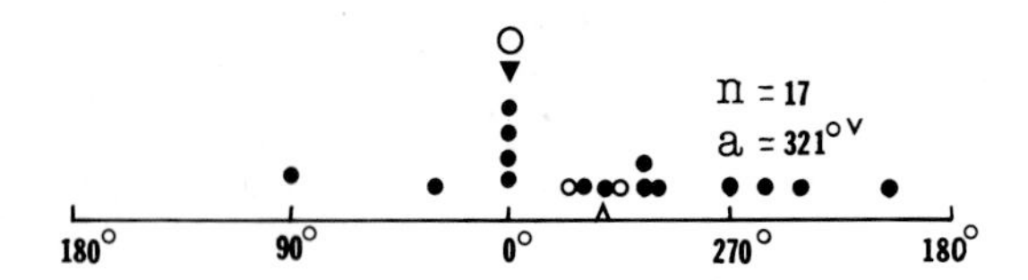

Fig. 7. Effect of removing queue leaders. Queues tend to continue travel in the original heading (0; V test, <.05). *Closed circles*, leader removed by diver; *open circles*, queue separated due to natural disruption

each situation and can occur under a wide array of environmental conditions characteristic of the range of variability within the biotope of the species. We cannot specify the guidepost(s) directing the migratory orientation although hydrodynamic cues including wave surge and monodirectional water currents are suspected to be especially important (Walton and Herrnkind, 1977).

Theoretically, surge and currents can provide essential directional information during migration. One or the other or both cues are nearly omnipresent in the migration pathway and their detection requires no visual component. Surge is directionally conservative since swells consistently approach from the deep water to the west or northwest (at Bimini). Also, surge is detectable both in the absence of current and when superimposed upon it. Currents may be more variable than surge and can be masked by heavy surge. However, strong currents run along the edge of the Gulf Stream and also occur as tidal flow at the edge of the shallow portion (5 - 8 m depth) of the Bahama Bank. The chemical signature of oceanic and bank waters are certainly different and, if detectable to lobsters (Ache et al., 1976; Laverack, 1974), could provide a discriminable directional cue as in the case of estuarine animals (Creutzberg, 1975).

The paths of migrant lobsters indicate that orientation to hydrodynamic cues prevalent in the area might not be a simple positive or negative taxis; i.e., the paths are at some angle to the primary surge and current directions. We propose to further test an hypothesis of hydrodynamic orientation and to examine the influence of other stimuli, such as visual, acoustic, slope, and magnetic, both singly and in combination. Especially, we hope to do long-range tracking of migrants both during spontaneous movements and under conditions of experimental manipulation. The obstructions posed by the marine medium make this a challenging prospect.

References

Ache, B., Fuzessery, Z., Carr, W.: Antennular chemosensitivity in the spiny lobster, *Panulirus argus:* Comparative tests of high and low molecular weight stimulants. Biol. Bull. 151, 273-282 (1976)

Batschelet, E.: Statistical methods for the analysis of problems in animal orientation and certain biological rhythms. Am. Inst. Biol. Sci. 1-57 (1965)

Batschelet, E.: Recent statistical methods for orientation data. In: Animal Orientation and Navigation. Galler, S. et al. (eds.): NASA Stock No. 3300-0392, 61-91, 1972

Berrill, M.: Gregarious behavior of juveniles of the spiny lobster, *Panulirus argus* (Crustacea: Decapoda). Bull. Mar. Sci. 25, 515-522 (1975)

Bill, R., Herrnkind, W.: Drag reduction by formation movement in spiny lobster. Science 193, 1146-1148 (1976)

CLIMAP: The surface of the ice-age earth. Science 191, 1131-1137 (1975)

Creutzberg, F.: Orientation in space: Animals (8.1) invertebrates. In: Marine Ecology II, Part 2, Physiological Mechanisms. Kinne, O. (ed.), 555-655, 1975

Dorgelo, J.: Salt tolerance in crustacea and the effect of temperature upon it. Biol. Rev. 51, 255-290 (1976)

Herrnkind, W.: Queuing behavior of spiny lobsters. Science 164, 1425-1427 (1969)

Herrnkind, W.: Spatio-temporal attributes of movement patterns in palinurid lobsters: Review, synthesis and prospectus. In: Workshop on Lobster and Rock Lobster Ecology and Physiology. Cobb, S., Phillips, B. (eds.). CSIRO Division of Fisheries and Oceanography Circular No. 7, 209-244, 1977

Herrnkind, W., Cummings, W.: Single file migrations of the spiny lobster *Panulirus argus* (Latreille). Bull. Mar. Sci. Gulf Carib. 14, 123-125 (1964)

Herrnkind, W., Kanciruk, P., Halusky, J., McLean, R.: Descriptive characterization of mass autumnal migrations of spiny lobster, *Panulirus argus*. Proc. Gulf Caribbean Fish. Inst. 25, 78-98 (1973)

Herrnkind, W., McLean, R.: Field studies of homing, mass emigration and orientation in the spiny lobster, *Panulirus argus*. Ann. N.Y. Acad. Sci. 188, 359-377 (1971)

Herrnkind, W., VanDerwalker, J., Barr, L.: Population dynamics, ecology and behavior of spiny lobster, *Panulirus argus*, of St. John, U.S. Virgin Islands: Habitation and pattern of movements. Results of the Tektite Program, Vol. 2. Bull. Nat. Hist. Mus. L. A. County 20, 31-45 (1975)

Kanciruk, P., Herrnkind, W.: Preliminary investigations of the daily and seasonal locomotor activity rhythms of the spiny lobster, *Panulirus argus*. Mar. Behav. Physiol. 1, 351-359 (1973)

Kanciruk, P., Herrnkind, W.: Autumnal reproduction in *Panulirus argus* at Bimini, Bahamas. Bull. Mar. Sci. 26, 417-432 (1976)

Kanciruk, P., Herrnkind, W.: Mass migration of spiny lobster, *Panulirus argus* (Crustacea: Palinuridae): behavior and environmental correlates. Bull. Mar. Sci. (in press)

Laverack, M.: The structure and function of chemoreceptor cells. In: Chemoreception in Marine Organisms. Grant, P., Mackie, A. (eds.). Academic Press, N. Y. 1-48, 1974

Olla, B., Studholme, A.: The effect of temperature on the activity of the bluefish, *Pomatomus saltatrix* L. Biol Bull. 141, 337-349 (1971)

Sutcliffe, W.: Effect of light intensity on the activity of the Bermuda spiny lobster, *Panulirus argus*. Ecology. 37, 200-201 (1956)

Walton, A., Herrnkind, W.: Hydrodynamic orientation of spiny lobster, *Panulirus argus*. Wave surge and monodirectional currents. Memorial Univ. of Newfoundland Marine Sciences Research Lab. Tech. Rept. No. 20, 184-211 (1977)

Witham, R.: Preliminary thermal studies on young *Panulirus argus*. Scientist 36, 154-158 (1973)

Migration and Homing of Marine Invertebrates: A Potpourri of Strategies

JAMES T. ENRIGHT, Scripps Institution of Oceanography, University of California, La Jolla, CA 92093, USA

Abstract

Marine invertebrates have a diversity of behavioral strategies for finding and remaining in a suitable habitat. The cases considered here on a comparative basis include vertical migration of zooplankton, the homing of talitrid amphipods (*Orchestoidea*), and the tidal migrations of a burrowing crab (*Emerita*) and a tiny amphipod (*Synchelidium*). The behavioral mechanisms and sensory systems underlying these performances vary with the ecologic situation, and range from simple phototaxis (plankton), straightforward geotaxis (plankton; *Emerita; Synchelidium*), and thigmotaxis (*Orchestoidea* in sea water); through finely tuned responses to hydrostatic pressure (*Emerita; Synchelidium*) and critically timed reversals of phototaxis (*Synchelidium*); to time-compensated celestial orientation (both sun and moon) and visual recognition of form (*Orchestoidea*).

Because of the emphasis at this Symposium on the truly remarkable homing and migratory performances of birds, it would be easy to overlook the fact that the behavior of many invertebrates involves clear if usually more modest counterparts. A few striking examples have, of course, already been brought to your attention here, and it is my objective to broaden the coverage, based on some examples from the behavior of marine invertebrates. In my opinion, the most important kind of animal migration, which has otherwise apparently been overlooked here, is the diurnal vertical migration of zooplankton. Whether viewed in terms of the number of species involved, the number of animal-kilometers traveled per year, or simply the gross biomass that is transported, these vertical migrations are probably the most impressive of the temporally coordinated mass movements of animals known to science. The German-language term for the migrations, *Aufwanderungen*, suggests a rather casual, liesurely sort of activity, not at all comparable with the *Zug* of birds, but that implication probably does the invertebrates an injustice. Some tiny oceanic crustaceans, only 2 to 3 mm in length, are now known to sustain surprisingly high average speeds of upward swimming, for hours at a time, daily covering vertical distances of hundreds of meters (Enright, 1977), with a single-mindedness that compares favorably with that of many migrant birds. Until very recently, it was widely accepted that the daily *timing* of these mass movements of plankton resulted from a direct response to the stimulus of prevailing light regimes, but that interpretation probably also does the invertebrates an injustice. Just as endogenous timing mechanisms play a significant role in the temporal coordination of bird migrations (Gwinner, 1968), biologic clocks clearly also underlie some instances of daily vertical migration (Enright and Hamner, 1967); similar mechanisms are strongly suspected in other cases as well (Enright, 1976; Enright and Honegger, 1977). In one significant way, however, the daily migrations of plankton are relatively easy to understand, perhaps justifying their neglect at this Symposium: The direction of migration is either upward or downward, and the stimuli which could serve as beacons to guide that sort of directional choice are self-evident. Responses to gravity, to light intensity, and to hydrostatic pressure constitute three obvious possibilities, and probably all three are often involved. In contrast, the directionality in the migration and homing of birds demands considerably greater complexity in terms of special behavioral and sensory equipment.

The other examples of migration and homing of marine invertebrates which I wish to discuss today in greater detail involve instances in which directionality of the movements is not so obviously dictated by clear, vectorial environmental stimuli. The cases involve several species of sand-beach crustaceans, all of them abundant members of the fauna on the Southern California beach in front of my office in La Jolla. Anyone who has taken more than a casual look at the intertidal fauna of a rocky shore must be impressed by the conspicuous vertical zonation of the attached animals and plants. This kind of zonation is, of course, an adaptive response to the fact that the intertidal region is characterized by extremely steep gradients in a whole array of important parameters of the physical environment. Although such zonation in the rocky intertidal zone is so familiar, it is less widely appreciated that an intertidal sand beach is often also characterized by a vertical zonation of the animals, which is no less clear-cut than that on the rocky shore. The phenomenon is simply less conspicuous because the animals are buried beneath the surface of the sand.

In temperate latitudes throughout the world, including the beaches of Southern California, talitrid amphipods, commonly known as beach-hoppers, inhabit the highest, most landward zone in the intertidal region. On the La Jolla beach, the talitrids consist of two, or sometimes three species of the genus *Orchestoidea*. These are strictly marine creatures, in the sense that their distribution is closely linked to the open-coast shoreline, and yet their adaptations to the semi-terrestrial habitat of the high-beach shoreline have gone so far that the animals cannot survive immersion in salt water for more than a few hours. At the other extreme, however, *Orchestoidea* cannot survive more than a few minutes of exposure to the heat and desiccation typical of the dry, supralittoral beach at midday in summer. It should not be surprising to find, therefore, that these amphipods inhabit a very narrow strip of the shoreline, usually just below the crest of the most recent high tide, where the subsurface sand remains moist but not wet. There they remain hidden beneath piles of seaweed or buried in the sand throughout the day, unless disturbed by a shorebird or a probing naturalist. At nighttime, they emerge to wander over the exposed beach in search of food, and return again to their burial zone before sunrise.

The behavioral mechanisms by which *Orchestoidea* finds its way back to this narrow habitable region on the shoreline constitute a very broad spectrum of responses. The most publicized of these is their ability to perform time-compensated solar orientation, a phenomenon initially explored by Pardi and Papi (1953) while working with European relatives of *Orchestoidea*, of the genus *Talitrus*. As clear-cut as this solar orientation usually is, also in *Orchestoidea* (Hartwick, 1976), it is apprently an escape mechanism only necessary for those rare and unpredictable occasions when the animals are exposed in the daytime, on a level, featureless beach. As has recently been shown by Hartwick (1976), the directional choice dictated by sun orientation can easily be overridden on the shore by a view of the local landmarks. A back-shore cliff or its facsimile is treated by the amphipods as a more reliable clue about the landward-seaward axis than the position of the sun. At close range, even a pile of seaweed on the moist beach can prove to be a stronger orienting stimulus than either sun or backshore cliff. The animals, when offered the chance, will ignore both solar azimuth and a view of cliffs, in order to dash for shelter beneath nearby piles of algae.

At night, *Orchestoidea* can, at least sometimes, use the moon as an orientational reference point for ecologically appropriate directional decisions: time-compensated lunar orientation (Enright, 1972). The question of how the animals can use one timing system, in their compensation for solar movement, and another, seemingly very different system, in compensating for lunar changes in azimuth, remains one of the most striking outstanding problems in talitrid behavior. In addition to these forms of visual orientation, some factor related to moisture of the environment or the substrate plays an important role in the behavior of *Orchestoidea*. The directional choice of the animals, given solar clues about the landward-seaward axis, is consistently seaward in a *dry* orientation chamber, and completely reversed to a landward direction, on a moist substrate. Still other factors that are known or suspected as orienting stimuli for the amphipods include slope of the substrate, when it is steep enough; direction of the wind, when it is strong; and texture of the substrate: On uncompacted moist sand, the animals tend to bury themselves immediately, rather than searching further.

With this large, complex, and flexible repertoir of orientational mechanisms at its disposal, *Orchestoidea* is not likely to lose its way on the beach itself, but what happens to the amphipods when they find themselves in sea water - as must occasionally happen when they are surprised by a wave? Their weak swimming ability would be of little use in the turbulent, nearshore waves; but in this situation, *Orchestoidea* shows a strong thigmotaxis, clinging tightly to any floating object. At first glance, this may not sound at all like orientation behavior, but in the nearshore zone, essentially all floating objects are very soon deposited as wrack on the high beach, at the top of the wave uprush: exactly where the amphipods would wish to be taken. Perhaps orientation is not the correct word for this kind of behavior; the movement of the animals is ultimately passive, and a thigmotaxis for floating objects simply gives the animals a means of "rectifying" the "alternating current" of the nearshore wave regime. As will be seen in other examples considered below, behavioral mechanisms to "rectify" transport by waves are key components of the orientational repertoir of other beach crustaceans as well.

These behavioral schemes by which *Orchestoidea* finds its way back to a suitable, restricted region of the beach can, in a very loose sense, be considered a sort of "homing," but there is an important distinction to be made here. Instead of returning to a particular *point* in space, as does a pigeon or a honey bee, the "home" to which the amphipods return is a *line*, a one-dimensional strip of habitat. Ultimately, then, the directional choice of the amphipods is simply "landward or seaward"; the complexity of the orientational mechanisms involved arises only because the environmental cues available to guide them in that choice are more subtle than those available to vertically migrant planktonic animals.

The other kinds of behavior of beach crustaceans to which I now direct your attention are related to talitrid orientation in the broad sense that the animals seek to remain in a narrow, optimal habitat zone of the shore, but that is about the full extent of the resemblances. The creatures involved, anomuran crabs of the genus *Emerita* and oedicerotid amphipods of the genus *Synchelidium*, seek their food in the uprush zone of the beach, that narrow region at the water's edge where sand is alternately inundated and exposed by essentially every wave. The position of the uprush zone on the beach moves landward and then seaward with the flow and ebb of the tides. To stay in their habitat zone, the animals migrate up and down the beach with the progression of the tides. In common with the familiar phenomenon of seasonal migration, these mass population movements of beach crustaceans are strongly coordinated in their timing, and clearly synchronized with predictable habitat fluctuations. I do not, therefore, hesitate to use the term "migration," even though the time scale involved is one of hours rather than months, and the distances traveled are tens of meters rather than hundreds of kilometers.

A first question of interest to a behavioral physiologist is how temporal coordination of the animals' tidal migrations is achieved. Is it simply a passive response to environmental fluctuations, or do the animals have some sort of endogenous timing capacity, to assist in the decision of whether landward or seaward movement is appropriate? Clear - cut experiments have demonstrated that endogenous timing plays a major role (Enright, 1963, 1974). The physiology of the animals, and with it, their behavioral repertoir, changes markedly, under the control of a "biological clock" which is synchronized by the tides on the shoreline. When confined in an aquarium with still water and sand, the crustaceans, both crabs and amphipods, remain buried in the sand throughout the times of rising tide. At, or shortly before tide crest, they spontaneously become extremely active, swimming in the water of the aquarium, and continue this "migratory restlessness" for several hours, at times corresponding to their seaward migration on the beach. My phrasing here is a deliberate misuse of the term "migratory restlessness," simply to call to your attention a functionally similar phenomenon, which is necessarily different in detail as well as in specific mechanism. My point is that, as in the case of long-distance migration of birds, these beach crustaceans are not simply at the mercy of relatively unreliable stimuli of the moment, for the temporal coordination of their directional movements; an endogenous biological clock gives them "inertia" in the time domain, so that they are less likely to be misled by occasional environmental anomalies.

Now what about the stimuli which determine the direction of movement? In sharp contrast with the high-beach talitrid amphipods, the directional decisions made by *Emerita* and *Synchelidium* are apparently independent of visual recognition of celestial or terrestrial landmarks. Instead, the directional choices are derived from the nature of wave movement in the animals' habitat. A landward-rushing wave in the uprush zone of a beach brings with it kinetic energy from distant sources. The flow is usally extremely turbulent, entraining sand (and animals!) from the substrate; the uprushing wave often advances as a "wall of water," with a very steep front. At the end of this landward surge, the direction of motion turns seaward, but (particularly on a gently sloping beach like that at La Jolla) the return flow only slowly picks up speed. The backrush tends to be much more a laminar flow, with a sheet of water several cm thick moving smoothly and gently seaward.

In the behavior that underlies the direction of migration, both *Emerita* and *Synchelidium* rely upon the hydrostatic pressure regimes which are associated with these differences in the direction of wave motion on the beach. Both species are exquisitely sensitive to small increases in hydrostatic pressure, of the sort associated with the landward-rushing, turbulent uprush; both respond to sudden pressure increase with hyperactivity that lasts from 5 - 30 s, with a threshold on the order of 5 millibars, equivalent to a 5-cm increase in water height (Enright, 1961, 1962, 1967). In *Emerita*, this hyperactivity seems to be exclusively an attempt by the animals to bury themselves deeper into the substrate (Enright, unpublished). Since turbulent, landward-rushing waves tear the crabs from the bottom, they seek to bury themselves again, as soon as possible, so as to avoid being cast up to the peak of the uprush. Following this short-duration response, the sand crabs then raise their long, setose antennae into the gentle, laminar flow of the returning wave, in order to filter out food particles from the water. On the ebbing tide, their spontaneous tendency to swim in the water, triggered by their tidal rhythm, leads to a net seaward migration as the tide falls. The animals thereby avoid the risk of being marooned on the high beach by the receding tide, since they cannot withstand the desiccation of the high intertidal zone between tides.

The amphipod *Synchelidium*, on the other hand, is a much more fragile creature; adult individuals are only some 2 mm in length, and they are able to bury into the sand only to a depth of a few mm, and that can be accomplished only in conditions of relatively still water. Their habitat zone on the beach, then, tends to be somewhat landward from that of the sand crabs. On the rising tide, *Synchelidium* is aggregated in the narrow zone of moist sand near the very peak of the uprush zone, where they feed on single algal cells and tiny bits of debris deposited by the waves.

Synchelidium, too, relies upon its respones to hydrostatic pressure, in its tidal migrations, but in a somewhat different way, which involves, in addition, a peculiar sort of phototactic response. During the rising tide, i.e., the inactive phase of the animals' tidal rhythm, the amphipods are negatively phototactic when in still water. When, however, one subjects the animals to vigorous water motion, of the sort which would arise from a landward-rushing wave, they become positive in their phototaxis, for an interval of some 5 - 15 s. Since the landward-rushing wave will also evoke their hyperactivity response to pressure increase, the amphipods can be expected to swim vigorously to the water surface and to be carried landward by the uprushing wave. At times of rising tide, both the positive phototaxis and the response to wave pressure have a similar duration: 5 - 15 s. This response-duration corresponds roughly to the duration of uprush by a wave, following which the animals will be in still water and can bury into the sand (Enright, 1962). This kind of behavior, by itself, would permit the animals to maintain a zonation near the highest uprush on the beach - during the rising tide; but it would also lead them to be stranded on the high beach, as soon as the tide turns. During the ebbing tide, however, the animals' endogenous activity rhythm comes into play: The amphipods are spontaneously more active at times of ebbing tide. Moreover, they are continuously positively phototactic, and, in addition, their responses to increases in hydrostatic pressure change markedly. Instead of a brief burst of randomly oriented hyperactivity, *Synchelidium* shows a response to pressure increase with a duration nearly twice as great: some 20 - 25 s; and the animals swim *upward* in the water (Enright, 1962). At times of ebbing tide, then, the long-duration, upwardly directed response to pressure increases, coupled with continual positive phototaxis, will lead the animals not only to follow an uprushing wave landward, but to continue to swim at the water surface for a sufficient time to be transported back seaward in the subsequent backrush of that same wave. During ebbing tide, then, the amphipods are no longer to be found at the peak of the uprush zone on the beach, but are essentially all in the turbulent, roiling water some tens of meters farther seaward.

These remarkable and integrated behavioral responses, which underlie the migrations of *Synchelidium*, do not involve "orientation," as the term is usually understood, in which an animal aligns its body axis with the desired direction of movement. Instead, the animals take advantage of wave transport in order to be carried in the "right" direction. Any single wave, on the beach, has both landward and seaward components of motion, with essentially no residual direction of net water transport. The behavioral interaction of the amphipods with the waves, however, enables them to "rectify" this "alternating current" system, so as to be carried landward during rising tide and seaward during ebbing tide. It is a much more complex strategy than that of *Orchestoidea* in sea water, which seeks only to be transported in a single direction, back toward the high beach.

A. Conclusions

I make no pretense that these aspects of orientation, homing, and migration among marine invertebrates can offer any direct clue toward the understanding of specific behavioral problems that arise in the orientation, homing, and migration of birds. On the other hand, I believe there are certain general messages to be derived from this sort of comparative approach. Such studies of crustacean behavior indicate in diagrammatic clarity, I believe, that the behavior which leads to proper habitat choices has been subject to strong selective pressures, which specify the *end* and not the means. These few examples suggest that there are dozens, and probably even hundreds of schemes, involving a broad spectrum of environmental parameters and sensory mechanisms, for accomplishing the ecologically important objective of finding and remaining in an appropriate region of the biotope. An inquisitive naturalist, having learned, for example, that talitrid amphipods can use time-compensated sun orientation to make an environmentally appropriate choice of directions, is open to the strong temptation to overgeneralize, to presume that celestial orientation ought to be a dominant directional beacon for other sand-beach crustaceans as well. Instead, we find other crustacean species, in an adjacent habitat, performing successfully, in terms of maintaining their position in a suitable ecologic zone, on the basis of fundamentally different sensory cues and environmentally keyed responses. Beyond this, a reexamination of the talitrid-amphipod case has revealed that the complex and highly evolved phenomenon of time-compensated sun orientation is only one of several orientational possibilities open to *Orchestoidea*, and by no means the dominant mechanism in a natural context. The implicit warning, then, is to be cautious in speculation about *the* mechanism by which, say, *birds* in general accomplish their goals in habitat selection, homing, and migration. As discouraging as it may seem to the generalist, the safer description of observations will be in terms of *one* of the mechanisms by which a *given species* finds its way in space. Perhaps, for example, with homing pigeons, which have been selected for so long as separate, inbred races, and in which selection has been only for the trait, "Find your way home!", we should even hesitate to generalize about homing pigeons as though all races were identical.

Lest this cautious approach lead to despair, however, it should not be overlooked that there are striking similarities in the migratory behavior of *Emerita* and *Synchelidium:* Both have strongly expressed endogenous tidal rhythms in spontaneous activity, and both show a strong, short-duration, hyperactivity response to minute increases in hydrostatic pressure, based on sensory systems which are, at this time, one of the major unsolved problems of sensory biology. The taxonomic distance between the crab and the amphipod is so great that one must presume that these behavioral similarities are the result of convergent evolution. This interpretation, then, gives us reason to *hope* that similar ecologic pressures may lead to related behavioral adaptations, despite the warning that one dare not prematurely assume this to be so.

If there is any other unifying principle underlying the performance of the invertebrates which I have discussed here, it is embodied in the self-evident concept that the right directional choice often depends upon time: time of day for vertical migration and celestial orientation, time of tide in the other cases considered here. The "where" and the "when" of habitat selection, homing, and migration are commonly intimately interrelated. For a determination of the "when" of orientational behavior, an endogenous timing mechanism, a biological clock synchronized by the environment, seems often to be a useful bit of accessory equipment. We should not, therefore, be surprised, in the future, to find that animals may make use of biological clocks in a variety of other, as yet unsuspected ways.

References

Enright, J.T.: Pressure sensitivity of an amphipod. Science 133, 758-760 (1961)

Enright, J.T.: Responses of an amphipod to pressure changes. Comp. Biochem. Physiol. 7, 131-145 (1962)

Enright, J.T.: The tidal rhythm of activity of a sand-beach amphipod. Zeit. vergl. Physiol. 46, 276-313 (1963)

Enright, J.T.: Temperature compensation in short-duration time measurements by an intertidal amphipod. Science 156, 1510-1512 (1967)

Enright, J.T.: When the beach-hopper looks at the moon: the moon-compass hypothesis. In: Animal Orientation and Navigation. Galler, S., Schmidt-Koenig, K., Jacobs, G., Belleville, R. (eds.) U.S. Govt. Printing Office, Wash., D.C., 1972, pp. 523-555

Enright, J.T.: Orientation in time: endogenous clocks. pp. 467-494. In: Marine Ecology, Kinne, O. (ed.). New York: Wiley-Interscience, 1974, Vol. II

Enright, J.T.: Copepods in a hurry: high-speed upward migration. Limnology and Oceanography 22, 118-126 (1977)

Enright, J.T., Hammer, W.M.: Vertical diurnal migration and endogenous rhythmicity. Science 154, 532-533 (1967)

Enright, J.T., Honegger, H.W.: Diurnal vertical migration: selective advantage and timing. Part II. Details of timing, a test of the model. Limnology and Oceanography, in press (1977)

Gwinner, E.: Artspezifische Muster der Zugunruhe und ihre mögliche Bedeutung für die Beendigung des Zuges im Winterquartier. Zeit. Tierpsych. 25, 843-853 (1968)

Hartwick, R.F.: Aspects of celestial orientation behavior in talitrid amphipods. In: Biological Rhythms in the Marine Environment. DeCoursey, P.J. (ed.) U. South Carolina Press, Columbia, S.C., 1976, pp. 189-197

Pardi, L., Papi, F.: Ricerche sull'orientamento di *Talitrus saltator* (Montagu) Zeit. vergl. Physiol. 35, 459-489 (1953)

Orientation Based on Directivity, a Directional Parameter of the Animal's Radiant Environment

F. J. VERHEIJEN, Laboratory of Comparative Physiology, State University of Utrecht, The Netherlands

Abstract

Orientation exhibited by animals that presumably cannot see the sun as a discrete source because of clouds or because of experimental interference with detailed vision, is often considered to be indicative of nonvisual cues. However, the remaining anisotropic radiance distribution may also be a directional cue. A vectorial parameter, the directivity D, is proposed for the directional quantification of a radiance field. Its computation resembles that of the test statistic of the Rayleigh test. Theoretically D ranges between 0 in an isotropic radiance field, and 1 in a parallel beam. Various examples are given of orientation in an anisotropic radiance field.

A. Introduction: Directional Orientation and Radiant Environment

Almost fourty years ago Lashley (1938) emphasized in his opening address as a president of a meeting of the Eastern Psychological Association in New York that "Psychological theories based upon the relations of stimulus and response remain sheer nonsense so long as the stimulus is defined only as whatever the experimenter puts in front of the animal." It would seem that many a psychologist has reasoned that this slashing criticism was passed on in 1938 on April Fool's Day. Up to the present Lashley's judgment still characterizes quite a few studies of the relations between photic stimulus and response, and, within the scope of this conference, more specifically studies of the relations between stimulus and response underlying photic orientation.

Animals have been allowed or forced to "orient" with respect to photic stimuli such as lamps, beams, or screens, whether or not presented in pairs with different brightness or color. The finishing touches to such apparatus are invariably given by the abundant application of flat-black paint. My already long blacklist of papers reporting such experiments is still supplemented by recent publications. However, the fivehundredth anniversary of this University does not seem to be an appropriate occasion for publishing a blacklist.

The American psychologist Gibson (1958) emphasized that light "fills" the environment so that rays will converge to any point: The environment is projected to this point. Thus the potential stimulus for an animal is all the light, whether or not via reflection "belonging" to an object, and converging to its eyes. On the basis of the available literature and of experimental evidence, I made at that time a similar assumption about phototaxis (Verheijen, 1958). Earlier the forgotten Russian physicist Gershun (1938) considered a light field as "a part of space studied from the standpoint of transmission of radiant energy within that space."

Photic directional orientation - as distinct from goal orientation, see Wallraff (1972, 1974) - must be assumed to be based on a mechanism tuned to extract certain vectorial features, which are related to the goal to be reached by the animal, out of the radiant[1] environment.

The radiant environment of an organism can be considered as a multiplicity of imaginary juxtaposed sources situated all around the organism as the measuring point. The radiance emitted by any given source in the direction of the measuring point (active regime) can be measured by a detector (passive regime) that can be provided with a radiance tube limiting the solid angle of acceptance to the angle occupied by that source. This tube or hood is termed Gershun tube after the Russian physicist mentioned previously. For the unrealistic purpose of measuring the "real" angular radiance distribution (ARD) at a given point it would be necessary to make an infinitive number of measurements using a detector with an infinitesimal solid angle of acceptance. Fortunately reality refuses to be measured, and consequently we have to be content with a model of the ARD obtained by a finite number of measurements made with a detector with a finite solid angle of acceptance. A three-dimensional display of the ARD, the ARD solid, is obtained by constructing a surface through the end points of the pseudovectors whose length are proportional to the radiances measured in various directions by a rotating meter. It is clear that the angle of acceptance of the detector is reflected in the shape of the radiance distribution solid (RDS) measured at a given point in a given radiance field: The larger the angle of acceptance of the detector, the more the shape of the RDS is smoothed toward a more or less prolate spheroid (egg-shape) (Fig. 1).

The shape of the "real" RDS is the result of the interactions of the radiation with the environment by which either the direction of the radiation is changed (scattering; reflection) or its intensity is decreased (absorption). Theoretically (Tyler and Preisendorfer, 1962) the RDS would shift to either a sphere (completely isotropic radiance field) or a needle (extremely anisotropic radiance field: a parallel beam) if, respectively, only scattering or only absorption were involved (Fig. 1). Within the theoretical range of RDSs between a sphere and a needle only a restricted class occurs on earth, and, more specifically, within the habitat of a given species, by the characteristic contributions of factors that influence beam direction and intensity. Previously (Verheijen, 1958, 1969) I have given theoretical arguments and experimental evidence for assuming that only radiance distributions within this natural class will allow a species to show orientation with interpretable biologic consequences, whereas in radiance distributions not included in the natural class, an organism will show either no orientation (if the degree of anisotropy of the radiance field is too low) or nonsense orientation (if the degree of anisotropy is too high). It would seem that we are badly in need of a vectorial measure to quantify the degree of anisotropy of a given radiance distribution. Unaware of the monumental work of Gershun (1938) I made a premature effort (Verheijen, 1958) to find such a measure, but I shall let that pass.

---

1 Because in this paper both the spectral distribution of a radiance field and the spectral sensitivity of an organism behaving within this stimulus field will be left out of consideration, I need not be concerned as to whether the stimulus should be termed light or radiance; polarization is also not included in this paper.

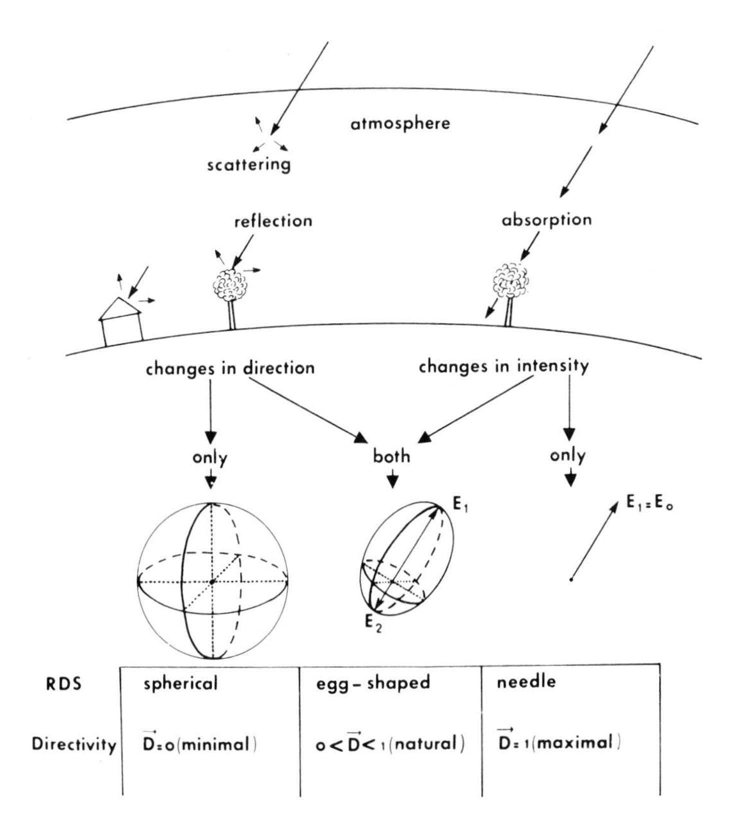

Fig. 1. Schematic diagram of the interactions between radiation and the environment, the resulting radiance distribution solid, and the directivity. Explanation: See text

B. Directivity of a Radiance Distribution

The directivity (D) proposed below is based on the work of Gershun (1938). Among all pairs of opposite irradiances E_1 and E_2 that can be measured at a given point in a given radiance distribution, there is one pair showing a maximal difference $E_1 - E_2$. Gershun (1938) proposed a true vectorial quantity for the characterization of the directional properties of a radiance field, the radiant vector $\vec{V}_R$ (this is a flux density vector; Gershun used the terms light field an light vector) such that $|\vec{V}_R| = \max\{E_1 - E_2\}$. $|\vec{V}_R|$ denotes the magnitude or modulus of the radiant vector. In an isotropic radiance field $\vec{V}_R = 0$, and consequently, the direction of the radiant vector is undetermined. This is in harmony with the fact that photic orientation is impossible in an isotropic radiance field. In a parallel beam the maximal difference $E_1 - E_2 = E_1$ since $E_2 = o$. Moreover E_1 now equals the scalar irradiance E_o, the total energy per unit area arriving at the measuring point from all directions (Tyler and Preisendorfer, 1962), so that in a parallel beam the radiant vector $|\vec{V}_R| = E_o$ (Fig. 1).

A physical measure, which, unlike the radiant vector, quantifies the directional properties of a radiance distribution independently of the amount of scalar irradiance, is obtained by dividing the radiant vector by the scalar irradiance. Thus the proposed directivity $|\vec{D}| \equiv |\vec{V}_R|/E_o$. This directional measure - Gershun's degree of diffusion r_o - is only sensitive to the "form" of the radiance distribution solid, and not to its "size". Obviously the magnitude of the directivity ranges between zero in an isotropic radiance field, and one in a parallel beam.

It is a fundamental problem in psychophysics to find some parameter of a stimulus situation that is reasonably related to the performance of an organism exposed to that stimulus situation. Which geometric aspects of a radiance field are most meaningful with respect to photically steered behavior in an organism - whether plant, animal, or man - is still anything but clear. The introduction of the directivity $\vec{D}$ as a quantitative measure of the degree of anisotropy of a radiance field - or, more correctly, the re-introduction of Gershun's obviously completely forgotten concept - might further the design of carefully controlled experimental conditions in order to unravel the relations between photically guided behavior and the anisotropy of the natural or the artificial radiant environment.

An attractive aspect of the proposed directivity is that its computation resembles the determination of the test statistic of the Rayleigh test, which is frequently used for determining the direction and degree of concentration of the angular orientation shown by animals released at a given point in a given stimulus field (Batschelet, 1965).

C. Visual Mechanisms for Directional Orientation in a Radiance Field

The directional features of an anisotropic ARD can be detected by means of different receptor systems. In the following considerations use is made of the classification proposed by Schöne (1973, 1975), namely, in one-input unit systems, two-input unit systems, and multiple-input unit systems (rasters). If the animal's only photoreceptor is a nondirectional simple irradiance detector (one-input unit system), then it can sequentially sample the stimulus intensity in different angular directions ("klinotaxis", Fraenkel and Gunn, 1940) and thus choose the brightest, the darkest, or another direction. If the animal has two of such, often bilaterally arranged, receptors (two-input unit system), it can detect directional properties of the ARD by the simultaneous comparison of the stimulation intensities of both eyes ("tropotaxis", Fraenkel and Gunn, 1940). Many eyes are multiple-input unit systems ("rasters") and many animals have two of such eyes. Because of the directional sensitivity of these eyes, the animal can obtain information about directional properties of the ARD from the stimulus gradient along the retina of each of these eyes, and turn in a brighter or a darker direction. The animal might additionally use the differences between the stimulation intensities of its two eyes. In man, however, a reduction of the stimulation intensity of one eye by a neutral density filter is not perceived as the introduction of a "dark direction" but as a darkening of the complete field of view, because both of his eyes function as one cyclopean eye (Julesz, 1971).

In a detailed discussion of the taxis problem, Precht (1942) emphasized that moving in a bright or in a dark direction (phototaxis) must be distinguished from moving with respect to a configurational stimulus ("visible thing"). It would seem that this distinction shows similarities with Wallraff's (1972, 1974) directional or goal orientation. Moreover, I suggest that these two orientational systems harmonize with the accumulating evidence that, at least in vertebrates, there are two visual systems. These two systems would be separately located both in the receptor and in the central nervous system. Spatial orientation would be a function of ambient vision by the peripheral retina, with central mechanisms in the phylogenetically older brain parts, whereas focal vision, located in the fovea and parafovea, and with central mechanisms in the newer brain parts, would represent vision tuned to identifying forms, (Trewarthen, 1968; Schneider, 1969; Ingle, 1973; Leibowitz and Dichgans, 1977).

D. Orientation of Various Species of Animals in a Radiance Field Without a Discrete Source

Because the directivity concept was developed very recently it was not applied in our experiments reviewed below.

In an imitated natural angular radiance distribution without a discrete sun-like source, positive, negative, and menotactic orientation was demonstrated in the collembolan *Podura aquatica* (Verheijen and Brouwer, 1972). Sea finding in hatchlings of the sea turtle *Chelonia mydas* was found to be in part based on certain directional features of the angular radiance distribution at the beach (Verheijen and Wildschut, 1973; F.A. van Rhijn, in prep.). The suggestion (Verheijen and Brouwer, 1972) that birds might show compass orientation with respect to the azimuthal radiance distribution of the sky exclusive of the sun, was confirmed in training experiments with starlings in a simple planetarium (Verheijen, 1975). Recently we started directional training by means of cardiac conditioning in the European eel, *Anguilla anguilla* - silver stage with enlarged eyes - as was done previously with pigeons by Wallraff (1968) and by Schlichte (1973). The animal was fastened in a transparent perspex tube of which the "upper" half had been removed in the "head region". The head of the animal was positioned in the center of a white opaline plastic globe (diameter 45 cm), which, in turn, was placed underwater in a large white opaline polyethylene tub. The anisotropic underwater radiance field (see Tyler, 1960) was produced by a 100 W lamp outside the tub. For a human observer the lamp was invisible inside the globe. An underwater meter for detailed measurements of the ARD is planned. In the radiance field the tube with the globe and the eel could be rotated with variable velocity up to about one revolution in 3 min. Electric shocks (50 Hz, 12 V) were given via two electrodes in the "tail region" of the tube. The response - the delay of one or a few heart beats - was registrated (see Kalmijn, 1966) via two electrodes embedded close to the heart and fixed to the skin. The distribution of the directions of the longitudinal body axis in which the longest heart-beat interval was found during test rotations was tested for randomness (Rayleigh test, Batschelet, 1965). Figure 2

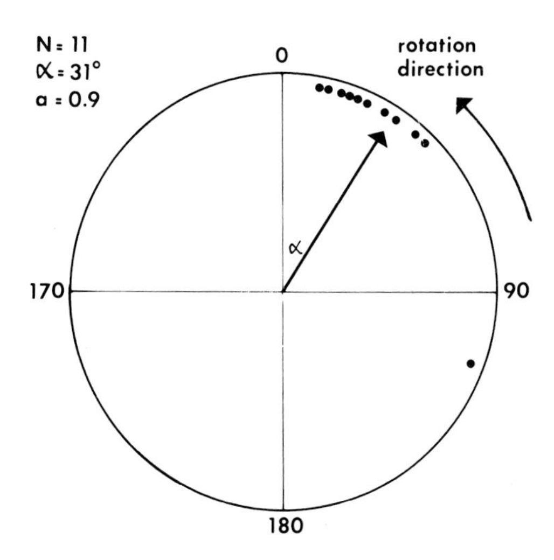

Fig. 2. Directions of body axis (● = head side) of a heart rate-conditioned eel showing the longest heart-beat interval per revolution during rotation in an anisotropic radiance field. Training direction: 0^{0} (brightest azimuthal direction)

shows one of the first results. The training direction was 0°, in this case the constant direction of the major horizontal vector in the radiance field. Obviously the animal perceived the training direction and anticipated the electric shocks. It remains to be seen whether the eel can show a compass orientation, and if so, what will be the directivity threshold in relation to the ambient radiance level. Moreover silver eels with enlarged eyes might show a better performance than nonmigratory eels.

I shall conclude with some data that show that the negative phototaxis of sand hoppers wetted with sea water (Debenedetti, 1963) is only manifest if the directivity of the radiance field is not too high, and that the sign of phototaxis reverses to positive in a highly directive radiance field.

The animals were studied in the apparatus diagrammed in Figure[2] 3. It consisted of the two halves of a circular arena, diameter 20 cm and

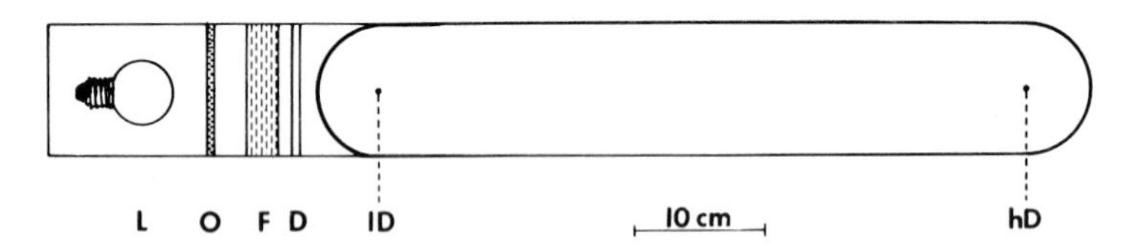

Fig. 3. Top view of alley for testing orientation of sand hoppers in anisotropic radiance distributions. *lD*, test place with low directivity; *hD*, test place with high directivity. The vector diagrams at these two places are given in Figure 4. Further explanation: See text

height 15 cm, between which a rectangular 100-cm long corridor was placed. The walls of the complete alley, inclusive of floor and ceiling, were flat gray. Light of an incandescent lamp penetrated via a plate of opaline glass O, a heat filter F (Plexiclass cuvet with water), and a horizontal slit diaphragm D into the alley through the translucent white perspex window forming one of the circular endings of the alley. Different levels of radiation were produced by an incandescent lamp of either 200 W or 15 W, and by changing the horizontal slit of diaphragm D. The construction assured that the directivity of the ARD increased the further away from the perspex window the measuring point was. The animals were tested at the points lD (low directivity) and hD (high directivity). The photic directional information of these two points differed only in the pattern of the horizontal vector diagram (Fig. 4). Twelve vectors of these diagrams were measured by a selenium cell with a hood giving it an acceptance angle of about 30°. The cell was calibrated by means of a calibrated thermopile. The animals were kept at 18 - 20° C in an isotropic radiance field and tested at about the same temperature in a hollow half-sphere of clear plastic placed either at lD or at hD in the alley. The methodology developed by Pardi and Papi (1953) was adopted. The directional distributions of the animals were tested by the Rayleigh test (Bat-

2 The experiments were carried out by the graduate student J. Groten in 1963. I did not publish the results until now because I was in doubt about their correct place in my ideas about photo-orientation, and because results that are only "odd" tend to be soon forgotten. The experiments are dated by methodologic and statistical aspects that do not quite meet present standards. Nevertheless I have every confidence in the results and in the proposed interpretation.

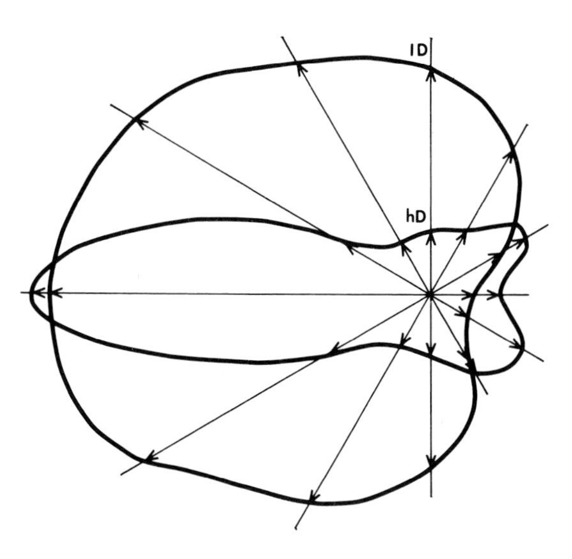

Fig. 4. Horizontal vector diagrams at places *lD* and *hD* of radiance field in alley diagrammed in Figure 3. Further explanation: See text

schelet, 1965) although one might dispute the required independence of the collected data (up to 10 animals recorded up to about ten times at intervals of a few minutes; the vessel was shuttled after every observation).

In the ARD at lD (15 W lamp; diaphragm open) dry animals showed a positive phototaxis, whereas animals wetted with sea water showed a negative phototaxis (Fig. 5). This confirms the motivational change of the sign of phototaxis observed by Debenedetti (1963). In Figure 6

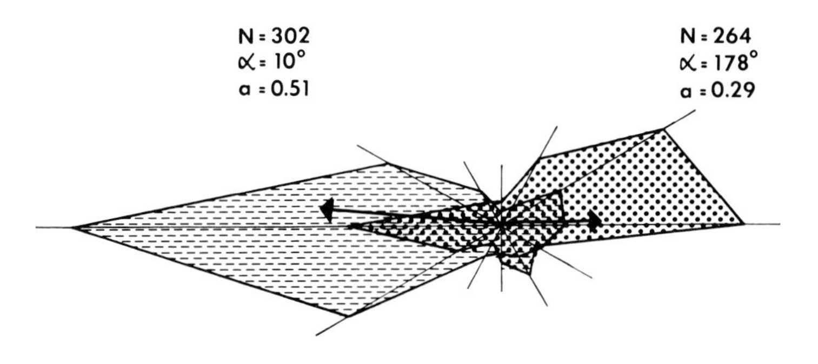

Fig. 5. Circular distribution, mean vector direction, and vector length of sea water-wetted (*stippled polygon*) and dry (*hatched polygon*) *Talitrus* at *lD* in test alley. Medium level of ambient radiation

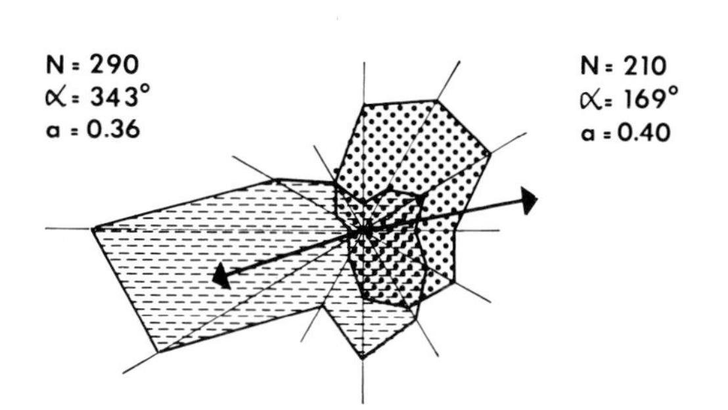

Fig. 6. Circular distribution, mean vector direction, and vector length of sea water-wetted *Talitrus* at *lD* (*stippled polygon*) and at *hD* (*hatched polygon*) in test alley. Medium level of ambient radiation

(15 W lamp, diaphragm open) it can be seen that in the more directional radiance field at hD (V_{maj} = 0.7 erg cm^{-2} s^{-1}) the animals wetted with sea water showed a positive phototaxis whereas at lD (V_{maj} = 64 erg cm^{-2} s^{-1}) the phototaxis was again negative. According to the model of phototaxis recently proposed by Hailman and Jaeger (1976) this reversal of phototactic sign might be the result of the lower ambient level of radiation at hD (ratio of major vectors at the two places of the order of 10^2 with a given lamp and a given diaphragm opening). However, even at the lowest level of radiation (15 W lamp, diaphragm maximally closed) the wetted animals showed a negative phototaxis at lD (V_{maj} = 0.7 erg cm^{-2} s^{-1}) (Fig. 7), whereas at the highest level of radiation (200 W lamp; diaphragm open) they showed positive phototaxis at hD (V_{maj} = 11.1 erg cm^{-2} s^{-1}). With this lamp and open diaphragm the animals were, of course, negatively phototactic at lD (V_{maj} = 910 erg cm^{-2} s^{-1}) (Fig. 8). If we assume with Hailman and

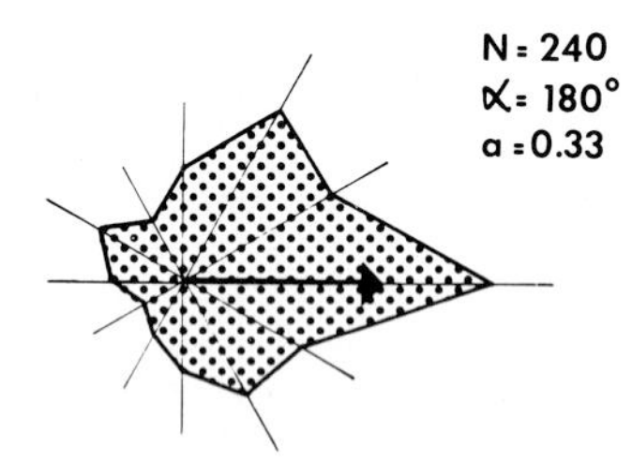

Fig. 7. Circular distribution, mean vector direction, and vector length of sea water-wetted *Talitrus* at lD in test alley. Low level of ambient radiation

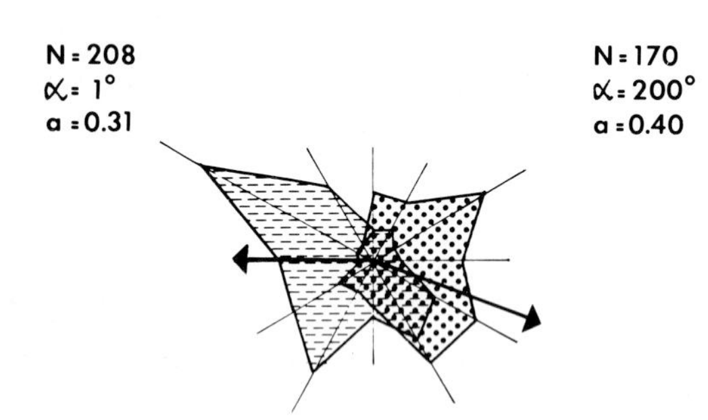

Fig. 8. Circular distribution, mean vector direction, and vector length of sea water-wetted *Talitrus* at *lD* (*stippled polygon*) and *hD* (*hatched polygon*) in test alley. High level of ambient radiation

Jaeger (1976) that the higher or lower the level of ambient illumination than "the" optimum ambient illumination - as determined, however, by otherwise identical motivational conditions, cf., e.g., Precht (1942) - the more an animal will show negative or positive phototaxis, then it is clear that the positive phototaxis observed at point hD in sand hoppers wetted with sea water cannot be explained along these lines of thought. It must be concluded that the highly directional character of the radiance field in the alley at the point hD interferes with the negative phototaxis, which is motivationally induced in the sand hoppers by another environmental factor, namely, sea water (cf. Verheijen, 1958).

E. Discussion

It bears emphasizing that, in photic orientation, the angular radiance distribution may play a considerable role, which has as yet been insufficiently formulated and which therefore has remained obscure. An animal may move in some direction because it perceives a discrete and configurational stimulus source (entaxis, Precht, 1942; goal orientation, Wallraff, 1972, 1974) or, alternatively, because it perceives features of the angular radiance distribution (ARD) that are directionally related to the goal (phototaxis, Precht 1942; directional orientation, Wallraff, 1972, 1974). Therefore the animal must have a more complex internal model of its photic environment - whether or not accompanied by conscious awareness (Griffin, 1976) - than is generally assumed. The place of the celestial bodies among these two types of orientational stimuli is not obvious. With respect to the required dioptric apparatus, and probably also the neural apparatus, I am inclined to attribute orientation by means of celestial bodies to the "focal" visual system. However "visible" the celestial bodies may be, they are no goal to be approached or to withdraw from. How animals tell the sun and moon apart is not clear (see Schmidt-Koenig, 1975, pp. 11-12, for a discussion about *Talitrus*).

Herrnkind's (1972) "simple phototaxis" of fiddler crabs in an arena with a one sided illuminated translucent white wall was no doubt oriented with respect to the arena's ARD. Tesch (1975) concluded that in migrating eels "sun-compass orientation appears entirely impossible because several of the field experiments were conducted during the night, and during long periods, heavy cloud cover obscured the sun during the day." Our first success in directional training of the eel in an anisotropic underwater radiance field indicates that one should hesitate before deciding that visual orientation in this animal is entirely impossible because of some not yet quantifiable photic condition. The theoretical ability of a fish to derive the bearing of the sun in azimuth from the underwater light distribution was discussed in detail by Harden-Jones (1968). Homing in pigeons with reduced vision because of frosted contact lenses has during prolonged series of experiments been attributed to nonvisual information, until it was recently admitted that "they can certainly see the sun as a bright area in their blurred, milky white visual field, and this is apparently sufficient to provide compass information" (Schmidt-Koenig and Keeton, 1977). In my opinion this interpretation is unnecessarily blurred and milky, and it emphasizes how strong the tendency still is to reduce the animal's photic environment to discrete sources (cf. Loeb, 1918). Verheijen and Brouwer (1972) assumed that "the birds might orient menotactically by a detector system with a large field of view. In the azimuthal angular radiance distribution of the sky the maximal weighted vector then indicates the azimuthal position of the sun." This assumption proved to be true for starlings (Verheijen, 1975).

In evaluating the homing performances in pigeons in relation to distance, Keeton (1970) emphasized that poor results may be of two different types. At the release site the vanishing directions may show a high degree of scatter, however with a small deviation of the - small - mean vector from the true home direction; or the vanishing directions may show a low degree of scatter, however with a great deviation of the - long - mean vector from the true home direction. Without pretending any relation to the distance problem as discussed by Keeton, I suggest that more generally these two types of "poor" performance may be the result of definite photic stimulus conditions to which the two photic orientation mechanisms pictured above are tuned.

Orientational scatter around the correct direction may increase as a result of the difficult perceptibility of discrete visual stimuli - inclusive of the celestial bodies - or as a result of the low degree of anisotropy of the ARD. Therefore the species-specific sensitivity to the anisotropy of a radiance field, or, in other words, the threshold for directivity, is a crucial sensory-perceptual ability to be studied. It would seem that in man the threshold for directivity must be closely related to the threshold data for spatial luminance gradient collected recently by van der Wildt et al. (1976). A simple egg-shaped radiance field consists essentially of two stimulus gradients over 180°. The method and the horizontal angular dimensions of the gradients applied by van der Wildt et al. (up to 10°) suggest that the nonfocal orientational visual system was involved.

Orientation may strongly deviate from some expected direction when a discrete source - for instance a lamp used as an artificial sun or moon - is introduced in an otherwise impoverished visual environment, because of derangement of the focal system (for a detailed discussion of this "trapping effect" see Verheijen, 1958, 1969). The resulting "positive phototactic tendency" is not only observed under experimental conditions (Ercolini and Scapini, 1974; and many others) but also, however to a moderate degree, with respect to a natural celestial body (Pardi et al. 1958; and many others). A highly directional ARD may also result in a "strong but wrong" orientation, as was reported in this paper for the sand hopper.

I was initially surprised about the advice of Larkin et al. (1975) to reduce bird-aircraft collisons by equipping the aircraft with a lamp producing a very narrow beam forward, because beams are notorious for their attractive power to birds, e.g., around lighthouses, and especially on dark (no moon!) nights. Whether or not the authors reckoned with this aspect, their advise was, however, correct, "because the airspeeds of airplanes are much greater than those of birds." Thus the problem is not whether birds are attracted to an aircraft, but whether they can avoid collision by an evasive alarm maneuvre.

Man-made illumination conditions appear to be disastrous to sea turtle rookery. McFarlane (1963) collected some data about the Atlantic loggerhead turtle *Caretta caretta*. It would seem that "strong but wrong" orientation - up to 95 % of hatchlings were prevented from reaching the surf - can be caused by the more or less natural angular radiance distribution, though with a wrong brightest direction, of the artificially illuminated sky over resort areas, and by the "trapping effect" (Verheijen, 1958) of streetlights. It might appear that hatchlings can more easily be protected against streetlights (scrub vegetation between beach and lights) than against an artificially illuminated sky in a wrong direction. Many more examples of photic orientation could have been discussed. The few examples given above show that both from a fundamental and from an applied point of view we still have much to learn about photic guidance systems and about the used directional stimuli.

Acknowledgments. I am indebted to my colleagues Mr. M.C. Arendse and Dr. Ir. A. Schuijf for the discovery of the abundance in original concepts of Gershun's paper. More details about the directional properties of a radiance distribution will be published elsewhere.

References

Batschelet, E.: Statistical methods for the analysis of problems in animal orientation and certain biological rhythms. Am. Inst. Biol. Sci., Washington, D.C., 1965

Debenedetti, E.F.: Preliminary observations on the orientation of *Talitrus saltator* in fresh and seawater. Naturwissenschaften 50, 25-26 (1963)

Ercolini, A., Scapini, F.: Sun compass and shore slope in the orientation of littoral Amphipods (*Talitrus saltator* Montagu). Monitore Zool. Ital. (N.S.) 8, 85-115 (1974)
Fraenkel, G.S., Gunn, D.L.: The orientation of animals. Kineses, taxes and compass reactions. (Oxford Univ. Press, London, republished 1961 by Dover, New York) (1940)
Gershun, A.: The lightfield. J. Math. Phys. 17, 51-151 (1938)
Gibson, J.J.: Visually controlled locomotion and visual orientation in animals. Brit. J. Psychol. 49, 182-194 (1958)
Griffin, D.R.: The question of animal awareness.(Rockefeller Univ. Press, New York) (1976)
Hailman, J.P., Jaeger, R.G.: A model of phototaxis and its evaluation with anuran amphibians. Behaviour 56, 215-249 (1976)
Harden-Jones, F.R.: Fish Migration. London: Arnold, 1968
Herrnkind, W.F.: Orientation in shore-living arthropods, especially the sand fiddler crab. In: Behaviour of Marine Animals. Winn, H.E., Olla, B.L. (eds.) Invertebrata. New York: Plenum Press, 1-59, 1962, Vol. 1
Ingle, D.: Two visual systems in the frog. Science 181, 1053-1055 (1973)
Julesz, B.: Foundations of cyclopean perception. (Univ. of Chicago Press, Chicago) (1971)
Kalmijn, A.J.: Electroperception in sharks and rays. Nature London, 212. 1232-1233 (1966)
Keeton, W.T.: "Distance effect" in pigeon orientation: an evaluation. Biol Bull. 139, 510-519 (1970)
Larkin, R.P., Torre-Bueno, J.R., Griffin, D.R., Walcott, Ch.: Reactions of migrating birds to lights and aircraft. Proc. Nat. Acad. Sci. USA 72, 1994-1996 (1975)
Lashley, K.S.: Experimental analysis of instinctive behavior. Psychol. Rev. 45, 445-471 (1938)
Leibowitz, H., Dichgans, J.: Zwei verschiedene Seh-Systeme. Neue Untersuchungsergebnisse zur Raumorientierung. Umschau 77, 353-354 (1977)
Loeb, J.: Forced movements, tropisms and animal conduct. (Lippincott, Philadelphia; republished 1973 by Dover, New York) (1918)
McFarlane, R.W.: Disorientation of loggerhead hatchlings by artificial road lighting. Copeia 1963, 153 (1963)
Pardi, L., Ercolini, A., Marchionni, V., Nicola, C.: Ricerche sull'orientamento degli Anfipodi del litorale: il comportamento degli individui allevati in laboratorio sino dall'abbandono del marsupio. Atti Acc. Sci. Torino, Cl. Sci. Fis. Mat. Nat. 92, 308-316 (1958)
Pardi, L., Papi, F.: Ricerche sull' orientamento di *Talitrus saltator* (Montagu) (Crustacea, Amphipoda). Z. Vergl. Physiol. 35, 459-489 (1953)
Precht, H.: Das Taxis-Problem in der Zoologie. Z. Wiss. Zool. 156, 1-128 (1942)
Schlichte, H.-J.: Untersuchungen über die Bedeutung optischer Parameter für das Heimkehrverhalten der Brieftaube. Z. Tierpsychol. 37, 257-280 (1973)
Schmidt-Koenig, K.: Migration and homing in animals. (Springer, Berlin - Heidelberg - New York) (1975)
Schmidt-Koenig, K., Keeton, W.T.: Sun compass utilization by pigeons wearing frosted contact lenses. Auk 94, 143-145 (1977)
Schneider, G.E.: Two visual systems. Science 163, 895-902 (1969)
Schöne, H.: Raumorientierung, Begriffe und Mechanismen. Fortschr. Zool. 21, (2/3): 1-19 (1973)
Schöne, H.: Orientation in space: Animals. General introduction. In: Marine Ecology. Kinne, O. (ed.). London: Wiley 499-553, 1975, Vol. II Part 2
Tesch, F.-W.: Orientation in space: Animals. Fishes. In: Marine Ecology. Kinne, O. (ed.). London: Wiley 657-707, 1975, Vol. II Part 2
Trevarthen, C.B.: Two mechanisms of vision in primates. Psychol. Forsch. 31, 299-337 (1968)
Tyler, J.E.: Radiance distribution as a function of depth in an underwater environment. Bull. Scripps Inst. Ocean. 7, 363-412 (1960)
Tyler, J.E., Preisendorfer, R.W.: Light. In: The Sea. Hill, M.N. (ed.). Physical Oceanography, London: Interscience, 397-451, 1962, Vol. 1
Verheijen, F.J.: The trapping effect of artificial light sources upon animals. Arch. Néerl. Zool. 13, 1-107 (1958)
Verheijen, F.J.: Some aspects of the reactivity of fish to visual stimuli in the natural and in a controlled environment. FAO Fish. Rep. 62 (2), 417-429 (1969)

Verheijen, F.J.: Sky radiance features representing orientational cues to birds. Paper 14th Int. Etholog. Conf. Parma, Italy (1975)

Verheijen, F.J., Brouwer, J.M.M.: Orientation of *Podura aquatica* (L.) (Collembola, Insecta) in a natural angular radiance distribution. Neth. J. Zool. 22, 72-80 (1972)

Verheijen, F.J., Wildschut, J.T.: The photic orientation of hatchling sea turtles during water finding behaviour. Neth. J. Sea Res. 7, 53-67 (1973)

Wallraff, H.G.: Über das Orientierungsvermögen von Vögeln unter natürlichen und künstlichen Sternenmuster. Dressurversuche mit Stockenten. Verh. Dtsch. Zool. Ges. Innsbruck 1968, 348-357 (1968)

Wallraff, H.G.: Fernorientierung der Vögel. Verh. Dtsch. Zool. Ges. 65, 201-214 (1972)

Wallraff, H.G.: Das Navigationssystem der Vögel. (Oldenbourg, München) (1974)

Wildt, G.J. van der, Keemink, C.J., Brink, G. van den: Gradient detection and contrast transfer by the human eye. Vision Res. 16, 1047-1053 (1976)

Subject Index

K. Schmidt-Koenig

Migration and Homing in Animals

1975. 64 figures, 2 tables. XII, 99 pages
(Zoophysiology and Ecology, Vol. 6)
ISBN 3-540-07433-3

The phenomena of animal migration and homing are described here, clearly and interestingly as an introduction of students and laymen, but with thorough discussion of specialized experiments. The vast material ranges from the sand hopper Talitrus, moving only a few meters, to the immense global migration of whales and birds. Each descriptive section is followed by a section discussing the problems and possible solutions of the orientation and navigation involved. There is still no conclusive answer to the mystery: how do animals direct their movements? Here particular emphasis has been placed on exact description of methods which have been invented and employed in the search for solutions. The answers are as varied and fascinating as the fact of animal migration and homing itself, the many relevant figures adding greatly to the interest of the book.

Contents: Field Performance in Orientation; Experimental and Theoretical Analysis in Crustaceans and Spiders, Locusts, Bees, Butterflies, Fishes, Amphibians, Reptiles, Birds, Mammals: Bats, Whales and Terrestrial Mammals-Conclusion; Appendix – Some Statistical Methods of Analysis of an Animal Orientation Data; References.

Springer-Verlag
Berlin
Heidelberg
New York